HANDBOOK OF
EMERGENCY
RESPONSE

A Human Factors and
Systems Engineering Approach

Industrial Innovation Series

Series Editor

Adedeji B. Badiru

Department of Systems and Engineering Management
Air Force Institute of Technology (AFIT) – Dayton, Ohio

PUBLISHED TITLES

Carbon Footprint Analysis: Concepts, Methods, Implementation, and Case Studies,
  Matthew John Franchetti & Defne Apul

Computational Economic Analysis for Engineering and Industry, *Adedeji B. Badiru &*
  Olufemi A. Omitaomu

Conveyors: Applications, Selection, and Integration, *Patrick M. McGuire*

Global Engineering: Design, Decision Making, and Communication, *Carlos Acosta, V. Jorge Leon,*
  Charles Conrad, and Cesar O. Malave

Handbook of Emergency Response: A Human Factors and Systems Engineering Approach,
  Adedeji B. Badiru & LeeAnn Racz

Handbook of Industrial Engineering Equations, Formulas, and Calculations, *Adedeji B. Badiru &*
  Olufemi A. Omitaomu

Handbook of Industrial and Systems Engineering, *Adedeji B. Badiru*

Handbook of Military Industrial Engineering, *Adedeji B.Badiru & Marlin U. Thomas*

Industrial Control Systems: Mathematical and Statistical Models and Techniques, *Adedeji B. Badiru,*
  Oye Ibidapo-Obe, & Babatunde J. Ayeni

Industrial Project Management: Concepts, Tools, and Techniques, *Adedeji B. Badiru, Abidemi Badiru,*
  & Adetokunboh Badiru

Inventory Management: Non-Classical Views, *Mohamad Y. Jaber*

Kansei Engineering - 2 volume set
  * Innovations of Kansei Engineering, *Mitsuo Nagamachi & Anitawati Mohd Lokman*
  * Kansei/Affective Engineering, *Mitsuo Nagamachi*

Knowledge Discovery from Sensor Data, *Auroop R. Ganguly, João Gama, Olufemi A. Omitaomu,*
  Mohamed Medhat Gaber, & Ranga Raju Vatsavai

Learning Curves: Theory, Models, and Applications, *Mohamad Y. Jaber*

Modern Construction: Lean Project Delivery and Integrated Practices, *Lincoln Harding Forbes &*
  Syed M. Ahmed

Moving from Project Management to Project Leadership: A Practical Guide to Leading Groups,
  R. Camper Bull

Project Management: Systems, Principles, and Applications, *Adedeji B. Badiru*

Project Management for the Oil and Gas Industry: A World System Approach, *Adedeji B. Badiru &*
  Samuel O. Osisanya

Quality Management in Construction Projects, *Abdul Razzak Rumane*

Quality Tools for Managing Construction Projects, *Abdul Razzak Rumane*

Social Responsibility: Failure Mode Effects and Analysis, *Holly Alison Duckworth &*
  Rosemond Ann Moore

Statistical Techniques for Project Control, *Adedeji B. Badiru & Tina Agustiady*

STEP Project Management: Guide for Science, Technology, and Engineering Projects, *Adedeji B. Badiru*

Systems Thinking: Coping with 21st Century Problems, *John Turner Boardman & Brian J. Sauser*

Techonomics: The Theory of Industrial Evolution, *H. Lee Martin*

Triple C Model of Project Management: Communication, Cooperation, Coordination, *Adedeji B. Badiru*

FORTHCOMING TITLES

Communication for Continuous Improvement Projects, *Tina Agustiady & Adedeji B. Badiru*

Cellular Manufacturing: Mitigating Risk and Uncertainty, *John X. Wang*

Essentials of Engineering Leadership and Innovation, *Pamela McCauley-Bush & Lesia L. Crumpton-Young*

Sustainability: Utilizing Lean Six Sigma Techniques, *Tina Agustiady & Adedeji B. Badiru*

Technology Transfer and Commercialization of Environmental Remediation Technology, *Mark N. Goltz*

HANDBOOK OF
EMERGENCY RESPONSE

A Human Factors and
Systems Engineering Approach

Edited by
ADEDEJI B. BADIRU
LEEANN RACZ

CRC Press
Taylor & Francis Group
Boca Raton London New York

CRC Press is an imprint of the
Taylor & Francis Group, an **informa** business

CRC Press
Taylor & Francis Group
6000 Broken Sound Parkway NW, Suite 300
Boca Raton, FL 33487-2742

© 2014 by Taylor & Francis Group, LLC
CRC Press is an imprint of Taylor & Francis Group, an Informa business

No claim to original U.S. Government works

Version Date: 20130410

International Standard Book Number-13: 978-1-4665-1456-0 (Hardback)

Library of Congress Cataloging-in-Publication Data

Handbook of emergency response : a human factors and systems engineering approach / editors, Adedeji B. Badiru and LeeAnn Racz.
　　pages cm. -- (Industrial innovation series)
　　ISBN 978-1-4665-1456-0 (hardcover : alk. paper) -- ISBN 978-1-4665-1457-7 (ebook)
　　1. Emergency management--Technological innovations. 2. Rescue work. 3. Survival and emergency equipment. 4. Incident command systems. 5. Human engineering. I. Badiru, Adedeji Bodunde, 1952- II. Racz, LeeAnn.

HV551.2.H355 2014
363.34'8068--dc23   2013012924

Visit the Taylor & Francis Web site at
http://www.taylorandfrancis.com

and the CRC Press Web site at
http://www.crcpress.com

*Dedicated to our children,
who shall carry on the legacy.*

Contents

Preface

The April 2013 Boston Marathon bombing and the fertilizer plant explosion in West Texas emphasize the need to be prepared for emergency response by everyone at all times and in all places. Rapid response, widespread information flow, social media, and first-responder coordination are essential for providing much-needed medical services to victims, restoring order to chaos, and bringing perpetrators to justice, where applicable.

This handbook presents an integrative guide for emergency response project management following disasters resulting from accidental, man-made, or natural disasters. The collection of chapters provides a mix of human factors and systems engineering approaches. The guides contained in this book are useful as a "first-aid" tool to mitigate logistical problems that often follow disasters or extreme events, whether inadvertent or deliberate. The core of the integrative guide is the managerial processes needed to apply science, technology, and engineering tools to the challenges of emergency response. The managerial approach involves a systematic structure for communication, cooperation, and coordination. Managing a project in the wake of a tragedy is complicated and involves various emotional, sentimental, reactive, and chaotic responses. This makes it difficult to clearly pursue the traditional methods of planning and execution. This is the time that a structured communication model is most needed. Unfortunately, practical models are often not readily available. Conventional wisdom and assumptions, in lieu of direct communication, portend failure for an emergency response project. At the time of a disaster, reactive rhetoric may not coincide with action or reality. Having a guiding model can help put things into proper focus. This is what the *Handbook of Emergency Response* provides. This handbook highlights what must be done, by whom, where, how, and when. It can also help to identify the resources (personnel, equipment, facilities, etc.) required for each effort. The handbook is organized into three parts:

Section I: Technical challenges
Section II: Human factors issues
Section III: Managerial models

Acknowledgments

We thank the multitude of individuals who contributed to this handbook, either as chapter authors, support staff, or editorial assistants. We particularly recognize the editorial and administrative support of Rebecca Taylor and Air Force Institute of Technology students Aaron Zorn, Kelsey Smith, Kalyn Tung, Robert Poisson, and Jacob Petter. Colleagues and friends also provided valuable suggestions and operational guidance. We appreciate and thank everyone. We thank the editorial and professional staff members at CRC Press for their dedication to this project.

Editors

Adedeji Badiru is a professor and head of Systems Engineering and Management at the Air Force Institute of Technology. He was previously a professor and department head of Industrial and Information Engineering at the University of Tennessee in Knoxville. Before that, he was a professor of industrial engineering and dean of University College at the University of Oklahoma. He is a registered professional engineer, a certified project management professional, a fellow of the Institute of Industrial Engineers, and a fellow of the Nigerian Academy of Engineering. He holds a BS in industrial engineering as well as MS degrees in mathematics and industrial engineering from Tennessee Technological University, and a PhD in industrial engineering from the University of Central Florida. His areas of interest include mathematical modeling, project modeling and analysis, economic analysis, systems engineering, and efficiency/productivity analysis and improvement.

Dr. Badiru is the author of more than two dozen books and scores of book chapters, refereed journal articles, conference proceedings, and presentations. He has also published more than two dozen magazine articles, editorials, and periodicals. He is a member of several professional associations and several scholastic honor societies. He has won multiple awards for his teaching, research, and professional accomplishments. Dr. Badiru is the recipient of the 2009 Dayton Affiliate Society Council Award for Outstanding Scientists and Engineers in the Education category with a commendation from the 128th Senate of Ohio. He also won the 2010 IIE/Joint Publishers Book-of-the-Year Award for co-editing *The Handbook of Military Industrial Engineering*. He also won the 2010 the ASEE John Imhoff Award for his global contributions to industrial engineering education; the 2011 Federal Employee of the Year Award in the managerial category from the International Public Management Association, Wright Patterson Air Force Base; the 2012 Distinguished Engineering Alumni Award from the University of Central Florida; and the 2012 Medallion Award from the Institute of Industrial Engineers for his global contributions in the advancement of the profession. He has served as a consultant to several organizations around the world. He holds a leadership certificate from the University of Tennessee Leadership Institute. He is also a program evaluator for ABET.

His professional accomplishments are coupled with his passion for writing about everyday events, interpersonal issues, and socially responsible service to the community. Outside of the academic realm, he writes self-help books, motivational poems, editorials, and newspaper commentaries, engages in painting and crafts, and manages a STEM (Science, Technology, Engineering, and Mathematics)-and-Sports education website. He is also the founder of the Association of Military Industrial Engineers.

LeeAnn Racz is an assistant professor of environmental engineering and director of the Graduate Environmental Engineering and Science Program in the Department of Systems Engineering and Management at the Air Force Institute of Technology. She currently holds the rank of Lieutenant Colonel in the U.S. Air Force. Her previous assignments have taken her to the U.S. Air Force School of Aerospace Medicine, Texas; the U.S. Air Force Academy, Colorado; Peterson Air Force Base, Colorado; Osan Air Base, South Korea; and Cannon Air Force Base, New Mexico. She is a registered professional engineer and holds a BS in environmental engineering from California Polytechnic State University (San Luis Obispo), an MS in biological and agricultural engineering from the University of Idaho, and a PhD in civil and environmental engineering from the University of Utah. Her areas of interest include characterizing the fate of chemical warfare agents and pollutants of emerging concern in natural and engineered environments, environmental health issues, and the use of biological reactors to treat industrial waste.

Dr. Racz has authored dozens of refereed journal articles, conference proceedings, magazine articles, and presentations. She is a member of several professional associations and honor societies. She received the 2012 Southwestern Ohio Council for Higher Education Teaching Excellence Award and was the 2011 Faculty Scholar of the Year for the Department of Systems Engineering and Management. She has received numerous other honors, such as the 2010 American Water Resources Association Utah Section Student Conference and Scholarship Competition, as well as being named 2006 U.S. Air Force School of Aerospace Medicine Company Grade Officer of the Year, 2004 Colorado Air Force Association Company Grade Officer of the Year, 2004 U.S. Air Force Academy Company Grade Officer of the Year, and the 2002 810th Medical Operations Squadron Company Grade Officer of the Year. She is also the recipient of the Air Force Meritorious Service Medal and Air Force Commendation Medal (three oak leaf clusters).

Contributors

Julie A. Adams
Vanderbilt University
Nashville, Tennessee

L. Erwin Atwood
The Pennsylvania State University
University Park, Pennsylvania

Adedeji B. Badiru
Air Force Institute of Technology
Dayton, Ohio

Erdinc Bakır
Yeditepe University
Istanbul, Turkey

Budhendra L. Bhaduri
Oak Ridge National Laboratory
Oak Ridge, Tennessee

Hal Brackett
Virginia Polytechnic Institute and
    State University
Blacksburg, Virginia

Robert K. Campbell
Alliance Solutions Group, Inc.
Newport News, Virginia

John B. Coles
University at Buffalo
The State University of New York
Buffalo, New York

Harley M. Connors
U.S. Air Force
Panama City, Florida

Onur Demir
Yeditepe University
Istanbul, Turkey

Stephen A. Devaux
Analytic Project Management, Inc.
Swampscott, Massachusetts

Vinayak Dixit
University of New South Wales
Sydney, Australia

Leily Farrokhvar
Virginia Polytechnic Institute and
    State University
Blacksburg, Virginia

Steven J. Fernandez
Oak Ridge National Laboratory
Oak Ridge, Tennessee

John M. Flach
Wright State University
Dayton, Ohio

Susan Gaines
University of Central Florida
Orlando, Florida

Jeffery M. Gearhart
Air Force Research Laboratory
Dayton, Ohio

James R. Gruenberg
Wright State University
Dayton, Ohio

Glenn C. Hamilton
Wright State University
Dayton, Ohio

William H. Holbach
Virginia Polytechnic Institute and
 State University
Blacksburg, Virginia

Bernard B. Hsieh
U.S. Army Corps of Engineers
Vicksburg, Mississippi

Mario E. Ierardi
Environmental Protection Agency
Washington, DC

David R. Jacques
Air Force Institute of Technology
Dayton, Ohio

Tay W. Johannes
Air Force Institute of Technology
Dayton, Ohio

Mark R. Jourdan
U.S. Army Corps of Engineers
Vicksburg, Mississippi

Gino J. Lim
University of Houston
Houston, Texas

Michael K. Lindell
Texas A&M University
College Station, Texas

Ali Haskins Lisle
Virginia Polytechnic Institute and
 State University
Blacksburg, Virginia

Ann Marie Major
The Pennsylvania State University
University Park, Pennsylvania

Tom Martin
Virginia Polytechnic Institute and
 State University
Blacksburg, Virginia

Pamela McCauley-Bush
University of Central Florida
Orlando, Florida

David A. McEntire
University of North Texas
Denton, Texas

Heather Nachtmann
University of Arkansas
Fayetteville, Arkansas

Olufemi A. Omitaomu
Oak Ridge National Laboratory
Oak Ridge, Tennessee

Linet Özdamar
Yeditepe University
Istanbul, Turkey

Brian J. Phillips
Science Applications International
 Corporation (SAIC)
Waldorf, Maryland

Helen W. Phipps
Booz Allen Hamilton (contractor
 for U.S. Air Force)
Panama City, Florida

Paul S. Pirkle III
Battelle Memorial Institute
Atlanta, Georgia

Edward A. Pohl
University of Arkansas
Fayetteville, Arkansas

Tasha L. Pravecek
U.S. Air Force
Dayton, Ohio

LeeAnn Racz
Air Force Institute of Technology
Dayton, Ohio

Mukesh Rungta
University of Houston
Houston, Texas

Gregory G. Seaman
U.S. Marine Corps
Oceanside, California

Valerie L. Shalin
Wright State University
Dayton, Ohio

Michael L. Shelley
Air Force Institute of Technology
Dayton, Ohio

Katrina Simon-Agolory
Transformational Communities
Atlanta, Georgia

David A. Smith
U.S. Air Force
Falls Church, Virginia

Jack E. Smith II
Wright State University
Dayton, Ohio

Tonya L. Smith-Jackson
North Carolina Agricultural and
   Technical State University
Greensboro, North Carolina

Robin R. Sobotta
Embry-Riddle Aeronautical
   University
Prescott, Arizona

Michelle L. Spencer
MLSK Associates LLC
Millbrook, Alabama

Debra Steele-Johnson
Wright State University
Dayton, Ohio

Michael J. Surette Jr.
Booz Allen Hamilton (contractor
   for U.S. Air Force)
Panama City, Florida

Anuradha Venkateswaran
Wilberforce University
Wilberforce, Ohio

Kera Z. Watkins
Wilberforce University
Wilberforce, Ohio

Arturo Watlington III
University of Central Florida
Orlando, Florida

Woodrow W. Winchester
Virginia Polytechnic Institute and
   State University
Blacksburg, Virginia

P. Brian Wolshon
Louisiana State University
Baton Rouge, Louisiana

Samet Yılmaz
Yeditepe University
Istanbul, Turkey

Jun Zhuang
University at Buffalo
The State University of New York
Buffalo, New York

Introduction

Emergency response is a topic that touches every facet of a community. The case examples of the 2010 earthquake in Haiti, Hurricane Katrina in the Gulf States' region, and other recent disasters have shown that there is need for an integrated approach to emergency response. A risk to one is a risk to all. Thus, everyone must be prepared, even though the level of preparedness will differ depending on the specific roles and functions involved. The present availability of technology for social networking and communication has heightened the awareness of various disasters occurring in different parts of the world. This increased awareness has raised the sensitivity to the need for emergency response. In spite of the various guides available in organizations, when a disaster occurs, there is a higher potential for victims to run helter-skelter without a coordinated emergency response plan. To this end, a handbook such as this provides one more avenue for a ready guide that everyone can place on a reference shelf for quick access when needed.

The handbook specifically considers a systems approach coupled with the human factors considerations in emergency response, touching on both technical and human behavioral issues. We cannot solve today's complex problems with yesterday's conservative tools and approaches. The interdependencies of factors make a modern systems approach quite imperative for a coordinated and effective emergency response. This is aptly conveyed by the elements provided in Figure I.1. Infrastructure dependency refers to a linkage or connection between two infrastructures, through which the state of one infrastructure influences or is correlated to the state of the other. The figure shows six major attributes or dimensions of interrelationships:

1. Type of failure
2. Infrastructure characteristics
3. State of operation
4. Types of interdependencies
5. Environment
6. Coupling and response behavior

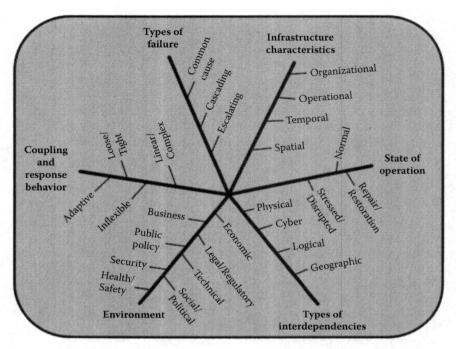

Figure I.1 Interdependency diagram for critical infrastructure in emergency response. (Adapted from Rinaldi, S.M., et al., *IEEE Control Syst.*, 21, 11–25, 2001.)

Each attribute is defined by a set of factors, such as economic, legal, social, political, and geographic. In conventional emergency response, such interdependencies are often ignored, neglected, or unknown. This may be due to a lack of knowledge or an omission due to the pressure of the emergency situation, but to get an effective result, interdependencies must be considered by using a robust systems architecture. Consider a specific and individual connection between two infrastructures (Rinaldi et al. 2001), such as electricity used to power a telecommunications switch. In this case, the relationship is usually unidirectional. That is,

Infrastructure *i* depends on infrastructure *j* through a specific link.
However, infrastructure *j* does not depend on infrastructure *i* through the same link.

The emergency response going in one direction may, thus, differ from the emergency response going in the other direction. This simple consideration, if not adequately recognized and considered, may lead to response failure that will end up adversely affecting everyone.

Communication is the common basis for executing emergency response and is best approached from a systems perspective considering all the directional flows of information, instructions, and announcements. The Triple C model, first introduced by Badiru (1987), has been adapted for application in a variety of communication, cooperation, and coordination scenarios. An adaptation for emergency response is shown in Figure I.2.

Situational awareness is a key element of an effective emergency response. Barriers to complete situational awareness include the following:

- Panic and anxiety
- Stress
- Complacency
- Insufficient communication
- Misinterpretation of information
- Conflicting information
- Response overload
- Poor health
- Excessive haste
- Denial
- Group effect (going with the flow of response without full awareness of what is going on)

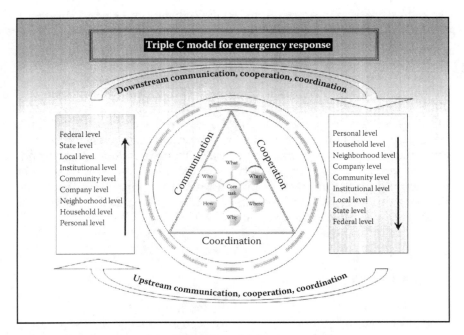

Figure I.2 Adaptation of the Triple C model for emergency response.

A good example of executing a communication network for situational awareness for emergency response was demonstrated by the Virginia Tech Police Department in 2011. That case example is illustrated in the following section.

Crisis communication: Virginia Tech case example

Social media, including Facebook, Twitter, and Blogs, is emerging as an effective tool for facilitating communication, cooperation, and coordination of emergency response. Issues such as situation awareness, broadcast message, and notifications can quickly reach a large population via social media tools. The brevity of these tools can ensure that they are quickly received, understood, and acted upon. Communication should be for the purpose of understanding and responding rather than just pushing data through.

A key component of Virginia Tech's (VT) Emergency Management Plan is how they communicate with the university community during an emergency. The VT Alerts system is one important component of this communication; it includes a prewritten message that requires only the date, time, and specific location to be keyed in before sending. The message is then sent via the Emergency Notification System (VT Alerts), which consists of multiple notification channels, including blast e-mail, homepage post, VT Phone Alerts (SMS, phone messaging, or non-VT e-mail), VT Desktop Alerts, campus digital signage, and social media posts to Twitter and Facebook. More than 30 people are trained and have access to VT Alerts, including VT police dispatchers, enabling the rapid response following the shooting on December 8, 2011. Another valuable feature of the communication plan is that university officials can replace the university website with a stripped down version that requires less bandwidth and is less likely to go down with increased website traffic. The "light" page, already designed and held in reserve, holds only basic navigation and linkages, reserving remaining space for notices, statements, or amplifying information. The normal website experienced an 80-fold increase in demand within 40 minutes of the first notice, prompting usage of both the light page and additional file servers. Figure I.3 illustrates the timeline for the Virginia Tech case example for emergency responsiveness.

Chapter Synopses

Section I: Technical challenges

Chapter one—Robotic technologies for first response: A review
After introducing the progressions and capabilities of today's robotics, this chapter discusses avenues for robotics in emergency situations.

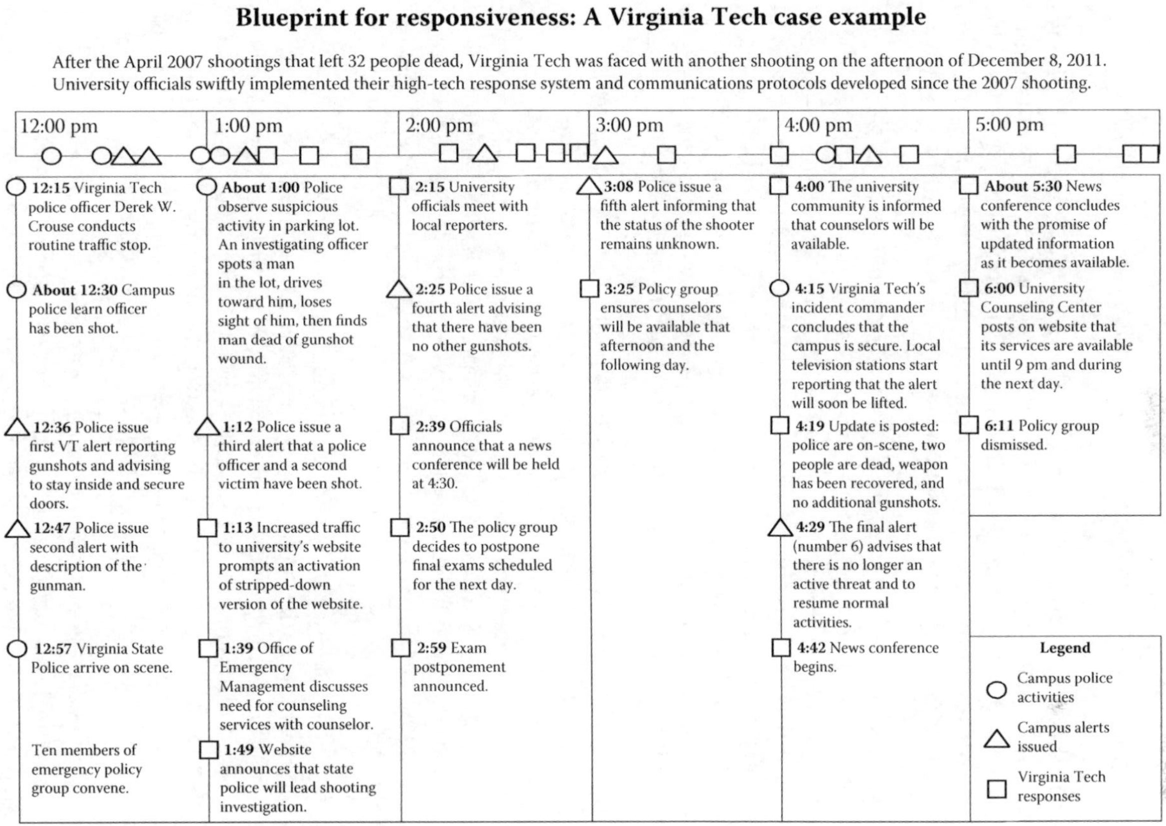

Blueprint for responsiveness: A Virginia Tech case example

After the April 2007 shootings that left 32 people dead, Virginia Tech was faced with another shooting on the afternoon of December 8, 2011. University officials swiftly implemented their high-tech response system and communications protocols developed since the 2007 shooting.

12:00 pm

12:15 Virginia Tech police officer Derek W. Crouse conducts routine traffic stop.

1:00 pm

About 1:00 Police observe suspicious activity in parking lot. An investigating officer spots a man in the lot, drives toward him, loses sight of him, then finds man dead of gunshot wound.

About 12:30 Campus police learn officer has been shot.

12:36 Police issue first VT alert reporting gunshots and advising to stay inside and secure doors.

12:47 Police issue second alert with description of the gunman.

12:57 Virginia State Police arrive on scene.

Ten members of emergency policy group convene.

1:12 Police issue a third alert that a police officer and a second victim have been shot.

1:13 Increased traffic to university's website prompts an activation of stripped-down version of the website.

1:39 Office of Emergency Management discusses need for counseling services with counselor.

1:49 Website announces that state police will lead shooting investigation.

2:00 pm

2:15 University officials meet with local reporters.

2:25 Police issue a fourth alert advising that there have been no other gunshots.

2:39 Officials announce that a news conference will be held at 4:30.

2:50 The policy group decides to postpone final exams scheduled for the next day.

2:59 Exam postponement announced.

3:00 pm

3:08 Police issue a fifth alert informing that the status of the shooter remains unknown.

3:25 Policy group ensures counselors will be available that afternoon and the following day.

4:00 pm

4:00 The university community is informed that counselors will be available.

4:15 Virginia Tech's incident commander concludes that the campus is secure. Local television stations start reporting that the alert will soon be lifted.

4:19 Update is posted: police are on-scene, two people are dead, weapon has been recovered, and no additional gunshots.

4:29 The final alert (number 6) advises that there is no longer an active threat and to resume normal activities.

4:42 News conference begins.

5:00 pm

About 5:30 News conference concludes with the promise of updated information as it becomes available.

6:00 University Counseling Center posts on website that its services are available until 9 pm and during the next day.

6:11 Policy group dismissed.

Legend

○ Campus police activities

△ Campus alerts issued

▢ Virginia Tech responses

Figure I.3 Virginia Tech case example of emergency responsiveness.

Specifically, this chapter displays the roles and implementation of robotics in fires, emergency services, search and rescue, natural disasters, law enforcement, and hazardous incidents.

Chapter two—A practical, simplified chemical agent sensor placement methodology

The possibility of the release of a chemical or biological agent in buildings has become a growing concern, intensified by the difficulty of managing such a weapon in heating, ventilation, and air-conditioning units. This chapter addresses the need for highly effective, easily operable chemical agent sensor networks and presents system design algorithms.

Chapter three—Harnessing disruptive technologies: Quick systems engineering for emergency response

Technology has experienced an exponential growth over the years, and the use of different machinery is now essential in aiding those involved in emergency management. This chapter explores different technologies that are currently being implemented in emergency situations. The chapter also addresses the use of "disruptive" technologies or technologies that disrupt the former way in which users performed tasks.

Chapter four—An emerging framework for unmanned aircraft systems in the national airspace system: Opportunities and challenges for first responders and emergency management

The role of unmanned aircraft systems (UASs) has increased dramatically as technological advances have been made. In this chapter, the use of the UAS in emergency situations is analyzed through its history along with its transition of use from traditional military operations to disaster applications.

Chapter five—Optimization in evacuation route planning

Our increasing population has caused a progressively more complex need for evacuation planning. The movement of a mass amount of people is no easy task, but this chapter uses the implementation of operations research models to provide more adequate evacuation planning. Throughout the chapter, main concerns in evacuation methods are addressed, such as timing, establishment of routes, and proper infrastructure.

Chapter six—Evacuation planning, analysis, and management

Although most evacuation studies concentrate on the evacuation of people, this chapter focuses on the evacuation of vehicles from environmental hazards and is designed as a guide for those managing and planning evacuation needs. Additionally, this chapter addresses preimpact evacuations and the proper methods of analyzing and managing an evacuation.

Chapter seven—Hierarchical optimization for helicopter mission planning in large-scale emergencies

Helicopters are frequently used in the course of responding to emergencies. This chapter discusses past methods and provides a new adaptation to optimizing the use of helicopters in disaster response.

Chapter eight—Planning and management of transportation systems for evacuation

Transportation is a key part of disaster relief. Although this chapter introduces all aspects of transportation management in a disaster situation, it also concentrates on research and other information that has been gathered in recent years.

Chapter nine—Riverflow prediction for emergency response and military applications using artificial neural networks

Emergency and military planning often requires accurate predictions of riverflow and flooding. This chapter presents a method using similar watersheds and artificial neural networks to predict these events even when the data are missing.

Section II: Human factors issues

Chapter ten—Understanding the influence of the "cry wolf" hypothesis" and "false alarm effect" on public response to emergency warnings

The "cry wolf" hypothesis, just as noted in the fable, may be affecting the public's reaction to natural disaster false alarms. This chapter addresses the issue of reaction to disasters using case studies and provides methods for improving the public's sensitivity to disaster response after "cry wolf" scenarios.

Chapter eleven—Accessible emergency management: A human factors engineering approach

Although the presence of a sufficient disaster relief system is imperative to relief success, the accessibility of such a system is even more crucial. This chapter focuses on disaster relief management for individuals with disabilities. It introduces possible changes that could be made to existing disaster relief in addition to pooling ideas of how modern methods of communication could be applied in a disaster situation.

Chapter twelve—Framework for preparing for the unprepared
Preparedness in a disaster situation will primarily influence the results of the solution. In this chapter, the authors introduce ways of integrating different entities to work quickly and effectively together. Specifically, the chapter focuses on the use of social networking in relief management and emphasizes the importance of individuals preparing themselves before a disaster strikes.

Chapter thirteen—Lest we forget: A critical analysis of bioterrorist incidents, national exercises, and U.S. prevention, response, and recovery strategies
Concern about possible biological attacks has been increasing in past years. This chapter addresses the history of biological events and prevention and discusses the U.S. policy on biological weapon anticipation and avoidance. This chapter also illuminates strategic possibilities to avoid confrontations involving biological weapons.

Chapter fourteen—Resilience to WMD: Communication and active participation are key
This chapter examines the evolution of information and communications through the lens of six case studies. This discussion provides valuable lessons in public resilience following a disaster.

Chapter fifteen—First responders: A biomechanical evaluation of supply distribution
In emergency management situations, those involved in disaster response must consider victim extraction, supply distribution, and the movement of those who were injured in the disaster. This chapter focuses on supply distribution activities and how emergency responders themselves can prevent injuries.

Chapter sixteen—Dynamics and dangers of therapeutic strategies for organophosphate poisoning: A physiologically based model
This chapter addresses the issue of organophosphate poisoning and conventional antidotes. The authors have developed a model that shows the physiological effect of antidotal atropine and oximes in response to organophosphate poisoning.

Section III: Managerial models

Chapter seventeen—Coordinated systems approach to emergency response
Proper project management is essential for providing adequate disaster relief. Through use of a systems approach, this chapter emphasizes the importance of a proactive response in emergency situations while also

providing various frameworks, models, and considerations to take into account when organizing disaster relief.

Chapter eighteen—Overcoming obstacles to integrated response among incongruent organizations

Using Japan's 2011 earthquake and tsunami tragedy as an example, this chapter on integrated response illuminates the issues of implementing successful reactions to disasters when dealing with disconnected organizations. Through studies of the malfunctions that occurred in Japan's postdisaster actions, the author discusses how greater organizational communication and compatibility lend themselves to more successful responses.

Chapter nineteen—Decisions in disaster recovery operations: A game theoretic perspective on organization cooperation

With every disaster comes the issue of organizing the separate parties involved in the relief and support in the area. This chapter gives a new perspective on establishing these interactions between different organizations.

Chapter twenty—Integrating Department of Defense response with nongovernmental organizations during a disaster

Although most incidents are handled at the lowest levels possible, disasters such as Hurricane Katrina in 2006 show the importance of proper communication between government and nongovernment entities. This chapter discusses the major players that operate in disaster response within the Department of Defense and nongovernmental organizations while also advising how these players can achieve better results through the implementation of more integrated responses.

Chapter twenty-one—Time is a murderer: The cost of critical path drag in emergency response

The optimization of time and resources is essential in any emergency response. This chapter presents project management techniques for improving response times and capabilities.

Chapter twenty-two—Coordination and control in emergency response

Through the interviews conducted with individuals affected by the devastation caused in Ohio after Hurricane Ike in 2008, this chapter uses theories of complex adaptive organizations to establish what happened during Hurricane Ike restoration. Additionally, the authors use these theories to establish possible alternative actions to be implemented during a future disaster.

Chapter twenty-three—Managing the complexities of incident command

To establish more effective response management within both civilian and military sectors, the National Incident Management System (NIMS) was created. This chapter serves to introduce the Incident Command System (ICS), a sector of NIMS. After addressing the history of ICS, the chapter identifies the system's challenges and offers potential solutions to alleviate issues in major responses.

Chapter twenty-four—Begin with the end in mind: An all-hazards systems approach to waste management planning for Homeland Security incidents

With Homeland Security incidents and spills of national significance in the foreground, this chapter explores waste management techniques that concentrate on an "all-hazards systems approach." This chapter provides guidelines for establishing a systems approach for reacting to a hazardous disaster.

Chapter twenty-five—Creating effective response communications

Emergencies require an integrated response that is supported by open communication. This chapter concentrates on the use of communication in a disaster situation. It specifically focuses on communication plans, the use and application of information systems, and the impact of communication via social networking and new media.

Chapter twenty-six—Assessing the state of knowledge about emergency management in other countries

In order to more effectively manage disasters, the Federal Emergency Management Agency developed a management plan to better respond to future disasters. This chapter addresses this plan in addition to illustrating how the international community can better recognize the difficulties of disaster management and quickly respond to issues.

Chapter twenty-seven—Framework for real-time all-hazards global situational awareness

In order for adequate disaster relief to occur, several aspects of control must fit together. Organizations should be able to communicate and share their information with one another, update their information as the disaster progresses, and possess an all-hazard framework in order to make analyses. This chapter addresses these aspects through an introduction to the Energy Awareness and Resiliency Standardized Services framework.

Chapter twenty-eight—All-hazards response team preparation: Planning and training concepts

The use of an "all-hazards" application in dealing with disasters is an important aspect of dealing with disaster management. This chapter concentrates on how to train, coordinate, and use organizations that are tasked with dealing with a multitude of different dangers.

Chapter twenty-nine—Medical supply chain resiliency in disasters

This chapter introduces several strategies for accurately distributing medical supplies and expertise in an emergency situation. Comparisons are made to commercial supply chain management with an emphasis on proper disaster management.

Chapter thirty—Decision support for inland waterways emergency response

There are times during contingencies that roads and bridges may not be available. Inland waterways may be a viable alternative to the highway system as a transportation mode. This chapter offers decision support in order for emergency managers to design inland waterway-based emergency response systems.

In summation

As evidenced by the breadth and depth of the contents summarized above, the contents of this handbook will add to and enhance the overall repertoire of tools and techniques for a comprehensive emergency response.

References

Badiru, A. B. (May 1987), "Communication, Cooperation, Coordination: The Triple C of Project Management," in *Proceedings of 1987 IIE Spring Conference*, Institute of Industrial Engineers, Washington, DC, pp. 401–404.

Rinaldi, S. M., J. P. Peerenboom, and T. K. Kelly (2001), "Identifying, Understanding, and Analyzing Critical Infrastructure Interdependencies," *IEEE Control Systems* 21(6): 11–25.

Virginia Tech Police Department (2011), "Virginia Tech Emergency Response Timeline," Case Example Report.

section I

Technical challenges

chapter one

Robotic technologies for first response
A review

Julie A. Adams

Contents

1.1 Introduction

Robotics has the potential to transform first response capabilities. Today robots are typically deployed by municipality bomb squads, who sometimes improvise to assist with other incidents and personnel, such as Special Weapons and Tactics (SWAT) teams (e.g., Barrett 2010; Edwards 2011). The widespread use of ground and aerial robots in recent military conflicts has resulted in a large number of first responders who understand the benefits of such technology. However, much of that technology still resides with the military or in the research laboratory. Recently, the United States Congress mandated that the Federal Aviation Association provide a plan for integration of unmanned aerial vehicles into the national air space by 2015. This mandate will facilitate the deployment of unmanned aerial vehicles by first responders significantly and transform the manner in which first response activities are conducted.

The use of robots for hazardous situations dates back to the Three Mile Island Nuclear incident (e.g., Pavelek 1985; Whittaker and Champeny 1987; Geifer et al. 1988). More recently, the impetus for first response robots was spurred by the 1995 bombing of the federal building in Oklahoma City (Blitch 1996). This incident essentially resulted in a new field of robotics focused on urban search and rescue (USAR) that then splintered into additional fields such as response for natural disasters, mining incidents, and CBRNE (Chemical, Biological, Radiological, Nuclear and Explosive) incidents. The first well-publicized deployment of robots occurred after the World Trade Center incident on September 11, 2001. Since that time, both ground and aerial robots have been deployed by the thousands in recent military conflicts and have seen growing deployments for civilian incidents.

For more than a decade, robotics for incident and disaster response has seen vibrant growth with a large focus on developing capabilities, primarily in the research laboratory. One of the prominent individuals in this area of research is Dr. Robin Murphy who has deployed robots at the World Trade Center, the La Conchita mudslides, and after Hurricanes Katrina, Ike, and Wilma.

While robotic technology continues to advance, a number of difficult challenges exist. One of the most difficult challenges is the development of the capabilities for the human responder to interact with the robots. There are a number of relationships that can occur between human and robotic entities. Scholtz (2003) and Goodrich and Schultz (2007) provided the commonly accepted interaction roles: supervisor, operator, mechanic, peer, bystander, mentor, and information consumer. While victims are representative of bystanders and future emergency response robotic systems will incorporate a majority of these roles, current systems primarily incorporate humans as operators or as information consumers (e.g., a sensor operator). An operator is defined as an individual who directs the behaviors and actions of the robot either by modifying parameters or through remote control (e.g., teleoperation). An information consumer does not directly interact with the robots, but rather uses information that originates, at least partially, from the robots.

The remainder of this chapter focuses on presenting current capabilities from both a research perspective and a commercial perspective. Specifically, the chapter presents capabilities for response to fires (Section 1.2), including wild fires; emergency management services (Section 1.3); search and rescue (Section 1.4), including urban and wilderness search and rescue; natural disasters (Section 1.5); law enforcement (Section 1.6); and CBRNE device incidents (Section 1.7). Robotics technology has the potential to keep thousands of first responders safe while performing dull, dangerous, and repetitive tasks, as well as facilitating the response activities.

1.2   Firefighting

Robots face a number of challenges when deployed for firefighting. The response concerns include keeping human responders safe, saving victims, and minimizing damage (Humphrey and Adams 2011). Robots must be able to navigate dangerous environments, understand the environment, interact with human responders, and respond appropriately in these highly uncertain and dynamic environments. As a result, there are a number of necessary capabilities that must be understood and developed.

The ability to navigate and understand the response environment when fighting a fire is one aspect that must be provided in a fire response robot. In particular, robots must be able to autonomously, or at least at a high level of semiautonomy, navigate the potentially unknown hazardous environment. Fires present unique challenges, including smoke and charring that can hinder visual sensing, heat and bright light (e.g., fire itself) that can hinder thermal sensing, and changing landscapes that can hinder accurate mapping and understanding of the environment. Oftentimes, the fire incident area is unknown and thus requires "exploration" to locate victims and minimize the hazards from the fire. A common means of understanding a new area with robots is called simultaneous localization and mapping (SLAM) (Thrun et al. 2005), which is considered a necessary technology for rescue robots (Birk and Carpin 2006). SLAM typically relies on a combination of laser range finders, visual cameras, odometry, internal measurement units, and/or Global Positioning System (GPS) to develop a map of the environment.

While a large number of SLAM algorithms exist, some algorithms have been specifically designed for first response. Calisi et al. (2005) designed their SLAM algorithm to incorporate two algorithms. One algorithm focuses on the SLAM problem, while the other plans the robot's motions. Specifically, this algorithm has been designed for cluttered environments but does not consider smoky conditions. Brunner et al. (2011) focus on developing localization capabilities in a smoky environment. Brunner et al. used visual and infrared cameras for their SLAM application and demonstrated that multiple sensors are important for smoky environments. A limitation of both these techniques is their two-dimensional (2-D) nature. Nüchter et al. (2005) provided the first three-dimensional (3-D) SLAM algorithm for this domain. The result of applying their algorithm is a 3-D map of the rescue environment. Pellenz et al. (2010) expanded on Nüchter et al.'s results by creating a 3-D map with terrain classification using a 3-D laser range finder, GPS, an internal measurement unit sensor, and color vision in real time for outdoor environments. Šakėnas et al. (2007) use time-of-flight image sensors and a stereo camera to segment 3-D point clouds into planar semantic floor plans. This mapping and segmentation does not occur locally on the robot, but the algorithm and robot can be used to generate floor maps for first responders about to enter the building. Multiple robots can be used to develop a centralized 3-D laser–based digital elevation map of the environment (Nagatani et al. 2009).

Tran et al. (2011) analyzed the performance of four common SLAM algorithms using the Microsoft Kinect sensor to reconstruct 3-D scenes and determined that each algorithm has advantages and disadvantages. A limitation of Tran et al.'s analysis was that it was for common, cluttered office environments. Tretyakov and Linder (2011) analyzed the performance of three common laser range finders, a time-of-flight camera, and the Microsoft Kinect for developing 3-D maps and analyzing terrain traversability of radiant surfaces in smoky environments. All sensors were tested under the same conditions using a fog machine that emits particles similar to real smoke. All sensors' performance degraded as smoke density increased, with the Kinect performing particularly well, although the Kinect has a short sensing range compared to the other sensors. Capitan et al. (2007) suggested vision-based algorithms for recognizing and tracking first responders (e.g., firemen) using a combination of a ground sensory network (e.g., cameras) and an unmanned aerial vehicle. While this system is not focused on developing a map of the environment, it does focus on scene understanding and segmentation.

SLAM, terrain traversability, and scene reconstruction not only permit the robots to navigate first response domains but can also be used to allow personnel from team leaders through incident commanders to know the location of their personnel during an incident. The maps can be provided to the remote response personnel monitoring the personnel within the incident scene as well as to the response personnel in the incident scene (Saarinen et al. 2005). This information is particularly important in smoky and dark environments for allowing responders to be aware of the spatial orientation of the environment and providing a means, other than tactile, to locate one's self in the environment (Saarinen et al. 2005).

SLAM technologies are also an important component of the GUARDIANS project, which employs a swarm of robots to guide human first responders during warehouse fires, while maintaining communications with external entities (Penders et al. 2011). The swarm of robots is intended to overcome the environmental sensing limitations mentioned previously that are common in smoky environments. The human first responder is able to see feedback from the robots on the protective headgear visor. The robot swarm reacts to the human responder's navigational motions, and the robots can assist the human first responder with navigating the environment, for example, through a heavily smoky environment to the exit.

Another means of monitoring first response personnel is through imagery provided by unmanned aerial vehicles. Maza et al. (2010) presented a multiple unmanned helicopter approach for monitoring first response personnel based on visual imagery. While this is a theoretically sound approach, it is not practical given current unmanned aerial power limitations (typically on the order of 20–45 minutes) and the number of highly trained human operators required to supervise each aircraft. Rao et al. (2005) also suggest unmanned aircraft, but focus on airships (e.g., balloons) due to their stability, limited power usage once they are in place for monitoring purposes, and typically higher payloads that permit a broader range of sensor capabilities. In addition to monitoring first responders, Rao et al. suggest that airships can be used to provide surveillance, visual-based search and rescue, and a communications link between incident command and ground-based first responders. While this approach incorporates only a single unmanned aerial vehicle, it also relies on ground-based cameras, which either must be in place before the incident or can be transported into the incident environment through ground robots. Of course, a limitation of deploying multiple ground robots, particularly into buildings, is related to managing the initial congestion as multiple robots attempt to enter the building (Ma et al. 2009). This deployment congestion problem can be managed in the same manner as human responder deployment is managed, but a better approach is to let the robots autonomously determine how to manage such congestion on their own, as proposed by Ma et al. (2009).

Sometimes specialized systems are required for particular types of fires. For example, fighting fires in traffic tunnels where it is necessary to quickly deploy firefighting capabilities, but traffic may block access to the incident area. Celentano et al. (2005) developed a roof-mounted monorail robot system to solve the accessibility issues often encountered in tunnel fires.

Wildfires also present different challenges than building or structure fires. Wildfires often encompass large geographical areas across varying terrain and weather conditions. Murphy et al. analyzed appropriate functions for robotic wildfire responders with corresponding technologies, including an outline of necessary research areas (Murphy et al. 2009a, 2009b; Murphy 2010b). Specific recommended tasks include moving items such as cargo or people, moving through unburned areas in front of the fire to determine potential direction and speed of the fire; providing information regarding fire suppression and the location of hot spots; investigation of hazardous areas; and acting as a mobile weather station. Murphy (2010b) also investigated the use of unmanned aerial vehicles for application to wild fires and noted the immediate benefit from deploying such vehicles for wilderness fires. Specifically, she outlined criteria for selecting a platform for specific tasks. NASA (2005) has investigated the potential tasks for unmanned aerial vehicles in response to wildfires. While NASA outlines similar tasks as Murphy et al., NASA's tasks are broader and include provision of real-time communications between ground personnel and the field command center, and the application of fire retardants (NASA 2005). Control algorithms for tracking the front line of a fire and fire suppression by multiple unmanned aerial vehicles have been evaluated in simulation for several hot spots and heterogeneous fire conditions (Kumar et al. 2011).

One focus of wildfire unmanned aerial vehicle research has been computer vision techniques for characterizing the fire to determine appropriate response strategies (Martinez-de Dios et al. 2008). In particular, Martinez-de Dios et al. have employed visual and infrared imaging along with GPS capabilities to develop 3-D representations of the location of the front of the fire, the height of the flames, the flame inclination angle, and the base width of the fire. The captured and fused information can be georeferenced to support fire command planning.

1.3   EMS

Assisting victims is a very demanding aspect of first response and one that many have focused on for robotic applications. The ultimate intent is to assist the human first responders in such a way as to keep these individuals safe, allow them to rescue more victims, and minimize the injuries to themselves and the victims. This section reviews some of the burgeoning technologies in this area of robotics for first response.

A key area in which robots can assist is in identifying and locating victims. Robots in general are able to enter response areas and locate victims. Ground robots can use images/video (Castillo and Chang 2005), hyperspectral imagery (Trierscheid et al. 2008), and thermal cameras (Hahn et al. 2011), or a combination (Aziz and Mertsching 2010) to locate victims. Vision-based systems are still faulty and must rely on specific techniques for identifying victims within the images. For example, Castillo and Chang developed a system based on templates incorporating aspects such as detecting the silhouette of the victim and skin to identify victims. Hahn et al. (2011) combine thermal sensors with SLAM technologies to create a 2-D heat map of victim locations. The combination of the detection with localization should allow responders to go to the victim's location and allow the robot to map and explore the environment, while simultaneously searching for victims. Trierscheid et al.'s (2008) spectral imaging approach permits heat detection through infrared, thus overcoming limitations of visual systems that rely on color detection as well as common limitations with fog or dust. Additionally, hyperspectral imaging can detect victims independent of body temperature.

Once victims are identified, it may be appropriate for the robot itself to triage or to instruct an uninjured victim or bystander to triage the victim. Chang and Murphy (2007) identified how robots can assist with triage, in addition to the technology required to complete these tasks.

The specific tasks incorporating the robot sensors and the remote robot operator include encouraging movement and assessing respiratory frequency, perfusion, and mental status. Harriott et al. (2011a, 2011b) investigated the use of a robot to instruct an untrained civilian in triaging victims for incidents involving contaminants before human responders are permitted into the contaminated hot zone. This scenario assumes that the robot can be used to locate the position of the victim, that the civilian agrees to assist the robot with triage, that the human follows the instructions provided by the robot, and that remote EMS personnel are overseeing the triage results. The intent is that victims that require immediate care can be identified, and when first responders are permitted into the hot zone, they can go directly to the victims requiring the most urgent care.

While Harriott et al.'s approach does not require the robot to actually conduct triage steps, others have focused on providing such capabilities. Specifically, Murphy et al. (2004a) investigated medical reachback capabilities for area denial situations in which the robot was equipped with imagery, auditory communication, a triage sensor, and a fluid delivery mechanism. The system allows remote medical personnel to interact and assess the victim. Bethel and Murphy (2008) explored this concept further by investigating how the robot should approach the victim, how close the robot should position itself to the victim, and how environmental constraints can impact interaction with the victim. Asaoka et al. (2008) developed electrocardiogram and arterial blood oxygen saturation degree sensors that robots can deploy during triage to detect a victim's vital signs.

Another important task robots can assist EMS personnel with is transporting victims post-triage. Vecna Robotics™ Battlefield Extraction-Assist Robot (BEAR) has been designed for this task, among others. The BEAR carries the rescued victims in its arms. However, Iwano et al. (2011) have focused on developing a robotic stretcher capable of climbing stairs. Iwano et al.'s robot requires the first responder to direct the stretcher robot, but the robot is able to position the victim on the stretcher autonomously and is able to transport average victims up and down stairs. Both the BEAR and Iwano et al.'s systems are quite large, making them difficult to use when transporting victims from small confined spaces. Alternatively, Yim et al. (2011) focused on victim manipulation using small, man-portable robots to move the victim by grasping the wrists or ankles and stabilizing a victim's head using foam.

1.4   Search and rescue

Search and rescue is a prominent first response activity, and the necessary capabilities vary depending on the type of search and rescue to be performed. USAR typically focuses on rescuing victims from collapsed structures, wilderness search and rescue seeks to locate and rescue victims lost in the wilderness, while water search and rescue focuses on victims in the water. USAR became a prominent robotics research topic after the 1995 Oklahoma City bombing. Blitch (1996) was the first to outline the search and rescue challenge for robotics. Murphy took up the challenge and has led the research field in this area since (Casper and Murphy 2000; Blitch et al. 2002; Murphy 2005; Murphy et al. 2008a). This section provides a brief overview of USAR robotics technologies, one of the most studied areas of robotics for first response, as well as wilderness search and rescue.

1.4.1 Urban search and rescue

USAR focuses on the problem of locating and rescuing victims in rubble piles, such as collapsed structures. Murphy and Blitch founded the field of USAR robotics, and Murphy's team was one of the five teams deployed to ground zero after the incident on September 11, 2001 (9-11). After 9-11, the area of USAR robotics saw a surge of research funding and activity; thus, it is one of the first response areas that has received the most attention. It is also an area where there have been significant advances in robotic technology (Micire 2007).

USAR is a very challenging problem for robotics. As such, a number of competitions, simulators, and test environments have been established for evaluating robotic capabilities before deployment. USARSim is a simulation tool for developing capabilities (Lewis et al. 2007; Balakirsky and Kootbally 2012). The RoboCup Rescue competition was designed to provide challenge problems for robotics research and development in this domain (Murphy et al. 2002; Osuka et al. 2002; Balakirsky et al. 2007). While the competition began as a simulation-based competition, it has transformed into a physically situated environment developed as a USAR test bed by the National Institute of Standards and Technology (NIST) (Jacoff et al. 2003). The competition has also inspired the development of test beds outside of the United States, such as the Japanese House (Shimaoka et al. 2011). The past few years have seen the testing of robots at locations such as Disaster City (Birk et al. 2009; Disaster City 2012), and new locations are coming online that will be useful for testing robotic first response technologies, such as Calamityville (Calamityville Technical Laboratory 2012).

While the fidelity and breadth of testing options continues to grow and first responders have become involved in system tests, there is no substitute for actual incident deployments. Murphy is one of the most experienced individual researchers worldwide in deploying robots for actual incidents. Murphy's team was one of five teams who deployed robots in the aftermath of September 11, 2001 (Murphy 2002, 2003, 2004a; Micire 2007). While the robots generally performed well, the deployments at the

World Trade Center realized a number of technological limitations that served as research motivation. Since 2001, robots have been deployed in response to the La Conchita mudslides (Murphy and Stover 2008); Hurricanes Katrina, Ike, and Wilma (Murphy et al. 2008b, 2009c); and building structure collapses (Pratt et al. 2008; Linder et al. 2010); however, the technology has evolved. Tread and wheel-based navigation were cited as limitations in early deployments; thus, many new navigation technologies have been developed (Voyles and Larson 2005; Tesch et al. 2009; Linder et al. 2010; Hatton and Choset 2011). However, these new capabilities do not always mean that robots can be deployed in response to an incident. The deployment of robots is dependent on the type of building collapse or rubble pile. Pancake, lean-to, and v-shape voids are typically searchable by robots. However, the voids must be large enough for the robots to be inserted, and the rubble pile structure must be stable enough to support humans on the pile to deploy the robots. For example, it was intended that robots would be deployed for the collapse of the municipal building in Cologne. The resulting small voids would have permitted the caterpillar-like robot to enter the voids. However, the structure could not support humans standing near the point of interest to deploy the robot (Linder et al. 2010).

While the primary focus has been on ground-based robots for USAR, more recently aerial vehicles have also been incorporated. For example, the Berkman Plaza II collapse incorporated an unmanned aerial vehicle for the purpose of conducting a structural forensic inspection of the collapsed structure (Pratt et al. 2008).

Many technological issues persist for deploying robots for USAR. The majority of current vehicles are tracked vehicles that are teleoperated and require a trained operator to maneuver the robot. Issues persist in relation to navigation capabilities, sensing of the remote environment, autonomy, and human interaction with the system.

Navigation and sensing have been cited as a large problem for USAR because every incident results in a different and challenging navigational environment (Casper and Murphy 2000; Blitch et al. 2002; Murphy 2010a; Linder et al. 2010). Tracked vehicles provide needed traction and are power efficient, but incidents can require the robot to rappel down into the void (Murphy 2010a) and large variances in structure types can impact mobility (Murphy 2004a). As a result, many researchers have focused on developing unique and interesting robotic capabilities.

One area of development has been basic robot architectures for this domain. Murphy has used the American Standard Robotics Inuktun VGTV (Variable Geometry Tracked Vehicle) Xtreme series ground-based tracked platforms (Micire 2007). Tracked platforms are common because of their mobility on uneven surfaces, traction, and speed (Kang et al. 2005; Carnegie 2007; Garcia-Cerezo et al. 2007), which typically use a skid-steer system. An extension of tracked vehicles is crawler vehicles (Arai et al. 2008). Crawler vehicles are tracked bodies that are connected; typically two to three bodies are connected using active joints. Recent developments have resulted in an independently actuated double-sided crawler and a mono-tread crawler (Arai et al. 2008) that permit rotation about a point, improve balance on rough terrain, permit lateral movement, and include the ability for the robot to turn over. This type of structure allows the robot to change its body posture. An extension of the crawler robots is platforms that provide snakelike locomotion (Wolf et al. 2005; Tesch et al. 2009; Maruyama and Ito 2010; Simodate and Ito 2011). The high-degree-of-freedom (DOF) snake robots can access narrow and small spaces by manipulating their bodies around obstacles and so on.

Alternatives to tracked platforms are hybrid combinations of wheels with legs, spokes, lobes, and so on (Eich et al. 2008; Herbert et al. 2009; Hunt et al. 2011). Eich et al. (2008) developed a robot platform that uses four wheels and each wheel has five compliant legs. The resulting vehicle can move quickly on flat ground, while also being able to navigate rough terrain and stairs. Wheeled robots are a common alternative in most domains and have been considered for USAR. Hunt et al.'s (2011) USAR Whegs™ is able to move at the speed of a person walking with the treads, but can locomote faster than average walking speed with its wheel-legs. While wheeled robots have been suggested (Ruangpayoongsak et al. 2005; Sato et al. 2007), wheeled robots suffer limitations when attempting to navigate rubble or uneven terrain. Carnegie (2007) suggested a wheeled platform with a split chassis to permit traversal of rough terrain.

Robots that jump also provide a unique alternative design that allows a robot to traverse rough terrain and obstacles that are larger than the robot (Dunwen et al. 2011). Robots that locomote by jumping can be smaller than the tracked robots, thus allowing jumping robots to access smaller cavities in the rubble. An advantage of jumping robots is the ability to adjust the height and distance of the jump, but a limitation of such an approach is the potential difficulty of controlling the locomotion accuracy of jumping robots. While the TerminatorBot reconfigurable robots do not jump, they crawl and can reconfigure themselves to access small spaces (Voyles and Larson 2005; Voyles et al. 2010). Reconfigurable, modular robots are able to change their shape to a larger extent than robots such as the Inuktun micro-VGTV. Li et al.'s (2005) three-module reconfigurable robot can transform from a shape similar to the above-cited crawler robots to a standard tread robot (similar to the Inuktun micro-VGTV). Similar in concept is the CUBIC-R robot, which uses crawlers to transform into an omnidirectional hexahedron (Tabata et al. 2005).

One element of interest is providing platforms with manipulation capabilities. Garcia-Cerezo et al. (2007) designed a vehicle that provides a 4-DOF arm with a dual 3-DOF manipulator to provide dexterous manipulation at the ground level and areas above the robot base. While not strictly focused on manipulation, another approach provides a robot jack for lifting rubble to access victims (Tanaka et al. 2005).

Many of the cited robots employ standard image cameras, infrared cameras, or laser range finders. Murphy has cited the difficulty of identifying victims (and remains) using visual imagery (Casper and Murphy 2003). Testing by Bostelman et al. (2005) determined that the CSEM SwissRanger-2 3-D range camera could be used to detect large objects in a USAR environment, but that the sensor provided false positives when attempting to detect human body components (e.g., arms, legs, and torsos). Thus, their conclusion was that the sensor can be useful for general navigation, but it is an inappropriate sensor for detecting victims. Craighead et al. (2006) evaluated the Canesta EP200 series range cameras for application to USAR. While they did not specifically test for identification of humans, they did determine that the sensor allowed an operator to better understand the environmental layout, but the sensor was unable to detect certain types of materials and has limited capabilities for outdoor daytime deployments.

An important component of USAR is the identification of the location of buried victims. Akiyama et al. (2007) focused on developing a ground-penetrating radar system that detects surviving victims' location using the victims' respiration frequency. Their technology was intended to locate victims up to 5 m from the sensor antenna. While the authors demonstrated their approach, it requires multiple antennas to detect the victim, a limitation of the proposed approach.

Irrespective of navigation mode, the ability to navigate and identify victims is a challenging one. Typically, the environment must be explored to locate victims. Carbone et al. (2005) developed a model-based algorithm to permit interactive exploration of the USAR environment that also identified unexplored areas. Saeedi et al. (2009) developed more generic algorithms for exploring USAR environments by permitting the assessment of performance metrics that include effectiveness of coverage and the number of victims located. Specifically, the authors developed an algorithm that allows look-ahead analysis to determine where to search next.

One of the most challenging aspects of USAR robotics is the interaction between the human and the robot. Most such systems are teleoperated; thus, the human operator is a necessary component to a successful deployment. A limitation for interacting with remotely located robots is that they provide limited environmental information, such as images to the operator (Casper and Murphy 2003; Burke et al. 2004; Murphy 2004b; Murphy and Burke 2005; Murphy 2010a). A significant limitation remains in that typically raw sensor data is presented to the operator for interpretation. As Murphy (2010a) indicates, the environmental circumstances, image displays, and the basic user interface frequently hinder the human's ability to fully interpret the circumstances presented in the image. As a result, there has been a large research focus on developing interaction techniques to assist human operators.

The prior efforts have focused on developing an interface for a single operator to a single robot or multiple robots (Burke and Murphy 2004; Lewis et al. 2009; Lee et al. 2010). However, research has shown that multiple humans in the team can prove beneficial. Burke and Murphy (2004) indicate that USAR "currently requires a minimum of 2:1 human-to-robot ratio," meaning that at least two humans are required to operate a single robot. Their field deployment results suggested that a team of humans, with defined roles who communicate in a goal-directed manner and develop a shared mental model of the task, robot, and environment, proves more effective. In fact, later research demonstrated that two rescuers working in a cooperative manner with a robot were "nine times more likely" to locate victims than a team consisting of a single human and a single robot (Murphy 2005).

Lewis et al. investigated using multiple robot and multiple human teams for USAR (Lee et al. 2010; Lewis et al. 2010). This effort assumes that the individual robots have some autonomous capabilities, thus relieving the operators of teleoperating the individual robots all the time. This work focuses on understanding if particular robots should be assigned to particular operators or if a shared pool of operators can effectively supervise and direct robots assigned to them as needed. In other words, the pool of operators each have temporary assignments to particular robots. The dedicated operators were each assigned 12 robots and the corresponding interface provided 12 video feeds, one from each of the assigned robots. The pool of operators used the same interface, but instead had 24 video fields, one from each robot in the system. The authors' conclusion was that the shared pool of operators encountered difficulty monitoring all the video feeds, which actually served as a distraction. As a result, the operators were less able to focus on particular robots and locate victims. The operators became confused when multiple robots found similar looking victims and sometimes recorded the same victim as a new victim multiple times. While these results suggest that more humans and robots may prove more efficient, these results combined with Murphy's suggest that well-defined roles may better serve operators in USAR situations.

Existing results show that operators rely heavily on video streams and such reliance can be a determent (Baker et al. 2004); however, video streams still play a significant role in interface design, particularly for USAR. Baker et al. (2004) developed an interface intended to better integrate necessary information on the interface with the video stream. The primary intent was to enhance the operator's awareness and efficiency and assist with selecting robotic modality, while reducing cognitive load. More recent results from this collaboration have resulted in further integration of the sensory information (Yanco et al. 2007). Specifically, the team has integrated two video channels (one of which is intended to be the primary focus) with a representative map of the environment and a distance panel. The map provides an indication of the robot's current position in the environment, along with a representative trail of where the robot has been. This map can be zoomed so that obstacles in front of the robot are more apparent. The distance panel provides an indication of how close obstacles and objects are relative to the robot on all sides of the robot. The interface also provides a display of the robot's operating mode and status, such as battery level. The research shows the progression of the interface design and the final design was demonstrated to provide increased capacity for navigating narrow spaces and resulted in fewer collisions. However, the authors suggest a number of important improvements. For example, the authors acknowledge that the distance display does not facilitate the operator's ability to improve overall environmental awareness.

The ability of operators to have an awareness of the remote USAR environment is critical for mission success. As a result, many have focused on developing appropriate interface support capabilities. Scholtz et al. (2005) investigated alternative methods of supporting the operator's situation awareness. One method relied on automatic mapping of the environment by onboard robot sensors, while the other relied on an overhead camera for providing a frame of reference. The environment for their assessment was the NIST Orange test arena. They found that both methods resulted in fewer critical navigation incidents and that there was virtually no difference in the awareness provided to the operators. However, both approaches suffer from limitations. The onboard sensing and mapping is expensive computationally, while most real USAR environments will not be realistically outfitted with overhead cameras.

Standard efforts have been pursued related to USAR robots and human–robot interaction (Murphy 2010c; Jacoff et al. 2003, 2010). While the standards have progressed over the years, there will need to be a continued effort to refine and develop new standards as the response robotic technology continues to evolve.

A final interesting concept is the coupling of a search and rescue canine with a small ground-based robot (Tran et al. 2010). The canine carries the robot in a harness and the robot can be deployed from the canine's position. This approach allows the robot to be deployed from a location that humans may be unable to access.

1.4.2   Wilderness search and rescue

Wilderness search and rescue is focused on locating and rescuing individuals lost in the wilderness. The search techniques and environment vary dramatically across different incidents and from USAR. In particular, search personnel may have to cover large areas over rugged and sometimes inaccessible terrain. The focus of robot-enabled wilderness search and rescue has been on the use of unmanned aerial vehicles to provide assistance with the search process. Goodrich and colleagues at Brigham Young University have conducted the majority of the research in this area (Goodrich et al. 2009; Lin et al. 2010).

An important issue to be addressed is how to integrate the unmanned aerial vehicle into the search response. Unmanned aerial vehicles can assist with the search, can fly over areas inaccessible to human searchers, and can be used to guide human personnel to critical signs or the missing person. However, it was unclear how to incorporate the unmanned aerial vehicle and the unmanned aerial vehicle operator into the search and rescue process. Thus, research provided guidelines on team structure and personnel roles when unmanned aerial vehicle support for wilderness search and rescue is provided (Goodrich et al. 2007; Cooper and Goodrich 2008; Goodrich et al. 2009; Adams et al. 2009; Lin et al. 2010).

The most common unmanned aerial vehicle sensor is imagery. Goodrich and Morse have focused on techniques to support the unmanned aerial vehicle sensor operator (e.g., information consumer) with detecting signs of the missing person. The team has completed fundamental work focused on fusing, or mosaicking, individual images from the unmanned aerial vehicles, along with various stabilization algorithms that keep the images on the screen longer to support the human sensor operator's detection of signs of the missing person (Goodrich et al. 2008). The team has used video georegistration to develop see-ability coverage maps to prioritize areas to be investigated by the human sensor operator (Morse et al. 2010). The team has also combined visual and infrared video for identification of the missing person and found that the fused view resulted in users identifying as many potential victims as was possible with a side-by-side presentation of the same information, with lower workload than a method that placed two sets of video side by side (Rasmussen et al. 2009). Most recently, the team has focused on developing a spectral anomaly detector to augment the presentation of objects that differ in color from the expected background (Morse et al. 2012). Their results found that operators were able to significantly increase detection of objects and make the overall task easier.

Another important element of the research is the development of unmanned aerial vehicle–specific capabilities. To that end, Lin and Goodrich (2009a, 2010) have studied methods for modeling expected behavior of lost persons that account for the terrain features in the specific area. These models can be incorporated into intelligent path planning algorithms that account for the unmanned aerial vehicle's limited flight time and the most likely area in which the missing person may be located (Lin and Goodrich 2009b).

1.5   Natural disasters

Natural disasters are frequently in the news across the globe, from tornados in the Midwest of the United States, earthquakes worldwide, tsunamis in Asia, hurricanes and typhoons, and so on. These types of disasters can result in collapsed structures that require ground-based search and rescue robots (see Section 1.4.1) or other natural incident areas, such as mudslides, but these types of disasters also provide unique opportunities for aerial and water-based robots.

The La Conchita mudslide that buried 15 homes with known survivors provided a unique opportunity to apply ground-based robot technologies developed for USAR (Murphy and Stover 2008). The deployment used commercially available USAR robots, namely the American Standard Robotics Inuktun VGTV Xtreme robot. The incident emphasized the need to be able to easily transport the vehicles and associated equipment through backpack, since the search locations were only accessible on foot. One home's interior was inaccessible to animals and humans, and it was unclear if the robot could access the necessary areas. Two minutes after insertion, the robot's track was displaced, rendering the robot incapable of navigation (Murphy and Stover 2008). The second home searched also presented mobility challenges for the robot, in the form of a shag carpet that caused the track to come off. A similar issue had been encountered during deployment at the World Trade Center, but the robot had been successfully deployed in other response efforts. The differences between the incidents included the environment surface that the robot had to navigate. While tracks are a preferable navigation method, they can encounter problems.

Hurricanes have also provided unique opportunities to deploy robotic technology. Hurricane Katrina provided an opportunity to understand how small-unmanned aerial vehicles can be employed in disaster response. Specifically, the iSensys IP3 vertical takeoff and landing vehicle was deployed to assist with structural surveys of multistory commercial structures in Biloxi and Gulfport (Pratt et al. 2009). A concept of operations was developed for deployment of such vehicles in cluttered urban environments. The deployment also resulted in a recommendation that flight crews consist of at least three people to one vehicle: one pilot, one mission specialist, and one flight director. The team also cited the need to develop semiautonomous capabilities to assist with flight control and data collection as well as a need to focus on human–system interaction.

Murphy's Center for Robot-Assisted Search and Rescue also deployed after Hurricane Wilma to survey damage at the Marco Island Yacht Club (Murphy et al. 2008b). In addition to deploying the iSensys VTOL aerial vehicle, the team also deployed an AEOS-1 unmanned surface vehicle with a dual-frequency identification acoustic camera below the waterline and a video camera mounted on the mast. The aerial vehicle was employed to check the roof structure of the club, while the surface vehicle was employed to assess the level of damage to the seawall bounding the yacht club dock. Additionally, the two vehicles were deployed to inspect a nearby bridge and fishing pier. Murphy et al. determined that the small surface vehicle's advantages over manned surface and unmanned underwater vehicles included a smaller size that permitted investigation of areas inaccessible to larger vehicles, the ability to provide simultaneous above and below surface imagery, and the ability for real-time control and video imagery.

The combination of the aerial and surface vehicles also provided unique advantages (Murphy et al. 2008a). Specifically, the aerial vehicle assisted with positioning the surface vehicle relative to the inspection target enhanced the safety of operating the surface vehicle and acted as a communication relay between the surface vehicle and the human responders.

There remain open challenges for deploying aerial and surface vehicles for natural disaster response. The ability to provide SLAM based on above and below surface imagery for the surface vehicles is pertinent. The structure of the environment also poses a number of challenges, such as intermittent GPS and communications, fog, lighting, waves, tides, and currents.

Following Hurricane Ike, the Center for Robot-Assisted Search and Rescue deployed unmanned marine vehicles for purposes of inspecting the Rollover Pass bridge in Texas (Murphy et al. 2009c). The team deployed the Sea-RAI surface vehicle, the VideoRay tethered remotely operated vehicle, and the YSI Ecomapper autonomous underwater vehicle. The Sea-RAI vehicle is similar to the AEOS-1 system deployed after Hurricane Wilma, but with the addition of video cameras to provide forward, backward, and hemispherical coverage above the water. This vehicle was able to autonomously navigate to the bridge and was then manually controlled. The YSI Ecomapper provided a side-scan sonar and was able to map the debris field. The Ecomapper suffered a 40% deployment incident rate due to debris in the water until a map of the channel was provided by the Sea-RAI. The VideoRay's tether became entangled in a pipeline, illustrating the importance of tether management.

As aerial and water-based unmanned technology continues to develop, additional deployments and uses for this technology will emerge for response to natural disasters and other types of incidents.

1.6   Law enforcement

Law enforcement has seen usage of ground-based bomb squad robots over the past decade. Bomb squad robots have very specific capabilities and are remotely controlled (see Section 1.7 for further discussion). While it is true that these robots have been used for other law enforcement tasks (Basich 2012), the size and configuration of these robots limits their general applicability to a broad spectrum of law enforcement tasks. As ground robot capabilities continue to improve and marine and aerial robots become more prevalent, there will be expanded use of robots for future law enforcement tasks.

Patrol and surveillance activities are a commonly cited law enforcement task cited as dull and boring that will be good to assign to robots (Humphrey and Adams 2009). Ground robots can be tasked with such capabilities if the environment is known (Vig and Adams 2007). However, unknown environments become more challenging due to uncertainty regarding terrain, features to be monitored, and so on.

Randelli et al. (2011) have developed a system for patrol and surveillance of complex indoor environments. The system relies on SLAM (see Section 1.2 for detailed SLAM discussion), semantic mapping, image recognition, and human intelligence to add items to the map and to recognize anomalous situations and humans.

Future surveillance and patrol tasks will require multiple robots or teams composed of humans and robots. Such deployments will require the human to move through the environment with the robot indicating items of interest and potentially providing a semantic mapping to the item so that the robot can incorporate the items into the map. The resulting map can be used by the robot to patrol a building autonomously. Such a system has limited application to a particular building. While similar systems have been developed (Chien et al. 2005; Mullins et al. 2007), these systems still rely heavily on a human operator, even when they incorporate a broad set of sensor capabilities (Chien et al. 2005) and autonomy. While these systems are single-robot systems, there is also a focus on developing autonomous patrol and surveillance capabilities for multiple ground robots (Portugal and Rocha 2011). Of course, such systems must be autonomous to deploy them effectively and minimize the manpower required to supervise the activity.

As aerial vehicles become available, there is great discussion regarding their use by law enforcement for patrol and surveillance activities, particularly in relationship to privacy concerns (Brookings 2012). The U.S. Border Patrol currently deploys larger aerial vehicles for patrol and surveillance activities. However, many smaller scale systems have been developed both at research institutes and at corporations that permit persistent surveillance (e.g., Belloni et al. 2007). A limited number of local law enforcement organizations have obtained permission to deploy aerial vehicles in limited situations (Dean 2011). Such systems can be easily deployed from any location and can collect high-resolution video and images. As the air space opens up, it is expected that the number of aerial vehicles deployed by law enforcement will increase significantly to provide persistent surveillance of events that involve large numbers of attendees, hostage situations, and so on. There is little question that this technology will arrive sooner, rather than later (Singer 2010).

1.7 CBRNE incidents

Most large metropolitan areas, and many smaller metropolitan areas, have at least one teleoperated ground robot as part of their bomb squad. The common bomb squad robot platforms include the Northrop Grumman Andros Mark and F6 series, iRobot Packbot® series, and QinetiQ® Talon and Dragon Runner™ series robots. Robots for bomb and improvised explosive device detection and defusion became popular over the past decade as they saw increasing usage during the United States' conflict in the Middle East. As such, these robots have become more prevalent in the civilian sector as well.

Unfortunately, these robots suffer from a number of limitations. The teleoperative capabilities do not provide tactile feedback to the operator regarding the force being applied to the gripper. Related to the inability to obtain a good grasp, the standard robots often do not provide depth sensors to allow the operator to understand how well the gripper is positioned for grasping an object. The operator must rely on the various independent camera views provided through the robot. While work has been done to develop more gamelike interfaces (e.g., iRobot's hand controller for the PackBot), the operator control units are still difficult to use, do not provide consistency across different robots, and typically provide raw data rather than information that the operator can easily interpret. For example, a typical robot will have two or more cameras, but the operator has no means of integrating the images from the cameras into a single representative view. The operator must switch between each camera or look across multiple sub-windows providing the different views. These robots also tend to be difficult to maneuver into position due to their kinematic constraints and the barren sensor feedback landscape provided by the robots to the operator.

Even given these limitations, most bomb squads will not retire their robots. In fact, we continue to see incidences of bomb squad robots being deployed to assist in other police matters. For example, it has become common for SWAT teams to use the bomb squad robot in stand-off situations and to breach entry to buildings (e.g., Basich 2012).

To date, we do not find bomb squad robots that exhibit autonomous capabilities; thus, these systems continue to be highly demanding on their operators. Research has attempted to address some of these cited limitations. For example, Day et al. (2008) demonstrated the addition of a small depth sensing camera mounted on the manipulator arm with an associated interface display for the Remotec Mini-Max robot. The resulting system was evaluated by a local bomb squad and was found to improve their ability to grasp objects. Ryu et al. (2005) also attempted to assist the operator by developing a multimodal interface coupled with a semiautonomous controller. The multimodal interface required the operator to wear a head-mounted display through which augmented reality was provided by overlaying information onto the display of images from the robot. The operator used simple speech commands to interact with the robot. Additionally, the system provided a semiautonomous capability that freed the operator from navigating and driving the robot to the target location.

The design and development of bomb squad robots is a challenge. However, as progress continues, new uses of these robots arise to include response to incidents involving chemical, biological, radiological, and nuclear hazards. Such systems require more extensive sensors and capabilities to handle the range of contaminants associated with such incidents. As well, it is necessary to provide robots that can be decontaminated using liquids. Initial work focused on augmenting existing systems with appropriate sensors and an associated user interface (Neilsen et al. 2008; Jasiobedzki et al. 2009; Rohling et al. 2009). In many cases, the standard bomb squad robots can be augmented to use hazard appropriate sensors.

Since the robot platforms for these conditions are the same base systems as bomb disposal, they suffer from the same limitations and open problems. Cecchini et al. (2011) designed a control architecture intended to support the operator of a teleoperated robot by providing warning messages of dangerous situations and potential harmful operations. The evaluated system incorporated a Geiger counter, and the evaluation incorporated a number of system and communication failures. Typically, sensors available for first response robots include MultiRAE detector, x-ray, and hazardous materials detection capabilities. Recent robotics research in response to the Fukushima Daiichi nuclear incident has resulted in new advances (Guizzo 2011; Nagatani et al. 2011a, 2011b). However, significant limitations persist. An indication of these limitations is represented by the core research challenges in DARPA's 2012 Grand Challenge.

1.8   Discussion

One of the largest limitations to deploying many of the technologies described in this chapter is the lack of high-fidelity deployments (e.g., Disaster City) and real deployments. Robotics researchers and developers need to partner with first responders and their associated training activities at the local, state, and regional levels (e.g., FEMA task force or civil support team exercises). Without deployments in such situations, the technology will not be ready for response to real incidents.

Many technical challenges persist. While this chapter presented numerous technologies for exploring and understanding unknown, dynamic environments, the environmental conditions and changing circumstances in the first response domain provide unique challenges for robots of all types. Smoke, fire, and charred buildings all create environmental sensing challenges, requiring additional research to develop exploration, route planning, SLAM, and (semi)autonomous navigation capabilities.

Navigation challenges also persist for the type of navigation capabilities to provide ground-based robots that can quickly and safely traverse differing terrains. Tracked vehicles can have their tracks fall off, while wheeled vehicles can encounter problems navigating uneven terrain and rubble. While many consider the general ground robotics navigation problem to be solved, first response domains offer challenges that remain unsolved. The navigation challenges are not limited to ground vehicles. Urban environments provide navigational challenges for aerial robots, while debris-filled bodies of water create navigation challenges for surface and underwater robots.

Sensing for environmental understanding and robotic capabilities is still a challenge for all types of robots. Fires, water, and smoke all provide difficult sensing challenges. While some results exist, there is still a need to further develop these capabilities to provide a robust and understandable interpretation of the incident environment.

Most robotic technology deployed to date is fully teleoperated (e.g., remotely operated), which is very demanding on the human responders and operators. To facilitate future deployments and seamlessly integrate robotic technologies, it will be necessary to develop semiautonomous and autonomous capabilities for all types of robots. Each of the open challenges described above will impact the ability to develop robots capable of semiautonomous and autonomous deployments. Another aspect of this challenge is ensuring that first responders are incorporated into the design and development process so that autonomous capabilities will be developed to actually provide a benefit to the response activities. There are numerous examples in the literature of ideas that sounded "great," but when tested provided little value to the first responders or actually hindered their response activities.

Current systems typically incorporate humans as an operator of the robot or as an information consumer (e.g., sensor operator). Existing interfaces suffer a number of limitations for providing the remote human responder with an adequate understanding of the incident area, as sensed by the robots. Continued research is necessary to provide improved information displays and decision support capabilities.

As robotic technology capabilities continue to improve, new relationships will form between the robots and human responders, thus creating additional human–robot interaction relationships. Technologies are needed to allow the entire command hierarchy to receive and understand information provided by robotic technologies. The development of such capabilities will incorporate the command hierarchy, from team leaders to incident commanders, as supervisors of the robotic systems. It is unlikely that these individuals will directly interact with the robots, but just like their human responders, the command hierarchy will require an understanding of the robots' health and activities. Additionally, at this level of command, the humans will likely require transparency regarding the source of the information provided in support of their tasks. In other words, the information provided by the robotic technology will likely need to match the types of information provided by the human responders.

Another anticipated human–robot interaction relationship is that of the peer, which will occur when humans and robots are deployed as teammates. The human will require easily accessible information regarding the robots' activities, intentions, status, and so on as will the robot of the human as they collaborate in loosely or tightly coupled team relationships. Supporting this type of relationship for the human will require more naturalistic interaction capabilities with the robots than that exists today.

Coupled with the interaction challenges is the ability to provide communications between the deployed robotic and human systems. Many of the robots deployed today are tethered to provide direct, secure, and continuous communications between robots and humans. Future deployments may require robots to access spaces that cannot easily be reached when tethered; thus, it will be necessary to develop capabilities to provide communication infrastructure. It is common after disasters for cellular communications to be limited or unavailable. It is uncommon to find wireless networks in most locations. However, some cities do provide citywide wireless networks. Even with citywide wireless networks, the limitation is that after a disaster, such wireless networks may not function as intended. While some systems exist to provide wireless communications in the incident response area (SkyFiber 2010), these systems have limitations such as getting the technology into position for deployment.

Ground, surface, and underwater robots have not received the level of public scrutiny that the incorporation of unmanned aerial vehicles into the national airspace and first response deployment has received. While efforts are under way for incorporating unmanned aerial vehicles into the national airspace, there are counterefforts to limit the use of such technologies by first responders due to the privacy concerns of citizens. The results of these efforts will determine the trajectory of unmanned aerial vehicles and arguably ground, surface, and underwater vehicles (which can be argued as being below the public's awareness level at this time) for integration into the first responder toolbox.

While the field of robotics has made great strides in developing potential technologies to assist first responders in saving property and lives, there are still many challenges to be addressed. To overcome these challenges, it is critical for first responders and the robotics developers to work collaboratively in a tightly coupled relationship. The first responders must recognize that robots are not perfect or as capable as the robots presented by Hollywood movie directors. It is critical the first responders provide feedback to the robotics developers regarding the unique response challenges, needs, and benefits robots can address and provide. At the same time, it is absolutely critical that robotics developers directly involve first responders in their development efforts, listen to what first responders say they need and want, and test their technologies with first responders in high-fidelity testing environments, such as Disaster City and FEMA task force exercises. Murphy has demonstrated the importance of developing such a relationship, a relationship that must be embraced by all individuals involved to provide future robotic capabilities that will seamlessly integrate and provide benefit for incident response.

1.9   Conclusion

The purpose of this chapter has been to provide an overview of some of the existing deployed and research-based ground, aerial, surface, and underwater robotic technologies and the associated technical challenges for use in first response. Beyond bomb squad teleoperated ground robots, there have been limited real-world first response deployments of these technologies. At the same time, there has been significant research activity related to first response robotics. While challenge competitions, such as the RoboCup Rescue Challenge, and test beds, such as the National Institute of Standards and Technology USAR test bed, have been developed, there is no substitute for high-fidelity simulation test beds and real deployments. The Disaster City facility and others that are coming online provide excellent resources to test real robot systems in representative environments. Such facilities need to be employed to address the remaining research challenges outlined in this chapter to ensure that robotic systems are developed that will provide cost-effective benefits to first responders.

References

Adams, J. A., C. M. Humphrey, M. A. Goodrich, J. L. Cooper, B. S. Morse, C. Engh and N. Rasmussen. 2009. Cognitive task analysis for developing UAV wilderness search support. *Journal of Cognitive Engineering and Decision Making* 3(1): 1–26.

Akiyama, I., N. Yoshizumi, A. Ohya, Y. Aoki and F. Matsuno. 2007. Search for survivors buried in rubble by rescue radar with array antennas—Extraction of respiratory fluctuation. In *Proceedings of the IEEE International Workshop on Safety, Security and Rescue Robotics.* IEEE: Piscataway, NJ.

Arai, M., Y. Tanaka, S. Hirose, H. Kuwahara and S. Tsukui. 2008. Development of Souryu-IV and Souryu-V: Serially connected crawler vehicles for in-rubble searching operations. *Journal of Field Robotics* 25(1): 31–65.

Asaoka, T., Y. Kanaeda and K. Magatani. 2008. Development of the device to detect human's bio-signals by easy sensing. In *Proceedings of the 30th Annual International IEEE Engineering in Medicine and Biology Society Conference,* 526–529. IEEE: Piscataway, NJ.

Aziz, M. Z. and B. Mertsching. 2010. Survivor search with autonomous UGVs using multimodal overt attention. In *Proceedings of the IEEE International Workshop on Safety, Security and Rescue Robotics.* IEEE: Piscataway, NJ.

Baker, M., R. Casey, B. Keyes and H. Yanco. 2004. Improved interfaces for human-robot interaction in urban search and rescue. In *Proceedings of the 2004 IEEE International Conference on Systems, Man and Cybernetics,* 2960–2965. IEEE: Piscataway, NJ.

Balakirsky, S., S. Carpin, A. Kleiner, M. Lewis, A. Visser, J. Wing and V. Ziparo. 2007. Towards heterogeneous robot teams for disaster mitigation: Results and performance metrics from RoboCup rescue. *Journal of Field Robotics* 24(11): 943–967.

Balakirsky, S. B. and Z. Kootbally. 2012. USARSim/ROS: A combined framework for robotic control and simulation. In *Proceedings of the AMSE 2012 International Symposium on Flexible Automation.*

Barrett, J. 2010. Friends made in low places: New robot line, more useful than predecessors, are being adopted by SWAT teams. *The Wall Street Journal,* October 18.

Basich, M. 2012. Robots at the ready: Using remote-controlled machines to breach doors and gather intel keeps officers out of harm's way. *Police: The Law Enforcement Magazine,* March 2.

Belloni, G., M. Feroli, A. Ficola, S. Pagnottelli and P. Valigi. 2007. An autonomous aerial vehicle for unmanned security and surveillance operations: Design and test. In *Proceedings of the IEEE International Workshop on Safety, Security and Rescue Robotics,* 1–4. IEEE: Piscataway, NJ.

Bethel, C. L. and R. R. Murphy. 2008. Survey of non-facial/non-verbal affective expressions for appearance-constrained robots. *IEEE Transactions on Systems, Man, and Cybernetics, Part C—Applications and Reviews* 38(1): 83–92.

Birk, A. and S. Carpin. 2006. Rescue robotics—A crucial milestone on the road to autonomous systems. *Advanced Robotics Journal* 20(5): 595–695.

Birk, A., S. Schwertfeger, K. Pathak and N. Vaskevicius. 2009. 3D data collection at disaster city at the 2008 NIST Response Robot Evaluation Exercise (RREE). In *Proceedings of the IEEE International Workshop of Safety, Security and Rescue Robotics.* IEEE: Piscataway, NJ.

Blitch, J. G. 1996. Artificial intelligence technologies for robot assisted urban search and rescue. *Expert Systems with Applications* 11(2): 109–124.

Blitch, J. R., R. R. Murphy and T. Durkin. 2002. Mobile semiautonomous robots for urban search and rescue. In *Encyclopedia of Library and Information Science.* vol. 72, A. Kent and J. G. Williams, Eds., CRC Press: New York, NY, 261–275.

Bostelman, R., T. Hong, R. Madhavan and B. Weiss. 2005. 3D range imaging for urban search and rescue robotics research. In *Proceedings of the IEEE International Workshop of Safety, Security and Rescue Robotics,* 164–169. IEEE: Piscataway, NJ.

Brookings Institute. 2012. Drone surveillance in the US. Panelists: K. Anderson, C. Crump, P. Rosenzweig, J. Villasenor and B. Wittes. April 4.

Brunner, C. J., T. Peynot and T. A. Vidal-Calleja. 2011. Combining multiple sensor modalities for a localisation robust to smoke. In *Proceedings of the 2011 IEEE/RSJ International Conference on Intelligent Robots and Systems*, 2489–2496. IEEE: Piscataway, NJ.

Burke, J. L. and R. Murphy. 2004. Human-robot interaction in USAR technical search: Two heads are better than one. In *Proceedings of the 13th IEEE International Workshop Robot and Human Interactive Communication*, 307–312. IEEE: Piscataway, NJ.

Burke, J., R. R. Murphy, M. Coovert and D. Riddle. 2004. Moonlight in Miami: An ethnographic study of human-robot interaction in USAR. *Human-Computer Interaction, Special Issue on Human-Robot Interaction* 19: 85–116.

Calamityville Technical Laboratory. 2012. Wright State University, School of Medicine. http://med.wright.edu/medicalreadiness/calamityville.

Calisi, D., A. Farinelli, L. Iocchi and D. Nardi. 2005. Autonomous navigation and exploration in a rescue environment. In *Proceedings of the IEEE International Workshop on Safety, Security, and Rescue Robotics*, 54–59. IEEE: Piscataway, NJ.

Capitan, J., D. Mantecon, P. Soriano and A. Ollero. 2007. Autonomous perception techniques for urban and industrial fire scenarios. In *Proceedings of the IEEE International Workshop on Safety, Security and Rescue Robotics*. IEEE: Piscataway, NJ.

Carbone, A., A. Finzi, A. Orlandini, F. Pirri and G. Ugazio. 2005. Situation aware rescue robots. In *Proceedings of the IEEE International Workshop on Safety, Security and Rescue Robotics*, 182–188. IEEE: Piscataway, NJ.

Carnegie, D. A. 2007. A three-tier hierarchical robotic system for urban search and rescue applications. In *Proceedings of the IEEE International Workshop on Safety, Security and Rescue Robotics*. IEEE: Piscataway, NJ.

Casper, J. and R. R. Murphy. 2000. Issues in intelligent robots for search and rescue. In *SPIE Ground Vehicle Technology II*. SPIE Proceedings, Volume 4024 SPIE: Bellingham, WA pp 292–302.

Casper, J. and R. R. Murphy. 2003. Human-robot interaction during the robot-assisted urban search and rescue response at the World Trade Center. *IEEE Transactions on Systems, Man and Cybernetics Part B* 33: 367–385.

Castillo, C. and C. Chang. 2005. A method to detect victims in search and rescue operations using template matching. In *Proceedings of the IEEE International Workshop on Safety, Security and Rescue Robotics*, 201–206. IEEE: Piscataway, NJ.

Cecchini, T., P. Villella and F. Rocchi. 2011. System control architecture for a remotely operated platform for S&R, IED-EOD and NBC applications. In *Proceedings of the IEEE International Symposium on Safety, Security and Rescue Robotics*, 259–264. IEEE: Piscataway, NJ.

Celentano, L., B. Siciliano and L. Villani. 2005. A robotic system for fire fighting in tunnels. In *Proceedings of the IEEE International Workshop on Safety, Security and Rescue Robotics*, 253–258. IEEE: Piscataway, NJ.

Chang, C. and R. R. Murphy. 2007. Towards robot-assisted mass-casualty triage. In *Proceedings of the 2007 IEEE International Conference on Networking, Sensing and Control*, 267–272. IEEE: Piscataway, NJ.

Chien, T. L., K. L. Su and J. H. Guo. 2005. The multiple interface security robot—WFSR-II. In *Proceedings of the IEEE International Workshop on Safety, Security, and Rescue Robotics*, 47–52. IEEE: Piscataway, NJ.

Cooper, J. and M. A. Goodrich. 2008. Towards combining UAV and sensor operator roles in UAV-enabled visual search. In *Proceedings of ACM/IEEE International Conference on Human-Robot Interaction*, 351–358. ACM: New York, NY.

Craighead, J., B. Day and R. R. Murphy. 2006. Evaluation of Canesta's range sensor technology for urban search and rescue and robot navigation. In *Proceedings of the IEEE Workshop on Safety Security Rescue Robots*. IEEE: Piscataway, NJ.

Day, B., C. Bethel, R. R. Murphy and J. Burke. 2008. A depth sensing display for bomb disposal robots. In *Proceedings of the IEEE International Workshop on Safety, Security, and Rescue Robotics*, 83–92. IEEE: Piscataway, NJ.

Dean, S. 2011. New police drone in Texas could carry weapons. *Officer.com*, October 29.

Disaster City, Texas Engineering Extension Service. 2012. http://teexweb.tamu.edu/teex.cfm?pageid=USARprog&area=usar&templateid=1117.

Dunwen, E., G. Wenjie and L. Yiyang. 2011. The concept of a jumping rescue robot with variable transmission mechanism. In *Proceedings of the IEEE International Symposium on Safety, Security, and Rescue Robotics*, 99–104. IEEE: Piscataway, NJ.

Edwards, D. 2011. Naked man opens fire on SWAT robot in Florida. Raw Story Media, April 7.

Eich, M., F. Grimminger and F. Kirchner. 2008. A versatile stair-climbing robot for search and rescue applications. In *Proceedings of the IEEE International Workshop on Safety, Security, and Rescue Robotics*, 35–40. IEEE: Piscataway, NJ.

Garcia-Cerezo, A., A. Mandow, J. L. Martinez, J. Gomez-de-Gabriel, J. Morales, A. Cruz, A. Reina and J. Seron. 2007. Development of ALACRANE: A mobile robotic assistance for exploration and rescue missions. In *Proceedings of the IEEE International Workshop on Safety, Security, and Rescue Robotics*, 1–6. IEEE: Piscataway, NJ.

Geifer, D. L., R. Fillnow and P. Bengel. 1988. Criteria development for remotely controlled remote devices at TMI-2. In *Proceedings of the 1988 American Nuclear Society Winter Meeting*.

Goodrich, M. A. and A. C. Schultz. 2007. Human-robot interaction: A survey. *Foundations and Trends in Human-Computer Interaction* 1(3): 203–275.

Goodrich, M. A., J. L. Cooper, J. A. Adams, C. Humphrey, R. Zeeman and B. G. Buss. 2007. Using a mini-UAV to support wilderness search and rescue practices for human-robot teaming. In *Proceedings of the IEEE International Conference on Safety, Security and Rescue Robotics*. IEEE: Piscataway, NJ.

Goodrich, M. A., B. S. Morse, D. Gerhardt, J. L. Cooper, M. Quigley, J. A. Adams and C. Humphrey. 2008. Supporting wilderness search and rescue using a camera-equipped mini UAV. *Journal of Field Robotics* 25(1–2): 89–110.

Goodrich, M. A., B. S. Morse, C. Engh, J. L. Cooper and J. A. Adams. 2009. Towards using unmanned aerial vehicles (UAVs) in wilderness search and rescue: Lessons from field trials. *Interaction Studies* 10(3): 455–481.

Guizzo, E. 2011. Fukushima robot operator writes tell-all blog. *IEEE Spectrum.* http://spectrum.ieee.org/automaton/robotics/industrial-robots/fukushima-robot-operator-diaries.

Hahn, R., D. Lang, M. Haselich and D. Paulus. 2011. Heat mapping for improved victim detection. In *Proceedings of the IEEE International Symposium on Safety, Security, and Rescue Robotics*, 116–121. IEEE: Piscataway, NJ.

Harriott, C. E., T. Zhang and J. A. Adams. 2011a. Evaluating the applicability of current models of workload to peer-based human-robot teams. In *Proceedings of the 6th ACM/IEEE International Conference on Human-Robot Interaction*, 45–52. ACM: New York, NY.

Harriott, C. E., T. Zhang and J. A. Adams. 2011b. Predicting and validating workload in human-robot teams. In *Proceedings of the 20th Conference on Behavior Representation in Modeling and Simulation,* 162–169.

Hatton, R. L. and H. Choset. 2011. Geometric motion planning: The local connection, Stokes' Theorem, and the importance of coordinate choice. *International Journal of Robotics Research* 30(8): 988–1014.

Herbert, S., N. Bird, A. Drenner and N. Papanikolopoulos. 2009. A search and rescue robot. In *Proceedings of the 2009 IEEE International Conference on Robotic and Automation,* 1585–1586. IEEE: Piscataway, NJ.

Humphrey, C. M. and J. A. Adams. 2009. Robotic tasks for CBRNE incident response. *Advanced Robotics* 23: 1217–1232.

Humphrey, C. M. and J. A. Adams. 2011. Analysis of complex, team-based systems: Augmentations to goal-directed task analysis and cognitive work analysis. *Theoretical Issues in Ergonomics* 12(2): 149–175.

Hunt, A. J., R. J. Bachmann, R. R. Murphy and R. D. Quinn. 2011. A rapidly reconfigurable robot for assistance in urban search and rescue. In *Proceedings of the 2011 IEEE/RSJ International Conference on Intelligent Robots and Systems,* 209–214. IEEE: Piscataway, NJ.

Iwano, Y., K. Osuka and H. Amano. 2011. Development of rescue support stretcher system with stair-climbing. In *Proceedings of the IEEE International Symposium on Safety, Security and Rescue Robotics,* 245–250. IEEE: Piscataway, NJ.

Jacoff, A., H. Huang, E. R. Messina, A. M. Virts and A. J. Downs. 2010. Comprehensive standard test suites for the performance evaluation of mobile robots. In *Proceedings of the 2010 Performance Metrics for Intelligent Systems Workshop,* 116–173.

Jacoff, A., E. Messina, B. A. Weiss, S. Tadokoro and Y. Nakagawa. 2003. Test arenas and performance metrics for urban search and rescue robots. In *Proceedings of the 2003 IEEE/RSJ International Conference on Intelligent Robots and Systems,* 3395–3403. IEEE: Piscataway, NJ.

Jasiobedzki, P., H.-K. Ng, M. Bondy and C. H. McDiarmid. 2009. C2SM: A mobile system for detecting and 3D mapping of chemical, radiological, and nuclear contamination. In *Proceedings of Sensors, and Command, Control, Communications, and Intelligence (C3I) Technologies for Homeland Security and Homeland Defense VIII.*

Kang, S., W. Lee, M. Kim and K. Shin. 2005. ROBHAZ-rescue: Rough-terrain negotiable teleoperated mobile robot for rescue mission. In *Proceedings of the IEEE International Workshop on Safety, Security and Rescue Robotics,* 105–110. IEEE: Piscataway, NJ.

Kumar, M., K. Cohen and B. HomChaudhuri. 2011. Cooperative control of multiple uninhabited aerial vehicles for monitoring and fighting wildfires. *Journal of Aerospace Computing, Information and Communication* 8: 1–16.

Lee, P., H. Wang, S. Chien, M. Lewis, P. Scerri, P. Velagapudi, K. Sycara and B. Kane. 2010. Teams for teams: Performance in multi-human/multi-robot teams. In *Proceedings of the 54th Annual Human Factors and Ergonomics Society Annual Meeting,* 438–442.

Lewis, M., J. Wang and S. Hughes. 2007. USARSim: Simulation for the study of human-robot interaction. *Journal of Cognitive Engineering and Decision Making* 1(1): 98–120.

Lewis, M., H. Wang, P. Velagapudi, P. Scerri and K. Sycara. 2009. Using humans as sensors in robotic search. In *Proceedings of the 12th International Conference on Information Fusion,* 1249–1256.

Lewis, M., H. Wang, S.-Y. Chien, P. Scerri, P. Velagapudi, K. Sycara and B. Kane. 2010. Teams organization and performance in multi-human/multi-robot teams. In *Proceedings of the 2010 IEEE International Conference on Systems, Man and Cybernetics,* 1617–1623. IEEE: Piscataway, NJ.

Li, B., S. Ma, J. Liu and Y. Wang. 2005. Development of a shape shifting robot for search and rescue. In *Proceedings of the IEEE International Workshop on Safety, Security and Rescue Robotics,* 31–35. IEEE: Piscataway, NJ.

Lin, L. and M. A. Goodrich. 2009a. A Bayesian approach to modeling lost person behaviors based on terrain features in wilderness search and rescue. In *Proceedings of the 18th Annual Conference on Behavior Representation in Modeling and Simulation.* BRIMS Society: Charleston, SC.

Lin, L. and M. A. Goodrich. 2009b. UAV intelligent path planning for wilderness search and rescue. In *Proceedings of IEEE/RSJ International Conference on Intelligent Robots and Systems.* IEEE: Piscataway, NJ.

Lin, L. and M. A. Goodrich. 2010. A Bayesian approach to modeling lost person behaviors based on terrain features in wilderness search and rescue. *Computational and Mathematical Organization Theory* 3: 300–323.

Lin, L., M. Roscheck, M. A. Goodrich and B. S. Morse. 2010. Supporting wilderness search and rescue with integrated intelligence: Autonomy and information at the right time and the right place. In *Proceedings of the Twenty-Fourth AAAI Conference on Artificial Intelligence, Special Track on Integrated Intelligence.* AAAI: Palo Alto, CA.

Linder, T., V. Tretyakov, S. Blumenthal, P. Molitor, D. Holz, R. Murphy, S. Tadokoro and H. Surmann. 2010. Rescue robots at the collapse of the municipal archive of Cologne city: A field report. In *Proceedings of the IEEE International Workshop on Safety, Security and Rescue Robotics.* IEEE: Piscataway, NJ.

Ma, Z., P. Lee, Y. Xu, M. Lewis and P. Scerri. 2009. An initial evaluation of approaches to building entry for large robot teams. In *Proceedings of the IEEE International Workshop on Safety, Security, and Rescue Robotics.* IEEE: Piscataway, NJ.

Martinez-de Dios, J. R., B. C. Arrue, A. Ollero, L. Merino and F. Gomez-Rodriguez. 2008. Computer vision techniques for forest fire perception. *Image and Vision Computing* 26: 550–562.

Maruyama, H. and K. Ito. 2010. Semi-autonomous snake-like robot for search and rescue. In *Proceedings of the IEEE International Workshop on Safety, Security, and Rescue Robotics.* IEEE: Piscataway, NJ.

Maza, I., F. Caballero, J. Capitan, J. R. Martinez-de Dios and A. Ollero. 2010. Firemen monitoring with multiple UAVs for search and rescue missions. In *Proceedings of the IEEE International Workshop on Safety, Security, and Rescue Robotics.* IEEE: Piscataway, NJ.

Micire, M. 2007. Evolution and field performance of a rescue robot. *Journal of Field Robotics* 25(1/2): 17–30.

Morse, B. S., C. H. Engh and M. A. Goodrich. 2010. UAV video coverage quality maps and prioritized indexing for wilderness search and rescue. In *Proceedings of the 5th ACM/IEEE International Conference on Human-Robot Interaction.* ACM: New York, NY.

Morse, B. S., D. Thornton and M. A. Goodrich. 2012. Color anomaly detection and suggestion for wilderness search and rescue. In *Proceedings of the 7th ACM/ IEEE International Conference on Human-Robot Interaction,* 455–462. ACM: New York, NY.

Mullins, J., B. Horan, M. Fielding and S. Nahavandi. 2007. A haptically enabled low-cost reconnaissance platform for law enforcement. In *Proceedings of the IEEE International Workshop on Safety, Security and Rescue Robotics*. IEEE: Piscataway, NJ.

Murphy, R. R. 2002. Rats, robots, and rescue. *IEEE Intelligent Systems* 17: 7–9.

Murphy, R. R. 2003. Rescue robots at the World Trade Center. *Journal of the Japan Society of Mechanical Engineers, Special Issue on Disaster Robotics* 102: 794–802.

Murphy, R. R. 2004a. Activities of the rescue robots at the World Trade Center from 11–21 September 2001. *IEEE Robotics and Automation Magazine* 11(3): 50–61.

Murphy, R. R. 2004b. Human-robot interaction in rescue robotics. *IEEE Systems, Man and Cybernetics Part A, Special Issue on Human-Robot Interaction* 34: 138–153.

Murphy, R. R. 2005. Humans, robots, rubble, and research. *Interaction* 12: 37–39.

Murphy, R. R. 2010a. Navigational and mission usability in rescue robots. *Journal of the Robotics Society of Japan* 28: 142–146.

Murphy, R. R. 2010b. Potential uses and criteria for unmanned aerial systems for wildland firefighting. In *AUVSI Unmanned Systems North America*. AUVSI: Arlington, VA.

Murphy, R. R. 2010c. Proposals for new UGV, UMV, UAV, and HRI standards for rescue robots. In *The 2010 Workshop on Performance Metrics for Intelligent Systems*. ACM: New York, NY.

Murphy, R. R. and J. L. Burke. 2005. Up from the rubble: Lessons learned about HRI from search and rescue. In *Proceedings of the 49th Annual Meetings of the Human Factors and Ergonomics Society*.

Murphy, R. R. and S. Stover. 2008. Rescue robots for mudslides: A descriptive study of the 2005 La Conchita mudslide response. *Journal of Field Robotics* 25: 3–16.

Murphy, R. R., J. Blitch and J. Casper. 2002. AAAI/RoboCup-2001 urban search and rescue events: Reality and competition. *Artificial Intelligence Magazine* 23(1): 37–42.

Murphy, R., D. Riddle and E. Rasmussen. 2004. Robot-assisted medical reachback: A survey of how medical personnel expect to interact with rescue robots. In *Proceedings of the 13th IEEE International Workshop on Robot and Human Interactive Communication*, 301–306. IEEE: Piscataway, NJ.

Murphy, R. R., S. Tadokoro, D. Nardi, A. Jacoff, P. Fiorini and A. Erkmen. 2008a. Rescue robotics. In *Handbook of Robotics*, B. Sciliano and O. Khatib, Eds., Springer-Verlag: New York, 1151–1174.

Murphy, R., E. Steimle, C. Griffin, C. Cullins, M. Hall and K. Pratt. 2008b. Cooperative use of unmanned sea surface and micro aerial vehicle at hurricane Wilma. *Journal of Field Robotics* 25: 164–180.

Murphy, R. R., R. Brown, R. Grant and C. T. Arnett. 2009a. Preliminary domain theory for robot-assisted wildland firefighting. In *IEEE Workshop on Safety Security Rescue Robotics*. IEEE: Piscataway, NJ.

Murphy, R. R., R. Brown, R. Grant and C. T. Arnett. 2009b. Requirements for wildland firefighting ground robots. In *AUVSI Unmanned Systems North America*.

Murphy, R. R., E. Steimle, M. Lindemuth, D. Trejo, M. Hall, D. Slocum, S. Hurlebas and Z. Medina-Cetina. 2009c. Robot-assisted bridge inspection after hurricane Ike. In *IEEE Workshop on Safety Security Rescue Robotics*, 1–5. AUVSI: Arlington, VA.

Nagatani, K., S. Kiribayashi, Y. Okada, S. Tadokoro, T. Nishimura, T. Yoshida, E. Koyanagi and Y. Hada. 2011a. Redesign of rescue mobile robot Quince: Toward emergency response to the nuclear accident at Fukushima Daiichi Nuclear Power Station on March 2011. In *Proceedings of the 2011 IEEE International Symposium on Safety, Security and Rescue Robotics*, 13–18.

Nagatani, K., S. Kiribayashi, Y. Okada, K. Otake, K. Yoshida, S. Tadokoro, T. Nishimura, T. Yoshida, E. Koyanagi, M. Fukushima and S. Kawatsuma. 2011b. Gamma-ray irradiation test of electric components of rescue mobile robot Quince: Toward emergency response to the nuclear accident at Fukushima Daiichi Nuclear Power Station on March 2011. In *Proceedings of the 2011 IEEE International Symposium on Safety, Security and Rescue Robotics*, 56–60.

Nagatani, K., Y. Okada, N. Tokunaga, K. Yoshida, S. Kiribayashi, K. Ohno, E. Takeuchi, S. Tadokoro, H. Akiyama, I. Noda, T. Yoshida and E. Koyanagi. 2009. Multi-robot exploration for search and rescue missions: A report of map building in RoboCupRescue. In *Proceedings of the IEEE International Workshop on Safety, Security and Rescue Robotics*.

NASA, Civil UAV Assessment Team. 2005. Civil UAV Capabilities Assessment, Interim Status Report. NASA Online publication.

Neilsen, C. W., D. I. Gertman, D. J. Bruemmer, R. S. Hartley and M. C. Walton. 2008. Evaluating robot technologies as tools to explore radiological and other hazardous environments. In *Proceedings of American Nuclear Society Emergency Planning and Response, and Robotics and Security Systems Joint Topical Meeting*.

Nüchter, A., K. Lingemann, J. Hertzberg, H. Surmann, K. Pervölz, M. Hennig, K. R. Tiruchinapalli, R. Worst and T. Christaller. 2005. Mapping of rescue environments with Kurt3D. In *Proceedings of the IEEE International Workshop on Safety, Security, and Rescue Robotics*, 158–163.

Osuka, K., R. R. Murphy and A. C. Schultz. 2002. USAR competitions for physically situated robots. *IEEE Robotics and Automation Magazine* 9(3): 26–33.

Pavelek, M. D. 1985. Use of teleoperators at Three Mile Island 2. In *Proceedings of the American Nuclear Society Exec. Conference on Remote Operations and Robotics in the Nuclear Industry*.

Pellenz, J., D. Lang, F. Neuhaus and D. Paulus. 2010. Real-time 3D mapping of rough terrain: A field report from Disaster City. In *Proceedings of the IEEE International Workshop on Safety, Security and Rescue Robotics*.

Penders, J., L. Alboul, U. Witkowsi, A. Naghsh, J. Saez-Pons, S. Herbrechtsmeier and M. El Habbal. 2011. A robot swarm assisting a human fire-fighter. *Advanced Robotics* 25(1–2): 93–117.

Portugal, D. and R. P. Rocha. 2011. On the performance and scalability of multi-robot patrolling algorithms. In *Proceedings of the IEEE International Symposium on Safety, Security and Rescue Robotics*, 50–55.

Pratt, K., R. Murphy, J. Burke, J. Craighead, C. Griffin and S. Stover. 2008. Use of tethered small unmanned aerial system at Berkman Plaza II collapse. In *IEEE International Workshop on Safety, Security, and Rescue Robotics*, 134–139.

Pratt, K., R. Murphy, S. Stover and C. Griffin. 2009. CONOPS and autonomy commendations for VTOL SUASs based on Hurricane Katrina operations. *Journal of Field Robotics* 26: 636–650.

Randelli, G., L. Iocchi and D. Nardi. 2011. User-friendly security robots. In *IEEE International Symposium on Safety, Security and Rescue Robotic*, 308–313.

Rao, J., Z. Gong, J. Luo and S. Xie. 2005. Unmanned airships for emergency management. In *Proceedings of the IEEE International Workshop on Safety, Security and Rescue Robotics*, 125–130.

Rasmussen, N., B. Morse, M. A. Goodrich and D. Eggett. 2009. Fused visible and infrared video for use in wilderness search and rescue. In *Proceedings of the IEEE Workshop on Applications of Computer Vision*, 1–8.

Rohling, T., B. Bruggemann, F. Hoeller and F. E. Schneider. 2009. CBRNE hazard detection with an unmanned vehicle. In *Proceedings of the IEEE International Workshop on Safety, Security and Rescue Robotics*.

Ruangpayoongsak, N., H. Roth and J. Chudoba. 2005. Mobile robots for search and rescue. In *Proceedings of the IEEE International Workshop on Safety, Security and Rescue Robotics*, 212–217.

Ryu, D., C.-S. Hwang, S. Kang, M. Kim and J.-B. Song. 2005. Wearable haptic-based multi-modal teleoperation of field mobile manipulator for explosive ordnance disposal. In *Proceedings of the IEEE International Workshop on Safety, Security and Rescue Robotics*, 75–80.

Saarinen, J., S. Heikkila, M. Elomaa, J. Suomela and A. Halme. 2005. Rescue personnel localization system. In *Proceedings of the IEEE International Workshop on Safety, Security and Rescue Robotics*, 218–223.

Saeedi, P., S.-A. Sorensen and S. Hailes. 2009. Performance-aware exploration algorithm for search and rescue robots. In *Proceedings of the IEEE International Workshop on Safety, Security and Rescue Robotics*.

Šakėnas, V., O. Kosuchinas, M. Pfingsthorn and A. Birk. 2007. Extraction of semantic floor plans from 3D point cloud maps. In *Proceedings of the IEEE International Workshop on Safety, Security and Rescue Robotics*.

Sato, N., F. Matsuno and N. Shiroma. 2007. FUMA: Platform development and system integration for rescue missions. In *Proceedings of the IEEE International Workshop on Safety, Security and Rescue Robotics*.

Scholtz, J. B. 2003. Theory and evaluation of human robot interactions. In *Proceedings of the 36th Annual Hawaii International Conference on System Sciences*, 125.1.

Scholtz, J., B. Antonishek and J. Young. 2005. A field study of two techniques for situation awareness for robot navigation in urban search and rescue. In *Proceedings of IEEE International Workshop on Robots and Human Interactive Communication*, 131–136.

Shimaoka, K., K. Ogane and T. Kimura. 2011. An evaluation test field design for a USAR robot related to a collapsed Japanese house. In *Proceedings of the IEEE International Symposium on Safety, Security and Rescue Robotics*, 221–225.

Simodate, Y. and K. Ito. 2011. Semicircular duplex manipulator to search narrow spaces for victims. In *Proceedings of the IEEE International Symposium on Safety, Security and Rescue Robotics*, 7–12.

Singer, P. W. 2010. *Wired for war: The robotics revolution and conflict in the 21st century*, Penguin Books: New York.

SkyFiber, Inc. 2010. SkyFiber's Optical Wireless Broadband. http://www.skyfiber.com/solutions_ern.php. Downloaded June 10, 2012.

Tabata, K., A. Inaba and H. Amano. 2005. Development of a transformational mobile robot to search victims under debris and rubble. 2nd report: improvement of mechanism and interface. In *Proceedings of the IEEE International Workshop on Safety, Security and Rescue Robotics*, 19–24.

Tanaka, J., K. Suzumori, M. Takata, T. Kanda and M. Mori. 2005. A mobile jack robot for rescue operation. In *Proceedings of the IEEE International Workshop on Safety, Security and Rescue Robotics*, 99–104.

Tesch, M., K. Lipkin, I. Brown, R. L. Hatton, A. Peck, J. Rembisz and H. Choset. 2009. Parameterized and scripted gaits for modular snake robots. *Advanced Robotics* 23: 1131–1158.

Thrun, S., W. Burgard and D. Fox. 2005. *Probabilistic robotics*, The MIT Press: Cambridge, MA.

Tran, J., A. Ferworn, M. Gerdzhev and D. Ostrom. 2010. Canine assisted robot deployment for urban search and rescue. In *Proceedings of the IEEE International Workshop on Safety, Security and Rescue Robotics*.

Tran, J., A. Ufkes, M. Fiala and A. Ferworn. 2011. Low-cost 3D scene reconstruction for response robots in real-time. In *Proceedings of the IEEE International Symposium on Safety, Security and Rescue Robotics*, 161–166.

Tretyakov, V. and T. Linder. 2011. Range sensors evaluation under smoky conditions for robotics applications. In *Proceedings of the IEEE International Symposium on Safety, Security and Rescue Robotics*, 215–220.

Trierscheid, M., J. Pellenz, D. Paulus and D. Balthasar. 2008. Hyperspectral imaging for victim detection with rescue robots. In *Proceedings of the IEEE International Workshop on Safety, Security and Rescue Robotics*, 7–12.

Vig, L. and J. A. Adams. 2007. Coalition formation: From software agents to robots. *Journal of Intelligent Robotic Systems* 50(1): 85–118.

Voyles, R. M. and A. Larson. 2005. TerminatorBot: A novel robot with dual-use mechanism for locomotion and manipulation. *IEEE/ASME Transactions on Mechatronics* 10(1): 17–25.

Voyles, R. M., S. Povilus, R. Mangharam and K. Li. 2010. RecoNode: A reconfigurable node for heterogeneous multi-robot search and rescue. In *Proceedings of the IEEE International Workshop on Safety, Security and Rescue Robotics*.

Whittaker, W. L. and L. Champeny. 1987. Capabilities of a remote work vehicle. In *Proceedings of the American Nuclear Society Topical Conference on Robotics and remote Handling*, 254.

Wolf, A., H. Choset, H. B. Brown Jr. and R. Casciola. 2005. Design and control of a mobile hyper-redundant urban search and rescue robot. *International Journal of Advanced Robotics* 19(8): 221–248.

Yanco, H. A., B. Keyes, J. L. Drury, C. W. Nielsen and D. J. Bruemmer. 2007. Evolving interface design for robot search tasks. *Journal of Field Robotics* 24(8–9): 779–799.

Yim, M. and J. Laucharoen. 2011. Towards small robot aided victim manipulation. *Journal of Intelligent Robotics and Systems* 64(1): 119–139.

chapter two

A practical, simplified chemical agent sensor placement methodology*

David A. Smith and David R. Jacques

Contents

* The material contained in this chapter was prepared as an account of work sponsored by an agency of the U.S. government. Neither the U.S. government nor any agent thereof, nor any of their employees, makes any warranty, express or implied, or assumes any legal liability or responsibility for the accuracy, completeness, or usefulness of any information, apparatus, product, or process disclosed, or represents that its use would not infringe upon privately owned rights. Reference herein to any specific commercial product, process, or service by trade name, trademark, manufacturer, or otherwise does not necessarily constitute or imply its endorsement, recommendation, or favoring by the U.S. government or any agency thereof. The views and opinions of authors expressed herein do not necessarily state or reflect those of the U.S. government or any agency thereof.

2.1   Introduction

Concerns regarding the intentional release of a biological, chemical, or radiological agent are growing. The scenario of a release has been identified by some as a "when" situation rather than an "if." The recently released Report of the Commission on the Prevention of WMD Proliferation and Terrorism (Graham et al., 2008) reiterates this point. For high-value targets, the associated vulnerability to these types of threats are more rigorously and frequently assessed, and plans for response are developed and implemented. For lower value targets, analyses may be conducted intermittently, if at all. Conducting a vulnerability assessment and implementing recommended actions requires, in most cases, expenditure of a significant commitment of resources, which must be balanced with the perceived risk.

It is generally agreed that the earlier a release of a chemical or biological agent is detected, the higher the probability of saving lives. Placement of sensors for detection of chemical and/or biological agent release has been studied recently by others (Sohn et al., 2002; Sohn and Lorenzetti, 2007; Chen and Wen, 2008). The focus of many has been on the optimal placement of the sensor based on airflow parameters. Chen and Wen (2008) noted that achieving maximum protection based on sensor placement requires sensors to not be placed in areas where the contaminant concentration would be low. Furthermore, Sohn and Lorenzetti (2007) contended that the sensor attributes are the key to determining a probabilistic solution to sensor placement. The common thread between each of the previous studies is that placement is typically based on a complicated algorithm requiring extensive knowledge of heating, ventilation, and air-conditioning (HVAC) parameters; a thorough understanding and application of fluid dynamics; and significant computational effort. The expertise required to implement these approaches puts optimum sensor placement beyond the resource means of those whose buildings are not considered to be high-value targets.

What has been missing, to date, is an easily implemented sensor placement solution based on the likelihood of the release location and the basic volumetric flow balance of a building. The intent here is to determine prioritized sensor locations that will be able to detect the most likely release scenarios, given budget constraints on the number of sensors. The authors analyzed the concepts needed to develop such an algorithm and determined that the requisite information needed for sensor placement can be based on a greatly simplified airflow model combined with a risk assessment of where a perpetrator might seek to release an agent. Given that an individual intent on releasing a chemical agent will seek to maximize the effect of the release, that is, enhance the consequence and do so from an accessible location, the solution is then based on the vulnerability of the release location and the subsequent consequence(s) that can be achieved.

The proposed detect-to-warn methodology presented herein was developed specifically for softer, that is, lower value, targets with limited resources to support threat analyses and sensor implementation. The purpose is to provide building facility managers with limited to no knowledge of terrorism or chemical agent sensor placement and a limited budget to develop a plan for sensor placement within their building. Specifically, the goal of this project was to develop a chemical/biological sensor wizard (CBW) to collect user-provided data with respect to the HVAC system, building type and facility vulnerability, and available budget to develop a detect-to-warn sensor placement algorithm.

With an innovative algorithm for developing chemical agent sensor networks in place, such as the CBW, it is feasible to perform risk analysis and determine recommended sensor placement locations in a timely manner. The CBW does not require expert operators, system architects, scientists, or engineers with advanced skill sets, but rather the system encapsulates basic knowledge and provides an expert-based CBW to automate the key decision-making processes. The typical system user is a building maintenance manager (hereafter referred to as the building manager) with knowledge of the building's HVAC system and security features.

The scope of this chapter is focused on detection of an intentional release of a chemical agent within a variety of building types.* Within any building type, the quickest means of dispersing an agent throughout the building is through the HVAC system. Furthermore, the vulnerability of the HVAC system is assumed proportional to the ease with which the system may be accessed by unauthorized personnel.

2.2 Proposed approach

The approach presented herein is based on an evaluation of a building's vulnerability/security attributes and basic building airflow parameters assuming adherence to indoor air quality standards and design practices. It assumes that the dominant driver for airflow within the building is the HVAC system. The information that feeds the CBW is provided by a building manager who has basic information regarding the building's structure, security, and HVAC system. The CBW uses this information to provide a relative risk rating (RRR) for all facilities in the building, an estimation of the agent concentration–response to standard release scenarios in the various facilities within the building, and a prioritized list

---

* The scope of the project was limited to a chemical agent release after the project was titled. Real-time, independently operated, biological agent sensors were not commercially available at the time this study was conducted.

of sensor placement locations based on the relative risk (RR) associated with the release locations and the ability of the sensors to detect releases from these locations.

It is widely accepted that distribution of a chemical agent or other toxic, gaseous material throughout a building would be accomplished via the HVAC system. Therefore, it is assumed that the point of attack is within or near an HVAC system node, that is, air return duct (ARD), air handling unit (AHU), or outside air vent. In general, particles or molecules of a chemical agent released indoors will pass through an ARD and will eventually be mixed in the AHU with outside air. In the event of a release external to the building, the agent particle or molecule will enter the system through the outside air intake, mix with the return flows at the AHU, and be distributed through the air supply. An HVAC system provides a means whereby a majority of building occupants may be exposed to a released agent generally on the order of 10–15 minutes or less, depending on air recirculation.

Each facility within a building (kitchen, restroom, office, etc.) is associated with a facility use designator (FUD), which identifies the principal use of the facility. The FUD influences occupancy ratings for a given facility size, and it also affects recommended volumetric flow rates and refresh (fresh air mixing) rates. An AHU zone is defined as the collection of facilities serviced by a specific AHU. Consequence (C), for the purposes of this discussion, is defined as human morbidity and/or mortality and results from the occupation of the facilities affected by the release of a chemical agent. Economic, social, or other disruptions are not considered for this investigation. The RRR for all facilities and AHUs within a building is determined by the product of consequence and vulnerability, as assessed based on the security/accessibility of the various facilities. This information is then fed into a linear time-invariant (LTI) concentration model (Jacques and Smith, 2010) whereby the RRR, chemical agent release amount, and unique HVAC parameters are analyzed to determine prioritized sensor placement locations for a range of release scenarios.

2.3   System overview

The CBW system is unique in that it does not require the user to have extensive knowledge of the HVAC systems, chemical agent release characteristics, or airflow dynamics. Figure 2.1 depicts a notional flow diagram for the CBW concept, and Figure 2.2 provides an external systems diagram, in integrated definition for function modeling (IDEF0) format, depicting the process by which the user interacts with CBSWEEP (the CBW graphic user interface as presently packaged and as referred to for the remainder of the chapter). IDEF0 is a graphical technique for depicting the data, functional flow, and control of system and/or organization processes (FIPS, 1993).

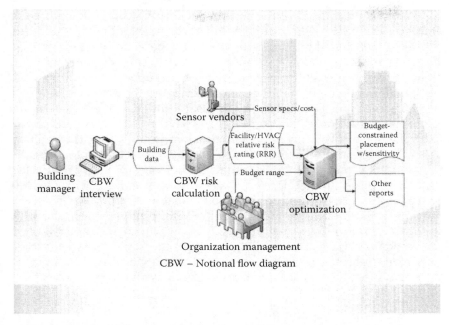

Figure 2.1 Notional flow diagram for chemical/biological sensor wizard.

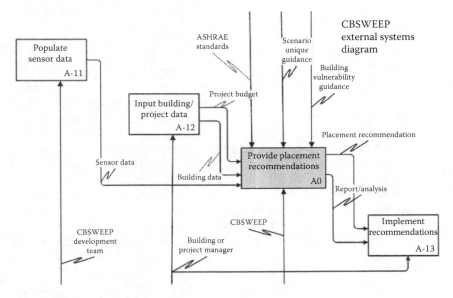

Figure 2.2 External systems diagram.

CBSWEEP is preloaded with American Society of Heating, Refrigerating and Air-Conditioning Engineers (ASHRAE) indoor air quality standards (ASHRAE, 2007), building vulnerability guidance, and predefined release scenarios by the CBSWEEP development team. The development team would also be responsible for keeping the wizard up to date with available sensor information. CBSWEEP, deployable as a desktop personal computer or Web-served application, "walks" the user through a series of questions, termed the "user interview," to gather the building information required for evaluation of the vulnerabilities of the building and associated facilities, as well as the estimated flow between facilities. The latter point is essential as it will determine the extent to which adjacent facilities are affected by a given release. The building manager provides budgetary data, which are applied as a constraint on the number of allowable sensors. CBSWEEP provides a risk analysis and placement recommendation consistent with the building information and project budget, which the building manager can use for implementation of a detect-to-warn strategy or as part of a more comprehensive active response approach. A description of each of the steps in determining the sensor placement evaluation is given later.

2.3.1 Assumptions and limitations

The proposed methodology is not intended to provide a rigorous evaluation of the probability of an attack on/in a given facility. In fact, for the purposes of CBSWEEP, an assumption is made that a release will occur and will be distributed through the HVAC system. Other assumptions include the following:

1. The zone air is well mixed.
2. CBSWEEP does not address actions to be taken after a release, but merely presents a method for determining optimal placement of sensors based on the occupancy of the building and vulnerability/consequence, with respect to a release within the HVAC system.
3. CBSWEEP is not intended for high-value (from a terrorist's perspective) military or government targets, or venues that are completely open access (carnivals, fairs, etc.) with limited or no restrictions to attendance.
4. CBSWEEP is intended to be used by nontechnical, nonscientific/engineering personnel who have a working knowledge of a building's HVAC system, that is, the building manager.
5. It is assumed a chemical sensor exists that can detect the agent(s) of concern.
6. The HVAC system is the point of release.
7. Airflow losses and absorption of the agent within building materials are not considered.

Additional assumptions and limitations associated with the airflow/concentration model are discussed in Section 2.4.3 and the references therein.

2.4   CBSWEEP functionality

An IDEF0 process diagram depicting the internal functionality for the wizard is shown in Figure 2.3. There are five main components comprising CBSWEEP:

- Interview/data collection module (A1): This module gathers user input and creates building model.
- Risk assessment module (A2): It assigns an RRR for each facility, AHU, and outside air intake.
- Agent concentration module (A3): This performs simulation of agent release scenarios to estimate agent concentration–time–response within affected facilities.
- Optimization module (A4): This provides prioritized placement of sensors, given detection capability and budget constraints.
- Reporting module (A5): This module provides reporting capability as an output to the user.

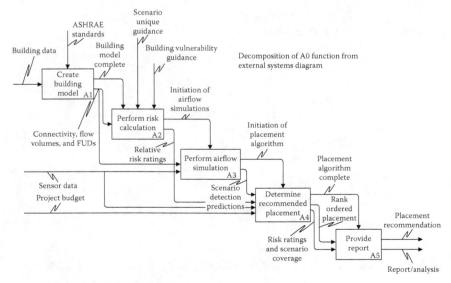

Figure 2.3 Wizard functionality.

2.4.1   Interview/data collection module

Beyond the summary information used to identify the project, CBSWEEP begins by querying the user regarding various aspects of the building and its associated facilities. The user chooses a building type (office, education, healthcare, etc.), and the wizard initially assumes a single floor and AHU. The user can add floors and/or AHUs as necessary. CBSWEEP establishes a zone associated with each AHU, and the user identifies which facilities (rooms, corridors, etc.) are associated with each AHU zone. Figure 2.4 shows an example of a data input screen associated with adding a new facility to an added AHU zone.

For each facility entered into CBSWEEP, the user is prompted for information regarding the use of the facility, or FUD, the square footage of the facility, and the number of supply and return air ducts servicing the facility. The square footage can use default ceiling heights to calculate room air volume, or the ceiling height for a facility can be specified by the user. The user provides adjacency information, both facility to facility within a zone and interzone adjacency, and the type of adjacency (closed door, open bay, etc.) is used to establish interzone flow rates within the concentration module. The facility size and FUD are then used to establish rated

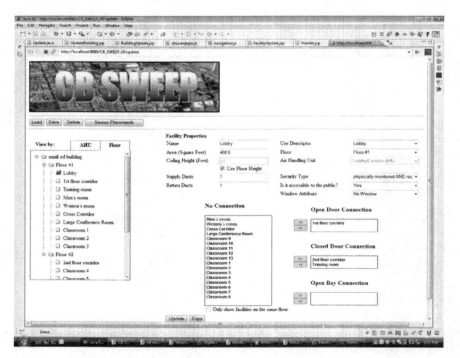

Figure 2.4 CBSWEEP data entry screen.

occupancy, volumetric flow (both supply and return) for each facility, and an overall refresh rate for the AHU zone based on ASHRAE standards (ASHRAE, 2007). An example of this implementation is a restroom that has a recommended volume of supply air based on the number of stalls, direct exhaust flow greater than the supply flow, and no return air going back to the AHU.

In addition to the information required to establish volumetric flow rates, the user provides security aspects of each facility and security aspects of the AHU and fresh air intake (which allows an evaluation of the associated vulnerability of each facility). The user is prompted to indicate whether a facility is open, locked, and/or monitored by security personnel or other means. The interview process also requests information regarding attributes for windows (e.g., nonoperable or open); these are used to assess the vulnerability of lower floor facilities to an attack from outside the building. Similarly, the user provides information regarding accessibility/security associated with the fresh air intakes and the mechanical rooms housing the AHUs. For roof-mounted AHUs/intakes, roof height and accessibility/security features of the roof are provided. The vulnerability and security parameters are preloaded into the wizard and are offered as a "pull-down" feature, so the user can easily and accurately enter the needed information. These parameters have values assigned to them, so the user does not have to determine the risk values associated with the various security and vulnerability features. This information is used by the risk assessment module to establish the RRRs for each facility, AHU, and outside air intake, as will be discussed in Section 2.4.2.

The CBSWEEP interview module allows the user to input all data at one time or over some course of time (a helpful feature since some buildings may contain several hundred facilities requiring, possibly, hours of input). The interview module also has an autocheck function to ensure data are correctly entered.

2.4.2   Risk assessment module

The RRR is determined following a series of inputs from the user of CBSWEEP based on HVAC vulnerability and building security, relative to critical nodes of the HVAC system. Typically, the RR is a function of the following:

$$RR = Consequence \times Vulnerability \times Probability \times Threat \qquad (2.1)$$

This type of approach is not without precedent. In fact, the Department of Homeland Security has used a similar scheme (DHS, 2007) in their risk-based approach to preparedness. For the purposes of this chapter, the Probability and Threat parameters are assigned a value of "1" as it is assumed the user of the wizard is concerned their facility may be attacked and they are taking proactive actions to prepare for such by placing sensors within the facility to improve the chance of detection. The resultant value, that is, the RRR, is then estimated as follows:

$$RRR = Consequence \times Vulnerability \qquad (2.2)$$

An RRR value is calculated for each serviced facility within a unique building. The potential consequence(s) (C) is based on potential human mortality and/or morbidity, which is itself based on occupancy. The user is allowed to input unique occupancy values for each of the facilities or a value will be estimated, based on the user defined. The occupancy is estimated based on ASHRAE Standard 62.1-2007, *Ventilation for Acceptable Indoor Air Quality* (ASHRAE, 2007). This guidance was used as it provides required, minimum, flow rates in cubic feet per minute (cfm) per square foot as a function of the designated use of the facility. Since the CBW captures the facility use and square footage, the occupancy can be estimated. In the present version of the model, each occupant is assigned an unweighted value of "1," indicating each occupant's mortality is equally considered.

The vulnerability of a facility, again with respect to the HVAC system, is based on the following factors and is designated as n in Equation 2.3.

1. Building accessibility (e.g., level of access controlled, if so how is it controlled)
2. Window operability
3. AHU distribution (e.g., is air recirculation shared between AHUs or is air exhausted to the outside)
4. Access to the fresh air intake (e.g., are louvers present, is the access screened, at what height is the access located)
5. General access control of various nodes (e.g., existence of locks, type of locks)
6. Access to specific HVAC nodes (e.g., who has access to the mechanical room with direct access to the AHU, are in-house personnel used for maintenance)
7. Access to the roof of the building (e.g., fixed external access, secured, height)

The general equation is as follows:

$$0 \leq u_{ij} \leq 1, \quad \sum_{j=1}^{n} u_{ij} = 1 \qquad (2.3)$$

where each factor (indexed by j) per serviced facility (indexed by i) receives an integer value between 0 (least vulnerable) and 5 (most vulnerable). The integer value is designated u.

Vulnerability of the *i*th serviced facility is then determined by the following equation:

$$Vf_i = \sum_{j=1}^{n} \frac{u_{ij} \cdot x_{ij}}{5}, \quad 0 \leq Vf_i \leq 1 \tag{2.4}$$

Vf identifies the vulnerability of facility *i*. The factor *x* is a weighting factor used to modify each vulnerability parameter based on predetermined values. In other words, each of the above parameters is assigned a weighted value relative to its security level. The value is predetermined but assigned based on answers to questions posed to the user of the wizard.

Stated another way, each of the above parameters is assigned a weighted value relative to its security level. Likewise, a similar calculation is performed for the fresh air intake. As an example, a fresh air intake located at less than 12 ft. from ground level, with no louvers (positioned at 45° from the building face) or screens, and located near a potential outdoor release site would receive a maximum value of "5" assigned for each of the two variables. These values are then summed and divided by the total number of possible points. The resultant value is then calculated and tabulated. In this case, the value assigned for the fresh air intake is [(5 + 5 + 5)/15] = 1.0. The same fresh air intake located at a distance of greater than 24 ft. from ground level, with a louver and screen, and not near or accessible for an external release would receive a value of [(1 + 0 + 0)/15] = 0.07. Each of the other parameters identified earlier are evaluated in a similar manner. It should be noted that the individual, assigned values are subjective in nature; however, since this model is intended to provide only a comparison of the RR within a specific building, the assignment of the values is not an issue as long as the values are consistently applied/evaluated throughout the process.

Once all individual parameter scores are determined, a cumulative vulnerability score is calculated. This value is then multiplied by the C value and an overall RRR score is achieved. As noted previously, the C score is currently unweighted. However, future versions of the model may allow the user to make a judgment on the assigned value of C. As an example, consider a large, complex building that houses several hundred employees and a day care facility. It may be the judgment of the user that the value of the children's lives is significantly greater than that of the adults. The user could then override the default C value to weigh the consequence for loss or impairment of a child's life greater.

As noted earlier, the resultant RRR value is not intended for use in comparing risks between buildings; it is intended to identify the RR of the various facilities within a single building to provide a framework from which to base sensor placement. It provides a reasonable estimation of the

RR between identified HVAC components and building security features so that a prioritization matrix can be developed. The placement of the sensors is then based on the available budget of the user, the HVAC characteristics, the building type/use, and the identified RRR as determined by the CBW.

2.4.3   *Agent concentration module*

The agent concentration module provides an estimate of the time-indexed concentration response within a building based on specific release scenarios. An LTI model is used to estimate concentration response based on an airflow connectivity model populated with the facility flow rates as described in the user interview module in Section 2.4.1. The LTI concentration model (Jacques and Smith, 2010) assumes steady-state flow rates and well-mixed air within a facility. The HVAC system is responsible for the movement of air within a building (it ignores pressure-driven effects because of environmental conditions). The form of the LTI model (Jacques and Smith, 2010) is as follows:

$$\frac{d\overline{x}}{dt} = \dot{\overline{x}} = A\overline{x} + B\overline{u}$$

$$\overline{y} = C\overline{x} + D\overline{u}$$

(2.5)

where the vectors $\overline{x}$, $\overline{u}$, and $\overline{y}$ represent the model states, inputs, and outputs, respectively. For this problem, the contaminant (agent) concentrations in each facility will be chosen as the state variables, and the inputs will be an agent release profile resulting from an internal release within a facility or an external release that enters the building through the outside air intake. An instantaneous release within a facility will be treated as an initial condition problem ($u = 0$) where the initial state associated with the release location will be the mass of agent released divided by the volume of the facility (room or corridor) where the release occurs. An additional state variable is added to track the amount of agent exhausted from the building to verify conservation of agent mass. The outputs of the model are the states (facility concentrations), the mass of agent remaining in the building, and the mass of agent exhausted from the building. Additional output variables are easily constructed to track agent concentrations in the supply and/or return air.

The coefficient matrices are functions of the flow rates and adjacency information describing flow paths between facilities or between a facility and its servicing AHU. A simple example serves to show how the model is assembled from a complete, balanced set of supply, return, and exhaust flows. Assume a two-room building (offices 1 and 2) with a corridor connecting the two rooms. For this simple illustration, we will assume that cross-flow occurs from the corridor to the offices. Defining the elements of the state vector x_1, x_2, and x_3 as the concentrations in offices 1, 2, and the corridor, respectively, we obtain the following differential equations for the model as follows:

$$V_1\dot{x}_1 = Q_{1s}c_s - Q_{1r}x_1 + Q_{13}x_3$$
$$V_2\dot{x}_2 = Q_{2s}c_s - Q_{2r}x_2 + Q_{23}x_3 \qquad (2.6)$$
$$V_3\dot{x}_3 = Q_{3s}c_s - Q_{3r}x_3 - Q_{13}x_3 - Q_{23}x_3 + u$$

V_i refers to the volume of the ith facility, and Q_{is} and Q_{ir} refer to the supply and return flow to/from the ith facility, respectively. Q_{ij} refers to the cross-flow from the jth to the ith facility. The sign for the cross-flow terms must be positive to properly associate with the facility where the concentration is being drawn from. The input, u, can be used to specify a time-varying agent mass release within a facility (shown here to indicate a corridor release), or it can be set to zero if an instantaneous agent release is assumed as discussed previously. The term c_s refers to the concentration in the supply air. For this simple example, we will assume that a specified fraction of the supply air is "fresh" coming from the outside air intake. This fraction is designated as μ, and it is assumed that this outside air exchange occurs in the AHU after the return flows have been mixed. With this assumption, c_s can be expressed as follows:

$$c_s = (1-\mu)\sum\left(\frac{Q_{ir}x_i}{\sum_j Q_{jr}}\right) + \mu c_{oa} \qquad (2.7)$$

where the term c_{oa} can be either set to zero, for clean outside air, or set according to an assumed outside air contaminant concentration for an external release scenario. For this latter case, c_{oa} can be set equal to u denoting a driving input to the system model. Substituting for c_s, and assuming an external release, we obtain

$$\dot{x}_1 = \frac{Q_{1r}}{V_1}\left(\frac{(1-\mu)Q_{1s}}{\sum_j Q_{jr}} - 1\right)x_1 + \frac{(1-\mu)Q_{2r}Q_{1s}}{V_1\sum_j Q_{jr}}x_2 + \frac{1}{V_1}\left(\frac{(1-\mu)Q_{3r}Q_{1s}}{\sum_j Q_{jr}} + Q_{13}\right)x_3 + \frac{Q_{1s}}{V_1}\mu u$$

$$\dot{x}_2 = \frac{(1-\mu)Q_{1r}Q_{2s}}{V_2\sum_j Q_{jr}}x_1 + \frac{Q_{2r}}{V_2}\left(\frac{(1-\mu)Q_{2s}}{\sum_j Q_{jr}} - 1\right)x_2 + \frac{1}{V_2}\left(\frac{(1-\mu)Q_{3r}Q_{2s}}{\sum_j Q_{jr}} + Q_{23}\right)x_3 + \frac{Q_{2s}}{V_2}\mu u \qquad (2.8)$$

$$\dot{x}_3 = \frac{(1-\mu)Q_{1r}Q_{3s}}{V_3\sum_j Q_{jr}}x_1 + \frac{(1-\mu)Q_{2r}Q_{3s}}{V_3\sum_j Q_{jr}}x_2 + \frac{Q_{3r}}{V_3}\left(\frac{(1-\mu)Q_{3s}}{\sum_j Q_{jr}} - 1 - Q_{13} - Q_{23}\right)x_3 + \frac{Q_{3s}}{V_3}u$$

The state differential equations are now in the desired form. Although this illustrative example assumed a specified fresh air fraction and no direct exhaust from the facilities, terms can be analogously defined for direct exhaust and/or infiltration/exfiltration in any facility.

To complete the LTI model, the output equations can be set equal to the state values, $y_i = x_i$, where $i = 1$–3, the concentration in each of the facilities. Additional outputs for agent mass in the building and supply air concentration can be constructed as follows:

$$y_4 \equiv V_1 x_1 + V_2 x_2 + V_3 x_3 \tag{2.9}$$

$$y_5 \equiv (1-\mu)\sum\left(\frac{Q_{ir}x_i}{\sum_j Q_{jr}}\right) \tag{2.10}$$

A complete description of the model, its assumptions and limitations, and its validation can be found in Jacques and Smith (2010).

CBSWEEP uses the concentration model to run a series of simulations by varying the agent release location and recording the time-varying agent concentration response within all facilities in the building, and as seen by each of the AHUs. The release amount is a function of the building volume and lethal amount of the given agent. For example, the approximate lethal dose to 50% of the population for sarin (gas/inhalational exposure) is approximately 100 mg/m^3. The current implementation of the model uses a fixed time step of 1 minute, which is considered sufficient given the modeling assumptions; however, the model is easily modifiable for other sample rates or continuous time. Information available from the model includes peak concentration seen at each facility/AHU, time to reach peak concentration, time to reach a given concentration threshold, and amount of time that agent concentration remains above a given threshold. Not all of this information is currently being used to determine sensor placement recommendations (see Section 2.4.4) but future versions of CBSWEEP may use this and/or provide it as additional reports to the user.

2.4.4 Optimization module

CBSWEEP uses the results from the agent concentration module simulations and the risk assessment module assigned RRR values to provide recommended sensor placements to the user. Based on the relatively sparse information available from the sensor manufacturers, a simplistic detection model using only a stated detection threshold is used in CBSWEEP. A concentration response exceeding this threshold by 20% is considered detectable and full value (unity) is given to detection by a sensor placed at that location for the given release location. A concentration response between 80% and 120% of the threshold is considered a marginal detection and receives a value of 0.5; a concentration response below 80% of the threshold value is considered undetectable by a sensor placed at that location, and receives zero value. The simplicity of this detection model is acknowledged; a more detailed detection model could be accommodated within the CBW, given sufficient data from testing of the candidate sensors. At present, only limited data are readily available from the manufacturers of the sensors. This limits the capability of the CBW.

Given the detectability of the release scenarios at the candidate sensor placement locations, CBSWEEP provides a prioritization of the placement location based on composite risk coverage. For each candidate sensor placement location, the detectability (0, 0.5, or 1) is recorded for each possible release location. The assessed value for a given placement location is calculated as a weighted sum of the detectability across all release locations, with the weights set equal to the RRR provided by the risk assessment module for the respective locations. Recall that the possible release locations are the facilities themselves, the AHUs, or the outside air intakes. The first recommended sensor placement is the facility, AHU, or air intake with the highest composite value. In general, a high-value sensor placement would likely be able to detect a number of release scenarios; an exception could be a vulnerable outside air intake capable of contaminating an entire building. Subsequent sensor placements will have the assessed value diminished based on the release scenarios (locations) that are already detectable based on the prior sensor placements. This is implemented by zeroing the RRR associated with the facilities/AHUs/intakes already covered (detectability greater than or equal to 1) based on all prior sensor placement coverage. This ensures adequate consideration for detecting releases from facilities with lower RRR values once the higher RRR release locations have been covered.

The model continues to "place" sensors in the manner above until either all release scenarios are covered, or the budget (constraining the number of sensors that can be placed) is reached or exceeded. To support "what if" budget scenarios, the CBW provides prioritized placements for up to 10 additional sensors beyond what is dictated by the budget constraint. Without the need to modify the building data, the user can also elect to rerun the wizard using a different budget value. As the run times for CBSWEEP once populated with the building information are on the order of minutes, it is easy for the user to obtain results for a wide range of budget amounts.

2.4.5 Reporting module

As discussed earlier, CBSWEEP is capable of generating a wide variety of information to support the building risk assessment and sensor placement decisions. The standard output currently includes the prioritized placement locations. The number of locations shown will be greater than the budget constraint allowed to support "what if" budgetary decisions. An exception is the case where a release from all nonzero RRR locations can be covered with fewer sensors than the budget allows. In this case, the sensor locations providing complete coverage will be reported. In all cases, the number and percentage of release scenarios covered by the budgeted number of sensors will be reported to the user. Other information available includes the time to detect for the release locations with the highest RRR values, and a report on the assessed RRR values for the various facilities, AHUs, and air intakes. This last information set can be used by the building manager to make decisions regarding modifications to the security features within the building. Although the CBW does not consider an active response strategy once detection has occurred, an important complement to using the CBW would be to incorporate it into an overall strategy for dealing with the threat of a chemical attack.

2.5   Example and results

CBSWEEP was run for an example building somewhat typical of a small two-story training location. Figure 2.5 shows a room layout for the first floor containing a lobby, training/conference rooms of various sizes, restrooms, and a corridor. There are three breakout rooms on the first floor; an additional 10 breakout rooms are located on the second floor (not shown), and they are accessed via stairwells and a central corridor. The lobby and central corridors for the first and second floors are serviced by AHU #1, and the training room, restrooms, and conference room are serviced by AHU #2. For this example building, each of the breakout rooms has a separate room air-conditioning unit.

Adjacency of the various facilities is as shown, noting that the second-floor corridor is connected to both the lobby and central corridor of the first floor via stairwells with closed doors. All of the second-floor breakout rooms have an adjacency to the central corridor of the second floor. Table 2.1 shows a sample of the building summary data provided by CBSWEEP.

The calculated RRR values are shown in Table 2.2. The training room adjacent to the lobby has the highest RRR value among all the rooms based on occupancy and vulnerability. Even though the large conference room has greater occupancy than the training room, the easier public access to the training room results in a higher RRR. The lobby also has high vulnerability based on ease of access, but a lower occupancy results in a much lower RRR than the training room. The breakout rooms on the first floor have a high value of RRR due in part to their occupancy, but primarily because of the accessibility of the room's air-conditioning units to unauthorized personnel. The upstairs breakout rooms have both lower occupancy and lower vulnerability because of the elevated location of the room air-conditioning units.

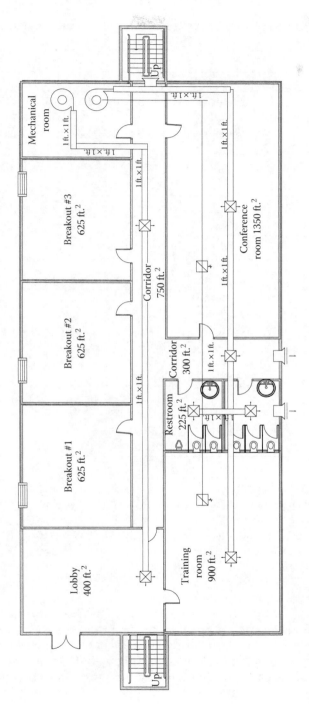

Figure 2.5 Small education building: first-floor layout.

Table 2.1 Sample Summary Screen for Small Education Building: Building Details

Zone name	Intake	Intake height
Lobby/corridor AHU	2	1

Facilities serviced by this zone

Name	Use descriptor	Area	Supplies	Returns	Floor
Lobby	Lobby	400.0	1	1	1
First-floor corridor	Corridor	750.0	1	1	1
Second-floor corridor	Corridor	1000.0	1	1	2

Zone name	Intake	Intake height
Training facility AHU	2	1

Facilities serviced by this zone

Name	Use descriptor	Area	Supplies	Returns	Floor
Training room	Classrooms	900.0	1	1	1
Men's room	Restroom	225.0	1	0	1
Women's room	Restroom	225.0	1	0	1
Cross corridor	Corridor	300.0	1	1	1
Large conference room	Conference rooms	1350.0	1	1	1

Table 2.2 Relative Risk Ratings for Small Education Building

Facility	RRR
Training room	85.5
First-floor breakout rooms	60.8
Large conference room	51.3
Second-floor breakout rooms	38.0
AHU #2	24.0
Lobby	19.2

Table 2.3 shows the results provided by CBSWEEP for the example building. It recommends placement of the first sensor in the central AHU based on its high RRR and the resulting prediction of being able to detect a release anywhere in the facilities serviced by AHU #2. A single sensor is not able to detect all considered release scenarios; detection of release

Table 2.3 Sample Output Screen for Small Education Building: Results

Top Release Locations/Time to Detect at This Location	
Location	Time to detect (min)
Training room	2
Large conference room	2

Top detection locations								
Scenarios covered								
Location	Floor	Zone	Rank	Cumulative	Cumulative%	Change	Change%	Budget
AHU #2	1	AHU #2	1	7	13.7	13.7	0	Within budget
Breakout room #1	1	AHU #3	2	10	19.6	3	5.9	Out of budget
Breakout room #2	1	AHU #4	3	13	25.5	3	5.9	Out of budget
Breakout room #3	1	AHU #5	4	16	31.4	3	5.9	Out of budget
Breakout room #4	2	AHU #6	5	19	37.3	3	5.9	Out of budget
Breakout room #5	2	AHU #7	6	22	43.2	3	5.9	Out of budget

within the facilities serviced by AHU #1 and any of the breakout rooms requires additional sensors. If the budget allowed placement of more than one sensor, the program recommends placement within the first-floor breakout rooms based on their higher RRR values. Other output provided by the wizard includes number and percent of release scenarios covered, and time to detect for the top (highest RRR) release locations.

Although this example building is rather simple, it shows the feasibility of the approach and the functionality of the CBW. Significantly more complex buildings, including an eight-story office building consisting of five AHU zones, have also been successfully run using the CBW.

2.6   Summary

The authors have developed an approach for an innovative wizard-driven system for quickly designing new sensor networks that can be used by inexperienced operators. The software system bases the recommended sensor placements on the building/facility usage, occupancy, and a high-level HVAC system architecture. The design involves identifying key components in the HVAC systems, such as AHUs and ARDs, and targeting these areas for placement of the sensors. An RRR is assigned to each facility, AHU, and outside air intake, which is subsequently used to prioritize sensor placements. An LTI model based on facility connectivity and standards-based flow rate calculations is used to predict the dispersion of a released agent throughout the building for given agent release scenarios. A simplified detection model then uses the concentration responses and a detection threshold to predict detectability for possible release locations, given a candidate sensor location. This produces a weighted value associated with each candidate sensor location, which is subsequently used to prioritize sensor placements consistent with assessed risk values and budgetary constraints. The approach was demonstrated using a two-story training building; larger and more complex buildings have been successfully modeled as well. Extensions of this work could address a wider variety of buildings and release scenarios involving biological agents, as well as possible adaptations to support shelter-in-place strategies.

Acknowledgments

This work was conducted while the authors were employed as consultants to Peerless Technologies Inc., Dayton, OH. The work was conducted under a Small Business Innovative Research contract awarded by the Department of Homeland Security to Peerless Technologies. The authors thank Peerless Technologies for the opportunity to contribute to this effort.

References

ASHRAE. ASHRAE Standard 62.1-2007. *Ventilation for Acceptable Indoor Air Quality*, American Society of Heating, Refrigerating and Air-Conditioning Engineers (ASHRAE), Atlanta, GA, 2007.

Chen, Y.L. and J. Wen. Sensor System Design for Building Indoor Air Protection. *Building and Environment*, 43, 1278–1285, 2008.

Department of Homeland Security (DHS). *National Preparedness Guidelines*, https://www.llis.dhs.gov/docdetails/details.do?contentID=26718, September, 2007.

Draft Federal Information Processing Standards (FIPS) Publication 183, 21 December 1993, 116 pp.

Graham, B., J. Talent, A. Graham, R. Cleveland, S. Rademaker, T. Roemer, W. Sherman, H. Sokolski, and R. Verma. *World at Risk: The Report of the Commission on the Prevention of WMD Proliferation and Terrorism*, First edition, Vintage Books, New York, 2008.

Jacques, D. and D. Smith. A Simplified Building Air Flow Model for Agent Concentration Prediction. *Journal of Occupational and Industrial Hygiene*, 7(11), 640–650, 2010.

Sohn, M. and D. Lorenzetti. Siting Bio-Samplers in Buildings. *Risk Analysis*, 27(4), 877–886, 2007.

Sohn, M., P. Reynolds, N. Singh, and A. Gadgil. Rapidly Locating and Characterizing Pollutant Releases in Buildings. *Journal of the Air and Waste Management Association*, 52, 1422–1432, 2002.

chapter three

Harnessing disruptive technologies

Quick systems engineering for emergency response

Brian J. Phillips

Contents

3.1 Introduction

This chapter describes relevant and existing technologies that can be used, adapted, or repurposed to support emergency management professionals. It presents a small collection of technologies that can support activities in the emergency response disciplines. These technologies can not only support current practices but can also act as a catalyst for changing those practices. These technologies are not in a research lab, and they are not "vaporware." They are actively being used in the commercial sector today.

    The emergency response community has difficult challenges in finding new technologies to address their real-world problems. These new technologies must be mature enough to be dependable. They must be affordable. They must interoperate with existing systems. The technical staff must be able to adopt and use these tools within their existing skill sets. This chapter lays out several new technologies that both support emergency professionals and match these restrictions. These technologies give responders new capabilities. They allow users to change the way they do their work and disrupts the way they perform their normal everyday tasks. Because of this, they are known as "disruptive" technology.

Disruptive technologies create new ways of doing things. In some cases, it allows people to perform their normal tasks in a better way. In other cases, it creates entirely new tasks and needs. Disruptive technologies change the basis of how users perform their tasks. Disruptive technologies are capable of changing what those user tasks are in the first place.

This chapter describes several of the highest value disruptive technologies commonly available (as of 2012). These technologies are new, but have rapidly reached a level of maturity that makes them useful to the emergency response community. All of these are the result of small increases in the capabilities of existing legacy technologies. One technology builds upon others. The technical underpinnings are not the "disruptive" part of them. A new technology becomes "disruptive" when it supplies capabilities that change what people do.

Ten years ago, it was hard to imagine that a huge portion of the population would be walking around town with cellular communication devices, small computers, global positioning system (GPS) tracking devices, gyroscopes, movie players, and Web browsers in their pockets. Today, it is considered odd if someone does not have those things. The emergency response community operates within this technical environment, where teenagers have more computing power in their back pocket than the entire Manhattan Project.

These computing platforms are not only for voice communication. They use small mobile applications that are quickly and easily downloaded from centralized repositories on the Web. An elementary school student can easily install, upgrade, or administrate these devices with new applications from these repositories. The Apple Inc. iStore is a popular repository that provides a way to automatically install and configure software. The Android platform uses a more open model through partners such as the Amazon Android Store (Amazon Technologies Inc.).

Technologies are no longer developed to stand alone. They inhabit the environment of other technologies and the users themselves. High-level system ideas (normally found in the realm of "Systems Thinking") can be used to describe the conceptual environment that these systems are engineered to inhabit (Boardman and Sauser 2008). Every system in the technology world has conceptual boundaries that define what is "inside" the scope of the system and what is "outside" that same scope. The modern era of technology is making those lines fuzzy though. The advent of a nearly universally adopted network protocol (Transmission Control Protocol/Internet Protocol [TCP/IP]) and the World Wide Web have created a baseline of interoperability that almost all modern consumer systems implement.

Web browsers have had a standard way to interoperate called HyperText Transfer Protocol (HTTP) since their inception. This is a standard that defines electronic documents with special tags embedded in the document (sometimes called "markup"). Web browsers read markup documents (called HyperText Markup Language [HTML]) and render them onto the computer screen. Modern Web browsers can partition their applications into pieces (sometimes called "portlets") so that a single view for a single user is actually formed from many sources at once.

This communication standard has found itself adopted by software developers as well. The technologies that make up the "back office" of the network now use "Web services" to communicate between each other. Adopting an open protocol allows systems to work together since one application can now use another as a "service." Developers can "stack" multiple services together quickly to write distributed applications that span the Web.

These services are all based on the concept of multiple tier content delivery. The fact that they are based on open standards allows different platforms to interoperate with them. Web browsers can interact with servers that interact with other servers, that interact with mobile processors, and that interact with other applications. These chains form a network of "togetherness" within which these applications can operate (Sauser and Boardman 2007). This network of applications can be thought of as a single larger application, made up of small applications that combined to form a new interconnected capability. This type of application is why websites such as expedia.com can easily search for airline and travel rates across dozens of websites. Technologies can easily connect together using standards and agreements. Organizations ensure interoperability through contracts and formal processes that adhere to them.

The previous generation of systems did not have this existing foundation of standards to rely upon. Every time a new system was built, a new communication and networking scheme would need to be built. In many cases, interoperatability was not considered in the design. Although the modern emergency responder has a set of legacy systems deployed back at their base stations, they have the next generation of technology with them on their hips and in their vehicles. If they have a cell phone or a laptop, they are connected to this vast network of interoperable systems.

3.1.1   Harnessing disruptive technology

The introduction of some technologies actually affects their end-users so drastically that it changes the tasks that those users are performing.

An example of this is the automobile, and its enabling technology, the assembly line. The assembly line made the automobile affordable. It did not directly change the lives of its users. The products of its work did. The assembly line was a disruptive technology that made other technologies more affordable, and thus more common and accessible.

Before the invention of the commonly affordable automobile, the world of transportation revolved around horses, railroads, and ships. Paved roads were designed for cities and towns only. No major petroleum industries existed; coal was the primary energy product of the day. After a generation of affordable automobiles appeared, the new automobiles were becoming more common than horses, railroads, and other transportation systems. Entire industries sprung up to support car users and their needs. A complex road network began growing in various parts of the world. In the modern age, the automobile has become so central to the industrialized way of life that it is considered "odd" to try to live without one in an industrialized country.

The automobile allowed the world to be more connected through road networks. It allowed each of its users to be autonomous. Drivers could pick their own destinations on their own schedules and even pick their own cars. Drivers still operated within a greater framework by adopting and adhering to standard laws and governances. This system created a basis to allow for new, emergent technologies. Automobiles got safer, faster, more luxurious, and more efficient over several generations of innovation, offering a diversity of transportation options for its user base.

Computers changed society in a similar way. Instead of being connected by roads, supported by an energy infrastructure and governed by laws, society uses information technology (IT) to connect us with cables and cellular towers. The IT realm is governed not just by laws, but also by standards that allow different platforms to operate together within the larger whole. A diverse set of systems and users inhabit this new information realm. Each of them operates autonomously, yet in harmony, as they inhabit this new environment of information. New systems can instantly "belong" to the current information infrastructure by supporting common and open standards. Users remain autonomous since the systems can operate without dependency on external processes.

The emergency response community is engaged in this arena as much as any other. The difficulty is not finding information, but understanding, filtering, and making sense of it. Deciphering the overload of data is a daunting but necessary task for emergency response professionals.

A special note about knowledge management: the practice of knowledge management is a combination of technology and process (Alavi and Leidner 2001). Embedding a process into an organization to collect, store, retrieve, reuse, and make sense of data is currently a human labor–intensive task. It is currently supported by technology, not solved by it.

The technologies required to truly automate that process are not yet mature enough for practical use. Some techniques such as keyword search, synonym recognition, and expert systems (Matkar and Parab 2011) are in the field and being used. They are a valuable part of the overall process, but they do not remove the reliance on the human expert. Current computers largely fail at understanding the context of information, which is exactly what human experts are good at. This limits the computers' ability to discern the importance and nuance of information when placed into greater context. Some technologies seem promising, namely statistical/kernel learning and adaptive systems (Bock 1988), but have a generation still to go before they are ready to take on the challenge of having a machine not only read information, but understand what it actually means.

New technologies and approaches are emerging faster than they can be incorporated into practice. The amount of time it takes to design, specify, and build a new system for a government or organization is so long that by the time a new technology is fielded, it is obsolete. Emergency professionals have an urgent need to fulfill many of their information processing requirements. Lives are literally on the line! This puts responders into the unenviable position of needing to use these new technologies without spending the time to allow them to mature or waiting on needed skills to emerge from the people on their teams. Emergency personnel need solutions in the short term.

Disruptive technologies can boost teams who have a requirement for accomplishing tasks immediately. These technologies are already out there and need only to be adopted and used to fulfill some user needs. Yes, it will change the way the users do their jobs. In the field of emergency response, chaotic and changing environments are normal and are simply part of the work.

It is a reality that by the time this chapter is published and disseminated, the current batch of new disruptive technologies will have changed. Many of the technologies discussed here will survive in one form or another, so they deserve some consideration when thinking about a capability for organizations that have little time, little money, and a lot at stake.

It is also important to delineate the difference between the technical capabilities within a modern industrialized society and the capabilities available in a limited or disadvantaged venue. Emergency responders are frequently required to travel into areas that do not have the infrastructure to support many of the technical capabilities mentioned in this chapter. In some cases, it may be advantageous (or necessary) to bring supporting technologies such as satellite communication links, mobile cellular networks, or even electrical power generation equipment.

3.2   *A tour of technologies*

Emergency response organizations stay busy trying to keep track of changes in the emergency response field. It can be strenuous to track those changes that are taking place in the IT field as well. Changes in IT can be translated to new capabilities if organizations (and their people) can be agile enough to adapt to them quickly.

New technologies can be coupled with existing technologies and processes to enhance the capabilities of organizations. Entirely new capabilities can be created by connecting new and legacy technologies together to form a new system-of-systems. This is sometimes called a "mash-up" application. The amount of technology available to the average user is greater than it has ever been. Understanding all that is available is a big challenge. Disruptive technologies discussed in this chapter have a few common features. They are useful for emergency response organizations. They provide an inexpensive capability that did not commonly exist five years ago. They are also within the skill range of most IT support staffs, and even some hobbyists. Each of these technologies has shown a real utility and could be readily repurposed to support emergency response, planning, command and control, or simple organization.

3.2.1   *The Web, Web 2.0, and netcentricity*

The Web arrived into the commercial world in the late 1990s. By the end of 2010, every facet of modern life was connected to it. Banking, shopping, and even medical consultations were being performed over the Web. The Web has not only become intrinsic to how people go about their daily business, it has also changed how they live their lives. The instant flow of information has become the expected norm.

At its base, the Web mainly relies on two technologies. The first is the communications protocol, called TCP/IP. This is a set of digital rules that allow the machinery that makes up the world's networks to reliably move information from one system to another through a network of intermediate points. The information is first broken up into smaller "packets." The packets are sent through the network to arrive at their final destination and then reassembled into the original.

The other technology is called HTML. This is a set of presentation rules. When information is presented using these rules, any application (such as a Web browser) can display and interact with it. The addition of scripting language support into HTML allowed the information to be dynamic so the user could interact with the information on the page.

The basic Web experience is transactional. A user requests information by using an application (such as the Microsoft Corporation's Internet Explorer® Web browser). The request is sent to another computer (normally referred to a "server"), and that server responds with either an error signal or the HTML document that the user requested. The application presents the requested data to the user using the interfaces specified in the HTML content.

This paradigm is very useful. Servers can use the information contained within requests to make changes to their internal states. This allows a large number of people to see data and also to make changes to that data set. People can query information and take action on that information without worrying about the communications pathway or (largely) the platform they are using. If users have an application that supports TCP/IP, and can process HTML according to the rules specified by the HTML standards, then it can work.

A purely transactional experience can be problematic. When multiple people are working on the same data set, they must constantly refresh their data to be current. This is highly difficult when there are many concurrent users, each making updates in real time. To support this, a new technology was invented. It allowed the scripts executing in browsers to make additional information requests through the Web, and update the user display in real time, with little or no interaction by the user.

There were several technologies that performed this. Now they are collectively referred to as Asynchronous JavaScript® (Oracle America Inc.) and Extensible Markup Language (XML), or Asynchronous JavaScript and XML (AJAX). It created a straightforward way to add a more granular transaction within the Web form. Requests for updates could occur in the background without users even being aware of it. Background requests are sent using a standard format using XML. Responses are informational rather than graphical, so only the data sets within the page needed to be changed. This technology and the highly interactive applications that it spawned were collectively known as Web 2.0. Web 2.0 has become completely embedded into the modern Web environment; it is embedded so deeply that people would only notice if it was not there. Web 2.0 applications include social media such as Facebook® (Facebook Inc.), real-time stock tracking with Google® Finance (Google Inc. n.d.), and tracking storms using live animations from the U.S. National Weather Service.

The Web (original and 2.0) supplies a technical basis for a huge number of new technologies and approaches to information-based problems. Alone, it reinvents the entire way knowledge is discovered, consumed, and published across the world. But more can be carried out with it. The Web can be placed at the center of operational organizations and used to address their complex problems. Using the Web as a central operational hub to not only display and collect data but also provide management on how people consume that data, and how data are transformed into knowledge is embedded in a concept called netcentricity.

A netcentric system connects people, machines, and data into a common information environment. A netcentric system is normally composed of different types of systems and users who collaborate to accomplish organizational goals. Examples of this type of system can be found in the U.S. Department of Defense (who originated this idea), corporate supply chain management systems, airline reservation systems, and many others. A netcentric system builds upon the existing Web (and Web 2.0) technology to create a management and data-sharing environment.

Emergency professionals who work for local, state, or national organizations will normally have some sort of Web-based capability. Some governments have already implemented a netcentric solution that attempts to organize and manage the emergency planning, management, and response functions. In the United States, the national system is called the National Incident Management System (U.S. Federal Emergency Management Agency).

These types of incident management systems have been built using large amounts of resources and effort. Their goal is to harness netcentricity and create management and collaboration layers so that emergency planners, managers, and responders can better execute their jobs. Systems of such size are seldom without gaps and faults and rely on user inputs and evaluation to steer the development of the system.

A distributed system-of-systems creates a challenging technical environment. The ability to perfectly define, build, test, and deploy such a system is almost nonexistent. It is extremely difficult to create a system perfectly when the environment it inhabits is rapidly changing. Instead, these large netcentric systems rely on organizational feedback loops. Emergency experts are not only given a system to use in the performance of their job, but they can also create suggestions (commonly referred to as a "use-case") so that the system builders can make improvements. This "creation—get input—upgrade—get input—repeat ..." loop acts to iteratively improve these systems piece by piece.

Those improvements may be simple repairs or enhancements, or they may be completely new, unforeseen capabilities. In some cases, those changes may completely change the way the system was originally envisioned to work. Input from the user community is key to evolving a better, more useful netcentric system. Emergency professionals have the ability to directly influence the system design by inserting new (and perhaps even disruptive) capabilities into the system through a formal system improvement program.

Emergency professionals do not need to be engineers or computer scientists to do this. The concept of a "use-case" can be captured in a simple story or scenario that the system needs to fulfill. Emergency professionals can define such use cases and participate with engineers to create, test, and field new capabilities as part of the greater system-of-systems.

3.2.2   Smart phones

Mobile phones, pads, and other "smart" devices have become common in the world. Ten years ago, they could only be described as science fiction. In the United States, almost half of the population has one of these devices (Neilsonwire 2011). Each of these devices is a mobile computer in its own right, with the power of a desktop just a few years ago. They are a cellular communications device, a GPS, an entertainment system, a camera, an accelerometer, and a connection from individual people to a network. These devices can be awake and active as long as the battery stays charged.

Each of these capabilities (phone, GPS, etc.) is a true advancement when considered alone. When they are considered together, it makes for a staggering leap in technology. Not only are people using these devices to accomplish their everyday tasks, but also new tasks are emerging because of these devices. People can check Foursquare® (Foursquare Labs Inc. n.d.) to determine the live locations of their friends. They can respond to social media alerts at a moment's notice. Schools use them to alert students of dangerous conditions or social events. They have become integrated into the fabric of our social and work lives to an extent that only the most visionary could have foretold.

These mobile computing platforms connect people to the network. They use the same basic technology that the Web (and Web 2.0) uses. This allows these mobile devices to be instantly integrated into the overall information community where people and servers communicate using standard methods and protocols. Applications can be written specifically for these platforms, or they can be written using legacy Web technologies. This means that it is possible for smart phones to interact with the Web, and with each other, in real time.

Many emergency professionals also carry these devices. They have become so inexpensive now that it is easy to imagine equipping entire organizations with smart devices (DePompa 2011). Such equipment would connect each member of an organization together in real time. It allows a netcentric system to be constructed that helps organize the combined efforts of many people at once.

Connecting people allows emergency professionals to create entirely new solutions to existing problems. People networked into a common information environment can accomplish tasks that go beyond those traditionally performed with legacy computing. Enabling this on a mobile platform brings this transformation out of the IT departments and into the response operation.

A smart phone is a communication device and a lightweight mobile computer. Either of these aspects is valuable. Both of these capabilities have been present in society for some time. The key difference with smart devices is that people always have them on their person and that device can be almost permanently continuously active. These devices can simultaneously receive information from cellular networks, collect data from their internal hardware (position, images, etc.), and process that information with applications that are hosted on the smart device. The smart device is not only sending and receiving user-generated information; it can process its own information flow and react according to its internal rules.

Instead of a passive device that responds to simple signals (such as receiving phone calls or checking maps), the device can become truly interactive. It may alert the owner to events happening within the network using either internal rules or a learning method based on what this user or others are doing with the information. These alerts can change the way people use their smart phones and change the way they actually do their jobs. Smart mobile devices can be embedded as a computing resource and as a thinking one as well. By coupling a computer, communications, physical location, and the desires of a user, a netcentric application can perform capabilities well beyond what systems traditionally perform.

Consider the idea of training. Emergency professionals routinely perform training on various tasks. These tasks can be orchestrated across the network so that each individual within a training scenario is connected to a central simulation. Training events can originate from the training managers, or an automated system, and allow the training to represent higher-level organizational challenges in addition to individual tasks. Training for events such as mass casualty triage and toxic chemical releases can be orchestrated across all participants at once. Exercise managers can have a central place to control the training activities. In addition, analytical data mining can be performed live during the exercise and afterward. This ability to quickly process data in real time while in the field allows exercise leaders to give participants feedback in real time.

In such a training scenario, an autonomous learning mobile application can "sense" simulated dangers and observe the users' actions in response to them. Such a system can share the actions of many exercise participants with an exercise management cell. Analytics can be performed in real time showing how exercise participants behaved when exposed to events.

Consider how emergency professionals perform command and control. Emergency leaders could use this technology to view maps that show the location of every person in each of their teams. Relating locations to resources (such as supplies or equipment) can give responders the ability to perform real-time queries on their material status. Interfacing these systems to hospital systems could allow responders to check hospital availabilities in real time and also allow the hospitals to get an early sense of impending crowds.

Use of cellular data networks can present a problem. Sometimes networks connections are weak, disabled, or nonexistent. More and more response teams are being equipped with mobile cellular communication vehicles. These are normally configured as a large utility vehicle with a robotic extending antenna and multiple power generators. These create a mobile cell tower that connects the communication devices in the local area to other towers or satellite links. Portable 3G cellular networks can be deployed across large areas when a disaster affects an area larger than what a single mobile antenna can support. Planning for communications, power, and data is already an important part of how response teams prepare and operate.

3.2.3   The cloud

"The cloud" is part marketing term and part revolution. The cloud is a term that tries to represent a new technology, but what it really represents is a new approach to technology. The new features of "the cloud" are focused on how applications are built, deployed, and used and not on what they actually do. According to the National Institute for Standards and Technology (NIST) draft definition of cloud computing (Badger et al. 2011), it is defined by the following characteristics:

- A cloud service is one that features on-demand servicing of users. The machines that make up the cloud do not wait endlessly for some predefined task. Instead they can be rapidly reconfigured to meet user demands.
- Cloud computing is connected to the Internet through high-speed broadband. Data transfer speed is very high.
- Cloud computing applications are elastic. This means that they can rapidly expand or decrease the resource requirements as needed.
- Cloud computing applications can pool resources, either conserving them or maximizing their use across multiple machines or users.
- Cloud computing platforms can measure their own usage and dynamically adjust the resources and services available to users as fits the demand.

These types of capabilities have been present for a long time. The cloud approach takes these existing technologies and packages them into Internet-focused capabilities. The NIST definitions presented above describe a specific vision of what the cloud "is," but not what it actually "does" for its users. The NIST definition shows a technical approach into an engineering mindset that ultimately affects users. It changes the basic rules of how software applications are acquired, deployed, and used.

To understand the cloud, these characteristics need to be tied to new ways of deploying software or data on servers. The cloud uses three new models of deploying software: Software as a Service (SaaS), Platform as a Service (PaaS), and Infrastructure as a Service (IaaS). Each of these deployment models have distinct differences with how computing systems have been traditionally built.

SaaS changes the way that software is sold and deployed to consumers. Instead of buying CDs, DVDs, or any media and installing software from them to a client machine, users simply subscribe to a service across the network. Applications that previously had required installation (and updating on a regular basis) now only require a connection. An example of this includes Google® Office (Google Inc.), a free (and very capable) office productivity suite that is accessed through a Web browser.

PaaS changes the model of how hardware platforms are acquired and deployed. Software engineers write computer programs to specific hardware platforms and operating systems. Combinations such as Intel® (Intel Corporation) and Microsoft Windows® (Microsoft Corporation) or Mac® OS X® (Apple Inc.) are common models for software, but many other combinations exist. By translating the hardware into a virtual machine, programs can be run on one hardware platform, but execute as if they were on another (sometimes called "virtualization"). A virtual system is simply a computer program that pretends it is a physical computer. PaaS allows engineers to digitize an entire computer so that every piece of data resident in the physical system can be transferred to a virtual one. Installed programs cannot tell the difference. PaaS allows people to move entire computers worth of software, hard drives, or even executing programs between hardware platforms where the applications (and the user) are unaware of the change. This creates a system that is fault tolerant, mobile, and easily replicated. It also allows engineers to build computing systems without being held hostage to legacy software and hardware requirements. Legacy systems can be moved into virtual platforms with little worry about their ability to work with new hardware.

IaaS is similar to PaaS. If a single computer can be moved from place to place by making a digital copy, then an entire data center can be moved as well. IaaS allows the construction of virtual data centers. A large number of machines can be configured, moved onto the virtual environment, and be available through the network. Entire organizations can be virtualized. IaaS allows the creation of new businesses that supply this virtual data center environment. Organizations can establish large enterprises or data centers without buying any hardware. In many cases, much of their software infrastructure can be rented as well.

These types of issues seem to be solely in the realm of IT professionals. The effects of these technologies create downstream opportunities. Many cloud technologies are freely available to anyone who seeks to use them. There are many applications within the cloud that are freely available and accessible across the Internet, where any type of platform (laptop, desktop, smart phone, or tablet) can access and use them.

Most emergency professionals are part of local, state, federal, or corporate organizations that have active network security policies. These policies can make the installation and use of software an onerous task. Using cloud applications normally removes the need to install software on a local machine. This means that the capabilities are available to try out quickly, with no difficult installation process. If you can reach the Internet site, then the application can be used.

Another facet of the cloud model of application delivery is that the data sets generated and used by end-users are stored within the cloud, not on the local machine. The data that users need are disconnected from the actual machine they are using. In a cloud, users can work with their application, save their data, and then go to another computer or mobile device to access that same data.

The cloud provides some real-world advantages to emergency professionals. The cloud removes the risk that the loss of a piece of hardware (laptop, phone, etc.) will result in the loss of data. In that case, it is easier and cheaper to buy inexpensive consumer-grade devices than to invest in hardened devices. After all, why spend thousands on a Panasonic Corporation Toughbook® instead of hundreds of dollars on a business-level Hewlett-Packard Company laptop? Why store data in a format that is difficult to access from mobile devices? Why continuously spend money and energy on tracking installation licenses for various office products?

Because cloud applications tend to be usable across platforms and operating systems and are openly available across the Internet, it becomes a straightforward exercise to construct a set of capabilities that are useful across the extended enterprise. A minimal (even hobbyist level) knowledge of IT is sufficient to collect enough cloud applications together to form some ability to meet mission requirements. Cloud applications are well suited to either formal or ad hoc systems engineering and integration efforts.

Consider the following example. A shared directory using Dropbox® (Thru LLC) (https://www.dropbox.com) can store documents so planners, managers, and responders can access them. During a planning phase, Google® Docs (Google Inc.) (https://docs.google.com) can be used to edit planning documents. When emergencies arise, managers can place situational data such as geospatial maps and images in a shared Dropbox directory (http://maps.yahoo.com) for access by team members. Responders can use Foursquare (https://foursquare.com) to get a sense of where the responders are within a large area and quickly create a map that managers and responders can use to discover where their people are. There is an abundance of tools available on the cloud that can be included into a responder enterprise with little to no customization. All of this can be carried out without paying for or installing any software.

3.2.4   The social network

Social networks are about people. They seek to provide an open environment where people can interact with each other based on friendship, subject areas, professions, or any other imaginable reason. The idea behind social networks is that technology can be used to connect people, and those connections extend across society.

The general idea of a social network can be represented by the Faberge® (Faberge Services Limited) shampoo commercials broadcast around 1982. In the commercial, Heather Locklear shows the power of an extended network of friends. As each friend tells two more friends, a large community begins to emerge. Social network applications attempt to give people tools that help them connect with others based on some common factor and help facilitate communications across the social network. It simply allows you to tell more than two friends at once.

If that were all there was to it, then it would quickly result in information overload. Social networks not only allow you to communicate to your friends but also to receive feedback from them. Asking a crowd for advice will result in some good answers and some not-so-good answers. Feedback allows a user to decide which is which. The only goal that the social networking applications have is to facilitate communication (and perhaps to display advertising in the process). The power of the content and feedback (provided by the users) makes the social network valuable.

Another key aspect to social networking tools is their lack of hierarchy. There is no concept of command and control between groups of cooperating friends. The removal of any pass-through or bottleneck to information flow makes this medium an extremely powerful tool. This is especially true in the hands of the emergency management and responder community. Traditional command and control applications attempt to regulate and govern all communications within the system. Social networks seek to allow unfettered access to information, relying on the fact that the social dynamics will by their nature self-organize the system. A totally open social system is well suited to chaotic and fast-moving situations where people are attempting to collaborate and interact in real time.

There are many real-world situations where emergency teams have set up ad hoc social networking applications to help organize their efforts (The Huffington Post 2011). The 2011 earthquake in Japan and Florida hurricane emergencies were a prime example of how social networking can be used to not only communicate within emergency teams but also to organize the response itself. The power of social networking is particularly useful in transcending organizational and institutional barriers so unstructured communications can be made directly between the public, emergency professionals, and community volunteers.

Social networking tools are normally free to use and available across the open Internet. Social networking tools have often been criticized for their misuse. Numerous cases of stalking, bullying, fraud, and abuse have been performed within existing social networks. This technology can be misused. Within the context of an emergency response activity, such misuse is a rather trivial concern given the short nature (hopefully) of a crisis and the self-organization found within society. When lives are at stake, such behavior will not likely be tolerated.

Social networking tools should be considered in any response effort. The open, widely scoped nature of their use and lack of formal structured communication make them ideal tools to quickly organize in an ad hoc manner.

3.2.5   Global positioning system

GPS provides the ability to determine a near-exact position for any point on the earth. The accuracy of that prediction is based on the ability of the receiver to detect its position using satellite signals (primarily) and other radio signals originating from the surface of the earth. There are two types of GPS; one for civilians and the other for military. The main difference is that a military GPS has two signals instead of one. More signal sources translate to better accuracy. The civilian GPS is built to be accurate to within 8 m (United States Government 2012). In real-world use, it is normally accurate within 3 m. Adding different signals allows GPS devices to enhance that accuracy.

When GPS data are tied to maps, they becomes quite valuable to emergency planners, managers, and responders. Tying location data to maps, and storing them in a computer system for later retrieval, is the vital to all geographic information systems (GIS). GPS data can be sent to GIS databases for later query, visualization, or playback.

The ability to locate positions of people, things, and areas on a map is a critical aspect to emergency management and response. Sharing that map with other people in real time transforms a set of data (knowing where something is) into knowledge (knowing what we can do with the data within the context of a specific emergency).

GIS mapping applications are available online. Applications such as maps.google.com will pinpoint the locations of items. When those maps are combined with GPS, a live picture of where things are begins to emerge. Emergency professionals can use geospatial maps to represent a planning area, an emergency area to be managed, or a tactical as-is representation of the emergency as it unfolds.

Overlaying the location data of responders, vehicles, inventories, and other equipment allows the construction of a real-time tactical picture.

Storing and rebuilding that data allows the construction of animations for after-event investigations. Applications such as Google Earth™ (Google Inc. 2012) are particularly suited for this task.

A primary source of GPS data can be found on modern smart devices. Smart phones, tablets, and other devices come configured to automatically track the location of the device. Emergency professionals can use these to capture data either in a basic file or to share that data within the larger, netcentric enterprise so that combined pictures of operations, readiness, and preparation can be created.

3.2.6   Security

Emergency professionals have valid concerns about the security and integrity of their data. Response plans, especially those that involve responding to criminal or terrorist events, need to be kept confidential and protected from distribution across the world. Communications and data generated during an emergency response may contain sensitive information that would be best kept from public view, at least until the response is finished. Financial records, personnel data, and equipment lists are all examples of data that could be subject to federal law (e.g., the Sarbanes-Oxley Act) or state law mandating their security.

The good news about security is that implementing some form of data encryption is not a difficult thing to do. Free resources abound on the Internet that allow for the encryption and decryption of individual files, entire directories, or entire computer systems. Most of these require very little knowledge of IT to use.

Free resources on the Web include applications such as TrueCrypt™ (TrueCrypt Developers Association, LC) (http://www.truecrypt.org). TrueCrypt is an open-source application that is ideal for encrypting entire directories or even entire hard drives. It has a long history of use and has the advantage of being small and portable. An entire directory can be encrypted with a password and copied on a small device (such as a DVD) along with the TrueCrypt application. Anyone who has possession of the device can read the data without installing additional software, as long as they know the password.

Email is another system where confidentiality is required. Organizations with a larger IT budget may have access to the Norton® PGP® (Symantec Corporation) application. Pretty Good Privacy (PGP) uses a public and private key system to encrypt email messages. If a user has the sender's key, then they can read the messages, if not, then it remains encrypted. PGP has the advantage of being integrated into email clients such as Microsoft Corporation's Outlook®. Many of the most popular email clients (such as Microsoft Outlook) already have similar capabilities in their latest versions, so PGP would not have to be purchased separately.

Information hosted on the Web has additional resources that can be used to encrypt and transmit securely. Almost every hosting company in the United States supports a technology called secure socket layer (SSL). SSL provides a two-part encryption mechanism to ensure messages are encrypted. Normally, one of these two parts (called a "public" key and a "private" key) is exchanged during an Internet request.

In some cases, system administrators can configure websites to force users to already have the public key installed in their browser, thus providing even better security. The effect of this is that anyone wanting to look at the data online must have a file that is not available unless they go through proper channels. The system administrators give those files out to individual people, so it creates an additional level of system control.

In addition to files and Web connections, devices also need security. Bluetooth® (Bluetooth SIG Inc.) is a common method of connecting devices wirelessly. It has the advantage of limited range, so any potential eavesdropper is required to be very close to the device. First-generation Bluetooth devices were open to several types of attacks that could compromise the data signal between them. The latest Bluetooth standard has moved to a public-key, private-key encryption method, so the security of those devices has been much improved.

Radio-frequency identification (RFID) is another common technology. RFID uses inexpensive "tags" placed on items. Those tags can be detected using radio signals. Items such as passports, corporate assets, and military identity cards (also known as common access cards) use this technology. The range of RFID is low enough to make it challenging to read data from the RFID but not impossible. The nature of the data stored in RFID devices makes them important to protect. The bad news is that RFID security has yet to be standardized across all vendors and products. The good news is that it is very easy to take simple precautions to protect information carried in RFID devices. The RFID signal is sufficiently weak so that a very thin material can be used to stop transmission of data. Many identity cards come with envelopes made of such material. Items such as U.S. passports remain vulnerable unless protected by some sort of special envelope or wallet.

RFID technology becomes valuable to emergency responders by providing an easy way to track items. A chaotic response event contains a massive amount of assets and resources that emergency personnel need to track. Connecting RFID tags to items makes it very simple to take inventory and locate items such as food supplies, emergency blankets and, in the event of catastrophic events, even tracking the location of body bags and their contents.

3.2.7   Tablet devices

In the near future, tablet devices such as the iPad® (Apple Inc.), the Kindle Fire® (Amazon Technologies Inc.), or the Samsung Galaxy™ (Samsung Electronics Co. Ltd.) will have the same performance capabilities as modern laptops, plus the ability to stay connected with advanced cellular networks, to use GPS, and to be configured with applications using "stores." Tablet devices will shortly be able to replace most common laptop functions in normal, everyday use (Jobs 2010). Tablet devices may not kill traditional laptop or desktop computers, but those legacy platforms will find themselves being used more and more for only specialty high-performance applications.

As these devices become more popular, the prices will drop even further. The original January 2010 price of the iPad was $499 (Keizer 2010). By 2012 similar devices such as the Kindle Fire were on the market for $199 (PC World Video 2011). Emergency professionals will be able to go to the office with the same portable device that they take to the field (IDG Connect 2012). That mobile device will be connected to the Internet and automatically backed up through the cloud, so when the device breaks or is lost, none of the data is lost.

Today's "out-of-the-box" tablet devices can replace most of what experts call "casual use" (Shanklin 2011). With the exception of writing large documents, developing computer programs, building intense graphics, and playing large gaming applications, a tablet device is more than capable. Hardware is maturing rapidly and limitations in data storage and speed will soon be overcome. Tablets give users all of the power of a lightweight laptop, but with much longer average battery life and improved portability.

3.2.8   Crowdsourcing

Crowdsourcing is more of a technique than a technology. The idea behind crowdsourcing is that information can be gathered and work tasks can be performed across a large, decentralized group of people with little or no command and control. This technique faces challenges in recruiting people, organizing a response, and maintaining some form of collective skill set. It does have a major advantage in that large numbers of people can be focused on tasks or needs without the necessity of large governing organizations or commitments.

Crowdsourcing techniques have long been used in emergency operations such as search and rescue. Emergent volunteers (Seana and Fothergill 2003) are gathered together and various geographic areas are searched with a large number of people.

Crowdsourcing is a very popular technique on the Internet (Doan et al. 2011). Internet projects such as the popular open-source community may be 100% reliant on crowdsourced and volunteer labor to accomplish their goals. The emergency response community can (and often does) engage the general public and volunteer groups using mass appeals for help. This type of approach allows response managers to steer the efforts of disconnected individuals toward some form of useful goal.

3.3 Science fiction in the near term

All of the technologies previously mentioned are in use today. Although they may not be completely embedded in our day-to-day lives, that level of involvement will occur sooner rather than later. Even more disruptive technologies are arriving into the marketplace. These technologies will, yet again, change the way that emergency professionals approach their jobs.

Voice technology is making huge leaps in capabilities. The latest Apple Inc. phone, the iPhone® 4s, has voice technology integrated throughout its system. Other device manufacturers are embedding this as well. Soon, voice recognition technology will be able to remove the need for responders to type, or perhaps even to have any computing device more powerful than a smart phone or tablet. Voice technology can enable encumbered rescue teams to use their systems without regard to their protective clothing or hand gear.

Technology is not just for small devices. New improvements in home technology systems allow the creation of mobile homes that contain the latest computing environments and medical systems. Companies such as N2Care build MEDCottage™ buildings. These portable homes can be used for medical shelters and redeployable housing for special needs populations.

Making all of these advances move forward even faster is multicore computing. Modern computing chips have progressed so more than one processor unit can execute commands within the system. New, advanced chips in commercial-grade computers (and even smart phones) can have four cores that work together in harmony, dividing up the work that the device needs to do so they can perform functions in parallel. Intel recently announced eight-core processors that will double the total chip speed yet again. It is easy to envision computing speeds on the smallest devices that operate far faster than anything we have on our largest workstations today. As these speeds increase, each of the technologies mentioned in this chapter will improve with it.

Robots and drones make up another new technology. These have been used for specialized activities (such as rescue or bomb removal) for some time (Scheiber 2003). The proven usefulness of drones in the Iraq and Afghanistan conflicts has encouraged industry to mature these technologies. Now robots and drones are being built that are less expensive and easier to use. This trend is expected to continue into the future. These autonomous devices will be used not just to safeguard lives but to make the work within the emergency response community faster and more efficient. Concerns about privacy and the role of civil authorities with such capabilities make a lively debate all but inevitable. The general population may worry about predator drones flying over their homes (Edwards 2011) when they associate that technology with warfare, but the implications and utility of using remote drones and robots to search for disaster victims and render assistance cannot be dismissed.

These debates will likely increase as new generations of robots become smaller and more capable (Wood 2008). Small insect-sized robots that can crawl into structures through small cracks to recognize buried victims and alert responders would be very valuable. The same technology that saves lives can be easily used to spy on a population. As more and more small robots are constructed, new "swarm" technologies that use more than one robot at a time will begin to emerge.

The application of these technologies to emergency management and response is still relatively new. Universities like Texas A&M have active programs that tie robot applications to emergency response needs. It is early in the development of this technology. Emergency responders can plan on it becoming very common in the future.

References

Alavi, M. and Leidner, D.E., 2001. Review: Knowledge management and knowledge management systems: Conceptual foundations and research issues. *MIS Quarterly*, 25(1), 107–136.

Badger, L., Grance, T., Patt-Corner, R., and Voas, J., 2011. Draft Cloud Computing Synopsis and Recommendations. Available at: http://www.google.com/url?sa=t&rct=j&q=&esrc=s&source=web&cd=1&ved=0CCMQFjA A&url=http%3A%2F%2Fcsrc.nist.gov%2Fpublications%2Fdrafts%2F800-146%2FDraft-NIST-SP800-146.pdf&ei=S-8qT5SWEIq3twfBh-DlDw&usg= AFQjCNGh4bgZ0LgIiao5uaCRs9I5T_RKDg (accessed February 2, 2012).

Boardman, J. and Sauser, B., 2008. *Systems Thinking: Coping with 21st Century Problems*. Boca Raton, FL: Taylor & Francis.

Bock, P., 1988. A perspective on artificial intelligence: Learning to learn. *Annals of Operations Research*, 16(1), 33–52.

DePompa, B., 2011 Productivity Spurs Smartphone and PDA Use in Federal Agencies: Washington Technology. *Washington Technology*. Available at: http://washingtontechnology.com/microsites/2009/smart-phone-pdas/pda-trends.aspx (accessed February 2, 2012).

Doan, A., Ramakrishnan, R. and Halevy, A.Y., 2011. Crowdsourcing systems on the World-Wide Web. *Communications of the ACM*, 54(4), 86.

Edwards, M., 2011. The Future Expansion of Unmanned Drones over the U.S. Alex Jones' Infowars: There's a War on for Your Mind! *Infowars.com*. Available at: http://www.infowars.com/the-future-expansion-of-unmanned-drones-over-the-u-s/ (accessed February 2, 2012).

Foursquare Labs Inc., n.d. foursquare.com. Available at: https://foursquare.com/ (accessed February 2, 2012).

Google Inc., 2012. Google Earth. *Google Earth.* Available at: http://www.google .com/earth/index.html (accessed February 19, 2012).

Google Inc., n.d. Google Finance: Stock Market Quotes, News, Currency Conversions & More. *Google Finance.* Available at: http://www.google.com/ finance (accessed February 6, 2012).

Huffington Post, The, 2011. Available at: http://www.huffingtonpost.com/2011/ 03/11/twitter-facebook-become-v_n_834767.html (accessed March 28, 2013).

IDG Connect, 2012. iPad for Business Survey 2012. Available at: http://www .idgconnect.com/download/8007/ipad-business-survey-2012?source = connect (accessed April 26, 2012).

Jobs, S., 2010. Apple CEO Steve Jobs at D8: The Full, Uncut Interview—Peter Kafka—D8—AllThingsD. Available at: http://allthingsd.com/20100607/ steve-jobs-at-d8-the-full-uncut-interview/?refcat = d8 (accessed April 26, 2012).

Keizer, G., 2010. Apple's iPad Profit: Breaking it Down. *PC World.* Available at: http://www.pcworld.com/article/188196/apples_ipad_profit_breaking_ it_down.html (accessed April 26, 2012).

Matkar, R. and Parab, A., 2011. Ontology Based Expert Systems—Replication of Human Learning. In S. J. Pise, ed. *Thinkquest~2010.* New Delhi: Springer India, pp. 43–47. Available at: http://www.springerlink.com/index/10.1007/ 978-81-8489-989-4_7 (accessed May 4, 2012).

Neilsonwire, 2011 Generation App: 62% of Mobile Users 25-34 Own Smartphones. *Nielsen Wire.* Available at: http://www.nielsen.com/us/en/newswire/2011/ generation-app-62-of-mobile-users-25-34-own-smartphones.html (accessed February 2, 2012).

PC World Video, 2011. Can Amazon's Kindle Fire Tablet Challenge the iPad? *PC World.* Available at: http://www.pcworld.com/article/240836/can_amazons_ kindle_fire_tablet_challenge_the_ipad.html (accessed April 26, 2012).

Sauser, B. and Boardman, J., 2007. Complementarity: In Search of the Biology of Systems. In *IEEE International Conference on System of Systems Engineering,* April 16–18, 2007, pp. 1–5. Available at: http://ieeexplore.ieee.org/lpdocs/ epic03/wrapper.htm?arnumber = 4304303 (accessed May 7, 2012).

Scheiber, D., 2003. Floridian: Robots to the Rescue. *St. Petersburg Times.* Available at: http://www.sptimes.com/2003/03/02/Floridian/Robots_to_the_rescue .shtml (accessed February 2, 2012).

Seana, L. and Fothergill, A., 2003 A Need to Help: Emergent Volunteer Behavior after September 11th. Available at: http://www.colorado.edu/hazards/ publications/sp/sp39/sept11book_ch11_lowe.pdf (accessed February 19, 2012).

Shanklin, W., 2011. Can an iPad Replace a Laptop? (2011 edition). *Gotta Be Mobile.* Available at: http://www.gottabemobile.com/2011/07/28/can-an-ipad-replace-a-laptop-2011-edition/ (accessed February 2, 2012).

United States Government, 2012. Available at: http://www.gps.gov/systems/ gps/performance/accuracy/ (accessed March 28, 2013).

Wood, R., 2008. Fly, robot, fly. *IEEE Spectrum,* 45(3), 25–29.

chapter four

An emerging framework for unmanned aircraft systems in the national airspace system

Opportunities and challenges for first responders and emergency management*

Robin R. Sobotta

Contents

4.1   Introduction

Recent legislative actions have paved the way for safe unmanned aircraft system (UAS) assimilation into the national airspace system (NAS) by September 30, 2015.[†] This congressionally mandated integration of manned

---

* This chapter is an expanded version of an original article, revised and republished herein with permission from the *Journal of Aviation and Aerospace Perspectives*; Vol. 1, No. 2: pp. 5–18, Fall 2011. The original article is available online at http://www.bga-aeroweb.com/JAAP.html.
† This was mandated via the FAA Modernization and Reform Act of 2012—legislation that is discussed at length later in this chapter.

and unmanned aircraft in the nation's nonrestricted airspace represents a significant shift in federal policy with regard to UAS operation in the NAS, as these systems move from extraordinary to everyday civil use within the United States. Although a number of public and private sector UAS applications are being envisioned, pressing public sector needs, such as first responder UAS applications for emergency response and management, was prioritized over less-emergent civil uses.

This chapter explores specialized needs, challenges, and potential public agency use of UASs for emergency response and management via presentation of the following: historical perspectives and evolution of UASs, an emerging civil UAS regulatory framework, legislative support for use of public UASs, viewing UASs as systems, nonmilitary UAS applications, successful first responder and emergency management UAS applications, potential barriers to public sector UAS use, and concluding guidance to public agencies considering UAS acquisition and use.

4.2   Evolution of UASs

The earliest recorded use of unmanned aviation devices reportedly involved an Australian attack on Venice, Italy, using hot air balloons armed with bombs (Scientific American 1849). Later, in 1915, Nikola Tesla authored a document discussing a fleet of unmanned aerial combat vehicles (Dempsey 2010). However, despite these early references to—and awareness of the value of—the so-called unmanned devices for aerial attacks, UASs did not encounter exponential growth in security and defense use until nearly a century later.

Following the tragic events of September 11, 2001, the users of unmanned aircraft witnessed a critical developmental surge, helping to create a solid foundation for the robust UAS industry that exists today. As is often the case, a sudden and extraordinary external event of some magnitude (i.e., an exogenous shock) can result in dramatic and rapid shifts in federal policy and related initiatives (Gesell and Sobotta 2007). In this case, the policy shift was driven by the U.S. Department of Defense (DOD), as supported by Congress and the Executive Branch, and resulted in rapidly increased development and acquisition of emerging unmanned technologies.

The war effort that followed the 9/11 attacks indeed created a unique window of opportunity for the rapid emergence of unmanned systems. UASs provide unique military solutions for accomplishing necessary intelligence, surveillance, and reconnaissance in hostile war environments, referred to as *theaters*. "UAS use has increased for a number of reasons. Advanced navigation and communications technologies [became available] just a few years ago, and increases in military communications satellite bandwidth have made remote operation of UAS more practical. The nature of the Iraq and Afghanistan wars has also increased the demand for UAS, as identification of and strikes against targets hiding among civilian populations required persistent surveillance and prompt strike capability, to minimize collateral damage. Further, UAS provide an asymmetrical—and comparatively invulnerable—technical advantage in these conflicts." (Gertler 2012).

Because of these advantages, the military rapidly increased its UAS acquisition, with notable support from Congress. "Reflecting a growing awareness and support for UAS, Congress has increased investment in unmanned aerial vehicles annually." "DOD's inventory of unmanned aircraft increased from 167 to nearly 7,500 from 2002 to 2010" with nearly $4 billion requested in 2012 for "procurement and development funding with much more planned for the outyears." (Gertler 2012). Clearly, the DOD has served a vital role in accelerating the development and expanded use of UASs over the past decade.

Despite the lack of earlier federal guidance on this issue, integration of manned and unmanned systems in the NAS is not simply a futuristic concept. In a recent *Airport Magazine* article, Kusy and Sobotta (2012) noted that "there are a number of U.S. locations where mixed operations are currently occurring. Sierra Vista (Arizona) Municipal Airport-Libby Army Airfield (home of the UAS Training Center at Fort Huachuca) boasts a long, safe history of combined (manned and unmanned) operations. In fact, over the past decade, more than 12,000 UAS military operators have been trained in Arizona. Remarkably, more than 33,000 UAS flight hours have been flown in the Grand Canyon State airspace since 2003, in support of military and homeland security needs." Although significant UAS military training has been occurring within the United States' restricted airspace areas since 1990, nonmilitary UAS operations have been strictly limited to a much smaller group of users (U.S. GAO 2008).

Aside from military training within the nation's restricted airspace, military UAS operators (based in the United States) have predominately operated their unmanned aircraft in remote, militarized locations outside the NAS. However, the Federal Aviation Administration (FAA) has permitted limited nonmilitary UAS operations in the NAS for selected domestic users including the following:

1. Those operating under special FAA *certificates of waiver or authorization* (COAs)* in designated areas with certain conditions/limitations.
2. Those operating under special airworthiness certificates in the experimental category (SAC-EC).
3. Those flying for recreational purposes, under FAA Advisory Circular (AC) 91-57, *Model Aircraft Operating Standards.*

* In Subtitle B, Section 331 of the Federal Aviation Administration Modernization and Reform Act of 2012 (H.R. 658.ENG), COAs are referred to as a "certificate of waiver" or a "certificate of authorization" and are described as "a Federal Aviation Administration grant of approval for a specific flight operation."

In a November 2011*, FAA Specialist Randy Willis reported 294 active COAs (already issued by the FAA), with 140 applications pending. By April 2012, in response to a Freedom of Information request, the FAA released a list of more than 60 UAS proponents granted COAs by the agency. Since that time, FAA has added several files with COA sponsor information on its UAS web page (FAA 2013a). Since July 2005, the FAA reports issuance of 94 SAC-EC (experimental certificates). COAs are limited to public-use aircrafts† such as those operated by federal, state, and local agencies and universities.

Written in 1981, FAA AC 91-57 permits recreational operation of model aircraft without FAA certification of the aircraft or the operator. However, these model aircraft must remain below 400 ft. above ground level (AGL) and away from spectators and noise-sensitive areas. Additionally, UAS operators must notify local airport operators, air traffic control officials, or flight service stations if operating within 3 miles of an airport and must yield to manned, full-scale aircraft (FAA 1981).

The AC 91-57 requirement to contact airport or air traffic control was extended to any airport within *five* statue miles of the proposed model aircraft activity, in the FAA Modernization and Reform Act of 2012. Additionally, model aircraft in this category were limited "to not more than 55 lbs unless certified through a design, construction, inspection, flight test, and operational safety program administered by a community-based organization." Since this AC applies to model or remote control aircraft "flown for hobby or recreational use" (U.S. House 2012, p. 67–68), it is important to note that AC 91-57 is *not* the proper guidance for commercial UAS users, first responders, or public agencies interested in UAS use for emergency response or management.

4.3 An emerging civil UAS regulatory framework

Today, UAS operations are emerging from their military and limited civil roots into a dramatic surge of airspace activity, as evidenced by the passage of two significant pieces of legislation that have called for the establishment of national UAS test ranges, the development of a federal regulatory structure for UAS civil operations, and ultimately, the full integration of UAS into the NAS by September 30, 2015.

---

* Presented at the FAA 27th Annual Great Lakes Regional Airports Conference.
† In Subtitle B, Section 331 of the FAA Modernization and Reform Act of 2012 (H.R. 658. ENG), a public UAS is defined as "an unmanned aircraft system that meets the qualifications and conditions required for operation of a public aircraft" (as defined in Section 40102 of title 49, United States Code). The term public aircraft or public-use aircraft is an important distinction from others that seek to operate in the NAS, as discussed later in this chapter and further defined in footnote * on page 85.

The technology does exist to safely integrate UASs into civil airspace, and accordingly, Congress is pressing the FAA to fully integrate unmanned systems into the NAS. The first official act of Congress (and the President) to usher UASs into civil airspace came via the National Defense Authorization Act (NDAA). Signed into law on December 31, 2011, the NDAA instructs the FAA administrator to "establish a program to integrate unmanned aircraft systems into the national airspace system at six test ranges" not later than 180 days after the NDAA enactment.

On February 14, 2012, President Obama also signed into law the FAA Modernization and Reform Act of 2012 to provide further guidance for integrating UASs into the NAS.* The Association for Unmanned Vehicles Systems International (AUVSI 2012) summarized the mandates and deadlines established in the FAA Modernization and Reform Act—guidance that instructs the FAA to swiftly but safely integrate civil UASs into the NAS, including the following:

- Setting a September 30, 2015, deadline for full integration of UASs into the national airspace
- Requiring a comprehensive integration plan be promptly drafted by the FAA's current Aviation Rulemaking Committee (AUVSI 2012)
- Requiring the FAA to create a 5-year UAS roadmap (to be updated annually)
- Requiring small UASs (sUASs; under 55 lbs) to be allowed to fly (in the NAS) within 27 months
- Requiring six UAS test sites within 6 months (however, due to subsequent FAA delays, the test ranges will not be operational until 2014)
- Requiring sUASs (under 55 lbs) be allowed to fly in the U.S. Arctic, 24 hours a day, beyond line-of-sight, at an altitude of at least 2000 ft within 1 year
- Requiring expedited UAS access for public users, such as law enforcement, firefighters, and emergency responders
- Allowing first responders to fly very small UASs within 90 days if they meet certain requirements; *the goal is to get law enforcement's and firefighter's immediate access to start flying small systems to save lives and increase public safety* (AUVSI interpretation in italics)
- Requiring the FAA to study UAS human factors and causes of accidents
- Exempting model aircraft operation in the nation's airspace, so long as the aircraft weighs less than 55 lbs and follows a set of community-based safety standards

---

* Congress also mandated that the test range program shall terminate 5 years after the date of enactment of the FAA Modernization and Reform Act.

In 2013, more than 50 applicants from 37 states had officially expressed an interest in seeking designation for one of the six UAS test ranges (FAA 2013b). States such as Arizona, Ohio, Oklahoma, New Mexico, New York, North Dakota, Texas, and many others had positioned themselves to be designated for UAS testing and research, as well as national UAS test range designation (FAA 2013b).

The economic stakes associated with UAS integration are very high. Initially, public safety and precision agriculture are expected to be the most promising civil markets. In 2012, AUVSI concluded that UAS civil integration into the nation's airspace will result in the following:

- An economic impact exceeding $13.6 billion in the first 3 years, and reaching $82.1 billion for the activity that will occur between 2015 and 2025
- Creation of more than 34,000 manufacturing jobs (40,000/annual wage), many of which will require technical baccalaureate degrees
- Tax revenue to the states exceeding $635 billion between 2015 and 2025

AUVSI (2012) further predicts that failure to integrate UAS into the nation's airspace could be very costly, resulting in a loss of *$10 billion* in annual economic benefit to the nation, or approximately *$27.6 million per day*. Following a call for public comment, the FAA (having consulted with the DOD and the National Aeronautics and Space Administration) announced the final criteria. UAS test range designations, to be announced in late 2013 or early 2014, are expected to extend for a period of 5 years from the time of FAA designation. The economic stakes are high. In 2013, AUVSI estimated $13.6 billion in economic impact in the first 3 years of civil UAS integration into the NAS. In the decade following civil integration, AUVSI projected an $82.1 billion economic impact and the creation of more than 103,000 related jobs (AUVSI, March 2013).

4.4   Legislative support for use of public UASs

The FAA Modernization and Reform Act also addresses the emerging need to facilitate the use of *public UASs* in the NAS (U.S. House 2012, p. 66)

> Not later than 270 days after the Act's enactment, the Secretary of Transportation is instructed to issue guidance regarding the operation of public unmanned aircraft systems to
>
> 1. Expedite the issuance of a certificate of authorization process;
> 2. Provide for a collaborative process with public agencies to allow for an incremental expansion

of access to the national airspace system as technology matures and the necessary safety analysis and data become available, and until standards are completed and technology issues are resolved;

3. Facilitate the capability of public agencies to develop and use test ranges, subject to operating restrictions required by the Federal Aviation Administration, to test and operate unmanned aircraft systems; and

4. Provide guidance on a public entity's responsibility when operating an unmanned aircraft without a civil airworthiness certificate issued by the Administration.

By December 31, 2015, the administrator is also tasked "to develop and implement operational and certification requirements for the operation of public unmanned aircraft systems in the national airspace system" (U.S. House 2012, p. 66).

This instruction was included in response to public agency claims of delays encountered when trying to gain permission to operate in civil airspace. On this point, AUVSI's Ben Gielow (2012a, p. 18) explains, "Anyone who has ever applied for a COA can attest to the difficulty and frustration with the process" (Gielow 2012a, p. 18). In 2012, the FAA responded to Congressional and industry concerns about COAs by establishing metrics to track COAs in the approval process, and the introduction of "an automated, web-based process to streamline steps and ensure a COA application is complete and ready for review." Additionally, the FAA announced expedited procedures to "grant one-time COAs for time-sensitive emergency missions such as disaster relief and humanitarian efforts" as well as doubling the length of a COA authorization from 12 to 24 months (FAA 2012).

Additional clauses of interest to public agencies in the FAA Act included the following:

- Expedited review and decision (60-day maximum response time) and expedited appeal of public sector agency applications for use of UAS in the NAS
- Allowance of a one-time approval of similar operations (missions) carried out during a fixed period of time (without the need to continually obtain additional FAA approvals)
- Permission for a government public safety agency to operate unmanned aircraft weighing up to 4.4 pounds, within operator line-of-site, up to 400 ft AGL, in daylight conditions, within Class G airspace and beyond five statute miles from airports [and other aviation activities] (U.S. House 2012)

The FAA also executed a 2012 agreement with the Department of Justice (DOJ) National Institute of Technology Program for an expedited law enforcement COA process for use of UASs *up to 25 pounds*.* Other items addressed in this progressive interagency agreement included the following (Gielow 2012b):

- Allowing for similar-type operations through an agency's jurisdiction (blanket COA) for operations (not just testing)
- Operating of UAS under 400 ft AGL, line-of-sight, and under visual flight rules or VFR conditions
- Permitting certain operations within class C, D, E, and G airspace
- Developing an online knowledge-based test in lieu of a pilot's license
- Creating a sample safety risk analysis plan (by the FAA) for agencies to use/adopt
- Delivering a DOJ-developed law enforcement conference to help educate agencies on these new rules

One of the most pressing concerns associated with UAS operation in the NAS relates to the operator's ability to yield the right-of-way to other aircraft operators. Under Federal Aviation Regulation Part 91-113, "vigilance shall be maintained by each person operating an aircraft so as to *see and avoid* other aircraft." This is one of the greatest challenges to UAS integration today.

Accordingly, many have suggested that a *sense and avoid* approach should be deemed the UAS equivalent to the Federal Aviation Regulation Part 91-113 *see and avoid* mandate (CFR 2004). Currently, to comply with this mandate, the FAA generally requires UASs operating under a COA to use a "chase" aircraft with visual contact or other constant visual observers. The FAA Modernization and Reform Act defines the term *sense and avoid capability* as meaning "the capability of an unmanned aircraft to remain a safe distance from and to avoid collisions with other airborne aircraft."

To address UAS-related challenges and provide recommendations for NAS integration, the FAA has previously established a UASs Aviation Rulemaking Committee (ARC). At an August 2011 AUVSI Conference in Washington, D.C., FAA officials discussed both their 2009 roadmap for both civil and public UAS NAS access, as well as their continued efforts to publish a notice of proposed rulemaking (NPRM) for sUAS domestic operation, in consultation with industry, government, academia, and federal committees (such as UAS ARC and RTCA). The FAA also created an Unmanned Aircraft Program Office (Aviation Safety) and an UASs Group (Air Traffic Organization) and recently announced its intent to establish UAS Center(s) of Excellence (Prosek and George 2011).†

---

* AUVSI's Government Relations Manager Ben Gielow reported the decision to include all UASs up to 25 pounds was made by the FAA, as at the time (Gielow 2012b), there were 146 UAS platforms weighing less than 25 pounds manufactured by 69 different North American companies.
† Paragraph drawn from Kusy and Sobotta (2012).

Although there has been much discussion about a pending sUAS rulemaking, the FAA has not yet issued the rule as of mid-2013. However, the passage of the many in the industry feel that sUAS rule has been largely drafted, but has been repeatedly delayed from public release due to federal political forces. Gielow (2012a, p. 17) notes, "for the past several years, the FAA has been working on a small UAS notice of proposed rulemaking (NPRM) to allow UAS weighing less than 55 pounds to fly in the (civil) airspace. This NPRM was originally supposed to be released in March 2011 but had been continuously pushed back." Per the FAA Act, the sUAS rulemaking process must be completed by the spring of 2014.

The FAA Modernization and Reform Act of 2012 also provides direction for desired UAS-related future research that will result in the development of the following:

- A better understanding of the relationship between human factors and UAS safety
- Technologies and methods to assess the risk of and prevent defects, failures, and malfunctions of products, parts, and processes for use in all classes of UASs that could result in a catastrophic failure of the unmanned aircraft that would endanger other aircraft in the NAS
- Dynamic simulation models for integrating all classes of UASs into the NAS without any degradation of existing levels of safety for all NAS users

4.5   Viewing UASs as systems

Despite the somewhat misleading name, *UASs are not without significant human control.* There have been numerous definitions offered for unmanned (or remotely piloted) aircraft over the past 20 to 30 years (U.S. DOD 2010 and many others). Most recently, the FAA Modernization and Reform Act provided a definition of *unmanned aircraft* as "an aircraft that is operated without the possibility of direct human intervention from within or on the aircraft." Further, the Act defines a UAS as an "unmanned aircraft and associated elements (including communication links and the components that control the unmanned aircraft) that are required for the pilot in command to operate safely and efficiently in the national airspace system." Generally there are two recognized types of UASs: remotely piloted vehicles and drones, as discussed later.

Over the years, UASs have lacked consistency in identification (naming) in the literature, reflecting different eras and user perceptions on the role of the operator/pilot, as follows (Haddal and Gertler 2010). A few descriptors are presented below:

- *Unmanned aerial vehicle (UAV)*: An older term, still widely used as synonymous with UAS. However, the term tends to focus on the platform (aircraft) as opposed to the entire system required to operate it.

- *Remotely piloted aircraft/vehicles (RPA/RPVs)*: UAS that are "actively flown—remotely—by a ground control operator" (Haddal and Gertler 2010). This descriptor is used widely by the U.S. military and also by organizations outside the United States.*
- *Remotely operated aircraft:* This term suggests an operator may not have (or need) traditional pilot credentials.
- *Drones:* Members of the media and others often use this term as synonymous with UAS. However, it is not an equivalent term. Haddal and Gertler (2010) define drones as UAS that "are programmed for autonomous flight." However, to those in the industry with great familiarity with UAS, drones are actually best characterized as *automated* air vehicles deployed on a preprogrammed mission with little or no capability for human intervention during the course of the flight.

The current favored terminology is UASs, with Congress having adopted this term throughout the final version of the FAA Modernization and Reform Act (after abandoning the use of unmanned *aerial* systems that appeared in earlier versions of the Act in 2011).

As noted earlier, UASs are often commonly thought of as just the aircraft or platform that accomplishes flight. However, UASs should be more accurately viewed as an entire system that includes several components: a platform or aircraft, the payload being carried, communications and data links, a ground control station, and the operator—all in the context of the national airspace and Next Generation Air Transportation System (NextGen) technology. One should not overlook the importance of the payload in determining the value of UAS use by first responders (Sobotta and Rayleigh 2011).

In selecting a system, first responders and other public sector users should consider a number of system elements and match the UAS to anticipated mission demands, including the following (Levy 2011; Miller and Bateson 2011; Olesen 2011):

- *Aircraft/platform*: Range, performance in a variety of environmental conditions, launch/recovery demands, acceptable payload size, aircraft noise, and operating and maintenance requirements

---

* At the Remotely Piloted Aircraft Systems (RPAS) Conference 2012, there was discussion of harmonizing FAA and ICAO unmanned system terminologies globally. The preferred term by regulatory representatives of multiple countries was actually agreed to be RPAS; however, changing the UAS terminology (currently in use in the United States) was made more challenging by the terminology definitions contained within the FAA Modernization and Reform Act of 2012. The RPAS Conference, held in Paris in 2012, is delivered annually by Unmanned Vehicle Systems International (UVI), which serves as an advocate for the global remotely piloted aircraft systems community (UVI International 2012).

- *Communications and data link*: Range, frequency use and congestion, sense and avoid capabilities, radio repeating (transmitting between vehicles)
- *Payload*: Cameras (electro-optical and infrared), sensing (chemical/biological)
- *Environment*: Meteorological, geographic, indoor/outdoor applications
- *Operator training*: Written and oral (pilot) testing, medical requirements, skill training and evaluation, partnerships with training organizations or academia

4.6   Potential nonmilitary UAS applications

UASs have long been successfully associated with provision of dull, dangerous, and dirty* missions and have been successfully deployed by the military for years. "UAS have become a transformational force multiplier for the DOD. When UAS's were introduced into the front-line DOD aircraft force structure over a decade ago, small numbers of aircraft were fulfilling niche capabilities. This is no longer the case. The numbers and roles of UAS have expanded dramatically to meet demand overseas, and in some categories, more unmanned aircraft are budgeted than manned" (Brooks 2011).

In much the same way that the Chair of the UAS Airspace Integration Product Team, Dallas Brooks (2011), referenced the use of UASs as a transformational force multiplier for the military, civil proponents of these systems are expecting the rapid adoption of UASs for civil use. A number of nonmilitary applications have already been envisioned for UASs including power line surveillance, agricultural applications, communications and broadcasting applications, real estate mapping, mining, freight transport, aerial photographing, weather research, movie production, and sporting event coverage (AUVSI 2013; Murphy 2011).

First responders such as law enforcement, firefighters, and emergency management representatives see UASs as potentially beneficial for two broad mission categories: *command and control* and *enhanced situation awareness*. Command and control applications may use UAS for enhanced management of search and rescue efforts, natural disasters (wildfires, floods, storms, tornados, hurricanes), crimes in progress, high-profile community events, HAZMAT incidents, and traffic accidents. Enhanced situational awareness or surveillance may be valuable in monitoring suspected criminal activity; community impacts from floods, storms, or wind; or in tracking the safety or security needs or concerns of isolated communities (AUVSI 2011; Murphy 2011).

---

* UASs are generally considered to be best suited for use in the support of dull, dangerous, or dirty missions; these missions are considered unappealing or unsafe for manned aircraft use. Sometimes a fourth "d" is added to this list: demanding (see Honeycutt 2008).

4.7   Successful first responder and emergency management applications

According to the AUVSI (2011), a number of first responders have successfully used UASs to support their following agency missions:

- *Border surveillance*: U.S. Customs and Border Patrol currently patrol portions of the U.S. border with General Atomics Predators and other unmanned assets.
- *Fire response*: The University of Alaska operated a ScanEagle UAS to map wildfire progress.
- *Flood plain surveillance*: A ScanEagle UAS was used by the University of North Dakota to monitor the Red River for flooding following heavy rains.
- *Search and rescue*: A team including the University of Colorado tested a Draganflyer UAS for this mission. Robyn Murphy (2011) also documented a number of other uses for "robots," such as during the aftermath of the Haitian earthquake.
- *Vehicle accident investigation*: The Utah Highway Patrol uses photos taken from a Leptron UAS to quickly recreate car accidents, aiding in the accident investigation process (AUVSI 2011).

In UAS use by the Miami-Dade Police Department (Toscano 2011), two well-publicized UAS law enforcement uses include the Mesa County (Colorado) Sheriff's Office (MCSO) and the Arlington (Texas) Police Department. The MCSO began its use of a UAS in an "18-month research project where extensive thought and comparison of UAVs, cost, application process with the FAA, and more were considered." Having recently received a "jurisdictional certificate of authorization" for a daily UAS operational program, the MCSO uses a Draganflyer (weighing just over 3 pounds, including a payload of infrared and digital cameras) for a variety of applications including hazardous material spills, crime scene photos, and other tactical uses where personnel could be at risk because of climate, terrain, health, and so on (Miller and Bateson 2011).

The MCSO used a deliberate, measured approach for acquiring, testing, training, and deploying its UAS asset in actual law enforcement applications. They considered both benefits and costs/risks associated with UAS use and tracked their UAS use and performance over several months. Today, they remain certain of the value of UAS use for law enforcement and first response applications. Sheriff's Office UAS Operator Ben Miller refers to UASs as "an applicable technology to law enforcement that I am convinced will save lives" (Benjamin 2009; Miller and Bateson 2011).

Another first responder agency using UASs in law enforcement operations is the Arlington Police Department. Originally obtained for

Super Bowl XLV security purposes, Arlington Police Deputy Chief Lauretta Hill says their Leptron Avenger UAS helicopter is now used for other purposes, including taking vehicle accident photos and measurements, which are then integrated in a "direct link download into a computer drawing software, to be able to clear accident scenes at a much quicker rate" (Levy 2011, p. 21).

In 2012, the U.S. Department of Homeland Security (DHS) Science and Technology (S&T) Directorate released details on an agency's initiative to support UAS testing and evaluation for first responders. In his presentation at *AUVSI's Unmanned Systems Review Program 2012*, John Appleby, PhD, introduced a DHS S&T concept for a *sUAS Center of Technology Transition for First Responders* with the following objectives:

- Develop sUAS sensor requirements, standards, [and] conops.
- Leverage DOD investments.
- Evaluate commercial off-the-shelf platforms in first responder and border security scenarios.
- Create a knowledge resource database consisting of test and evaluation reports, user testimonials, and guidelines for use on FirstResponder.gov.
- Guide future platform and sensor development to meet DHS and first responder operational requirements (Appleby 2012, p. 12–13).

DHS has since evaluated several small UAS for possible public safety use, following implementation of its Robotic Aircraft for Public Safety (RAPS) program in 2012 at Oklahoma State University's test site (within Fort Sill's restricted airspace). "Under the RAPS program, the DHS is conducting operational tests and generating reports on the performance of small UAS for law enforcement, disaster response and firefighting applications. The reports will be made available to public safety agencies that are considering using unmanned aircraft in their operations" (Carey 2013). This proactive support by DHS provides much needed guidance as increasing waves of first responders seek to acquire and safely (and efficiently) utilize UASs in a variety of future missions in the NAS.

4.8   Potential barriers to public sector UAS use

While many practical reasons exist for first responders to utilize UASs, serious barriers remain. Dallas Brooks (2011) identified three requirements for integrating UASs into the NAS, including the following: (1) aircraft must be certified as airworthy; (2) pilots or operators must be qualified to operate the aircraft in the appropriate classes of airspace; and (3) flight operations must be in compliance with applicable regulatory guidance.

As mentioned earlier in this chapter, public UAS users have historically expressed serious concerns about the slow, somewhat complicated COA approval process, which the FAA sought to improve in 2012. Lingon and Adelman (2010) also cite the lack of federal guidance relating to UAS for use as public aircraft as challenging, stating, "although not all FAA regulations apply to public aircraft,* some still do. If you cannot conduct a flight in conformance with all applicable regulations, you cannot legally conduct the flight or you must apply for a waiver from the FAA. ..."

Lingon and Adelman (2010) differentiate public aircraft (operated by or on behalf of a government) from being categorized as civil aircraft. "Public aircraft are not required to adhere to FAA's requirement for an aircraft airworthiness certificate, and the operator is ... not required to obtain an airman's certificate." On June 2, 2004, FAA Associate Administrator Nick Sabatini further testified to Congress that "with a public aircraft operation, ensuring that the government is meeting the safety standards falls to the government agency on whose behalf the operation is being conducted and not on the FAA." Tovar (2011) equally stresses the importance of knowing and adhering to the evolving FAA UAS requirements for civil airspace operation.

Although regulatory impediments and airspace approval delays (e.g., FAA COA process for UAS testing, training, and operations) are of great concern, other potential barriers to first responder use of UASs in the NAS must also be considered (U.S. GAO 2008). These include funding challenges, public acceptance and privacy concerns, lack of available testing or training ranges, and the operational risks associated with UAS use.

---

* It is the distinction of being designated as an agency operating "public aircraft" that enables first responders to seek permission to access the NAS (via the COA process mentioned earlier in this chapter) well in advance of most other civil UAS users. As defined in Section 40102 of title 49, United States Code, the term "public aircraft" includes any of following: (A) Except with respect to an aircraft described in subparagraph (E), an aircraft used only for the United States Government, except as provided in Section 40125(b). (B) An aircraft owned by the Government and operated by any person for purposes related to crew training, equipment development, or demonstration, except as provided in Section 40125(b). (C) An aircraft owned and operated by the government of a State, the District of Columbia, or a territory or possession of the United States or a political subdivision of one of these governments, except as provided in Section 40125(b). (D) An aircraft exclusively leased for at least 90 continuous days by the government of a State, the District of Columbia, or a territory or possession of the United States or a political subdivision of one of these governments, except as provided in Section 40125(b). (E) An aircraft owned or operated by the armed forces or chartered to provide transportation or other commercial air service to the armed forces under the conditions specified by Section 40125(c). In the preceding sentence, the term "other commercial air service" means an aircraft operation that (1) is within the United States territorial airspace; (2) the Administrator of the FAA determines is available for compensation or hire to the public, and (3) must comply with all applicable civil aircraft rules under title 14, Code of Federal Regulations.

During economically challenging times, law enforcement agencies are particularly concerned about the costs of acquiring, operating, and maintaining new technologies. UAS prices can vary widely, with small UASs ranging from tens to hundreds of thousands of dollars per system. Notably, larger platforms may not provide significantly more value than smaller systems for first responders.

Because of regulatory uncertainty, MCSO officials initially leased their very small Draganflyer UAS system, which reportedly had a purchase price of more than $25,000. MCSO greatly values their UAS acquisition, having conducted more than 72 UAS-facilitated search and rescue missions in the span of 10 months (Benjamin 2009). Interestingly, one law enforcement official reported that a UAS acquisition is easier to justify if the purchase price does not exceed that of a replacement patrol vehicle.

It is important to assure public safety in the operation of UASs, as well as respect citizens' rights to privacy in the operation of public UASs over communities. Safety concerns are generally a familiar item in the aviation community, and as such, public administrators are typically prepared for such inquires. However, it is the less familiar privacy issues that will be a challenge for UAS operators and public officials, as civil UAS use increases in the nation's airspace.

Kusy and Sobotta (2012) have examined some of the UAS-related issues of potential community concern, including privacy and safety. In response, they suggest, "communities and aviation planning agencies should share a commitment to assuring public privacy and safety as UAS are integrated into civil airspace." The Fourth Amendment and a few key court cases have previously established search and seizure parameters for privacy. Retired Port St. Lucie (Florida) Chief of Police Donald L. Shinnamon, Sr., states that items in a 'plain view observation' such as that seen from 'navigable airspace' (court-defined as 1,000' AGL for fixed-wing and as 400' AGL for helicopters) are not considered illegal searches (Kusy and Sobotta 2012, p. 32).

"Whether the courts will accept plain-view observations made below 400 feet [from UAS platforms] will be determined in the future," said Shinnamon in an August 2011 *Unmanned Systems* article. "However, it is clear that the warrantless use of technology to intrude into constitutionally-protected areas is unlawful." Shinnamon (2011) also notes, "The Fourth Amendment to the U.S. Constitution prohibits unreasonable searches and seizures, and requires warrants to be based on probable cause." Aircraft overflight-related search and seizure limits have been established by numerous court decisions, including those discussed later, which will also extend to UAS operations.

In *Katz v. United States*, 389 U.S. 347 (1967), the Supreme Court established a two-part test to determine when a search warrant is required. According to the court, a warrant is required when a person expects privacy in the property searched and when society believes that expectation is reasonable. Subsequent to Katz, the courts also distinguished between three types of public areas: businesses, open fields, and homes, with homes requiring a greater expectation of privacy over the first two areas (Shinnamon 2011). Later, in *Kyllo v. United States*, 533 U.S. 27 (2001), the court determined that police use of thermal imaging constitutes an unlawful search if done without a warrant.*

While there is no record of UAS-specific court decisions, appropriate precautions should be taken to anticipate and mitigate potential concerns associated with UAS operations. Shinnamon (2011) suggests that public agencies engage the community (including civil liberties groups) and the media by

- Including them in discussions about UAS acquisition and use.
- Emphasizing that "the objective of air support, whether manned or unmanned, is to provide incident commanders on the ground with information they need to bring an incident to a successful conclusion as quickly and efficiently as possible … Avoid terms such as gathering urban intelligence, surveillance, etc."
- Establishing a UAS policy/procedure manual including "the legal aspects of airport operations and state, in no uncertain terms, that the aircraft will be operated in accordance with the requirements established by the courts at all times."
- Holding agency personnel accountable for compliance with the rules.

Shinnamon (2011) also notes that, "as surveillance technology has made its way into the hands of the general public, many states have found it necessary to enact laws addressing video surveillance and voyeurism." The potential operation of UAS in civilian airspace will likely raise privacy concerns by citizenry. However, additional laws and court rulings may not be necessary if potential concerns are considered in advance and mitigated. Including citizens early in the UAS acquisition process—and conducting respectful UAS operations within communities—will hopefully avoid unnecessary community resentment and the potential legal pitfalls associated with improper UAS use.

If the privacy and safety issues are not adequately addressed in communities, it can be a "show stopper" for first responders or other public agencies attempting to acquire and use UASs. The FAA recognized the need to address this rapidly emerging issue by holding a special online session to invite public comment on a UAS test site privacy policy (FAA 2013c). Another approach to aiding community members to

---

* Further, see *California v. Ciraolo* 476 US 207 (1986) and *Florida v. Riley* 488 US 445 (1989) for additional overflight-related privacy rulings.

welcome local UAS operations is by establishing a proactive educational program. Informative websites can be established or community forums can be held to clarify UAS training/operating areas, discuss UAS operator training and procedures, and introduce agency measures to assure public privacy and safety concerns (such as those relating to "lost link" or lost communications with the platform).

Hosting a public education event in conjunction with industry, academia, or government officials (such as local airport operators)—like the *UAS Community Forum* recently held by Embry-Riddle Aeronautical University in Prescott, AZ—is one element of a proactive outreach campaign that can be used to gain increased community support for UAS operation near communities (Embry-Riddle 2012).

4.9 Conclusions

Recent congressional actions require full integration of UASs in the NAS by September 30, 2015. With this change in federal policy will come both opportunities and challenges for current and future national airspace users. In particular, first responders will find themselves in a unique role of being at the leading edge of the nation's introduction of UASs into civil airspace.

First responders are encouraged to work closely with partners in the aerospace industry, government, academia, industry associations, military, and public to safely integrate UASs into the national airspace and gain needed public acceptance. It is critical that public sector officials embrace this opportunity in a manner that balances the tremendous benefits associated with this emerging technology with the responsibilities and risks associated with its use. Below are several suggestions for first responders:

- Realize that bigger is not always better. Larger platforms may not provide significantly more value than smaller systems for first responders. Also consider leasing the UAS if the manufacturer provides this option.
- Establish reasonable standards and training to assure that operators understand UAS capabilities and limits and an approach to determine if UAS use is safe and appropriate for the selected mission.
- Select a technology integration pace that will assure proper training and testing, before wider use/deployment of a UAS.
- Consider and take active steps to anticipate and mitigate risks, costs, and potential privacy abuses that could be associated with improper UAS operation.
- Communicate agency efforts to the public in an active outreach campaign including Web-based education or community forums.

- Carefully track the emerging federal flight and operator requirements and remain in compliance at all times.
- Collaborate (or explore pilot projects) with internal stakeholders (existing aviation and SWAT units) as well as external stakeholders (government agencies, industry, UAS associations, academia, and the military or national guard, as appropriate) to train, test, and determine optimal missions for deploying a UAS.
- Work with regional and state officials to identify and establish safe, approved training and testing (range) facilities.
- Cooperate in establishment of UAS centers of excellence to promote research, testing, training, and sharing of best practices and lessons learned with other agencies.

References

Appleby, J. (2012). Unmanned aerial systems and airborne technology programs in DHS S & T. Presented at the *AUVSI's Unmanned Systems Program Review 2012*, February 8, 2012. Washington, DC: Omni Shoreham.

Association for Unmanned Vehicle Systems International (AUVSI). (2011). Unmanned aircraft system integration into the United States national airspace system: An assessment of the impact on job creation in the U.S. aerospace industry. Arlington, VA: AUVSI.

AUVSI. (March 2013). The economic impact of unmanned aircraft systems integration in the United States. Arlington, VA: AUVSI.

AUVSI (2012). Current legislative initiatives in the United States. Retrieved May 13, 2013 from http://www.auvsi.org/AUVSI/Advocacy/CurrentLegislative Initiatives.

Benjamin, H. (August 31, 2009). FAA gives sheriff green light to fly the Draganflyer X6 UAV RC helicopter. Mesa County Sheriff's Office press release. Retrieved May 26, 2012 from http://www.draganfly.com/news/2010/02/10/faa-gives-sheriff-green-light-to-fly-the-draganflyer-x6-uav-rc-helicopter/.

Brooks, D. (2011). DOD UAS plan for airspace integration. Available from AUVSI. Retrieved January 15, 2012 from http://www.space.com/missionlaunches/060719_falcon1_update.html.

Carey, B. (March 29, 2013). Companies demonstrate Small UAS for DHS Program. AIN Online. Retrieved April 7, 2013 from http://www.ainonline.com/aviation-news/2013-03-29/companies-demonstrate-small-uas-dhs-program.

Code of Federal Regulations (CFR) (2004). General Operating and Flight Rules, 14 CFR Federal Aviation Regulation (FAR) Part 91.113. Retrieved April 7, 2013 from http://rgl.faa.gov/Regulatory_and_Guidance_Library/rgFAR.nsf/0/934f0a02e17e7de086256eeb005192fc!OpenDocument.

Dempsey, M. (2010). Eyes of the Army: U.S. Army roadmap for unmanned aircraft systems 2010–2035. Retrieved January 2, 2012 from http://www.fas.org/irp/program/collect/uas-army.pdf.

Embry-Riddle Aeronautical University (March 2012). ERAU to host community forum on unmanned aircraft systems. Press release. Retrieved April 7, 2013 from http://prescott.erau.edu/news/community-forum-on-unmanned-aircraft.html.

FAA (April 2013a). Unmanned aircraft systems (UAS). Retrieved April 7, 2013 from http://www.faa.gov/about/initiatives/uas/.

FAA (March 2013b). 50 applicants from 37 states (Test Site Map). Retrieved April 7, 2013 from http://www.faa.gov/about/initiatives/uas/media/UAS_testsite_map.pdf.

FAA (March 27, 2013c). FAA to hold online session on UAS test site privacy policy. Retrieved April 7, 2013 from http://www.faa.gov/about/initiatives/uas/.

FAA (2012). FAA makes progress with UAS integration. Retrieved April 7, 2013 from http://www.faa.gov/news/updates/?newsId = 68004.

Federal Aviation Administration (FAA) (1981). Model aircraft operating standards. Federal Aviation Administration Advisory Circular, 91–57. Retrieved May 27, 2013 from http://www.faa.gov/documentLibrary/media/Advisory_Circular/91–57.pdf.

Gesell, L. and Sobotta, R. R. (2007). *The administration of public airports* (5th ed.). Chandler, AZ: Coast Aire Publications.

Gertler, J. (January 3, 2012). US Unmanned Aerial Systems. Congressional Research Service. pp. 1–2. Retrieved April 2, 2013 from http://www.hsdl.org/?view&did = 697556.

Gielow, B. (March 2012a). Into the wild blue yonder: U.S. Congress sets 2015 deadline for UAS to fly in the national airspace. *Unmanned Systems*; 30(3): 17–19.

Gielow, B. (May 2012b). Federal legislative and regulatory update. Association for Unmanned Vehicle Systems International presentation in May 2012. Retrieved August 12, 2012 from www.auvsi.org/.

Haddal, C. and Gertler, J. (July 8, 2010) Homeland Security: Unmanned Aerial Vehicles and Border Surveillance. Retrieved April 7, 2013 from http://www.fas.org/sgp/crs/homesec/RS21698.pdf.

Honeycutt, G. (2008). Unmanned systems down on the farm: Dull, dirty, dangerous, demanding. *Unmanned Systems*. Retrieved April 26, 2012 from http://www.m2mcomm.com/downloads/06.06.01-AUVSI-Unmanned-Systems-Down-on-The-Farm.pdf.

Kusy, L. and Sobotta, R. (2012). Unmanned aircraft systems: Coming soon to an airport near you. *Airport Magazine*; 24(1): 28–30, 32.

Levy, S. (2011). Arlington, Texas police ramp up UAS program. *Unmanned Systems*; 29(8): 21.

Lingon, L. and Adelman, T. (2010). What's stopping public safety agencies from operating UAS? *Air Beat*; September/October 2010: 42–44.

Miller, B. and Bateson, M. (2011). Mesa County: Onward and upward, now with fixed wings. *Unmanned Systems*; 29(8): 17–18.

Murphy, R. (2011). Use of unmanned aircraft systems for search and rescue in 2010. Proceedings from *AUVSI Unmanned Systems North America 2011 Conference*. San Diego, CA: Association Archives.

Olesen, A. (2011). How to start your own police UAS program in 176 simple steps. Proceedings from *AUVSI Unmanned Systems North America 2011 Conference*. San Diego, CA: Association Archives.

Prosek, R. and George, S. (2011). FAA UAS integration part 1. Proceedings from *AUVSI Unmanned Systems North America 2011 Conference*. San Diego, CA: Association Archives.

Scientific American (1849) More about balloons. *Remotely piloted aerial vehicles: An anthology*. Ed. Russell Naughton. Retrieved April 26, 2012 from http://www.ctie.monash.edu/hargrave/rpav_home.html#Beginnings.

Shinnamon, Sr., D. (2011). Law enforcement battles public acceptance with SUAS (Small Unmanned Aircraft Systems). *Unmanned Systems*; 29(8): 19–20.

Sobotta, R. R. and Rayleigh, S. (2011). Emerging civil applications for unmanned aircraft systems: Focus on first responders. Presented at the National Security and Intelligence Symposium. Prescott, AZ: Embry-Riddle Aeronautical University.

Toscano, M. (2011). Unmanned systems community leaping forward into the new year. *Unmanned Systems*; 29(12).

Tovar, E. (2011). Barriers for first responders. Proceedings from *AUVSI Unmanned Systems North America 2011 Conference*. San Diego, CA: Association Archives.

U.S. Department of Defense (U.S. DOD) (November 8, 2010, as amended through October 15, 2011). DOD dictionary of military terms. Joint publication I-02. Definitions retrieved April 26, 2012 from http://www.dtic.mil/doctrine/dod_dictionary/data/u/18955.html and http://www.dtic.mil/doctrine/dod_dictionary/data/u/18956.html.

U.S. General Accountability Office (U.S. GAO) (2008). Unmanned aircraft systems: Federal actions needed to ensure safety and expand their potential uses within the National Airspace System. FAO-08-511. Washington, DC: U.S. Government Printing Office. Retrieved from April 7, 2013 from http://www.gao.gov/assets/280/275328.pdf.

U.S. House of Representatives (U.S. House) (2012). Federal Aviation Administration (FAA) Modernization and Reform Act of 2012. (H.R. 658. ENR). 112th Congress. Retrieved April 7, 2013 from http://thomas.loc.gov/cgi-bin/bdquery/z?d112:H.R.658:.

UVI International (June 2012). Remotely Piloted Aircraft Systems (RPAS) Conference. Conference presentation slides, provided to attendees. Paris, France.

chapter five

Optimization in evacuation route planning

Gino J. Lim and Mukesh Rungta

Contents

5.1 Introduction

"Destruction, hence, like creation, is one of Nature's mandates." This famous quote by the Marquis de Sade signifies the inevitability of destruction as one of the laws of nature. News articles keep rolling in highlighting major and minor natural incidents that have resulted in casualties or are potential life threats. Besides nature's blow, there are numerous other major and minor incidents resulting from human deeds having short-term or long-term implications on the health and lives of people. The earthquake and tsunami in Japan and consequent nuclear accidents in Fukushima, wildfires in Nevada and Texas, and yearly hurricanes along the east coast of the United States are examples of the power of such disasters. Survival in these natural and man-made disasters depends on either how proactive or how reactive the response is in terms of scalability and effectiveness.

Urbanization has resulted in huge population concentrations in megacities of the world. Such cities as New York, Tokyo, London, Mumbai, and Shanghai all boast of populations in the tens of millions. All these cities have an excellent infrastructure to support their thriving population, but do these metropolitans have an efficient plan for evacuating their residing populations in case of an impending danger? The answer would probably be "no." This is because the dynamics of evacuation are highly complex resulting from behavior of people and unforeseen circumstances of the event. Attempts have been made to address this problem by applying the concepts of operations research (OR). Mathematical models have been designed that attempt to mimic the real evacuation scenario. These models are then solved for optimizing the evacuation process. This chapter focuses on applications of OR models in evacuation planning.

Evacuation involves the mass movement of a population in the wake of an impending danger from an impacted geographical region toward safer destinations. The U.S. federal government, through FEMA, requires all states to have a comprehensive emergency operations plan. These plans guide emergency operations for all types of hazards, from natural to man-made and technological. While the general evacuation issues faced by coastal states are similar, different strategies and plans have been developed to deal with variations in population, geography, and transportation system characteristics. States also differ in the way that they delegate authority, allocate people and resources, and enforce evacuations. They seek to maximize the efficiency of their emergency operation plans within these many constraints. Evacuation orders are issued by the local authorities after analyzing the severity and possible consequences of the disruptive event. The evacuation procedure is then carried out according to the devised plan. From a logistics viewpoint, the evacuation plan model answers the following basic questions:

1. How much time would be required for evacuation?
2. What are the ideal routes that should be used for evacuation?
3. How should the traffic flow be managed within limited infrastructure?

A critical issue in evacuations, particularly during hurricanes, is timing. The earlier the evacuation order is issued, the more time residents and tourists will have to evacuate. Unfortunately, the earlier it is issued, the greater the possibility the hurricane could change course before landfall, rendering the evacuation unnecessary. Emergency management centers would not want to "cry wolf" and issue an evacuation order in situations of false alarm. Therefore, they want to wait until the last minute before making such an important decision. The time required to evacuate is estimated from a combination of clearance times and the prelandfall hazard time (Florida Division of Emergency Management 2000) as shown in Figure 5.1. Clearance time is the time required to configure all traffic control elements on the evacuation routes, initiate the evacuation, and clear the routes of vehicles once deteriorating conditions warrant its end.

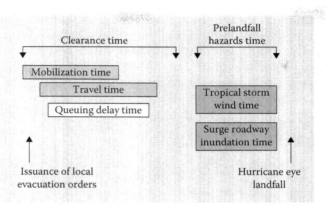

Figure 5.1 Components of evacuation time. (From Florida Division of Emergency Management "Evacuation and clearance times," 2000. Last accessed online June 4, 2012.)

Prelandfall hazards time is the time during which hazardous conditions exist before actual hurricane landfall. Hence, a nearly accurate estimate of the clearance time would arm the evacuation managers with a plan for the evacuation according to the horizon time before the danger hits the shores.

Clearance times are estimated using evacuation traffic models, which are dependent on data such as the population anticipated to evacuate, the number of lanes available for evacuation, and impacts from other areas that will affect the evacuation such as flooding and road closures. In an ideal scenario, the objective of a good evacuation plan is to minimize the evacuation time such that a maximum number of evacuees can be moved away from danger. In order not to overcharge the transportation infrastructure, it is essential to find an efficient set of routes to load the evacuating traffic and allocate a starting schedule and flow on those paths as per the priority. Due to the large number of people evacuating from a large geographical region, the problem of evacuation is very complex to solve. The amount of time required for clearance can be significantly lengthened by en route congestion and the setup time required for complex control features (such as those required for contraflow). Managing this enormous task effectively requires a similar scale of resources, thereby making the problem even more complex.

This chapter presents optimization planning problems for evacuation, and mathematical and computational optimization models that have been developed to solve them. The rest of this chapter is organized as follows. In Section 5.2, we introduce the evolution of optimization models in evacuation planning literature where we classify various problems and research areas in this field. We also review the routing and scheduling optimization models and the solution approaches that are developed specifically for evacuation planning in this chapter. There is a brief discussion on the software used for evacuation planning in Section 5.3, which can be a useful guide for emergency planners. In Section 5.4, we summarize the chapter and suggest directions for future research.

5.2   Bird's-eye view of evacuation route planning research

Evacuation planning has been a major research topic in the OR community. The research on this problem has evolved over the years to encapsulate various aspects of a real evacuation scenario and come up with a realistic evacuation plan. Mathematical models for evacuation planning are handled using network models. Initial research on this problem addressed building evacuations during emergency situations (Chamlet et al. 1982). Numerous other models were developed that handled building evacuations, and an exhaustive survey on these models is presented in the works by Aronson (1989) and Hamacher and Tjandra (2002). Many of the models designed for building evacuations are also applicable to regional evacuation. This section discusses the models that are applicable for both types of evacuation problems.

Designing an evacuation plan often involves a series of optimization problems with various objectives and constraints. Given the large number of research papers in this area, we classify the problem into broad categories to provide some structure for the rest of the chapter. Different aspects of the problem in evacuation planning can be classified as follows:

1. Deciding routes, assigning traffic, and finding clearance time
2. Efficient use of limited infrastructure
3. Reliable planning in stochastic conditions

Although we have offered this simple classification, it is found that many papers deal with problems that intersect two or more of the preceding categories. The models that we discuss in this chapter are classified under a common umbrella of macroscopic modeling. Macroscopic models do not consider any individual's behavior during the emergency situation but are useful to provide good lower bounds for evacuation time. Individual behaviors can be modeled using microscopic models, and simulations are used for their analysis.

Since time is a decisive parameter during the evacuation, an estimate of evacuation time or clearance time is the primary information required by the evacuation planners. Most of the literature in evacuation planning is therefore centered around minimization of the clearance time. As per the FEMA report (Dept. of Transportation and Dept. of Homeland Security

2006), one of the criteria for the evaluation of an evacuation plan is based on the effective implementation and ease of managing the routes loaded with the evacuating traffic. Therefore, deciding the best routes and assigning the traffic on the selected routes based on the evacuation schedule that leaves behind minimum evacuees is an important aspect of evacuation planning.

Highway network clearance times are greatly influenced by other factors such as location of shelters, number of intersections on the selected routes, and decision on the timing for start of contraflow on certain highway stretches. Overlooking these factors may result in a buildup of traffic on certain road sections. Efficient use of the limited transportation infrastructure is therefore required. Evacuation is a rare scenario and many of the parameters that are used to come up with a prescriptive evacuation plan are random. Parameters such as number of evacuees, travel time, and link capacity cannot be considered as deterministic. Therefore, stochastic models are essential to embed the uncertainty associated with the problem and come up with a reliable evacuation plan. A detailed discussion of the optimization models and solution techniques for each of the preceding classifications is provided in the following subsections.

5.2.1   Routing, traffic assignment, and clearance time

Routing evacuating traffic involves selecting a set of paths among the alternative paths between origin and destination pairs. Typically, the number of evacuees for a regional evacuation is huge and the limited number of paths cannot handle all the vehicles simultaneously. Therefore, evacuees have to be grouped according to a preallocated schedule of evacuation. Since the objective is to evacuate in minimum time, the vehicle routing and scheduling decisions are intertwined. At the macroscopic level, evacuation routing is a "many to many" routing problem with multiple origins and multiple destinations. From the perspective of an evacuation planner whose target is to evacuate the maximum number of people within a minimum time, Wardrop's traffic principle of system optimal (SO) flow is best suited to decide the routes, schedule, and traffic flow.

A precise estimate of the evacuation time is of primary importance to the evacuation planners. In this section, we will provide a detailed treatment of how the answer to this question is found and subsequently discuss the methods for deciding the routes and traffic assignment for the evacuation. The objective of finding the minimum time is modeled as a minimum cost network flow problem. Under the assumption of deterministic travel time and capacity of the arcs, the evacuation model minimizes the clearance time. Since time is a decisive parameter in such problems, dynamic networks are used instead of static ones and the evacuation planning problems are modeled under the discrete time dynamic network flow framework.

A discrete time dynamic network flow problem is a discrete time expansion of a static network flow problem. In this case, we distribute the flow over a set of predetermined time periods $t = 1, 2, \ldots T$. Consider a directed static network $G = (N, A)$ with N and A as the set of nodes and arcs, respectively. A constant transit time λ_{ij} is associated with each arc $(i, j) \in A$. The time expansion of G over a time horizon T defines the dynamic network $G_T = (N_T, A_T)$ associated with G. The time expansion essentially is the replication of the static network G at each discrete unit of time in T. Since there are multiple copies of source and sink nodes, a super source node s and sink node d are introduced to create a single source/ single sink network. The benefit of a time-expanded network is that it facilitates solving the flow over time problems by static flow computations. On the other hand, time-expanded networks are huge in practice and the size of the network increases linearly with the given time horizon T and, therefore, exponentially in the input size. This makes the problem difficult to solve and is proved to be a pseudopolynomial problem in the work by Hoppe and Tardos (1995).

Optimization Model: A number of dynamic network flow models were discussed by Tjandra (2003) in which the objective of the models was to push for maximum flow in minimum time. Given the total number of evacuees, the capacity of the road links, and the shelter destinations, the objective is typically to minimize the time for sending all the supply from source to destination for the underlying evacuation network. A mixed integer dynamic network flow model under the SO principle of traffic flow is used. One such model for minimizing the average travel time by all the evacuees is discussed below.

Decision Variables: In the dynamic network flow model, the flow $x_{ij}(t)$ is the number of vehicles that leave node i at time t and reach node j at time $t + \lambda_{ij}$. Flow variable $y_i(t + 1)$ represents the vehicles that move on holdover arcs (arcs connecting the same node from time t to $(t + 1)$), that is,

$$y_i(t+1) = x_{i(t),i(t+1)}$$

Let N_d denote the set of destination nodes, N_s the set of source nodes, and S_i the initial demand at source node i. We define objective function Z to model the average evacuation time required by an evacuee to leave the network, that is,

$$Z = \frac{\sum_{t=0}^{T} \sum_{i \in N_d} t x_{id}(t)}{\sum_{i \in N_s} S_i}$$

Objective function Z depends only on the flow variables since the denominator is constant. Therefore, Z can be redefined as follows:

$$Z = \sum_{t=0}^{T} \sum_{i \in N_d} t x_{id}(t)$$

Constraints:

1. *Arc capacity:* The flow $x_{ij}(t)$ on any movement arc $(i, j) \in A$ and flow $y_i(t)$ on any holdover arc should not exceed the capacity of the arc u_{ij} at any time t.
2. *Supply at source:* For complete evacuation, the total flow reaching the destination should be equal to the total number of initial evacuees present at the start of the evacuation.
3. *Capacity at destination:* Safe destinations have a limited capacity to hold evacuees (capacitated network). Therefore, the total flow reaching a particular destination $d \in N_d$ should be less than or equal to the capacity of the destination.
4. *Flow balance:* Total flow coming into a node should be equal to the total flow going out of the node.

Under assumptions of constant capacity and constant transit time on arcs, the evacuation model that minimizes the average evacuation time can be formulated as

$$\text{minimize} \sum_{t=0}^{T} \sum_{i \in N_d} t x_{id}(t) \tag{5.1}$$

$$x_{si}(0) = S_i, \quad \forall i \in N_s \tag{5.2}$$

$$\sum_{t=0}^{T} \sum_{i \in N_d} x_{id}(t) = \sum_{j \in N_s} S_j \tag{5.3}$$

$$\sum_{t=0}^{T} \sum_{k \in pred(i)} x_{ki}(t) \le C_i, \quad \forall i \in N_d \tag{5.4}$$

$$y_i(t+1) - y_i(t) = \sum_{k \in pred(i)} x_{ki}(t - \lambda_{ki}) - \sum_{j \in succ(i)} x_{ij}(t), \quad t = 0, \ldots, T; \forall i \in N \tag{5.5}$$

$$y_i(0) = 0, \quad \forall i \in N \tag{5.6}$$

$$y_i(t) = 0, \quad \forall i \in N_d, t = 0, \ldots, T \tag{5.7}$$

$$0 \le y_i(t) \le a_i, \quad t = 1, \ldots, T; i \in N \cup N_d \tag{5.8}$$

$$0 \le x_{ij}(t) \le u_{ij}, \quad t = 0, \ldots, T - \lambda_{ij}, \forall (ij) \in A \tag{5.9}$$

We can treat the time-expanded network as a static network and then apply any minimum cost static network flow algorithm to obtain the solution. Other variants of this problem can be obtained by changing the objective function to maximizing flow out of the network (maximum dynamic flow problem) during a given time horizon T, maximizing flow out of the network for any smaller time horizons $T' \in T$ (universal maximum flow problem), and minimizing the time horizon to clear the network. Since the problem is pseudopolynomial because of the large network size of a time-expanded network, a number of solution techniques have been proposed. Hoppe and Tardos (1995) came up with a first polynomial time algorithm for the quickest transshipment problem and provided an integral optimum flow that can send exactly the right amount of flow out of each source and into each sink in the minimum overall time.

Heuristic algorithms have been proposed to come up with routes and schedules for the evacuation. One such algorithm is Capacity Constrained Route Planning (CCRP) proposed by Lu et al. (2005). This algorithm divides evacuees into multiple groups and assigns a route and time schedule to each group. The CCRP algorithm employs a shortest path algorithm to find the shortest route from source nodes to destination nodes and assigns the evacuation schedule based on the available capacity on the route. The solution of the CCRP algorithm is within 10% of the optimal evacuation time in all test cases. Another heuristic approach by Lim et al. (2011) used the Evacuation Scheduling Algorithm (ESA) for the capacity constrained network flow optimization. ESA utilizes Dijkstra's algorithm for finding the evacuation paths and a greedy algorithm for finding the maximum flow of each path and schedule to execute the flow for each time interval.

In general, the mathematical models and the solution methods discussed in this section are only useful for finding a lower bound of evacuation time and the corresponding routes and schedules. A closer look at the transportation network (such as intersections), the dynamics of the evacuation (such as contraflow decisions and priority), and realistic assumptions (i.e., occurrences of incidents and congestion) would present a better picture and result in a realistic evacuation plan. Researchers and practitioners, therefore, consider the preceding factors carefully and incorporate them in the evacuation planning model (EPM). In Sections 5.2.2 and 5.2.3, we discuss the models for decision making in evacuation that take these factors into account.

5.2.2   Efficient use of the critical roadway segments

As transportation infrastructure is a limited resource in terms of both directional accessibility and capacity, it is worth pursuing a routing plan that makes optimal use of this infrastructure. Lane-based routing and contraflow of traffic are two such traffic engineering tools that can be very effective for routing the evacuating traffic during a catastrophic event. Apart from these two, decisions corresponding to the location of shelters are also very important so that the traffic is not biased toward a particular shelter location and does not saturate a specific route.

5.2.2.1   Lane-based routing

Most traffic delays during an evacuation occur at intersections. Lane-based routing is a valuable strategy for reducing these delays. In a lane-based routing plan, select turning options at intersections are restricted to improve traffic flow away from a hazardous area. A plan might require vehicles in the right lane of an intersection approach to turn right while requiring vehicles in the left lane to continue straight. Cova and Johnson (2003) model the evacuation routing problem as an integer extension of the min-cost flow problem. The primary objective is to route vehicles to their closest evacuation zone exit. A secondary objective is to minimize the number of intersection merging conflicts. Furthermore, the model prevents intersection crossing conflicts. The model can be specified as follows.

Decision Variables: As with the previous model for evacuation planning, variable x_{ij} represents the flow of vehicles on arc ij. Variable y_{ij} is a binary variable that is set if the flow on arc ij is positive; otherwise, it is set to 0. Variable z_i denotes the number of traffic streams that merge at node i. An upper bound on the number of merges allowed for the model is set to value M. For the evacuation network, parameter d_{ij} is the length of arc ij and b_i is the capacity of any node i.

Constraints: The objective of the model is to minimize the total travel distance with the following set of constraints.

1. *Flow balance constraint*: The flow out of a node minus the flow into the node must be equal to the net flow at the node. Typically, b_i is a constant that is positive for source nodes, zero for intermediate nodes, and negative for destination nodes. In this model, b_i is positive for source nodes, zero for intermediate nodes, but a variable for evacuation zone exit nodes. This allows the net outflow at exits to be resolved endogenously along with the routes.
2. *Intersection crossing conflict*: This constraint limits the flow on intersecting arcs by allowing a positive flow on one of the arcs and zero flow on the cross arc.
3. *Merge limiting*: Total number of merges at a node i are recorded for each traffic stream above 1 that terminates at the node and the total number of merges allowed in the routing plan is limited by an upper bound M.
4.

$$\text{Minimize } Z = \sum_i \sum_j d_{ij} x_{ij} \qquad (5.10)$$

$$\sum x_{ij} - \sum x_{ji} = b_i, \quad \forall i \tag{5.11}$$

$$x_{ij} \le u_{ij}y_{ij}, \quad \forall i \to j \text{ that cross } k \to l \tag{5.12}$$

$$x_{kl} \le u_{kl}(1 - y_{ij}), \forall k \to l \text{ crossed by } i \to j \tag{5.13}$$

$$\sum_j y_{ji} \le z_i + 1, \quad \forall i \text{ with a potential merge} \tag{5.14}$$

$$\sum_i z_i \le M \tag{5.15}$$

$$0 \le x_{ij} \le u_{ij}, \quad \forall i \to j \tag{5.16}$$

The central assumption underlying the model is that placing intersections in a state of uninterrupted flow and minimizing merging will result in fewer traffic delays and lower network evacuation clearing times under moderate to heavy volumes. This can require the cooperation of evacuees and many emergency personnel if there are a large number of intersections to control. The microsimulation results indicate that channeling flows at intersections to remove crossing conflicts can significantly decrease network clearing time over no routing plan (i.e., random destination choice).

5.2.2.2 Contraflow

Contraflow is emerging as an important and widely used tool to improve evacuation traffic capacity. Unlike normal traffic conditions where traffic is in both directions, evacuation events result in traffic that is directed in a single direction. As seen in Figure 5.2a, a road section is completely unutilized and can be used for traffic flow in the direction of evacuation as seen in Figure 5.2b. Contraflow is defined as the reversal of traffic flow

(a)   (b)

Figure 5.2 Contraflow during evacuation. (a) No contraflow. (b) With contraflow.

in one or more of the inbound lanes of a divided highway for use in the outbound direction, with the goal of increasing capacity. The increased capacity that contraflow provides can substantially reduce clearance times. Managerial questions that are posed pertain to starting times of contraflow, how long it will last, and which lanes in the network need to be reversed along with issues associated with safety, accessibility, convenience, enforcement, and cost.

From the optimization perspective, research on contraflow concentrates on deciding the network configuration, that is, coming up with the optimum network structure with lane directions that will result in minimum evacuation time. Heuristic approaches have been taken to solve this problem as the problem is nondeterministic polynomial (NP) complete (Kim and Shekhar 2005) and the solution will not be scalable for regional evacuation. We discuss here a greedy solution approach proposed by Kim and Shekhar (2005) to come up with a network configuration for contraflow.

The decision to start contraflow on a particular edge is influenced by the congestion history of that edge. The basic assumption of the greedy heuristic is that when the evacuation route planner is run over an original network configuration without contraflow, the edges having more congestion history are more influential in the determination of edge flips. To quantify the congestion history on each edge with the data from the evacuation route planner, the *FlowHistory* and *CongestionIndex* of an edge e is defined in the following way:

Definition 1 FlowHistory(e) = Total number of traveling units going through edge e during EvacuationTime.

Definition 2 CongestionIndex(e) = FlowHistory(e)/(Capacity(e) × EvacuationTime).

FlowHistory(e) is acquired from the result of the evacuation route planner. The denominator in Definition 2 refers to the maximally possible amount of flow of edge e during EvacuationTime. Thus, *CongestionIndex(e)* indicates the percentage of edge utilization during EvacuationTime. A higher *CongestionIndex(e)* value means that the edge e has been more congested during the evacuation process. The third definition used in the greedy approach is the Degree of Contraflow. The Degree of Contraflow in the reconfigured network is defined as follows:

Definition 3 Degree of Contraflow(DoC) = Number of Flipped Edges/ Total Number of Edges.

This percentage parameter indicates how many edges are flipped among all edges in the reconfigured network and is used to control unnecessary flips.

Run an evacuation route planner to produce *FlowHistory* and
EvacuationTime on $G_{original}$;
for all *edge* ∈ $G_{original}$ **do**
   CongestionIndex(e) = *FlowHistory*(e)/(*Capacity*(e)
    × *EvacuationTime*);
end for
Sort edges by *CongestionIndex*(e) in descending order;
$G_{reconfigured}$ = $G_{original}$;
for all (i,j) in the first *DoC*% edges in the sorted edges **do**
$G_{reconfigured}$ · *flip*((i,j));
Return $G_{reconfigured}$;

Algorithm 5.1 The greedy algorithm.

  The greedy algorithm shown in Algorithm 5.1 works in the following way: First, run any evacuation route planner to generate the flow
history and evacuation time of a given original network. Second, assign
a congestion index value to each edge. Third, the edges are sorted by
congestion index in descending order. Finally, flip edges in favor of a
higher congestion index value among the first *DoC*% of the sorted edge
set. The evacuation route planner must be run again over the reconfigured network to get the evacuation time of the reconfigured network.
The greedy heuristic designed for contraflow is quickly able to come up
with a good quality solution for network reconfiguration, minimizing
the evacuation time.

5.2.2.3 *Shelter locations and clearance time*

The location of shelters in a region threatened by a hurricane can greatly
influence the highway network clearance time, that is, the time needed
by evacuees to escape from origin points to safe areas. Sherali et al. (1991)
developed and solved a location-allocation model for determining a set
of viable shelter locations from among potential alternatives and, accordingly, simultaneously prescribing a traffic diversion strategy to minimize
the total evacuation time of automobiles from designated origins to the
shelters under some emergency conditions. The network model is based
on a directed graph $G = (N, A)$ with a node set N, and a set of directed
arcs A. Let $N_d \subseteq N$ denote a set of potential destination nodes and binary
decision variables y_i, which is set to 1 if a shelter is located at node $i \in N_d$.
  Constraints:

  1. The restriction on the set of chosen shelter locations is dictated by
   the maximum value S of any of these quantities:
   a. Total number of shelters to be opened
   b. Permissible number of new shelter locations

c. Number of shelters opened within designated subregions to avoid traffic bottlenecks on key roadways
d. Total staffing needs of all shelter locations

Such restrictions can be represented by the set

$$Y = \{y \equiv (y_i, i \in N_d): \sum_{i \in N_d} s_{ij} y_j \leq S_j, j = 1, \ldots, m, \text{and } y \text{ binary}\}$$

2. *Flow balance constraint*: Net rate of accumulation of demand occurring at node $i \in N_d$ during the planning horizon T is z_i. For other nodes $i \in N - N_d$, the flow balance equation would be equal to initial supply at that node. Note that for intermediate nodes, initial supply is zero.
3. *Shelter capacity constraint*: Total accumulation over the time horizon at any particular shelter location should not exceed its capacity.
4. *Arc capacity constraint*: The capacity of the arc connecting two nodes should not be exceeded.

Objective Function: The total time spent by the evacuees seeking shelters on the network system has to be minimized. This time is obtained using the equation for travel time on an arc proposed by U.S. Bureau of Public Roads, which is given by

$$T_{pq} = t_{pq}\{1 + \alpha(x_{pq} / u_{pq})^\beta\}, \quad \forall (p,q) \in A$$

The travel time T_{pq} is a function of steady-state flow rate x_{pq}, t_{pq} is the free-flow travel time, u_{pq} is the flow rate capacity of link (p, q), and α and β are nonnegative model parameters, typically 0.15 and 4.0, respectively. Assuming D_i denoting a constant rate of dissipation of evacuees originating at node $i \in N_s$ over a time horizon, T, that seek shelter locations, the total number of evacuees originating at node i is given by $D_i T$. The proposed model EPM1 is stated as follows:

$$\text{Minimize} \sum_{(p,q) \in A} [t_{pq} x_{pq} + \tau_{pq} x_{pq}(x_{pq} + \bar{x}_{pq})], \tag{5.17}$$

$$\sum x_{ij} - \sum x_{ji} = D_i - z_i, \ \forall i \in N_d \tag{5.18}$$

$$\sum x_{ij} - \sum x_{ji} = D_i, \ \forall i \in N - N_d \tag{5.19}$$

$$0 \leq z_i \leq C_i y_i, \ \forall i \in N_d \tag{5.20}$$

$$y \in Y, 0 \leq x_{pq} \leq \bar{u}_{pq}, \ \forall (p,q) \in A \tag{5.21}$$

Observe that problem EPM1 has the special structure that for a fixed $y \in Y$, the problem is a separable, convex cost, network flow programming problem that can be solved very efficiently. The objective seeks to minimize the total vehicle hours accrued during the evacuation procedure, and hence, the model adopts a systems-oriented approach. The objective function is nonlinear and represents the total system congestion-related travel time. The model is modified in case problem EPM1 is infeasible. One may wish to consider the case in which a portion of the evacuees originating at any of the nodes $i \in N$ are accommodated at makeshift locations at node i itself, possibly by overcrowding a shelter location if $i \in N_d$. To accommodate this situation, we can introduce an artificial variable x_{ai} in the left-hand side of constraint (5.2) to absorb excess flows for each $i \in N$ and, accordingly, penalize such a variable x_{ai} in the objective function using a suitable penalty factor M_i, for each $i \in N$. Incorporating this feature into EPM1, and rewriting the resulting problem in obvious matrix notation, the EPM can be formulated as follows:

$$\text{Minimize } F(x) + M \cdot x_a \tag{5.22}$$

$$Ax + Bz + x_a = D \tag{5.23}$$

$$0 \le z_i \le C_i y_i, \ \forall i \in N_d \tag{5.24}$$

$$y \in Y, 0 \le x \le \bar{u}, x_a \ge o \tag{5.25}$$

This problem is a NP-hard nonlinear mixed-integer programming problem and, by its nature, is most amenable for solution by some branch-and-bound scheme. The authors developed a heuristic and two versions of an exact implicit enumeration algorithm based on the generalized Benders' decomposition method to come up with the solution.

5.2.3   Stochastic models

The routing models do not properly capture the dynamic nature of transport risk factors at the tactical level (e.g., traffic conditions, population density, and weather conditions). Moreover, most of these risk factors cannot be known a priori with certainty. They are both time-dependent and stochastic in nature, that is, they are random variables with probability distribution functions that vary with time. Therefore, the routing and scheduling problem is best modeled as a path selection problem in a stochastic time-varying network.

There are numerous sources of uncertainty during evacuation and most are not easy to quantify or control. Factors such as severity of the disaster, human behavior during the evacuation, and the impacts of disasters on infrastructure are beyond our control. There are two major factors that affect evacuation planning that are being studied by the research community: the uncertain demand levels at the impact nodes and the degrading capacity of the road links during disasters. These unexpected changes in evacuation demand levels and capacity may result in significant differences in terms of the predictions of a model.

Developing a priori path sets for evacuation requires estimation of demand. The anticipated demand may deviate significantly from the actual number of people evacuating. If the solution consists of a set of paths, and more demand appears than was anticipated, there will be an insufficient number of paths to assign. Because the realized demand differs from the predicted demand, new paths must sometimes be calculated by suboptimal means. Similarly, estimating a static capacity value for the links is unacceptable. As pointed out by numerous works on transportation network reliability (Chen et al. 2002; Lo and Tung 2003), the capacity degrades as the number of vehicles on the link increases. This network has a higher probability of encountering a catastrophic event, such as extreme congestion or perhaps gridlock.

It becomes necessary to design approaches that account for demand and capacity uncertainty and develop more robust solutions that are less likely to fail under these extreme events and potentially reduce the variance of future costs. Robust optimization (RO) and chance constrained programming (CCP) are used to account for parameter uncertainty in cases when a mathematical program is formulated. The models are developed for dynamic traffic assignment (DTA) with the underlying principle of cell transmission modeling (CTM) (Daganzo 1994). The main advantage of using a stochastic programming technique is that they introduce reliability to the model.

For introducing the stochastic demand in the deterministic formulation, Waller and Ziliaskopoulos (2006) used CCP on a CTM-based model. In the deterministic linear programming (LP), there is a single set of constraints that generate the demand at each origin node for each time period. This demand constraint can be written as an inequality of the form

$$x_i^t - x_i^{t-1} + y_{ij}^{t-1} \geq d_i^{t-1}, \quad j \in \Gamma(i) \quad \forall i \in C_R, \quad \forall t \in T \qquad (5.26)$$

If d_i^{t-1} is a random variable with probability distribution $F_{d_i^{t-1}}$, the equivalent chance constraint for the demand constraint can be written as

$$Pr[x_i^t - x_i^{t-1} + y_{ij}^{t-1} \geq d_i^{t-1}] \geq \alpha, \quad j \in \Gamma(i) \quad \forall i \in C_R, \quad \forall t \in T \qquad (5.27)$$

By the definition of the distribution function

$$F_{d_i^{t-1}}(x_i^t - x_i^{t-1} + y_{ij}^{t-1}) \geq \alpha \qquad (5.28)$$

$$(x_i^t - x_i^{t-1} + y_{ij}^{t-1}) \geq F_{d_i^{t-1}}^{-1}(\alpha) \qquad (5.29)$$

Constraint 5.29 is the deterministic equivalent of the chance Constraint 5.27. Including Constraint 5.29 instead of the deterministic Constraint 5.26 in the model produces the equivalent deterministic LP for the CCP evacuation model. Parameter α in Constraint 5.29 specifies the reliability level for a single constraint and thereby relates to the network reliability level. By using the chance constraint and increasing the confidence level α, the linear program solves the model with demands greater than the expected value. A higher value of α, therefore, results in a more robust solution.

Often, it is impossible to know the exact probability distribution of the number of evacuees. Yao et al. (2009) proposed a robust CTM-based optimization model in which no infeasibilities were allowed. A robust optimal solution can be interpreted as the solution that is feasible for any realization of the uncertain data and achieves the best worst case objective value. In RO methodology, the demand uncertainty is represented using an appropriate uncertainty set U_d. An uncertainty set can be either a polyhedral set, ellipsoidal set, or a box set. The objective value of the model is guaranteed for any demand realization within an appropriate uncertainty set. The robust counterpart of the CTM with uncertain demand corresponds to a deterministic LP considering maximum possible demand in the uncertainty set.

Yazici and Ozbay (2010) employed a conventional chance constraint approach to model uncertainties in road capacities as well as traffic demand during evacuation. The model is based on CTM for DTA and can be written in compact standard form as follows:

$$\text{Minimize} \quad \sum_t \sum_i x_i^t \tag{5.30}$$

$$\text{Subject to} \quad A_{eq} v = b_{eq} \tag{5.31}$$

$$Rv = D \tag{5.32}$$

$$Av \le b \tag{5.33}$$

$$Tv \le C \tag{5.34}$$

$$x_i^t \ge 0, y_{ij}^t \ge 0, \quad \forall (i, j) \in \xi, \quad \forall t \in T \tag{5.35}$$

In the preceding formulation, Constraint 5.32 represents the demand constraints with demand vector D and Constraint 5.34 represents the capacity constraints with capacity vector C. Since the constraints are formulated for each time step, C and D vectors consist of time-expanded forms of all cell capacities and demands in the network. The SO DTA problem with probabilistic demand and capacity constraints can be modeled as given below.

$$\text{Minimize} \quad \sum_t \sum_i x_i^t \tag{5.36}$$

$$\text{Subject to} \quad A_{eq} v = b_{eq} \tag{5.37}$$

$$P\left(R_{\bar{l}} v = \psi_{\bar{l}}\right) \geq p_{\bar{l}}, \bar{l} = 1,\ldots l \tag{5.38}$$

$$A v \leq b \tag{5.39}$$

$$P\left(T_{\bar{k}} v \leq \phi_{\bar{k}}\right) \geq p_{\bar{k}}, \bar{k} = 1,\ldots k \tag{5.40}$$

$$x_i^t \geq 0, y_{ij}^t \geq 0, \quad \forall (i,j) \in \xi, \quad \forall t \in T \tag{5.41}$$

where $\bar{l}$ refers to the number of probabilistic demand constraints (having a total of l) and $\bar{k}$ refers to the number of probabilistic capacity constraints (having a total of k). Probability levels $p_{\bar{l}}$ and $p_{\bar{k}}$ are imposed on each probabilistic constraint individually and can be assigned different values based on the desired local reliability of the constraint at hand. Solutions for such a formulation should meet the reliability levels individually for each link and source cell at each time step. The values of $p_{\bar{l}}$ and $p_{\bar{k}}$ can be obtained by using their corresponding probability distribution function. The solution will follow substituting the $p_{\bar{l}}$ and $p_{\bar{k}}$ values as the right-hand side of the probabilistic equation for all probabilistic capacity and demand constraints, thus making them deterministic, and solve the problem through conventional LP solvers.

Probabilistic constraints ensure that the state of the system remains within a subset of all possible states where its functioning is undisturbed by major failures. Small capacity fluctuations (e.g., 1990 vehicles per hour per lane [vphpl] instead of 2000 vphpl) will not cause a significant change on the network traffic, but changes that correspond to a more significant percentage of the existing capacity (e.g., 1750 vphpl instead of 2000 vphpl) will affect network flows. A similar argument is also valid for demand fluctuations. Employing stochastic models with a reliability measure will result in a more realistic evacuation plan as compared to the results of the deterministic models that most likely would not be experienced during an actual evacuation process. Such probabilistic models will help decision makers to make a more informed decision.

5.3   Software and modeling tools

Modeling and analysis tools can provide emergency managers, engineers, and planners with a means to apply different disaster-related scenarios and make informed decisions about strategies to best accommodate evacuation demand. These tools can equip emergency managers to evaluate different alternative scenarios and give them the opportunity to develop alternative means to evacuate based on prevailing conditions associated with a particular evacuation event. There exist a number of modeling and analysis tools to aid the analysts or decision makers in dealing with evacuation. The major emphasis of such tools is traffic assignment, weather and assessment monitoring, and prediction.

In most of these evacuation models, traffic assignment is conducted using simulation. A few of the evacuation-specific software packages are Mass Evacuation (MASSVAC), Network Emergency Evacuation (NETVAC), Dynamic Network Evacuation (DYNEV), and Evacuation Traffic Information System (ETIS). A recent addition to the evacuation modeling packages is the Oak Ridge Evacuation Modeling System (OREMS), which can be used to model evacuation from various disasters. Sometimes, existing microsimulation transportation software packages such as NETSIM, VISSIM, CORSIM, and DYNASMART are used in evacuation modeling of the complete evacuation process or analysis of evacuation policies such as contraflow. ARENA, which is general simulation software, is also employed in evacuation modeling. Since those software packages are not specifically designed for evacuation modeling, they do not include demand generation models suitable for evacuation modeling. Customized assignment procedures have been developed specifically to model evacuation traffic. A summary of tools and software used are presented in Table 5.1. These models help emergency managers and planners in making evacuation decisions, but only a few of these models can be used to estimate hazards due to the disaster or event necessitating evacuation. Information regarding the scale (e.g., micro–macro) and employed demand models on selected evacuation-specific software packages are presented in Table 5.2.

Table 5.1 Models Used for Emergency Evacuation and Planning

Model	Primary purpose	Potential users	Developer/ funding	Release date
Clarus	Weather information	Transportation managers, weather providers, and travelers	USDOT and NOAA	2005/2006 (testing)
CATS/JACE	Estimate hazards from disaster	Emergency managers	DTRA and FEMA	2005 (version 6.0 for use with ArcView 9)
DYNASMART-P	Model traffic flow	Transportation managers	FHWA	2007 (version 1.3)

Table 5.1 (***Continued***) Models Used for Emergency Evacuation and Planning

Model	Primary purpose	Potential users	Developer/ funding	Release date
ETIS	Collect and disseminate transportation information	Emergency managers	FHWA	2002 (tested); online tool
ETDFS	Evacuation modeling and analysis	Emergency managers	FEMA, USACE, DOT	2002; online tool
HAZUS-MR3	Estimate potential losses from disaster	Emergency personnel and planners	FEMA	2007 (version 1.3)
HURREVAC	Evacuation decisions	Emergency managers	FEMA	May 2008
MASSVAC	Nuclear power plant and hurricane evacuation	Emergency planners	Hobeika and Kim	1985
NETVAC	Nuclear power plant evacuation/ analyze route choice	Emergency managers and planners	MIT	1982
OREMS	Evaluate large scale vehicular evacuation, develop evacuation plan	Emergency personnel and planners	CTA/ ORNL	2003 (version 2.6)
SLOSH	Estimate storm surge height and wind due to hurricane	FEMA, NOAA, USACE	NWS	2002 (version 1.31)

Apart from the list of models mentioned in the tables, DTA models are under development. These systems use simulation models combined with real-time traffic and origin-destination information to predict the effects of various management strategies, thus allowing more effective management and providing better traffic information than is currently possible. Two DTA traffic estimation and prediction system prototypes

Table 5.2 Evacuation-Specific Software Packages

Model	Developer	Brief information
NETVAC	Sheffi et al. (1982)	1. Macroscopic model developed for sites near nuclear plants 2. Insensitive to evacuees' behavior 3. Deterministic model 4. Time varying O-D tables are required as input
DYNEV	KLD and Associates	1. Macroscopic model developed for sites near nuclear plants 2. Employs static traffic assignment 3. Used to analyze the impacts of alternative traffic controls (traffic signals, stop signs, and yield signs) 4. Analyze network capacity and evacuation demand 5. Cannot deal with time-varying flows
EPP	PRC Voorhees	1. Dynamic and probabilistic model 2. Consider evacuees' behavior for determining loading and response rate
CLEAR	NRC	1. A microsimulation tool with static assignment 2. Cannot deal with time-varying flows
TEDDS	Hobeika (1985)	1. A microsimulation tool with quasi-dynamic assignment 2. Originally developed for power plants and is based on computer simulation model MASSVAC 3. Has a knowledge-based system that stores evacuation expert rules, disaster-related information, and area and transportation network characteristics 4. Event-based simulation designed to load evacuees, select routes, calculate clearance time, and identify bottlenecks
TEVACS	Anthony F. Han (1990)	1. Analyze large scale evacuation (cities in Taiwan) 2. Incorporates all modes of transport into the model
OREMS	ORNL	1. Based on ESIM (Evacuation SIMulation) 2. Combines trip distribution and traffic assignment with traffic flow simulation submodel 3. Analyze traffic management and control, and operational assessment 4. Dynamic model that allows for tracking evacuation at user-specified time intervals 5. Includes human behavior and weather information 6. Capable of modeling contraflow operations 7. Does not estimate timing of people's response and number of evacuees

*Table 5.2 (**Continued**)* Evacuation-Specific Software Packages

Model	Developer	Brief information
ETIS	PBS&J	1. Macroscopic model with static assignment
		2. Operates within GIS environment
		3. Incorporates human behavior and weather conditions
		4. Capable of modeling contraflow operations
HURREVAC	FEMA	1. Specifically for hurricane evacuation
		2. Operational tool, assisting decision makers in advance of and during an evacuation
		3. Estimates time required to evacuate an area

Source: M. Yazici, *Introducing uncertainty into evacuation modeling via dynamic traffic assignment with probabilistic demand and capacity constraints*, PhD thesis, Rutgers University-Graduate School-New Brunswick, New Brunswick, NJ, 2010.

are DynaMIT and Dynamic Network Assignment Simulation Model for Advanced Road Telematics (DynaSmart). There are two versions of DynaSmart being developed for real-time applications: DynaSmart-X and DynusT. DynaSmart has the developmental goal of serving as a real-time computer system for traffic estimation and prediction that supports both transportation management systems and advanced traveler information systems (ATIS). Similarly, DynaMIT is being developed as a real-time computer system for traffic estimation, prediction, and generation of traveler information and route guidance to support the operation of traffic management systems and ATIS at traffic management centers.

5.4   Conclusions

This chapter focused on providing a review of OR techniques for coming up with an a priori evacuation plan. An overview of mathematical models was provided that targeted various aspects of the evacuation problem. These models provided a comprehensive evacuation plan and answered important question of the total clearance time required in wake of the infrastructure limitations and the inherent uncertainties associated with regional evacuation. Deterministic as well as stochastic models were reviewed in this chapter and the solution techniques used for solving these models were introduced. A comprehensive list of the software developed for evacuation planning was also provided.

We believe that there are still many important OR problems in evacuation planning. Access to timely and accurate traffic information during evacuations is critical to the evacuation process. Information about traffic flow rates and speeds, along with lane closures, weather conditions, incidents, and the availability of alternative routes, is needed to effectively guide evacuees. We think the focus will shift from a priori optimization toward real-time adaptive decision making for several reasons, such as the availability of the necessary technology and data with the advent of Intelligent Transportation Systems equipment. OR techniques applied to evacuation planning significantly contribute to decision making during unfortunate disaster events. We expect that research will intensify in the near future and we hope that this chapter will be useful to practitioners and future researchers in this area.

References

J. Aronson, "A survey of dynamic network flows," *Annals of Operations Research*, vol. 20, no. 1, pp. 1–66, 1989.

L. Chamlet, R. Francis and P. Saunders, "Network models for building evacuation," *Fire Technology*, vol. 18, no. 1, pp. 90–113, 1982.

A. Chen, H. Yang, H. Lo and W. Tang, "Capacity reliability of a road network: An assessment methodology and numerical results," *Transportation Research Part B: Methodological*, vol. 36, no. 3, pp. 225–252, 2002.

T. Cova and J. Johnson, "A network flow model for lane-based evacuation routing," *Transportation Research Part A: Policy and Practice*, vol. 37, no. 7, pp. 579–604, 2003.

C. Daganzo, "The cell transmission model: A dynamic representation of highway traffic consistent with the hydrodynamic theory," *Transportation Research Part B: Methodological*, vol. 28, no. 4, pp. 269–287, 1994.

Dept. of Transportation and Dept. of Homeland Security, *Report to Congress on Catastrophic Hurricane Evacuation Plan Evaluation*, 2006.

Florida Division of Emergency Management, "Evacuation and clearance times," 2000. Last accessed online June 4, 2012.

H. Hamacher and S. Tjandra, "Mathematical modelling of evacuation problems— A state of the art," *Pedestrian and Evacuation Dynamics*, pp. 227–266, 2002.

A. Han, "TEVACS: Decision support system for evacuation planning in Taiwan," *Journal of Transportation Engineering*, vol. 116, no. 6, p. 821, 1990.

A. Hobeika and B. Jamei, "MASSVAC: A model for calculating evacuation times under natural disasters," *Emergency Planning*, pp. 23–28, 1985.

B. Hoppe and E. Tardos, "The quickest transshipment problem," in *Proceedings of the sixth annual ACM-SIAM symposium on discrete algorithms*, pp. 512–521, Society for Industrial and Applied Mathematics: Philadelphia, PA, 1995.

S. Kim and S. Shekhar, "Contraflow network reconfiguration for evacuation planning: A summary of results," in *Proceedings of the 13th annual ACM international workshop on geographic information systems*, pp. 250–259, ACM: New York, NY, 2005.

G. Lim, S. Zangeneh, M. Baharnemati and T. Assavapokee, "A Capacitated Network Flow Optimization Approach for Short Notice Evacuation Planning," tech. rep., University of Houston, Systems Optimization and Computing Lab Technical Report No. SOCL1111-01, 2011.

H. Lo and Y. Tung, "Network with degradable links: capacity analysis and design," *Transportation Research Part B: Methodological*, vol. 37, no. 4, pp. 345–363, 2003.

Q. Lu, B. George and S. Shekhar, "Capacity constrained routing algorithms for evacuation planning: A summary of results," *Advances in Spatial and Temporal Databases*, vol. 3633, pp. 291–307, 2005.

Y. Sheffi, H. Mahmassani and W. Powell, "A transportation network evacuation model," *Transportation Research Part A: General*, vol. 16, no. 3, pp. 209–218, 1982.

H. Sherali, T. Carter and A. Hobeika, "A location-allocation model and algorithm for evacuation planning under hurricane/flood conditions," *Transportation Research Part B: Methodological*, vol. 25, no. 9, pp. 439–452, 1991.

S. Tjandra, *Dynamic network optimization with application to the evacuation problem*, Universitatsbibliothek: TU Kaiserslautern, 2003.

S. T. Waller and A. Ziliaskopoulos, "A chance-constrained based stochastic dynamic traffic assignment model: Analysis, formulation and solution algorithms," *Transportation Research Part C: Emerging Technologies*, vol. 14, no. 6, pp. 418–427, 2006.

T. Yao, S. Mandala and B. Chung, "Evacuation transportation planning under uncertainty: A robust optimization approach," *Networks and Spatial Economics*, vol. 9, no. 2, pp. 171–189, 2009.

M. Yazici, *Introducing uncertainty into evacuation modeling via dynamic traffic assignment with probabilistic demand and capacity constraints*, PhD thesis, Rutgers University-Graduate School-New Brunswick, New Brunswick, NJ, 2010.

A. Yazici and K. Ozbay, "Evacuation network modeling via dynamic traffic assignment with probabilistic demand and capacity constraints," *Transportation Research Record: Journal of the Transportation Research Board*, vol. 2196, no. 1, pp. 11–20, 2010.

chapter six

Evacuation planning, analysis, and management

Michael K. Lindell

Contents

6.1   Introduction

Evacuation is a process intended to temporarily move people from a hazardous location to a place of greater safety. It takes place frequently; an evacuation involving 1000 people or more takes place every 2–3 weeks in the United States (Dotson and Jones 2005). Moreover, evacuation is effective; many people are removed from the threats of natural hazards such as floods, hurricanes, wildfires, and volcanic eruptions and of technological hazards such as explosions, toxic chemical spills, and radiological releases. Such movement can significantly decrease the number of deaths and injuries as well as the social impacts these casualties can cause (Lindell et al. 2006). However, despite its frequency and success, evacuation should not be assumed to be an unquestionably appropriate response to all hazardous situations. In some cases, characteristics of the hazard preclude evacuation. For example, earthquakes generally provide no forewarning, so preimpact evacuation is not possible. In addition, some tornadoes, toxic chemicals, and radiological incidents have such a rapid onset and short duration that evacuation would increase rather than decrease the danger to risk-area residents. Evacuation would also be inadvisable in any hazard when the risk of movement exceeds the risk of remaining in place as might be the case for evacuating intensive care patients. In other cases, evacuation is unnecessary because people are in structures that provide adequate protection. In such cases, shelter in-place will provide adequate protection. Finally, evacuation is inadvisable when the reduction in risk for residents on the periphery of the threat area might not be worth the expense and disruption involved. These evacuation expenses include lost income and temporary living expenses, as well as travel costs (Wu et al. 2012).

The logistics of evacuation are relatively simple when authorities can provide adequate forewarning (e.g., most floods on the lower reaches of major rivers) in areas where households have their own cars and have friends or relatives who live in safe areas nearby. However, significant logistical problems can arise in the course of warning, transporting, and providing accommodations for large numbers of residents, some of whom might be limited in their access to transportation or even in their personal mobility. In some cases, people lacking cars are sufficiently mobile that they can walk to nearby locations where they can board buses that will carry them to safety. In other cases, individuals have such limited mobility that they must be picked up from their homes and transported by vans. Finally, some individuals require intensive care and thus require ambulances for transport. Although many such individuals are in hospitals, nursing homes, or other institutions, some live in homes that are distributed throughout the community and must be identified through home health care organizations. Moreover, many households have pets and some have farm animals that must be evacuated. Emergency managers must take all these conditions into account when planning for and managing evacuations and public shelters.

The challenges of evacuation management increase substantially with the size of the risk area and its population. In this regard, it is noteworthy that many hazardous materials evacuations involve only a few square miles or less from well-defined release points. For example, the risk area for a small daytime release of many toxic gases is about 1/2 mile downwind (USDOT 2012). However, other hazards involve much larger areas. A nighttime spill of a large quantity of methyl isocyanate has a downwind protective action distance of 7 miles or more (USDOT 2012), and a nuclear power plant accident might require an evacuation of the entire 10 mile plume inhalation emergency planning zone (EPZ) (McKenna 2000). A major hurricane might only require evacuation of areas stretching a few miles inland, but uncertainty about the point of landfall could require hundreds of miles of coast to evacuate (e.g., Hurricane Floyd in 1999).

The purpose of this chapter is to guide emergency managers and transportation planners in developing plans for vehicular evacuation from environmental hazards. It does not discuss pedestrian evacuation from buildings, aircraft, or ships because the scale and dynamics of these evacuations is quite different (Aguirre et al. 2011; Bolton 2007; Nelson and Mowrer 2002; Peacock et al. 2011). Moreover, the focus is on preimpact rather than postimpact evacuations because preimpact evacuations face great challenges because of the time constraint of completion before the beginning of hazard exposure (in the case of natural hazards) or the accumulation of a lethal dose (in the case of toxic chemicals or radiological materials). The remainder of this chapter is divided into four major sections. Section 6.2 covers preimpact activities involved in evacuation planning. Section 6.3 covers evacuation analysis—the activities associated with assessing the time required to evacuate a risk area and the points on the evacuation route system (ERS) where bottlenecks are most likely to occur. Section 6.4 addresses evacuation management—the activities involved in actually implementing an evacuation plan. Of course, emergency managers should recognize that evacuation management, like any other emergency response activity, generally requires improvisation and planning (Kreps 1991; Mendonça and Wallace 2004, 2007). Improvisation is usually needed because the incidents that occur do not correspond exactly to the scenarios that guided plan development. Consequently, it is essential that emergency exercises be conducted so that responders learn how to adjust to conditions that they did not expect and that all incidents involving evacuations be thoroughly critiqued to ensure that lessons are learned, plans and procedures are revised, and training is adapted for a better response to future incidents (Lindell and Perry 2007). Section 6.5 presents the chapter's conclusions.

6.2 Evacuation planning

Disaster researchers (Lindell 2011; Lindell et al. 2006) and evacuation planners (Houston 2009) agree that a critical first step in evacuation planning is to establish continuing contacts among all the organizations that will be involved in implementing an evacuation—especially emergency management, transportation, police and fire departments, social services, and the American Red Cross (or other shelter operators). In the case of regional evacuations, this includes the affected organizations in other jurisdictions—counties or states—that will be involved in evacuation, shelter, and reentry. As will be discussed later, an effective planning process will identify a number of issues that need to be addressed in risk, transit, and host jurisdictions (Houston 2009). Critical activities during evacuation operations include interagency communications, provisions for accepting outside support personnel (e.g., common credentialing), multichannel/multilingual public information from joint information centers, media monitoring, support for emergency responder families, en route fuel operations, traffic monitoring, transportation support for transit-dependent populations, mobile response teams, and shelter coordination. Evacuation routes need to be coordinated to ensure that plans in adjacent jurisdictions do not conflict, as happened in the Three Mile Island incident (Chenault et al. 1980). Critical activities during reentry operations include damage assessment and documentation, debris removal, infrastructure restoration, and communication with returnees. Planning activities should include annual audits of evacuation plans to ensure plans and procedures remain current, continuing training to ensure that personnel understand their roles, periodic drills, tabletop exercises, field exercises to test the plan under different emergency scenarios, and critiques to ensure that problems are identified and changes made to correct those problems (Lindell and Perry 2007).

6.3 Evacuation analysis

Evacuation analysis is one component of population protection analysis, which in turn is one element in community preparedness analysis (Lindell and Perry 2007). There are four steps in evacuation analysis: (a) to identify risk areas/sectors, (b) to identify the ERS and assess its capacity, (c) to forecast evacuation demand, and (d) to model traffic behavior. The National Oceanic and Atmospheric Administration (NOAA) Coastal Services Center (csc.noaa.gov/hes/hes.html) provides access to technical guidelines for hurricane evacuation studies (USACE 1995) and examples of completed evacuation studies. Similarly, the U.S. Nuclear Regulatory Commission provides technical guidelines for nuclear power plant evacuation studies (Jones et al. 2011; USNRC/FEMA 1980) at www.nrc.gov/

reading-rm/doc-collections/#nuregs. Examples of evacuation studies for individual nuclear power plants can be found by searching for "evacuation time estimates (ETEs) for" that plant from the search engine at www .nrc.gov/.

6.3.1 Identify risk areas/sectors

A risk area is, as the name implies, the geographic areas at risk from a given hazard. For some hazards, equivalent terms are used such as special flood hazard areas (floods), vulnerable zone (VZ; toxic chemicals) or EPZ (nuclear power plants). For most hydrometeorological and geological hazards, specialized agencies conduct risk analyses and provide risk area maps—for example, hurricanes (www.nhc.noaa.gov/surge/risk/), landslides (landslides.usgs.gov/learning/nationalmap/), and tsunamis (www .conservation.ca.gov/cgs/geologic_hazards/Tsunami/Inundation_Maps/ Pages/Statewide_Maps). However, identification of risk areas for hazardous materials is more difficult because the size of the risk area depends on a number of factors, such as the quantity of material released and its volatility and, toxicity (or radioactivity), along with meteorological conditions such as wind speed, wind direction, and atmospheric stability. The USDOT (2012) provides simplified procedures for identifying risk areas for hazardous material transportation, and the U.S. Environmental Protection Agency (www.epa.gov/oem/tools.htm) provides procedures for fixed-site facilities. There is a standardized 10-mile plume inhalation EPZ for all U.S. commercial nuclear power plants (USNRC/FEMA 1980).

Whether for a natural (e.g., hurricane) or technological (e.g., nuclear power plant) hazard, it is unlikely that all parts of an extremely large risk area will be advised to take the same protective action in an emergency. In a nuclear power plant EPZ, for example, the some areas are likely to be advised to evacuate, others to shelter in-place, and still others to continue normal activities while monitoring the situation. Thus, large risk areas are often divided into smaller areas corresponding to recurrence intervals or event intensities. For example, the Federal Emergency Management Agency (FEMA 1998) primarily categorizes floodplains as A ("base" flood or 100-year floodplain), V (coastal floodplain subject to wave action), B (500-year floodplain or 100-year floodplain protected by a levee), or C (outside the 500-year floodplain). Hurricane risk areas are frequently subdivided according to their exposure to different Saffir–Simpson categories (Category 1–5).

The most effective way to implement such protective actions is by identifying Emergency Response Planning Areas (ERPAs)—often called sectors in hurricane planning—that are areas of relatively homogeneous risk in which residents will all be advised to take the same protective action. Since these ERPAs or sectors should also be readily identifiable by those within them, emergency planners sometimes define these ERPAs in terms of postal codes (see Figure 6.1). Clear definition of ERPAs is important because many people have difficulty in identifying their location on risk area maps (Arlikatti et al. 2006; Zhang et al. 2004). Defining ERPAs by postal codes can work well in urban areas because postal codes in such areas are relatively compact, but are likely to be problematic in rural areas because a single postal code can cover a very large area having heterogeneous levels of risk. In such cases, ERPA boundaries can be defined by well-known political boundaries (city lines), geographical features (rivers), and major roads. The identification of geographical features is especially important because these can serve as evacuation impediments. Mountains, peninsulas, islands, and rivers can all restrict the number, directions, and capacities of evacuation routes. For example, Figure 6.1 indicates there are multiple evacuation routes on Galveston Island and on the mainland portion of Galveston County. However, the bridge from the island to the mainland acts as a bottleneck that restricts evacuation route capacity below the levels on either side. In fact, careful analysis showed that the approach to the bridge limited capacity even more than the bridge itself (Urbanik 1979).

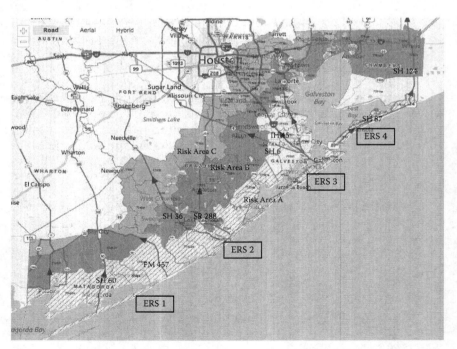

Figure 6.1 Risk areas and the evacuation route system for the Houston–Galveston study area.

6.3.2 Identify the ERS and its capacity

Evacuations typically result in substantially more vehicles attempting to use the ERS and, moreover, using those roads almost exclusively to travel in a single direction away from the hazard. Thus, continuing to use normal traffic control mechanisms such as stop signs and traffic signals—and even toll booths on tollways—can substantially delay vehicles attempting to leave the risk area. Consequently, it is necessary to designate the roads with the greatest capacity as the primary evacuation routes and to devise traffic management strategies that can be used to maximize their capacity. As noted in the section on ERS capacity management below, this typically requires staffing critical intersections with police officers and traffic department personnel to manually control signalized and signed intersections. This allows them to prevent inbound traffic from entering the risk area and to prevent cross traffic from disrupting outbound flows.

The capacity of the ERS can be estimated in three steps. First, the analyst must identify the distinct set of links (individual road segments) that provide routes out of the risk area. For example, ERS 4 in Figure 6.1 is defined by State Highway SH 87 (link 1) that traverses the Bolivar Peninsula and joins SH 124 (link 2) to leave the risk area. Second, the analyst must assess the capacities of the individual links considering each link's number of lanes, shoulder type and width, intersection characteristics (including turn lanes and signal phasing), speed limits, and surrounding land use (Dotson and Jones 2005; TRB 2010). Third, the analyst must examine the configuration of the links in each evacuation route to determine whether they form a serial, parallel, converging, or diverging pattern. Continuing the example of ERS 1, both SH 87 (link 1 running down the center of the Bolivar Peninsula to its base) and SH 124 (link 2 running from SH 87 inland) have capacities of 800 vehicles per hour (vph). Serial links generally are limited by the smallest capacity in the set of links for that evacuation route—as when a link with an 800 vph capacity overloads a link with a 600 vph capacity producing an overall capacity of 600 vph. It is also possible for the reverse to be true, as when a link with a 600 vph capacity underloads a link with an 800 vph capacity, also producing an overall capacity of 600 vph. However, there usually are additional vehicles attempting to access the downstream link—in this case, risk area B (dark shading inland from the coast, risk area A) through which SH 124 passes on the way inland from the coast—so a downstream link with greater capacity than the upstream link that feeds it usually becomes filled anyway.

In general, parallel links have additive capacities—as when two parallel links with capacities of 800 vph have a combined capacity of 1600 vph (see ERS 2 for an example of two parallel evacuation routes, SH 36 and SH 288). Diverging links (a single link splitting into multiple links, see SH 6 diverging from IH 45 on the mainland just across from Galveston Island in ERS 3) generally follow the same rule as serial links because, here too, the capacity of the overall evacuation route is determined by the capacity of the link with the smaller capacity. Converging links (multiple links feeding into a single link, see Route 457 converging into SH 60 in ERS 1) are problematic because traffic cannot merge seamlessly, so the actual flow into the downstream route is likely to be lower than its nominal level.

It is important to recognize that link capacity can be degraded by conditions that are controllable and uncontrollable. In some cases, the controllable traffic impediments can be corrected (e.g., traffic accidents) or temporarily suspended (e.g., road repairs) to return traffic capacity to its normal level. The uncontrollable impediments (e.g., adverse weather) may require the use of compensatory actions such as additional evacuation routes or earlier initiation of evacuation.

6.3.3   Forecast evacuation demand

Evacuation demand is the rate at which vehicles enter the ERS, which is a function of trip generation (the number of evacuating vehicles) and departure timing. As indicated in Table 6.1, forecasting trip generation begins with an assessment of the size and distribution of four broad population segments, which are the (a) vehicle-owning resident population, (b) transit-dependent resident population, (c) transient population, and (d) institutionalized population (the latter includes those with physical and mental disabilities as well as school children and prisoners). These population segments can be expected to differ in their distributions of time to warning receipt and time for evacuation preparation.

6.3.3.1   Vehicle-owning resident population

The basic unit of the vehicle-owning resident population is the household. The term *household* is important here because it is well documented that households tend to evacuate as a unit and will usually seek to reunite if separated before evacuating (Tierney et al. 2001). The speed at which such households can evacuate thus depends on whether members are united at the time of warning receipt, which is typical on evenings and weekends and, if not, how soon they can reunite. Moreover, depending on the immediacy of the threat (usually minutes of forewarning for transportation accidents or terrorist attacks but days for hurricanes) and the potential for property destruction (low for toxic chemicals and high for hurricanes), people might take time to protect property (e.g., install storm shutters) or to pack valuables before evacuating. Thus, a household's departure time is determined by adding its evacuation preparation time to its time of warning receipt. In many cases, there are multiple vehicles

Table 6.1 Evacuation Model Parameters and Data Sources

Parameter	Data source
Background data collection	
1. Risk area and emergency response planning area definition	Hazard analysis
2. Evacuation route system definition	Transportation department
Trip generation	
3. Number and distribution of vehicle-owning resident households	National census data
4. Number of evacuating vehicles per residential household	Behavioral research
5. Number of evacuating trailers per residential household	Behavioral research
6. Number of persons per residential household	National census data
7. Number and distribution of the transit-dependent resident households	National census data
8. Number and distribution of transient households	Local visitors' bureau
9. Number of evacuating vehicles per transient household	Behavioral research
10. Size and distribution of the institutionalized population	Local records
11. Size and distribution of the noninstitutionalized population with disabilities	Local records/ national census data
12. Percentage of residents' PAR compliance/spontaneous evacuation	Behavioral research
13. Percentage of transients' PAR compliance/spontaneous evacuation	Behavioral research
Departure timing	
14. Percentage of early evacuating residential households	Behavioral research
15. Percentage of early evacuating transient households	Behavioral research
16. Percentage of early evacuating institutions	Local agreements
17. Residential households' departure time distribution	Behavioral research
18. Transients' departure time distribution	Behavioral research
19. Institutions' departure time distribution	Local agreements
20. Noninstitutionalized population's departure time distribution	Local agreements
Destination/route choice	
21. Evacuees' ultimate destinations	Behavioral research
22. Evacuees' proximate destinations/route choices	Behavioral research
23. Evacuees' utilization of the evacuation route system	Behavioral research

Source: Adapted from Lindell, M.K. and C.S. Prater, *J. Urban Plan. Dev.*, 133, 18–29, 2007 and Lindell, M.K., *Transportation Res. A*, 42, 140–154, 2008.

and trailers per household. The number of vehicles and trailers is important because this variable, when divided by the ERS capacity in vehicles
per hour determines the number of hours to complete an evacuation. The
number of persons per household is important because this variable is
directly related to the required capacity of public shelters.

6.3.3.2 Transit-dependent population

The number of households without vehicles can be estimated by using the
data available from the National Household Travel Survey (NHTS) (nhts.
ornl .gov/tools.shtml) or other sources that identify the number of households that do not own a vehicle. Although the percentage of households in a
geographical area (derived from the NHTS) can be multiplied by the number
of households in that area to estimate the number of households that will
need public transit to evacuate, such estimates should be treated with caution. On the one hand, some households have vehicles that are in poor repair
and cannot provide reliable transportation over the long distances that are
required for hurricane evacuations. On the other hand, some households
that do not own vehicles are likely to obtain rides with other households that
do. These two effects tend to cancel each other, but when we do not know the
relative magnitude of the two effects, it is not possible to know whether they
exactly cancel each other or if there is a net bias in one direction.

A common procedure for evacuating transit-dependent populations
is to advise them to walk to the nearest elementary school where they can
obtain rides on buses. This procedure produces substantial evacuation
preparation times for this population segment because, unlike vehicle-
owning households that pack their car(s) after packing their bags, transit-
dependent households must walk to the nearest school, wait for their bus
to arrive, and (generally) wait on the bus as it travels to the next pickup
location before entering the ERS. Moreover, the time to evacuate transit-
dependent households increases substantially if there is insufficient bus
capacity to evacuate everyone in one trip. When buses must make multiple trips, the distribution of evacuation times for transit-dependent
households increases directly with the distance to the discharge point at a
reception center or public shelter.

6.3.3.3 Transients

Transients can be individuals visiting the area on business or households
vacationing in the area. Transient populations can vary substantially by
time of year, day of week (especially weekday vs. weekend), time of day,
and also because of special events such as festivals and athletic events.
Consequently, evacuation planners must be aware of this variation and
incorporate it into their plans. Aside from their effect on the size of the
evacuating population, transients are potentially problematic because
they are likely to differ from local residents in having a lower likelihood of

receiving informal warnings from peers, which is an important source of warnings in emergencies (Lindell et al. 2006). The lower level of informal warnings may delay their departure times. However, they have no need to protect property and need less time to pack, which tends to decrease their departure times. In most cases, transients have their own vehicles but some arrive by public transportation, especially commercial airlines, and might have difficulty in rescheduling their departures.

6.3.3.4   Special needs population

For purposes of evacuation, the special needs population comprises people with a variety of functional limitations in communication, medical health, functional independence, supervision, and transportation (FEMA 2010a, b; Kailes and Enders 2007). Some members of the special needs population are school children who must be evacuated directly to a safe location or who must be returned to their homes where they will evacuate with the other members of their households. Whichever option school officials adopt must be communicated to parents during the planning phase before any evacuation. Another important category of the special needs population is prisoners who must be evacuated under close supervision to another correctional institution—preferably before the rest of the risk area population begins to evacuate. The third major category in the special needs population comprises those who are institutionalized in hospitals, nursing homes, group homes, or other types of facilities. Like school children and prisoners, the size and distribution of institutionalized patients is relatively easy to identify. However, there is also approximately 16.7% of the noninstitutionalized U.S. population comprising people with disabilities that substantially limit one or more of activities of daily living (Brault 2008). Thus, there is likely to be a significant fraction of the population in any risk area that will need some type of assistance when evacuating but whose location and specific needs are difficult to identify in advance. Lack of information about the number, location, and types of disabilities presents problems for determining what types of vehicles will be needed to evacuate these people and to what types of facilities (and thus, the evacuation routes and destinations) they should be sent.

6.3.3.5   Time variation in ERPA population size

Evacuation analysts must recognize that the size of each population segment within each ERPA varies over time. There is variation by time of day as people travel to work, school, and other daytime activities and return home at night. This diurnal variation is most pronounced during the work week, but there can also be substantial variation by time of year in some communities, as tourists and people with second homes arrive in vacation communities. Finally, there can be substantial increases in population due to major festivals and sports events.

6.3.3.6   Evacuation compliance and shadow evacuation

As noted earlier, it is well established that there will be incomplete compliance with evacuation recommendations and that the level of compliance declines with people's perceptions of threat, which is a function of authorities' labeling of the evacuation (recommended vs. mandatory) and people's perception of their distance from that threat (see Lindell and Prater 2007). However, it is also well established that there will be *shadow evacuation*—evacuation from areas that authorities do not consider to be at risk (Zeigler et al. 1981). Both incomplete compliance and shadow evacuation have been observed in a wide variety of hazards, both technological and natural. For example, the nuclear power plant accident at Three Mile Island produced about 150,000 evacuees in response to an evacuation advisory that applied to 10,000 people at most (Lindell and Perry 1983). Furthermore, there was an estimated 47% evacuation rate from Harris County, Texas, for Hurricane Rita (Lindell and Prater 2008; Stein et al. 2010) even though less than 10% of the population was in the officially designated hurricane risk area. This shadow evacuation generated hundreds of thousands of additional evacuees.

6.3.3.7   ERS loading function

Vehicles enter (load) the ERS at a rate that is determined by the rate at which people are warned and the rate at which they prepare to leave. As Figure 6.2 indicates, the rate of warning reception and preparation times can each be represented as an exponential function, $p_t = 1 - \exp(-at^b)$, where p_t is the

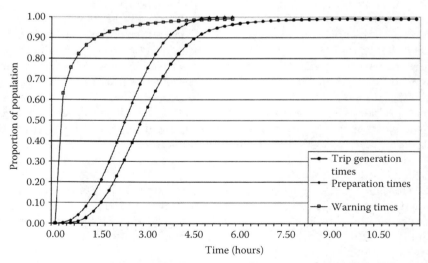

Figure 6.2 Generic trip generation time distribution function. (Reprinted from *Transportation Res. A*, 42, Lindell, M.K., EMBLEM2: An empirically based large-scale evacuation time estimate model, 140–154, 2008, with permission from Elsevier.)

proportion of the population prepared at time t and a and b are parameters that determine the shape of the curve. Lindell et al. (2002) used parameters $a = 2.00$ and $b = 0.50$ for a representative warning time distribution and $a = 0.085$ and $b = 2.55$ for a representative scenario preparation time distribution. These distributions would be appropriate for a rapid onset scenario such as an imminent flood, hazardous materials release, or nuclear power plant accident. However, they are inappropriate for hurricane evacuations in major metropolitan areas where evacuations take place over two or more days. In such cases, the departure time distributions *for each day* tend to take the form described in Figure 6.2 (Lindell et al. 2005).

When combined, these two distributions yield a loading function that describes the rate at which households enter the ERS (see the right-hand curve in Figure 6.2). It is important to note that the shape of the *warning distribution* depends on the nature of the warning system (Rogers and Sorensen 1988; Sorensen and Rogers 1989). The larger the percentage of the population that can be warned *simultaneously by broadcast systems* (e.g., siren and emergency alert system), the sooner the warning distribution will reach asymptote (the point at which everyone who can receive a warning does so). Conversely, the larger the percentage of the population that must be warned *sequentially by diffusion systems* (e.g., emergency responders going door-to-door), the later the warning distribution will reach asymptote. By contrast, the shape of the *preparation distribution* depends on warning recipients' perceptions of the urgency of the situation and their abilities to respond promptly. For example, people will begin their evacuations sooner for a chemical release in progress than for a hurricane that is 2 days away. In addition, able-bodied people whose families are united and who have emergency kits already packed will begin their evacuations sooner than those who lack these characteristics.

The estimation of the ERS loading function is even more complex for situations in which there is no forewarning before hazard onset, often called "no-notice" evacuations. In such cases, families are likely to be separated with, for example, two spouses working at different locations and multiple children at different schools. Families' attempts to reunify constitute an additional component of preparation time (before beginning their trip out of the risk area) that can produce inbound traffic into the risk area and crossflows through the risk area. This delays households' own departures and disrupts outbound flows of traffic from those seeking to leave directly after receiving a warning (Murray-Tuite 2007; Murray-Tuite and Mahmassani 2003, 2004).

6.3.4   Model traffic behavior

In a very simple case, the time required to evacuate a risk area could be estimated by dividing the number of evacuating vehicles by the roadway capacity (in vehicles per hour). Unfortunately, evacuations frequently produce demand that exceeds the available roadway capacity, thus producing queues of vehicles waiting for access to the roadway. The need to incorporate queuing time into the calculations means that modeling traffic behavior for any area other than a small neighborhood with an extensive road network requires the use of computer programs. Only these computer programs can adequately model the behavior of drivers as they leave their points of origin, drive to one of the principal evacuation routes, queue for access to the evacuation route, travel along that route to a proximal destination at the edge of the risk area, and travel from there to the household's ultimate destination where they will seek shelter. There are many traffic behavior models available (Hardy and Wunderlich 2007; Pel et al. 2012). These models require analysts to input the data from the previous stages of the evacuation analysis discussed in this section: identifying risk areas/sectors; identifying the ERS and its capacity; forecasting evacuation demand; and modeling traffic behavior. These traffic modeling programs can generate data on total traffic volume; average vehicle travel time; risk area clearance time; hourly volume, total volume, and average speed at each exit from the risk area; and queue lengths at selected intersections (Jones et al. 2011). It is important to conduct sensitivity analyses using a range of plausible values for each of the major input parameters to see how model output data, especially ETEs, are affected (Lindell 2008).

6.4   Evacuation management

The principal tasks involved in the management of an actual evacuation are threat detection and emergency classification, protective action selection, population warning, evacuation transportation support, evacuation traffic management, reception and care (shelter accommodations), and reentry management (Lindell et al. 2006; Perry and Lindell 2007). There are other population protection functions that must be performed in an emergency response but are not directly related to evacuation management. These include search and rescue (finding and transporting those who are lost or injured), impact zone access control and security (keeping people out of evacuated areas and protecting the property there), hazard exposure control (protecting emergency responders from danger), emergency medical care (providing definitive care to the sick and injured), and environmental surety (monitoring and remediating environmental contamination by hazardous materials). These tasks can be most effective if evacuation managers use the Incident Command System (ICS) to guide their response operations. However, ICS will be most effective when nonemergency mission agencies such as social services and American Red Cross have also been trained and participated in exercises using ICS (Lutz and Lindell 2008).

6.4.1   Incident detection and classification

Sometimes, as was the case in the 1978 Sumner, Washington flood, the risk area population needs no assistance in detecting and classifying environmental conditions as immediately dangerous (Perry et al. 1981). In other cases, such as hurricanes and nuclear power plant accidents, specialized agencies are needed to detect a hazard, forecast the areas it is likely to affect, and assess the severity of the threat to people and property, thus indicating the need for sheltering in-place or evacuation. The National Weather Service uses two emergency classes to define the certainty, severity, immediacy, and intensity of the threat at designated locations. A hurricane watch means hurricane conditions are possible in a defined area within the next 48 hours, and a hurricane warning means hurricane conditions are expected in a defined area within the next 36 hours. The U.S. Nuclear Regulatory Commission has established a set of four emergency classes, together with the specific plant and environmental conditions (emergency action levels) defining each class, and specific response actions expected for each class (USNRC/FEMA 1980). The problem of incident detection and classification is much more complex for chemical facilities because there are so many different types and sizes of facilities that produce, store, and dispose of chemicals. There appears to be no available data on chemical facilities, ability to promptly detect and accurately classify incidents requiring protective actions such as evacuation, but Sorensen and Rogers (1988) found that few communities even had adequate mechanisms for receiving notifications about incidents from chemical facilities within their jurisdiction.

6.4.2   Protective action selection

The goal of evacuation is to clear the risk area before the onset of hazard conditions. It is important to recognize that this deadline can vary from one hazard to another because some hazards are lethal as soon as exposure occurs (e.g., being overtaken by a tsunami wave), whereas other hazards may require an extended period of exposure before life-threatening doses occur (e.g., toxic chemicals or radiological materials). It is also important to realize that evacuation is unnecessary, and might be dangerous, for those who can achieve protection by sheltering in-place (Cova et al. 2009; Jetter and Whitfield 2005; Lindell and Perry 1992; Sorensen et al. 2004). Sheltering in-place within buildings is effective when these structures are elevated above the flood crest, are strong enough to withstand hurricane wind, can withstand wildfires, are dense enough to provide shielding from external radiation, or have sufficiently low air exchange rates that they delay the infiltration of contaminated air. If a hazard strikes without forewarning (e.g., railroad tankcar derailment), staying indoors with all ventilation shut off (e.g., doors, windows, heaters, and air conditioners) can often achieve lower exposures than would result from either pedestrian or vehicular evacuation through the airborne plume of chemical or radiological contamination.

All these different types of cases require an assessment of the relative risk of sheltering in-place compared to evacuating. For patients in intensive care wards in hospital, the risks of evacuation might be so high that sheltering in-place is the only feasible option. Advance analysis and planning, and perhaps facility retrofit, are needed to ensure that these patients and their caregivers remain safe while others evacuate.

An important issue that should be considered in the selection of a protective action is the government's authority to compel people to comply. In the case of hurricane evacuations, almost all Gulf and Atlantic coast states grant their governors the authority to order evacuations and most also grant that authority to one or more local officials, usually the highest elected local official at the county or city level (Wolshon et al. 2005a). Legal authority notwithstanding, no jurisdiction has enough police officers to arrest and jail those who refuse to evacuate. However, identifying an evacuation as "mandatory" rather than "recommended" significantly increases compliance (Baker 1991, 2000).

6.4.3 Warning

The findings of studies on individual response to environmental hazards and disasters can be depicted by the Protective Action Decision Model (Lindell and Perry 2004, 2012) (see Figure 6.3). The process of protective action decision making begins with environmental cues (sights such as

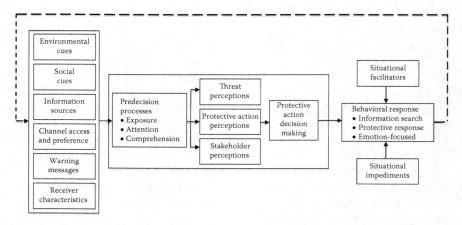

Figure 6.3 Information flow in the PADM. (From Lindell, M.K. and R.W. Perry, *Risk Anal.*, 32, 616–632, 2012.)

funnel clouds, sounds such as the roar of tornado wind, and smells such as chemical odors) and social cues such as observations of other households evacuating. However, protective response is usually initiated by warnings of environmental threats received from authorities and peers. These sources transmit warning messages via one or more channels to receiving individuals in the risk area. Warnings result in effects that depend on receivers' characteristics. The relevant effects are changes in receivers' beliefs and behaviors, whereas receivers' characteristics include their physical (e.g., strength), psychomotor (e.g., vision and hearing), and cognitive (e.g., primary and secondary languages as well as their mental models/schemas) abilities and their economic (money and vehicles) and social (friends, relatives, neighbors, and coworkers) resources.

Environmental cues, social cues, and socially transmitted warnings initiate a series of predecisional processes that, in turn, elicit core perceptions of the environmental threat, alternative protective actions, and relevant stakeholders. These perceptions provide the basis for protective action decision making, the outcome of which combines with situational facilitators and impediments to produce a behavioral response. In general, the response can be characterized as information search, protective response (problem-focused coping), or emotion-focused coping. In many cases, there is a feedback loop as additional environmental or social cues are observed or warnings are received. The dominant tendency is for such information to prompt protective response, but information seeking occurs when there is uncertainty at a given stage in the protective action decision-making process. Once the uncertainty is resolved, processing proceeds to the next stage in the process.

Warnings are crucial elements in initiating evacuation but emergency managers need to understand the best ways to construct and transmit them. First, a warning message should describe the threat in terms of its expected impact intensity, timing, and location, provide a recommended protective action, and indicate sources to contact for further information about the situation and to obtain assistance in taking protective action. Warnings should be transmitted repeatedly on each channel and disseminated via multiple channels so that people receive, attend to, and understand the warning message.

There are a number of different ways for disseminating warnings to those at risk (Lindell and Perry 1992). These are face-to-face warnings, mobile loudspeakers, sirens, commercial radio and television, NOAA weather radio, telephones, e-mail, Internet websites, social media such as Twitter, and newspapers. Each of these mechanisms can be evaluated with respect to eight characteristics. These are precision of dissemination (the ability to target only those at risk), penetration of normal activities (especially the ability to overcome ambient noise), specificity of the message (the ability to convey all the elements of an ideal warning message), susceptibility to message distortion (a problem that arises when a warning message must be passed through one or more intermediate communicators), rate of dissemination over time (the time between receipt by the first and last warning receivers), receiver requirements (especially expensive equipment or specialized knowledge and skill), sender requirements (also expensive equipment or specialized knowledge and skill), and feedback (sender verification that the receiver has received the warning). Communities need to choose which warning channels they will use based on the characteristics of the hazard (especially the amount of forewarning, the size of the impact area, and the likely impact intensity) and the community's financial resources.

6.4.4   Evacuation transportation support

As noted earlier, some segments of the risk area population lack vehicles or have other obstacles to evacuation. For the transit dependent, local authorities need to make arrangements with school districts or mass transit providers such as transit districts. Emergency managers need to verify that transit providers have not made arrangements with other jurisdictions that would need buses at the same time. Such duplicate demands could arise if the providers signed agreements with officials in two different counties on the assumption that these agreements would be activated only for very local evacuations, such as a hazardous materials incident or flooding that would not require simultaneous evacuations in multiple jurisdictions (Daines 1991). Similar caution must be taken to review concurrent service agreements for multiple hospitals and nursing homes, which have the added requirement of ensuring that there are available beds in receiving facilities before transporting patients (Auf der Heide and Scanlon 2007). Moreover, care must be taken to ensure that the number of buses, ambulances, and other vehicles is sufficient to accommodate the maximum capacity of these facilities. Criteria for transporting pets need to be made and publicized via the news media before and during an incident (Heath et al. 2001).

6.4.5   Evacuation traffic management

There are two fundamental strategies for evacuation traffic management: ERS capacity management and population demand management. ERS capacity management seeks to minimize or avoid queues that form when too many vehicles compete for space on the roads or to increase capacity through procedures such as contraflow. Population demand management seeks to regulate the number of vehicles attempting to use the ERS at any given point in time so that maximum capacity is maintained for a given lane use configuration.

6.4.5.1   ERS capacity management

The most common evacuation impediments mentioned in a recent survey of transportation officials in the nation's 26 largest metropolitan areas were ERS capacity and ERS condition (Vásconez and Kehrli 2010). Accordingly, maintaining, and if possible increasing, ERS capacity is a major issue in managing an evacuation. The primary methods for maintaining ERS capacity are intersection control and prompt clearance of traffic impediments such as vehicles blocking traffic lanes because of overheating, running out of fuel, and collisions (Ballard and Borchardt 2006; Jones et al. 2011; Wolshon et al. 2005b). Some traffic impediments can be avoided by, for example, stationing fuel tankers along evacuation routes to avoid allowing evacuating vehicles to run out of fuel. Some collisions can be cleared by other evacuating vehicles that push the incapacitated vehicle off the roadway, but others require authorities to station tow trucks along the evacuation routes. Since tow truck access to incident locations will be impaired if traffic is at a standstill and all lanes and shoulders are occupied, evacuation managers must designate specific lanes or alternate routes for emergency access.

The primary methods for increasing ERS capacity are utilization of highway shoulders as traffic lanes and implementation of contraflow, reversing the direction of inbound lanes or shoulders (Jones et al. 2011; Wolshon et al. 2005b). Utilizing shoulders as traffic lanes has some distinct limitations such as precluding their use for emergency vehicle access to accident sites (Ballard and Borchardt 2006). Moreover, they also increase capacity only by a small amount (estimated to be 8%) (Wolshon et al. 2005b). However, the cumulative effect of even this small capacity increase could be significant if implemented over multiple evacuation routes. For example, using shoulders on three evacuation routes could increase capacity by 24%. This increase in capacity could significantly decrease the number of hours required to evacuate areas with large populations such as the 34 hours required in some counties on the Texas coast (Lindell et al. 2002).

Planning for contraflow requires careful thought because a full contraflow configuration provides more outbound capacity, but partial contraflow configurations (reversing only one of two inbound lanes) provide an opportunity to maintain one or more lanes in their normal inbound configuration. Although such partial contraflow configurations provide one or more lanes for normal inbound traffic, returning buses, or emergency vehicles, a partial contraflow configuration obviously increases outbound capacity by a smaller amount (about 30%) than a complete contraflow configuration (about 70%). In addition, single lane contraflow configurations increase the potential for head-on collisions, and they must be examined carefully to ensure that the shoulder does not have a critically reduced width at a bridge or underpass. Regardless of which contraflow configuration is selected, evacuation managers need to correctly anticipate the length of time required to implement it. Estimates of the time required to staff traffic control points, install traffic control devices, and clear inbound lanes of inbound traffic range from 4 to 12 hours, depending on the length and complexity of the contraflow segment (Wolshon et al. 2005b).

6.4.5.2   Population demand management

The primary methods of demand management are to control the total number of people seeking to evacuate, control the allocation of vehicles to evacuation routes, and control the loading of the ERS. One important method of demand management is to avoid shadow evacuation from outside the officially designated evacuation zones. As noted earlier, shadow evacuation has added substantially to the number of vehicles in some past evacuations. In addition, the need for evacuation from inside the risk area can be reduced if local structures are adequate for sheltering in-place (e.g., Sorensen et al. 2004). As noted earlier, hazardous materials releases require that structures have low air infiltration rates, whereas hurricanes require that they are adequately strengthened against the wind and elevated above storm surge and inland flood levels. Moreover, the occupants must believe that these structures provide effective protection before they will comply with a shelter recommendation.

Another demand management strategy is to reroute evacuation traffic from overutilized routes to underutilized ones. This could be an effective strategy because people tend to use the most familiar routes to their evacuation destinations (Lindell and Prater 2007). Unfortunately, such habits tend to overload the freeways and underload minor highways (Dow and Cutter 2002). The challenge in redirecting traffic is for authorities to identify the alternative evacuation routes from specific originating points to specific destinations and to communicate in real time what average speeds are on these routes. This strategy would be extremely difficult to implement at present but is likely to be increasingly feasible with the development of intelligent transportation systems that are capable of monitoring and directing traffic.

In principle, evacuation timing can be controlled by phased (also called "staged") evacuations in which people in areas of lower risk are asked to delay their evacuations until after people in areas of higher risk have evacuated (Jones et al. 2007, 2011). Phased evacuations have been used many times in hurricanes and wildfires (Jones et al. 2008). However, conclusions about their desirability appear to be based only on indirect anecdotal evidence of success (i.e., an absence of spectacular failures) rather than collection and analysis of survey data. To implement phased evacuation effectively, evacuation managers would need the same capability for monitoring the traffic on evacuation routes as with traffic rerouting, but would also need high levels of evacuees' compliance with a phased evacuation recommendation and an ability to forecast accurately how soon people would begin to evacuate after a delayed evacuation advisory was issued. Unfortunately, there is little available evidence that people will comply with a request to delay their evacuation and no research on the amount of time people are willing to delay their evacuations while others are leaving. To the contrary, the available behavioral research indicates that people believe that sheltering provides less effective protection than evacuation for hazardous materials releases (Lindell and Perry 1992). Experts are also highly uncertain about people's willingness to comply with a shelter recommendation in a radiological release (Dombroski et al. 2006), and observation of others evacuating is a cue that prompts people to leave (Baker 1991). Notwithstanding these uncertainties, there is a natural tendency for those farther inland to begin their hurricane evacuations later than those near the coast (Lindell et al. 2011; Wu et al. 2012). This has the potential to create traffic jams if inland residents try to enter the ERS as the same time coastal residents are arriving there. Fortunately, this convergence of coastal and inland evacuees has the tendency to produce evacuation queues in inland areas where the risk of being overtaken by the hurricane surge is relatively low.

6.4.5.3   Evacuation safety

Lindell and Perry's (1992) review of the available evidence on mass evacuation during emergencies indicated that the accident rate for private vehicles is almost certain to be no higher and is quite likely to be lower than during normal driving periods. Traffic tends to be slow, moving at about 35 mph. Since traffic flows in a mass evacuation would be predominately one way (outbound), any accidents that would occur would be less serious and, therefore, less likely to produce deaths or injuries or to severely obstruct traffic flow.

There are insufficient data to determine whether this finding can be generalized to adverse road conditions. The finding by Bastien et al. (1985) that no traffic accidents were reported during an evacuation through a volcanic ashfall suggest that a moderate degradation of road conditions would not materially alter accident rates. One would have to assume that under some circumstances, road conditions could be so dangerous that the traffic accident risks of the evacuation might become an important consideration. Indeed, the risk of accidents owing to slick road surfaces (e.g., sleet, snow, rain), reduced visibility (e.g., fog, dust, ash), high winds, or fire (urban, forest, brush, grass) could exceed the risks from such hazard agents such as an imminent tornado, hurricane landfall, or hazardous materials release.

The available evidence on hospital and nursing home evacuation indicates that even mass evacuations of special susceptibility populations can be accomplished in short time periods without appreciable safety risks. There are, of course, important logistical barriers that must be overcome to assure that an evacuation is carried out in a timely and effective manner (Chenault et al. 1980). Historical evidence indicates that sound traffic management (for the evacuation of the general public by automobile) and a normal level of medical precaution (for hospital evacuations) would be sufficient to assure the safety of the evacuating population. The major planning element that emerges from these studies that may not be obvious to a trained professional would be the recommendation by Burton et al. (1981) to form a central registry of evacuated physicians, which lists telephone numbers at which they can be reached by their patients. Local radio and television broadcasts could be used to publicize the availability of the registry.

6.4.5.4   Real-time evacuation traffic management

Large-scale evacuations can generate traffic that crosses municipal, county, and state boundaries. Consequently, emergency management, transportation operations, and law enforcement can find themselves facing traffic flows that are unexpectedly large or even completely unexpected. This lack of coordination became a major problem during the 1999 response to Hurricane Floyd that generated 3.5 million evacuees in Florida, Georgia, South Carolina, and North Carolina. After the event, FEMA collaborated with the Federal Highway Administration to develop the *Evacuation Traffic Information System* (ETIS). ETIS is a computer system that supports the federal Evacuation Liaison Team in facilitating coordination among states affected by a major (Category 3 or higher) hurricane. ETIS provides information about contraflow implementation, traffic impediments (road closures and traffic accidents), traffic congestion, and shelter locations.

Some metropolitan areas have also adopted a number of features of intelligent transportation systems. For example, Houston TRANSTAR is a regional facility that is jointly operated by the Texas Department of Transportation, Harris County, City of Houston, and the Harris County Metropolitan Transit Authority for managing both normal and evacuation traffic. During normal operations, TRANSTAR receives traffic information from closed circuit television cameras and embedded highway traffic sensors so it can operate dynamic message signs and a highway advisory radio to communicate with motorists, control traffic signals, and dispatch emergency vehicles to remove traffic impediments. During emergency operations, TRANSTAR staff activates the Emergency Operations Center to accommodate representatives from other regional, state, and nongovernmental agencies. TRANSTAR supports a website (www.houstontranstar .org/) that provides motorists with hurricane information, including risk area maps, evacuation route maps, and brochures explaining the contraflow procedures for the major evacuation routes.

6.4.6 Shelter accommodations

When evacuations last overnight or longer, people need places to stay until it is safe to return. The types of shelter accommodations people choose is the best-studied aspect of evacuation logistics, with researchers finding consistently that only a small percentage of evacuees use public shelters. Mileti et al. (1992) reviewed evacuations and concluded that only 15% of evacuees go to public shelters, which are primarily occupied by ethnic minorities, the poor, and later departures. More recently, Baker (2000) reported that 15% of evacuees used public shelters during Hurricane Floyd. Whitehead (2003) found that only 6% of evacuees stayed in public shelters during Hurricane Bonnie, whereas 16% stayed in hotels/motels and 70% stayed with friends/relatives. Lindell et al. (2011) reported that only 3% of evacuees went to public shelters during Hurricane Lili, whereas 29% stayed in hotels/motels and 54% stayed with friends/relatives.

6.4.7 Reentry management

Managing evacuees' reentry after an evacuation can be a major challenge for emergency managers. Achieving high levels of compliance with return plans has been difficult following past hurricanes (Siebeneck and Cova 2008), so there is a need to better understand the factors that influence the timing of evacuees' reentry attempts. There have been several calls from within the research community for studies to be conducted on evacuation reentry (Sorenson et al. 1987; Stallings 1991). Disaster researchers' neglect of the reentry process is an important omission for practical and theoretical reasons. Specifically, evacuees who return too early could frustrate local officials' efforts to maintain security in the evacuation zone. Alternatively, evacuees might find themselves parked at traffic control points for hours or days waiting for police to let them into the evacuation zone (Dash and Morrow 2000). Moreover, those who enter the evacuation zone before damage assessment teams complete their work might become exposed to safety hazards from the collapse of unstable buildings and health hazards from moving back into homes that lack essential services such as water, sewer, electric power, and gas.

Siebeneck et al. (2012) found that Hurricane Ike evacuees substantially overestimated the extent to which there would be looting in their neighborhoods while they were away from home. They also overestimated the traffic problems they would encounter when returning but had relatively accurate expectations about lost income (which was expected to be and actually was relatively low) and loss of utilities such as phone, water, gas, and electricity (which were expected to be and actually were relatively high). Evacuees who expected greater problems being stuck in traffic were more likely than those who expressed less concern about this problem to comply with reentry plans. However, returnees expecting greater physical risk from damaged structures were less likely than returnees who expressed less concern with these issues to comply with reentry orders. There were no statistically significant relationships of looting expectations, income loss, and utility loss that were associated with reentry plan compliance. In addition, there were no statistically significant relationships between sociodemographic characteristics and likelihood of complying with reentry orders.

Because many evacuees rely on news media as a source of information about their communities while evacuated (Siebeneck and Cova 2008), it is important to gain a better understanding of what information, or misinformation, the news media are communicating and how this, in turn, shapes the expectations and behavior of the evacuated population. As yet, there is little information about the role social media play in evacuees' reentry timing, which is unfortunate because e-mail, Internet websites, and social media such as Twitter have the potential for disseminating information about conditions in the evacuation zone as well as official reentry plans. Recent research has found that social media can provide different types of information about incidents in progress (Sutton et al. 2008) by allowing users to readily access emergency information and forward (retweet) it to others (Vieweg et al. 2010). However, these channels have been found to vary in their use by different user types (private, government, and media), ages, and ethnic groups (Vultee and Vultee 2011; Wolshon et al. 2005b), and some evidence indicates that the overall use of these communication channels in emergencies remains rather limited (Lindell et al. 2011).

6.5   Conclusions

Evacuations can be managed effectively, but the larger the scale of the evacuation and the shorter the amount of forewarning, the greater is the need to plan, analyze, manage effectively. Effective planning requires coordination among diverse agencies at different levels of government (municipal, county, state, and federal), as well as with nongovernmental organizations such as the Red Cross. Effective analysis requires identifying risk areas/sectors, identifying the ERS and its capacity, forecasting evacuation demand, and modeling traffic behavior. Effective evacuation management requires incident detection and classification, protective action selection, warning, evacuation transportation support, evacuation traffic management, shelter accommodations, and reentry management. Evacuation managers have the best chance of protecting the risk area population from environmental hazards if they address all these issues successfully. In addition, they should work with local emergency managers to increase households' and businesses' levels of emergency preparedness (Lindell 2011, 2012; Lindell and Perry 2004).

Acknowledgments

This research was supported by the National Science Foundation under Grants SES-0527699, SES-0838654, CMMI-0826401, CMMI-1138612, and CMMI-1129998. None of the conclusions expressed here necessarily reflect views other than those of the authors.

References

Aguirre, B.E., El-Tawil, S., Best, E., Gill, K.B. and Fedorov, V. 2011. Contributions of social science to agent-based models of building evacuation. *Contemporary Social Science* 6:415–432.

Arlikatti, S., Lindell, M.K., Prater, C.S. and Zhang, Y. 2006. Risk area accuracy and hurricane evacuation expectations of coastal residents. *Environment and Behavior* 38:226–247.

Auf der Heide, E. and Scanlon, J. 2007. The role of the health sector in planning and response. In *Emergency management: Principles and practice for local government*, 2nd ed., eds. K.J. Tierney and W.F. Waugh, Jr., p. 183–206. Washington DC: International City/County Management Association.

Baker, E.J. 1991. Hurricane evacuation behavior. *International Journal of Mass Emergencies and Disasters* 9:287–310.

Baker, E.J. 2000. Hurricane evacuation in the United States. In *Storms*, Vol. 1, Ch. 16, eds. R. Pielke, Jr. and R. Pielke, Sr., p. 308–319. London: Routledge.

Ballard, A.J. and Borchardt, D.W. 2006. *Recommended practices for hurricane evacuation traffic operations*, FHWA/TX-06/0-4962-P2. College Station Texas: Texas Transportation Institute. Downloaded from tti.tamu.edu.documents/0-4962.P2.psf.

Bastien, M.C., Dumas, M., Laporte, J. and Parmentier, N. 1985. Evacuation risks: A tentative approach for quantification. *Risk Analysis* 5:53–61.

Bolton, P.A. 2007. *Managing pedestrians during evacuation of metropolitan areas*, FHWA-HOP-07-066. Washington, DC: Federal Highway Administration. Downloaded from www.its.dot.gov/its_publicsafety/evacuation.htm.

Brault, M. 2008. *Disability status and the characteristics of people in group quarters: A brief analysis of disability prevalence among the civilian noninstitutionalized and total populations in the American Community Survey*. Washington, DC: U.S. Census Bureau. Downloaded from www.census.gov/hhes/www/disability/GQdisability.pdf.

Burton, I., Kliman, M., Powell, D., Schmidt, L., Timmerman, P., Victor, P., Whyte, A. and Wojick, J. 1981. *The Mississauga evacuation: Final report*. Toronto ON: University of Toronto Institute for Environmental Studies.

Chenault, W.W., Hilbert, G.D. and Reichlin, S.D. 1980. *Evacuation planning in the TMI accident*. Human Sciences Research, Inc. Mclean, VA. Downloaded from oai.dtic.mil/oai/oai?verb=getRecord&metadataPrefix=html&identifier=ADA080104.

Cova, T.J., Drews, F.A., Siebeneck, L.K. and Musters, A. 2009. Protective actions in wildfires: Evacuate or shelter-in-place? *Natural Hazards Review* 10:151–162.

Daines, G.E. 1991. Planning, training, and exercising. In *Emergency management: Principles and practice for local government*, eds. T.E. Drabek and G.J. Hoetmer, p. 161–200. Washington, DC: International City Management Association.

Dash, N. and Morrow, B.H. 2000. Return delays and evacuation order compliance: The case of Hurricane Georges and the Florida Keys. *Environmental Hazards* 2:119–128.

Dombroski, M., Fischhoff, B. and Fischbeck, P. 2006. Predicting emergency evacuation and sheltering behavior: A structured analytical approach. *Risk Analysis* 26:1675–1688.

Dotson, L.J. and Jones, J. 2005. *Identification and analysis of factors affecting emergency evacuations: Main report.* NUREG/CR-6864, Vol.1, SAND2004-5901. Washington, DC: US Nuclear Regulatory Commission.

Dow, K. and Cutter, S.L. 2002. Emerging hurricane evacuation issues: Hurricane Floyd and South Carolina. *Natural Hazards Review* 3:12–18.

FEMA: Federal Emergency Management Agency. 1998. *IS-9 Managing floodplain development through the National Flood Insurance Program (NFIP).* Retrieved on May 29, 2012 from www.fema.gov/plan/prevent/floodplain/nfipkeywords/sfha.shtm.

Federal Emergency Management Agency. 2010a. *Developing and maintaining emergency operations plans: Comprehensive preparedness guide 101, Version 2.0.* Retrieved in April 2012 from http://www.fema.gov/pdf/about/divisions/npd/CPG_101_V2.pdf.

Federal Emergency Management Association. 2010b. *Guidance on planning for integration of functional needs support services in general population shelters.* Retrieved in April 2012 from http://www.fema.gov/pdf/about/odic/fnss_guidance.pdf.

Hardy, M. and Wunderlich, K. 2007. *Evacuation management operations (EMO) modeling assessment: Transportation modeling inventory.* Falls Church, VA: Noblis. Downloaded from www.its.dot.gov/its_publicsafety/evacuation.htm.

Heath, S.E., Kass, P.H., Beck, A.M. and Glickman, L.T. 2001. Human and pet-related risk factors for household evacuation failure during a natural disaster. *American Journal of Epidemiology* 153:659–665.

Houston, N. 2009. *Good practices in transportation evacuation preparedness and response: Results of the FHWA workshop series,* FHWA-HOP-09-040. Washington, DC: Federal Highway Administration, Office of Transportation Operations. Downloaded from www.ops.fhwa.dot.gov/publications/publications.htm#eto.

Jetter, J.J. and Whitfield, C. 2005. Effectiveness of expedient sheltering in place in a residence. *Journal of Hazardous Materials* A119:31–40.

Jones, J., Walton, F., Smith, J.D. and Wolshon, B. 2008. *Assessment of emergency response planning and implementation for large scale evacuations,* NUREG/CR-6981, SAND2008-1776P. Washington, DC: US Nuclear Regulatory Commission. Downloaded from www.nrc.gov/reading-rm/doc-collections/nuregs/.

Jones, J., Walton, F. and Wolshon, B. 2011. *Criteria for development of evacuation time estimate studies,* NUREG/CR-7002 SAND2010-0016P. Washington, DC: US Nuclear Regulatory Commission. Downloaded from www.nrc.gov/reading-rm/doc-collections/nuregs/.

Jones, J.A., Bixler, N., Schelling, F.J. and Sullivan, R. 2007. *Review of NUREG-0654, Supplement 3, "Criteria for protective action recommendations for severe accidents",* NUREG/CR-6953, Vol. I, SAND2007-5448P. Washington, DC: U.S. Nuclear Regulatory Commission. Downloaded from www.nrc.gov/reading-rm/doc-collections/nuregs/.

Kailes, J. I. and Enders, A. 2007. Moving beyond "special needs." A function based framework for emergency management and planning. *Journal of Disability Policy Studies* 17:230–237.

Kreps, G.A. 1991. Organizing for emergency management. In *Emergency management: Principles and practice for local government*, eds. T.S. Drabek and G.J. Hoetmer, p. 30–54. Washington, DC: International City/County Management Association.

Lindell, M.K. 2008. EMBLEM2: An empirically based large-scale evacuation time estimate model. *Transportation Research A* 42:140–154.

Lindell, M.K. 2011. Disaster studies. In *Sociopedia*, ed. B. Klandermans. Downloaded from www.sagepub.net/isa/resources/pdf/Disaster%20Studies.pdf.

Lindell, M.K. 2012. Response to environmental disasters. In *The Oxford handbook of environmental and conservation psychology*, ed. S. Clayton, p. 391–413. New York: Oxford University Press.

Lindell, M.K. and Perry, R.W. 1983. Nuclear power plant emergency warning: How would the public respond? *Nuclear News* 26:49–53.

Lindell, M.K. and Perry, R.W. 1992. *Behavioral foundations of community emergency planning*. Washington, DC: Hemisphere Press.

Lindell, M.K. and Perry, R.W. 2004. *Communicating environmental risk in multiethnic communities*. Thousand Oaks, CA: Sage.

Lindell, M.K. and Perry, R.W. 2007. Planning and preparedness. In *Emergency management: Principles and practice for local government*, 2nd ed., eds. K.J. Tierney and W.F. Waugh, Jr., p. 113–141. Washington, DC: International City/County Management Association.

Lindell, M.K. and Perry, R.W. 2012. The protective action decision model: Theoretical modifications and additional evidence. *Risk Analysis* 32:616–632.

Lindell, M.K. and Prater, C.S. 2007. Critical behavioral assumptions in evacuation analysis for private vehicles: Examples from hurricane research and planning. *Journal of Urban Planning and Development* 133:18–29.

Lindell, M.K. and Prater, C.S. 2008. *Behavioral analysis: Texas hurricane evacuation study.* College Station, TX: Texas A&M University Hazard Reduction and Recovery Center.

Lindell, M.K., Kang J.E. and Prater, C.S. 2011. The logistics of household hurricane evacuation. *Natural Hazards* 58:1093–1109.

Lindell, M.K., Lu, J.C. and Prater, C.S. 2005. Household decision making and evacuation in response to Hurricane Lili. *Natural Hazards Review* 6:171–179.

Lindell, M.K., Prater, C.S. and Perry, R.W. 2006. *Fundamentals of emergency management.* Emmitsburg, MD: Federal Emergency Management Agency Emergency Management Institute. Downloaded from www.training.fema.gov/EMIWeb/edu/fem.asp or archone.tamu.edu/hrrc/Publications/books/index.html.

Lindell, M.K., Prater, C.S. and Wu, J.Y. 2002. *Hurricane evacuation time estimates for the Texas Gulf coast.* College Station, TX: Texas A&M University Hazard Reduction and Recovery Center. Downloaded from www.txdps.state.tx.us/dem/downloadableforms.htm.

Lutz, L.D. and Lindell, M.K. 2008. The Incident Command System as a response model within emergency operation centers during Hurricane Rita. *Journal of Contingencies and Crisis Management* 16:122–134.

McKenna, T. 2000. Protective action recommendations based upon plant conditions. *Journal of Hazardous Materials* 75:145–164.

Mendonça, D. and Wallace, W.A. 2004. Studying organizationally-situated improvisation in response to extreme events. *International Journal of Mass Emergencies and Disasters* 22:5–29.

Mendonça, D. and Wallace, W.A. 2007. A cognitive model of improvisation in emergency management. *IEEE Transactions On Systems, Man, And Cybernetics. Part A: Systems and Humans* 37:547–561.

Mileti, D.S., Sorensen, J.H. and O'Brien, P.W. 1992. Toward an explanation of mass care shelter use in evacuations. *International Journal of Mass Emergencies and Disasters* 10:25–42.

Murray-Tuite, P. 2007. Perspectives for network management in response to unplanned disruptions. *Journal of Urban Planning and Development*, 133:9–17.

Murray-Tuite, P.M. and Mahmassani, H.S. 2003. Model of household trip-chain sequencing in emergency evacuation. *Transportation Research Record* 1831:21–29.

Murray-Tuite, P.M. and Mahmassani, H.S. 2004. Transportation network evacuation planning with household activity interactions. *Transportation Research Record* 1894:150–159.

Nelson, H.E. and Mowrer, F.W. 2002. Emergency movement, In *SFPE handbook of fire protection engineering*, eds. P.J. DiNenno, et al., Chapter 14. Quincy, Boston, MA: National Fire Protection Association.

Peacock, R.D., Kuligowski, E.D. and Averill, J.D. 2011. *Pedestrian and evacuation dynamics*. New York: Springer.

Pel, A.J., Hoogendoorn, S.P. and Bliemer, M.C.J. 2012. A review on travel behaviour modelling in dynamic traffic simulation models for evacuations. *Transportation* 39:97–123.

Perry, R.W. and Lindell, M.K. 2007. *Emergency Planning*. Hoboken NJ: John Wiley.

Perry, R.W., Lindell, M.K. and Greene, M.R. 1981. *Evacuation planning in emergency management*. Lexington, MA: Heath Lexington Books.

Rogers, G.O. and Sorensen, J.H. 1988. Diffusion of emergency warnings. *Environmental Professional* 10:185–198.

Siebeneck, L.K. and Cova, T.J. 2008. An assessment of the reentry process for Hurricane Rita 2005. *International Journal of Mass Emergencies and Disasters* 26:91–111.

Siebeneck, L.K., Lindell, M.K., Prater, C.S., Wu, H.C. and Huang, S.K. 2013. Evacuees' reentry concerns and experiences in the aftermath of Hurricane Ike, *Natural Hazards*, 65, 2267–2286.

Sorensen, J.H. and Rogers, G.O. 1988. Local preparedness for chemical accidents: A survey of U.S. communities. *Industrial Crisis Quarterly* 2:89–108.

Sorensen, J.H. and Rogers, G.O. 1989. Warning and response in two hazardous materials transportation accidents in the U.S. *Journal of Hazardous Materials* 22:57–74.

Sorensen, J.H., Shumpert, B.L. and Vogt, B.M. 2004. Planning for protective action decision making: evacuate or shelter-in-place. *Journal of Hazardous Materials* A109:1–11.

Sorensen, J.H., Vogt, B.M. and Mileti, D.S. 1987. *Evacuation: An assessment of planning and research*. Washington, DC: Federal Emergency Management Agency.

Stallings, R.A. 1991. Ending evacuations. *International Journal of Mass Emergencies and Disasters* 9:183–200.

Stein, R.M., Dueñas-Osorio, L. and Subramanian, D. 2010. Who evacuates when hurricanes approach? The role of risk, information, and location. *Social Science Quarterly* 91:816–834.

Sutton, J., Palen, L. and Shlovski, I. 2008. Back-channels on the front lines: Emerging use of social media in the 2007 Southern California wildfires. *Proceedings of the 2008 ISCRAM Conference.* Retrieved on October 5, 2011 from www.iscram.org/index.php?option=content&task=view&id=2236.

Tierney, K.J., Lindell, M.K. and Perry, R.W. 2001. *Facing the unexpected: Disaster preparedness and response in the United States.* Washington, DC: Joseph Henry Press.

TRB: Transportation Research Board. 2010. *Highway capacity manual.* Washington, DC: Transportation Research Board.

Urbanik, T. 1979. Hurricane evacuation demand and capacity estimation. In *Hurricanes and coastal storms: Awareness, education and mitigation,* ed. E.J. Baker. Tallahassee, FL: Florida State University.

USACE: US Army Corps of Engineers. 1995. *Technical guidelines for hurricane evacuation studies.* Downloaded March 28, 2013 from www.csc.noaa.gov/hes/hes.html.

USDOT: US Department of Transportation. 2012. *Emergency response guidebook.* Retrieved on March 29, 2013 from http://phmsa.dot.gov/staticfiles/PHMSA/DownloadableFiles/Files/Hazmat/ERG2012.pdf. from phmsa. dot.gov/hazmat/library/erg.

USNRC/FEMA: US Nuclear Regulatory Commission/Federal Emergency Management Agency. 1980. *Criteria for preparation and evaluation of radiological emergency response plans and preparedness in support of nuclear power plants,* NUREG-0654/FEMA-REP-1. Retrieved on May 29, 2012 from www.nrc.gov/reading-rm/doc-collections/nuregs/staff/sr0654/.

Vásconez, K.C. and Kehrli, M. 2010. *Highway evacuations in selected metropolitan regions: Assessment of impediments.* FHWA-HOP-10-059. Washington, DC: Federal Highway Administration, Office of Transportation Operations. Downloaded from www.ops.fhwa.dot.gov/publications/publications.htm#eto.

Vieweg, S., Hughes, A., Starbird, K. and Palen, L. (2010). Supporting situational awareness during emergencies using microblogged information. *Proceedings of the ACM 2010 Conference on Computer Human Interaction.* Retrieved 10/05/11 from www.cs.colorado.edu/~palen/Home/Crisis_Informatics.html.

Vultee, F. and Vultee, D.M. 2011. What we tweet about when we tweet about disasters: The nature and sources of microblog comments during emergencies. *International Journal of Mass Emergencies and Disasters* 29:221–242.

Whitehead, J.C. 2003. One million dollars per mile? The opportunity costs of hurricane evacuation. *Ocean and Coastal Management* 46:1069–1083.

Wolshon, B., Urbina, E., Wilmot, C. and Levitan, M. 2005a. Review of policies and practices for hurricane evacuation. I: Transportation planning, preparedness, and response. *Natural Hazards Review* 6:129–142.

Wolshon, B., Urbina Hamilton, E., Levitan, M. and Wilmot, C. 2005b. Review of policies and practices for hurricane evacuation. II: Traffic operations, management, and control. *Natural Hazards Review* 6:143–161.

Wu, H.C., Lindell, M.K. and Prater, C.S. 2012. Logistics of hurricane evacuation in Hurricanes Katrina and Rita. *Transportation Research Part F: Traffic Psychology and Behaviour* 15:445–461.

Zeigler, D., Brunn, S. and Johnson, J. 1981. Evacuation from a nuclear technological disaster. *Geographical Review* 71:1–16.

Zhang, Y., Prater, C.S. and Lindell, M.K. 2004. Risk area accuracy and evacuation from Hurricane Bret. *Natural Hazards Review* 5:115–120.

chapter seven

Hierarchical optimization for helicopter mission planning in large-scale emergencies

Linet Özdamar, Onur Demir, Erdinc Bakır, and Samet Yılmaz

Contents

7.1   Introduction

Disasters have occurred with increasing frequency and intensity during the last decade, which has required more transportation means to cover larger affected areas. With rising transportation costs and increased vehicle capacity requests, the logistics of disaster relief consumes an important amount of postdisaster resources. Consequently, the activities of relief materials distribution and wounded evacuation, which comprise "the last mile distribution and pickup problem" (Balcik et al. 2008), need to be carried out efficiently. It is conjectured that fleet size minimization and optimized vehicle routing in disaster relief could reduce relief costs by about 15% (Stapleton et al. 2009). Therefore, introducing more efficient models that deal with this set of activities is economically important.

The logistics problem considered here is described as follows. We consider emergencies where some of the delivery and pickup requests are satisfied by helicopters. For instance, in earthquakes that affect old cities with lots of building damage, land transportation would be very difficult, and helicopters are possibly the best alternative to reach people residing at inaccessible locations. Similarly, in flooded cities, stranded people can be safely and quickly evacuated by helicopters. In the aftermath of such emergencies, information about the locations of injured or stranded persons (e.g., GPS data) can be collected by reconnaissance flights conducted before evacuation to hospitals or camp areas begin. On the basis of the estimated number of survivors, urgent relief materials are prepared for distribution to survivors who cannot be immediately evacuated. We assume that the available relief material supplies are limited and stocked at different warehouses outside the disaster zone. These have to be picked up by helicopters and dropped off at designated locations. Noting that the available emergency patient capacities at hospitals are also limited, helicopters assigned to the evacuation task need to take these people to hospitals with free service capacity. Due to the growing size and intensity of emergencies, it is most likely that both the distribution and the pickup tasks at a given location will not be met by a single helicopter visit. There are also special aviation constraints pertaining to this problem, such as the altitude and temperature dependent takeoff cargo limit of the helicopter, and the range that the vehicle can fly without refueling.

In this chapter, we adapt and use the optimization-guided hierarchical planning and clustering (OHPC) procedure described in Özdamar and Demir (2012) to solve the logistics problem described earlier. The backbone of the OHPC is an efficient network flow model that optimizes vehicle and material flows over a given network of depots (air bases) and demand points. The network flow approach is more efficient than vehicle routing approaches in emergency logistics because of the larger scale of demand pickup or per drop-off location (Yi and Özdamar 2007). In the OHPC

(Özdamar and Demir 2012), the static network flow model discussed by Özdamar and Pedamallu (2011) is embedded within a parallel hierarchical network aggregation heuristic that subdivides the network recursively, grouping demand nodes into clusters and representing them with a single cluster center. The network partitioning process continues until the cluster sizes that are generated allow us to obtain an optimal solution for the network flow model. The heuristic preserves the consistency among solutions at different aggregation levels by passing on the upper-level solution to the lower level as a parameter.

To ensure every route is feasible with regard to fuel limits and available number of helicopters, a route management procedure (RMP) is proposed here. Once the optimal solutions of the cluster problems at the final level are obtained, they are processed by the RMP that first ensures every route's fuel feasibility by inserting refueling stops into itineraries. Then, the RMP recombines fuel-feasible routes into longer itineraries that are assigned to each helicopter that is available for the mission. Hence, the RMP can be used as a scenario generation tool where different numbers of helicopters are allocated resulting in varying mission completion times.

In the next section, we provide brief reviews on the previous models proposed in the area of emergency logistics and on the solution methods proposed for planning various helicopter missions. Then, we discuss the mathematical model (Özdamar 2011) that is utilized in the adapted OHPC and we describe the adapted OHPC and the RMP in detail. Finally, we describe two emergency scenarios to illustrate the performance of the OHPC.

7.2   Review of emergency logistics models

In this review, we discuss and compare the efficiency of the model structures previously proposed by researchers. There are four basic approaches used in modeling the last mile distribution and pickup problem in disaster relief. These approaches differ according to the manner in which they represent vehicle routes. In the first modeling approach, which we call the route construction vehicle routing (RCVR) model, each vehicle route is represented by a binary variable of four indices that define the vehicle and route identification and starting and ending nodes. The second modeling approach (denoted here as the route enumeration vehicle routing [REVR] model) enumerates all feasible routes between all pairs of supply and demand nodes. The decision of assigning a route to each vehicle is represented as a binary variable in the REVR and a capacitated assignment problem is solved rather than a vehicle routing problem. The third modeling approach (denoted the dynamic network flow [DNF] model) constructs a dynamic network flow model whose outputs consist of vehicle and material flows, rather than routes. The flows produced by the

DNF are parsed to construct vehicle routes and loads. Yi and Özdamar (2007) show that the DNF is more efficient than the RCVR, because the DNF has integer variables of three indices that represent the starting and ending nodes and the travel starting time of the vehicle. The size of the DNF does not depend on the number of routes or vehicles used; rather, it depends only on the number of nodes in the network and the length of the planning horizon. The DNF tends to be more compact than the REVR as well, especially in postdisaster activities, because of the large numbers of routes and vehicles needed in such operations.

Özdamar and Pedamallu (2011) and Özdamar (2011) point out that though the DNF is more compact than the RCVR and the REVR, it has a disadvantage that lies in the length of the specified planning horizon, T. If T is too small, then some of the goods are not delivered despite the availability of supplies. On the other hand, if T is too long, then the problem size becomes large and only small relief networks can be handled by the DNF. Obviously, the correct size of this parameter cannot be known a priori until the model is solved. A second disadvantage of the DNF is that it represents distances between pairs of nodes in terms of integral time periods. This is impractical, because each travel time has to be a multiple of the smallest travel time between all pairs of nodes in the network. Finally, in postdisaster situations, it might not be worth spending resources and time running dynamic models if the models cannot be solved fast enough, especially where large-scale relief operations are concerned. Because model inputs are roughly estimated at best, a static fast-response model that is rerun regularly on information updates is more meaningful from a practical point of view. Consequently, Özdamar and Pedamallu (2011) and Özdamar (2011) propose a fourth modeling category: the static network flow (SNF) model. The SNF eliminates the temporal structure of the dynamic model and presents a fast-response model that can solve problems with networks of up to 100 nodes within reasonable computation times.

Table 7.1 summarizes the number of vehicle-related integer variables found in the four model categories discussed earlier (we assume that a single vehicle type is used in every model category).

In the following review, we classify the models found in the emergency logistics literature in terms of model structure and discuss their capabilities and objective functions. Barbarosoglu et al. (2002) adopt the RCVR and develop a hierarchical decision support methodology for helicopter logistics planning. The first level of planning involves tactical decisions such as the selection of helicopter fleet size and the determination of the number of sorties to be undertaken by each helicopter. The second level addresses the operational decision of vehicle routing and tries to minimize the mission completion time. The RCVR can solve very small instances with relief networks of up to 10 nodes and 3 helicopters. This

Table 7.1 Model Categories and Numbers of Vehicle-Related Integral Variables

Model category	Number of vehicle-related integral variables
Route Construction Vehicle Routing	$v \cdot r \cdot C^2$ (binary variables)
Route Enumeration Vehicle Routing	$v \cdot R \cdot C^2$ (binary variables)
Dynamic Network Flow	$T \cdot C^2$ (integer variables)
Static Network Flow	C^2 (integer variables)

Key: v is the number of vehicles used in the operation; r is the maximum number of routes that a vehicle has to undertake to complete its task (has to be estimated before solving the problem); R is the number of all possible routes between all pairs of supply and demand nodes; C is the number of nodes in the relief network; and T is the length of the planning horizon (has to be estimated before solving the problem).

same type of model is reported in a NATO Research and Technology Organization (2008) report that describes military missions.

The RCVR is also adopted by DeAngelis et al. (2007) who consider the airplane routing and scheduling problem for transporting food to communities in Angola. In their model, airplanes can park at any depot that has available parking space. The authors calculate the weekly parking schedule of planes and maximize the total satisfied demand. The size of the relief network that they consider consists of 18 nodes. Both Barbarosoglu et al. and DeAngelis et al. consider refueling constraints in their models, making them very difficult to solve.

Mete and Zabinsky (2010) describe a two-stage stochastic program that solves the location problem of depots in the first stage and the transportation of aid materials in the second stage. Demand is random in the model. The authors adopt the REVR in the second stage and solve a small scenario with a 21-node relief network and 14 vehicles.

A number of DNF models are also proposed to coordinate disaster relief logistics. Haghani and Oh (1996) propose a multicommodity multimodal transportation model with time windows. The model explicitly expands the network into a time–space network to coordinate the transportation of critical items to affected areas. In this approach, vehicle flows are represented by integer variables with four indices. The objective function minimizes transportation and inventory costs. In the DNF category, Özdamar et al. (2004) consider the postdisaster aid distribution problem on an integrated multimode transportation network and introduce a multiperiod planning model that foresees future demand. Özdamar et al.'s model improves the Haghani and Oh model by introducing a time lag in equations and variables. Hence, the relief network is not expanded into the time–space and the number of integer variables is reduced significantly. Yi and Özdamar (2007) extend Özdamar et al.'s model by including the evacuation of injured persons. The authors consider patient queues and hospital space limitations in their model. In the last two models, the objective is to minimize unsatisfied demand. Yi and Özdamar (2007) can solve relief networks of up to 80 nodes with a planning horizon of eight time periods.

Özdamar and Pedamallu (2011) compare the DNF and the SNF and solve scenarios of up to 100 nodes with three objective functions as follows: minimize the number of vehicles, minimize the total delay in the arrival of goods, and minimize the total travel distance. The authors show that although the total delay objective can be solved within a CPU second, the total distance objective can take up to an hour. The authors also present the trade-offs between the three objectives. Campbell et al. (2008) and Huang et al. (2011) have also discussed different objective functions such as minimizing the maximum arrival time of supplies and minimizing the sum of arrival times of supplies.

Due to the computational complexity of the detailed vehicle routing models discussed earlier, some researchers (Tzeng et al. 2007; Yan and Shih 2009; Lin et al. 2011) prefer to represent the relief distribution problem as an uncapacitated network flow problem (as per the basic transportation model) without considering the vehicle links. Such models assume that the transportation cost between any pair of locations is known and that the specific route of transportation among these locations has already been selected. Despite the latter simplification, the multiobjective multiperiod transportation model proposed by Lin et al. (2011) can only solve very small scenarios of up to 10 delivery locations. There are also studies (Hodgson et al. 1998; Doerner et al. 2007; Jozefowiez et al. 2007; Nolz et al. 2010) where the commodity or service delivery problem is modeled as a multivehicle covering tour problem (e.g., the placement of water tanks or providing mobile health services at locations that are accessible to postdisaster survivors). The latter model type is more complex than the traveling salesman problem, where a set of existing locations have to be visited once by an unoccupied vehicle.

Some of the work found in the literature is dedicated to evacuation only and includes approaches other than the model categories mentioned earlier. For instance, Bakuli and Smith (1996) propose a queuing network model in designing an emergency evacuation network. Fiedrich et al. (2000) propose an evacuation model to minimize the number of fatalities. A linear programming approach is described in Chiu and Zheng (2007) who treat multipriority group evacuation in sudden onset disasters. The authors assume that important information such as population is known with certainty with a goal to minimize the total travel time. A similar objective is considered by Sayyady and Eksioglu (2008) who build a model for evacuating the population by public transit. Apte and Heath (2009) propose a model that is used for picking up people with disabilities in case of a hurricane warning.

For a more detailed discussion on different objectives and model properties in disaster relief routing problem, we refer the reader to de la Torre et al. (2011). For readers who are interested in all aspects of disaster response, we refer to the extensive surveys of Altay and Green (2006), Wright et al. (2006), Apte (2009), and Caunhye et al. (2011).

7.3   Review of helicopter mission planning approaches

An early study that describes a solution method for helicopter mission planning involves a larvicide treatment program in West Africa (Solomon et al. 1992). The authors try to minimize the cost of the larvicide dropped by calculating the river flow speed of breeding sites and then try to optimize the vehicle routes to minimize the flight time. To achieve the second objective, they group river segments together while considering the fuel and larvicide tank limitations.

Another study (Timlin and Pulleybank 1992) describes a helicopter mission between an oil company's air bases on land and offshore oil platforms whose crews are changed regularly. The goal is to minimize the flight time and capacitated and uncapacitated insertion heuristics are suggested to solve the problem. Sierksma and Tijssen (1998) describe a similar mission where the same number of personnel are delivered and picked up from 51 oil rigs in the North Sea and transported to the main base near Amsterdam, the Netherlands. During this mission, it is possible to make multiple visits to the same platform. The authors develop a mixed integer model (MIP) that contains information on all predefined feasible routes (the REVR approach). It is claimed that the model becomes intractable for missions with large demands. Therefore, the integrality constraints on variables are relaxed and a linear program (LP) is solved with a solution rounded into an integer solution. The authors generate 16 integer solutions from the LP solution by rounding different sets of real-valued variables and find that this is the best approach among other heuristics. In this study, the authors also develop a cluster formation heuristic that selects the most distant platform as the seed and greedily adds new platforms to the cluster until either the flight range or the vehicle capacity constraint is violated. The routes formed within each cluster are improved by 1-opt and 2-opt exchange methods. A similar MIP and a column generation technique are suggested by Menezes et al. (2010) for a network of 80 oil rigs and 4 air bases belonging to an energy company in Brazil. About 1900 workers are transported daily by 45 helicopters in 110 scheduled flights. This problem has constraints on the number of daily flights per helicopter, on the number of flight legs in routes, and available pilots with breaks. Similar to Sierksma and Tijssen (1998), split deliveries and pickups are allowed and the model needs to have a predefined set of feasible flights that abide these constraints before it is solved (the REVR approach).

A genetic algorithm is proposed by Armstrong-Crews and Mock (2005) to solve a helicopter routing problem that involves taking measurements at remote sites in Alaska. A single helicopter is involved and there is no transportation task; therefore, the problem becomes a traveling salesman problem with flight range constraints where refueling locations can be visited more than once. The authors' goal is to minimize the traveled distance to solve a 10-site problem.

7.4 Optimization guided Hierarchical Planning and Clustering: Routing model

The mathematical model discussed here is proposed by Özdamar (2011) and assumes the SNF approach. It solves the routing problem of a cluster network with given pickup and drop-off requests. The assumptions of the model are as follows:

- Refueling constraints are not taken into consideration by the model but rather carried out by a postprocessing algorithm, the RMP.
- It is assumed that the flight range of the helicopter is long enough to cover the two-way distance between any base (helipad) and any drop-off or pickup point.
- The takeoff and landing cargo weight limit for a helicopter at a given location is a function of the temperature and altitude of the location.
- Helicopters carry relief materials such as water on an external cargo hook, whereas injured persons or evacuees are transported as internal cargo. External and internal cargo carrying capacities are separate measures (in tons and number of persons, respectively), and the total internal–external cargo weight cannot exceed the takeoff and landing cargo limit defined on the helicopter's performance card.
- There is a single helicopter type used in the relief mission.
- A helicopter can start and end its tour at different locations.
- A helicopter may pay multiple visits to a site to satisfy demand for drop-off or pickup (split deliveries and pickup are allowed).
- The amount of available relief supplies are limited and stored at various warehouses.
- The number of available space at hospitals/evacuee camps is limited.
- The number and locations of injured/stranded persons are known.
- Tank refueling is carried out at available bases (helipads) near hospitals, warehouses, or evacuee camps.

The mathematical formulation of the problem is denoted as model P. Model P and the related notations are listed below. All sets and variables are indicated in capital letters whereas all parameters are in small letters.

Sets:

C: Set of all nodes including bases, drop-off/pickup locations
N: Set of edges connecting the nodes in the network
CD: Set of drop-off/pickup locations, $CD \subset C$
CS: Set of refueling bases (near hospitals, warehouses, and camps) where tours start and end, $CS \subset C$

Parameters:

d_{op}: Flight distance/time between locations o and p
lq_o: Number of persons to be evacuated from location $o \in CD$
uq_o: Amount of commodities to be dropped off at location $o \in CD$
cap_o: Takeoff/landing cargo weight limit (in tons) at location o depending on temperature and altitude
caplz: Maximum internal cargo (persons) that can be transported by a helicopter
capuz: Maximum external cargo weight (in tons) that can be transported by a helicopter
sup_o: Limited amount of relief supplies at base $o \in CS$
lsq_o: Limited amount of evacuee receiving space at base $o \in CS$
w: Average weight of a person

Decision variables:

UZ_{op}: Weight of commodity transported from location o to location p
LZ_{op}: Number of evacuees transported from location o to location p
Y_{op}: Integer number of helicopters traveling from location o to location p

Model P:

Minimize $\Sigma_{(o,p)\in N} d_{op}Y_{op}$ (7.1)

Subject to:

$\Sigma_{p\in C} UZ_{po} - \Sigma_{p\in C} UZ_{op} = uq_o$ ($\forall o \in CD$) (7.2)
$- \Sigma_{p\in C} LZ_{po} + \Sigma_{p\in C} LZ_{op} = lq_o$ ($\forall o \in CD$) (7.3)
$UZ_{op} + w\,LZ_{op} \le cap_o\,Y_{op}$ ($\forall (o,p) \in N$) (7.4a)
$UZ_{op} + w\,LZ_{op} \le cap_p\,Y_{op}$ ($\forall (o,p) \in N$) (7.4b)
$LZ_{op} \le caplz\,Y_{op}$ ($\forall (o,p) \in N$) (7.4c)
$UZ_{op} \le capuz\,Y_{op}$ ($\forall (o,p) \in N$) (7.4d)
$\Sigma_{p\in C} Y_{op} = \Sigma_{p\in C} Y_{po}$ ($\forall o \in C$) (7.5)

$$-\Sigma_{p \in C} UZ_{po} + \Sigma_{p \in C} UZ_{op} \leq \sup_o \quad (\forall\, o \in CS)\ (7.6)$$

$$\Sigma_{p \in C} LZ_{po} - \Sigma_{p \in C} LZ_{op} \leq lsq_o \quad (\forall\, o \in CS)\ (7.7)$$

$$Y_{op} \geq 0 \text{ and integer; } LZ_{op},\, UZ_{op} \geq 0\ (7.8)$$

The objective minimizes the total flight distance or time. Constraint sets 7.2 and 7.3 enforce the material balance for commodities and evacuees at the drop-off and pickup locations. We remark that some locations in the set CD can be only drop-off locations, some can be only pickup locations, and others can be both. Depending on the type of location, the uq_o and lq_o parameters will be zero or positive, that is, if it is a drop off–only location, uq_o is positive and lq_o is zero, and vice versa if it is a pickup-only location. Otherwise, both parameters are positive. Constraints 7.4a and 7.4b restrict the total cargo weight carried at take off and landing at every location according to its altitude and temperature. Constraints 7.4c and 7.4d impose internal and external cargo capacity limits. The next set of Constraints 7.5 balance the flow of vehicles at each location including the bases, ensuring that every location has equal indegree and outdegree in the optimal solution. In Constraints 7.6 and 7.7, we enable material balance of ingoing and outgoing flows at bases. The right-hand sides of these constraints may be limited numbers indicating limited supplies of commodities and limited evacuee receiving space. Again, some bases may represent warehouses, whereas some represent camps or hospitals.

7.5 Optimization guided Hierarchical Planning and Clustering algorithm

7.5.1 Brief overview

The OHPC is a parallel heuristic solution method that utilizes the solution of model P within a hierarchical network aggregation framework. The procedure always maintains feasibility among solutions at different levels of aggregation. As Bard and Jarrah (2009) point out, large-scale delivery and pickup operations are best dealt with by clustering demand nodes and downsizing the network. In the OHPC, the demand nodes in the relief network are partitioned into geographically dense clusters using the *k*-means clustering algorithm and then the routing problem of the top-level network is solved first. At the top level, the network is composed of demand cluster centers and bases (near hospitals, warehouses, and evacuee camps). Each cluster center represents the delivery and pickup requirements of demand locations that belong to the cluster. Model P is solved for the top-level network and provides the following information: the set of bases that send and receive material and people to and from each cluster center and the amounts sent and received. This information is passed on to the next-level problem. At the next level of the planning hierarchy, each cluster center is transformed into its own subnetwork that includes demand nodes and the specific bases that have served the cluster center at the previous aggregation level. Model P is rebuilt for this subnetwork and solved. All cluster problems originating from a parent aggregate problem are solved in parallel and their optimal routes are obtained. If a cluster has more demand nodes than desired, it is reaggregated into smaller subclusters and the whole procedure is repeated. Otherwise, the cluster's optimal vehicle routes are added to the final optimal solution constructed for the original network. A flowchart of the OHPC is illustrated in Figure 7.1.

7.5.2   Hierarchical planning

In the procedure described in Figure 7.1, clustering is carried out according to the k-means partitioning heuristic to obtain K disjoint clusters. The centroids of these clusters are selected as cluster centers denoted as cc_i, where $i = 1...K$. The original network is then condensed down to K cluster centers plus the set of bases, CS. This problem is defined as the aggregate zero-level problem, P^0. An optimal solution for P^0 is obtained by solving model P using CPLEX. This solution defines the vehicle routes and their loads moving between bases and cluster centers cc_i. Next, each cluster center cc_i is disaggregated into its own subnetwork that is defined as P^i. P^i consists of a set of demand nodes, D_i, that fall into cluster i and a set of bases, CS_i, that send and receive supplies to/from cc_i in the solution of P^0. The cluster center cc_i itself is not included in P^i. The parameter links between problems P^i and P^0 are organized as follows. The optimal commodity amount (UZ) sent to cc_i from a base in CS_i in the optimal solution of P^0 become the available supply (sup_o) of the base. The optimal number of evacuees (LZ) transferred from cc_i to the base in the solution of P^0 becomes the base's available evacuee space (lsq_o). Thus, these two sets of parameters are passed on to problem P^i from the optimal solution of P^0 and the consistency of solutions is preserved between different levels of network aggregation.

All cluster problems P^i are solved in parallel by CPLEX and detailed routes are obtained for each P^i. However, if a problem P^i has too many demand nodes, it is not solved immediately. Rather, the demand nodes in P^i are reaggregated into smaller clusters that make up the next level of subproblems. This hierarchical aggregation approach continues until a desirable final cluster size is obtained. Consequently, a problem P^i may consist of other smaller cluster centers at an intermediate level of aggregation, but it consists of original demand nodes at the final level. The details of the algorithm are found in Özdamar and Demir (2012).

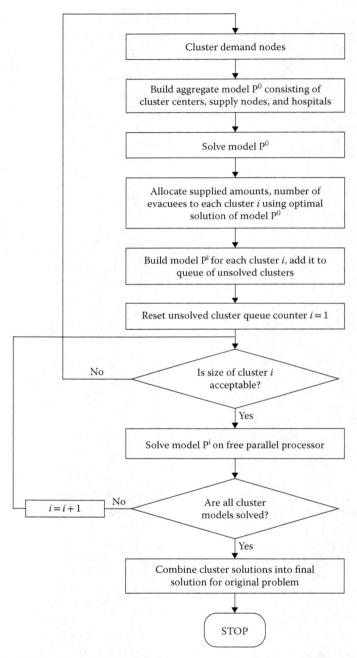

Figure 7.1 Flowchart of the Optimization guided Hierarchical Planning and Clustering (OHPC) procedure.

7.5.3   Route management procedure

The postprocessing algorithm RMP maintains the fuel feasibility of routes by comparing each route length to the helicopter's flight range. If a route length is longer, then the procedure inserts refueling stops into the route as needed. In a second step, the RMP unites fuel-feasible routes into longer routes that become helicopter itineraries. The longest helicopter itinerary defines the mission completion time. By specifying different mission completion times, the decision maker (DM) can obtain different numbers of itineraries (helicopters). A longer itinerary results in a lower number of utilized helicopters. The second step of the RMP is interactive and routes are recombined until the DM is satisfied both with the mission completion time and with the number of helicopters allocated to the operation.

7.5.3.1   Asserting the fuel feasibility of routes

This procedure inserts refueling stops into all routes that exceed the flight range of the helicopter. We assume that when a helicopter's tank is refueled, it is filled up completely and that refueling takes place only at bases.

If a specific route's length exceeds the flight range, the procedure breaks up the route into two by selecting the median node on the route and inserts a refueling base that is nearest to both the median node and the node next to it. Hence, two routes starting and ending with the refueling base emerge. Then, both of these routes are rechecked for fuel feasibility. If any one of them is found to be infeasible, it is subjected to the same procedure. This step of the RMP leads to an increase in total flight distance and time because inserted refueling stops result in additional flight distance and time.

7.5.3.2   Constructing helicopter itineraries

In this step, the DM is asked about the maximum length of a helicopter itinerary (i.e., aspired mission completion time). The goal is to merge fuel feasible routes into longer itineraries that will be assigned to the available helicopters. The merging procedure sorts the fuel-feasible routes in descending order of length. The longest route at the top of the list is assigned to the first helicopter. Then, the sorted list is checked for the route whose start node is nearest to the end node of the assigned route. Once the latter route is identified, it is appended to the itinerary of the first helicopter and deleted from the sorted list. The same procedure is repeated until the maximum itinerary length is reached on the first helicopter. Then, the itinerary of the next helicopter is constructed in the same manner. This merging procedure aims to balance the total transportation workload equally among all helicopters.

7.6 Emergency scenarios

7.6.1 An earthquake scenario

7.6.1.1 Description of the scenario

Here, we summarize a hypothetical postearthquake scenario whose details are found in Özdamar (2011). The scenario is generated based on the impact expectations of the Disaster Coordination Center in Istanbul, Turkey (AKOM: http://www.ibb.gov.tr/sites/akom/Documents/index.html). Due to the earthquake-related safety hazards of the viaducts connecting the main highways, AKOM perceives land transportation difficult. Therefore, a network of 72 helipads covering Istanbul has been recently built. If Istanbul is hit by an earthquake, AKOM will coordinate all daily helicopter missions and manage the supply chain for medical relief materials and injured evacuation.

In our scenario, we select 60 helipad locations in Istanbul as landing points. These locations cover most of the districts that are at risk, though some helipads are in safe regions nearby hospital clusters. We assume that light to moderate injuries are treated at mobile medical centers that are established at open spaces nearby the helipads. Heavily injured persons are picked up from these locations and transported to hospitals with vacant beds. AKOM has announced that the expected number of light to moderate injuries is about 600,000 persons, and that 130,000 people will be heavily injured.

We assume that medical aid materials such as intravenous materials, medicine, and vaccines are dropped to mobile medical centers near the helipads to help the moderately injured persons and vaccinate healthy individuals against epidemics. A minimal amount of drinking water is also transported for the injured. The data concerning material demand are based on the expected numbers of moderately injured persons in each district. This information is acquired from several geographical information system maps published in a consortium study led by the Kandilli Earthquake Research Institute. (Refer to "Earthquake risk analysis of Istanbul metropolitan area," Bogazici Universitesi Kandilli Rasathanesi ve Deprem Arastirma Enstitusu, Istanbul, Turkey, 2002.)

Our scenario data include 10,414 heavily injured persons waiting to be picked up at 60 temporary medical centers near the helipads. It is assumed that AKOM has 260,600 vaccines in stock. Sufficient water and medicine supplies exist for 116,560 light-to-moderately injured persons. The total external cargo to be distributed by helicopters weighs 1054.43 tons. The supplies are stocked at various airports, seaports, and warehouses near the two central train stations, one on each side of the Bosphorus Bridge. The total number of warehouses is eight.

We include 10 major state hospitals in the relief network, some of which represent hospital clusters. Most of these hospitals are chosen in safer zones to guarantee building security. The bed capacities of the hospital clusters are aggregate. The selected hospitals are able to treat a total of 18,272 heavily injured persons. The map of the relief network is provided in Figure 7.2. In Table 7.2, we provide statistical summaries for patient capacity of hospitals, number of heavily injured persons per demand node, and requested material weight per demand node.

We assume that the Turkish Air Force (TAF) dedicates most of its heavy lift helicopters to this mission due to the economic and strategic importance of Istanbul. Each TAF helicopter is able to carry 12.247 tons of external cargo and 55 persons as internal cargo. The flight range of the helicopter is given as 2 hours and 40 minutes at sea level. We assume refueling takes place only at helipads that are near hospitals or warehouses.

7.6.1.2   Scenario analysis

The OHPC is run with just one cluster because the relief network is small (78 nodes). We allow a CPU time of 20 minutes for the CPLEX run on a laptop with 445 MB RAM. In Table 7.3, we provide the properties of the obtained solution before the RMP is invoked: the total travel time, the total stopping time (assuming 1 hour per stop), the total number of routes constructed, the number of only pickup routes, the average and maximum number of stops per route, and the percentage relative gap from the best possible solution.

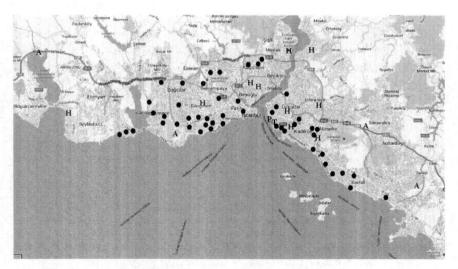

Figure 7.2 Map of the relief network (H: hospital, P: sea port; T: train station, black dot: delivery and pickup point).

Table 7.2 Statistical Summary of Scenario Parameters

	Bed capacity per hospital	Number of injured persons per demand node	Requested material weight per demand node (tons)
Average	1,827	173	17.57
Maximum	4,696	489	49.04
Minimum	100	21	2.22
Total	18,272	10,414	1054.43

Source: Özdamar, L. *OR Spektrum*, 33(3), 655–672, 2011.

Table 7.3 Properties of the Obtained Solution before the Route Management Procedure Is Invoked

Measurement	
Total travel time (hours)	9.69
Percentage relative gap from best possible solution	3.56
Total stopping time (hours)	634
Average number of stops per route	3.16
Maximum number of stops per route	5
Number of pickup-only routes	89
Total number of routes	217

Source: Özdamar, L., *OR Spektrum*, 33(3), 655–672, 2011.

We observe that the solution for the travel time objective is at most 3.56% above the best possible solution. When we analyze the solution, we observe that the optimal routes that are constructed are quite short with the average number of stops on a route being 3.16 stops. The reason why the optimal routes are so short is that helicopters fly with full internal cargo on almost all routes, carrying injured people to hospitals. The internal cargo capacity of the helicopter is 55, but the average number of injured people per demand node is three times this number. Therefore, helicopters have to fly short distances between temporary medical centers and hospitals multiple times. Eighty-nine of the routes start from hospitals and perform only the pickup function and the remaining routes start from warehouses and end at hospitals, performing both delivery and pickup functions. Because we can control the CPU time allowed for CPLEX to solve model P, we also try 10 CPU minutes and obtain a solution with a 6% gap from the best possible solution. We find that model P is quite efficient with regard to solvability, and it can be suitable for use in a dynamic environment where plans are updated every few hours.

Table 7.4 Results with Three Different Itinerary Lengths

Mission completion time	25 hours	10 hours	7 hours
Number of helicopters required to complete the mission	27	69	106
Flight hours due to additional arcs	2.62 hours	1.97 hours	1.81 hours

Source: Özdamar, L., *OR Spektrum*, 33(3), 655–672, 2011.

In the first step of the RMP, we confirm that all routes are fuel feasible. Next, we conduct an analysis on the maximum itinerary length so as to complete the mission in 25, 10, and 7 hours. In Table 7.4, the results are summarized. The additional flight times caused by the linkage of appended routes are indicated in the third row of Table 7.4. When the mission time is set to 7 hours, an average of two routes is appended into an itinerary, whereas about seven distinct routes are appended with a mission time of 25 hours. Therefore, the additional flight time caused by route union is larger in the 25-hour scenario as compared to the other two scenarios.

Table 7.4 shows that the number of helicopters required for each desired mission completion time are 27, 69, and 106, respectively. If the pilots are allowed to fly 10 hours per day, then 27 crews can complete this mission within 2.5 days. Considering the expected total number of heavy injuries, evacuation of all patients can be completed within a month. Obviously, more than 27 helicopters would be required to increase the survival rate of heavily injured patients.

7.6.2 A flood scenario

7.6.2.1 Description of the scenario

We partially reconstruct the Louisiana (the United States) flood scenario caused by Hurricane Katrina (2005), taking into consideration the evacuation of stranded citizens only. According to the 2006 report by Dr. D. L. Haulman ("The US Air Force Response to Hurricane Katrina," D. L. Haulman, Chief, Organizational Histories Branch, Air Force Historical Research Agency: http://www.afhra.af.mil/shared/media/document/AFD-070912-046.pdf), in the aftermath of the Katrina disaster, the USAF helicopters flew 648 sorties, 599 of which were on search and rescue missions that rescued a total of 4322 people. Between August 31 and September 10, 2005, 2836 people were rescued by 25 HH-60 helicopters, 1461 people by 5 MH-53 helicopters, and 25 people by 9 UH-1 helicopters. The helicopters conducted search and rescue missions, hoisting victims stranded on roofs in flooded areas of New Orleans to dry ground around the Superdome and the Convention Center.

In this scenario, we consider the rescue mission conducted by 25 HH-60 helicopters (that rescued 2836 people from roofs). The helicopters are assumed to start and end their tours at the Superdome and the Convention Center where refueling is also assumed to be carried out. The HH-60 has a flight range of 600 km and internal cargo capacity of 12 troops.

We created an evacuation demand from 620 pickup points (Figure 7.3) scattered around the flooded areas in New Orleans based on the flood map provided by FEMA (http://www.fema.gov/hazard/flood/recoverydata/ katrina/katrina_la_maps.shtm). The number of persons to be rescued from each pickup point is created randomly using a uniform distribution with lower and upper bounds (1, 8). We assume that both the Convention Center and the Superdome had the capacity to receive all survivors.

7.6.2.2   Scenario analysis

We solved this scenario using the OHPC allowing the number of clusters, K, to be 20, 30, and 40. We used a parallel server with 10 quad-core Xeon processors with 8 GB RAM, resulting in an overall number of 40 Xeon processors. In Table 7.5, we provide the total travel distances (excluding stopping times for picking up survivors) resulting from each solution as well as the CPU times, the number of routes, and the minimum, maximum, and average route lengths before the first step of the RMP is applied. We also provide the linear programming relaxation of the 622 node problem so that we can compare the solutions obtained against a weak lower bound (LB). In Table 7.5, we observe that none of the three solutions required invoking the fuel-feasibility procedure because routes are at most 253 km, which is well below the flight range of 600 km. However, the fuel spent while rescuing survivors is not considered. Also, the route lengths show great variation, the range between the minimum route length and the maximum is wide, and obviously, these routes should be merged for constructing balanced itineraries for 25 helicopters. Regarding the quality of the solutions, we obtain solutions that are within 7% above the weak LB, that is, our solutions should be much closer to the optimal integer solution. The CPU time taken for the 20 and 30 cluster solution is about 1 CPU hour, which makes the plans dynamically updatable in case new information arrives and plans need to be revised. When we compare the solutions obtained with different numbers of clusters, we observe that when $K = 40$, the top-level routing problem becomes more difficult and there are more cluster problems to solve, resulting in higher CPU time. In terms of solution quality, the lesser the clusters, the better the solution. The reason is that less clusters result in a lower degree of network aggregation at the top level.

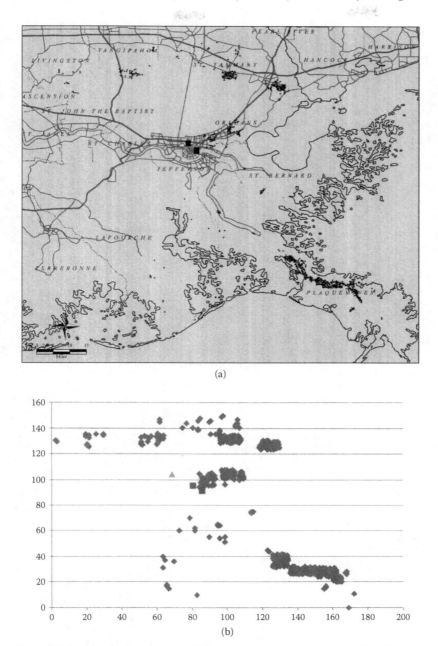

(a)

(b)

Figure 7.3 (a) Distribution of rescue request locations on Louisiana map in the Katrina scenario (♦: rescue request; ν: superdome and convention center). (b) Distribution of rescue requests in the Katrina scenario (♦: rescue request; ν: superdome and convention center; ▲: airport).

Table 7.5 Results for the Katrina Rescue Scenario with Different Numbers of Clusters

Number of clusters (K)	Distance traveled (km) (% above lower bound)	Number of routes	Average route length (km)	Maximum route length (km)	Minimum route length (km)	CPU (seconds)
20	31,515 (6.46)	241	131	253	7	3,786
30	31,541 (6.55)	246	129	253	10	3,860
40	32,711 (10.5)	252	129	253	6	8,625
Linear program relaxation (lower bound)	29,601	–	–	–	–	37,273

Table 7.6 Results for the Katrina Scenario with Cluster Number $K = 30$ Including Stop Times

$K = 30$	Distance traveled (includes stop km)	Number of routes	Average route length (km)	Maximum route length	Minimum route length
Before route management procedure is invoked	80,397	246	326	490	164
After second step of route management procedure	80,403	21	3,828	3,999	1,328

In Table 7.6, we indicate the same performance measures for the solution obtained with a cluster number of 30 before and after the second step of the RMP is invoked. Our goal is to observe the additional travel distance caused by introducing a stopping time of 15 minutes per visit, which is converted into a travel distance of 40 km per stop (due to fuel spent while hovering above a roof). We would also like to observe the additional flight distances caused by invoking both steps of the RMP.

In Table 7.6, we find that approximately 50,000 km worth of fuel is spent while hovering over roofs. The average route length extends by almost 200 km, implying that there are five stops on a route on the average. Here, the first step of the RMP need not be invoked because the longest route length is 490 km, which is below the 600 km limit.

The second step of the RMP is implemented with an aspired itinerary length of 4000 km. The resulting number of itineraries is 21, which is equivalent to the theoretical minimum number of helicopters necessary to complete the mission (the theoretical minimum is calculated by dividing the total distance traveled by the aspired itinerary length). Here, the length of the shortest itinerary is 1328 km (there is only one such short itinerary), whereas the average itinerary length is 3828 km. The additional flight distance caused by merging the routes is minimal because only two routes (out of the 246) start and end with a different node. Assuming a low flight speed, an itinerary of 4000 km might imply a flight duration of 15 hours per helicopter.

7.7 Conclusion

Our target is to coordinate the logistics of the postdisaster relief delivery and rescue/evacuation activities that are carried out by helicopters. With this goal, we describe an efficient OHPC that is based on clustering demand nodes and solving the aggregate problem first to find the optimal allocation of warehouses and hospitals among demand clusters. Once the aggregate solution is obtained, the detailed routing problem within each cluster's subnetwork is solved (on parallel computers) by letting relevant parts of the top-level solution become problem parameters. The hierarchical parameter links that are imposed between different levels of planning do not allow discrepancies between the solutions at consecutive aggregation levels. Once the routes are finalized, the OHPC makes them fuel feasible and merges them together into distinctive helicopter itineraries, enabling the DM to specify a desired mission completion time.

We implement the OHPC on two scenarios, the first concerning a potential Istanbul earthquake and the second involving the rescue mission conducted by helicopters after the flooding from Hurricane Katrina. Our results show that near-optimal solutions are obtained within reasonable CPU times making the method usable in a dynamic decision-making environment.

Acknowledgment

We thank the Scientific and Technological Research Council of Turkey (TUBITAK) who partially supported this study with research grant no. 110M578.

References

Altay, N., Green, W. G. 2006. OR/MS research in disaster operations management. *European Journal of Operational Research* 175: 475–493.

Apte, A. 2009. Humanitarian logistics: A new field of research and action. *Foundations and Trends in Technology, Information and Operations Management* 3: 1–100.

Apte, A., Heath, S. 2009. A plan for evacuation of disabled people in the face of a disaster. Working Paper, Naval Postgraduate School, Monterey, CA.

Armstrong-Crews, N. L., Mock, K. J. 2005. Helicopter Routing for Maintaining Remote Sites in Alaska using a Genetic Algorithm, 1586–1587. AAAI Press/ The MIT Press.

Bakuli, D. L., Smith, J. M. 1996. Resource allocation in state dependent emergency evacuation networks. *European Journal of Operations Research* 89: 543–555.

Balcik, B., Beamon, B., Smilowitz, K. 2008. Last mile distribution in humanitarian relief. *Journal of Intelligent Transportation Systems* 12: 51–63.

Barbarosoglu, G., Özdamar, L., Cevik, A. 2002. An interactive approach for hierarchical analysis of helicopter logistics in disaster relief operations. *European Journal of Operational Research* 140: 118–133.

Bard, J. F., Jarrah, A. I. 2009. Large scale constrained clustering for rationalizing pickup and delivery operations. *Transportation Research Part B* 43(5): 542–561.

Campbell, A. M., Vandenbussche, D., Hermann, W. 2008. Routing for relief efforts. *Transportation Science* 42(2): 127–145.

Caunhye, A. M., Xiaofeng, N., Shaligram, P. 2011. Optimization models in emergency logistics: A literature review. *Socio-Economic Planning Sciences* 46(1): 14–13.

Chiu, Y. C., Zheng, H. 2007. Real time mobilization decision for multi-priority emergency response resources and evacuation groups: Model formulation and solution. *Transportation Research: Part E* 43: 710–736.

DeAngelis, V., Mecoli, M., Nikoi, C., Storchi, G. 2007. Multiperiod integrated routing and scheduling of World Food Programme cargo planes in Angola. *Computers and Operations Research* 34–36(6): 1601–1615.

de la Torre, L. E., Dolinskaya, I. S., Smilowitz, R. K. 2011. Disaster relief routing: Integrating research and practice. *Socio-Economic Planning Sciences* 46(1): 88–97.

Doerner, K., Focke, A., Gtjahr, W. 2007. Multicriteria tour planning for mobile healthcare facilities in a developing country. *European Journal of Operational Research* 179: 1078–1096.

Fiedrich, F., Gehbauer, F., Rickers, U. 2000. Optimized resource allocation for emergency response after earthquake disasters. *Safety Science* 35(1–3): 41–57.

Haghani, A., Oh, S. C. 1996. Formulation and solution of a multi-commodity, multi-modal network model for disaster relief operations. *Transportation Research Part A-Policy and Practice* 30: 231–250.

Hodgson, M., Laporte, G., Semet, F. 1998. A covering tour model for planning mobile health care facilities in Suhum district, Ghana. *Journal of Regional Science* 38: 621–638.

Huang, M., Smilowitz, K., Balcik, B. 2011. Models for relief routing: Equity, efficiency and efficacy. *Transportation Research Part E: Logistics and Transportation Review* 48(1): 2–18.

Jozefowiez, N., Semet, F., Talbi, E. G. 2007. The bi-objective covering tour problem. *Computers & Operations Research* 34(7): 1929–1942.

Lin, Y. H., Batta, R., Rogerson, P., Blatt, A., Flanigan, M., Lee, K. 2011. A logistics model for emergency supply of critical items in the aftermath of a disaster. *Socio-Economic Planning Sciences* 45(4): 132–145.

Menezes, F., Porto, O., Reis, M. L., Moreno, L., de Aragao, M. P., Uchoa, E., Abeledo, H., do Nascimento, N. C. 2010. Optimizing helicopter transport of oil rig crews at Petrobras. *Interfaces* 40: 408–416.

Mete, H. O., Zabinsky, Z. B. 2010. Stochastic optimization of medical supply location and distribution in disaster management. *International Journal of Production Economics* 126(1): 76–84.

NATO Research and Technology Organization. 2008. Technical Report No. TR-SAS-045. Computer Based Decision Support Tool for Helicopter Mission Planning in Disaster Relief and Military Operations. NATO, Brussels, Belgium.

Nolz, P., Doerner, K., Gutjahr, W., Hartl, R. 2010. A bi-objective metaheuristic for disaster relief operation planning. *Advances in Multi-Objective Nature Inspired Computing*. Studies in Computational Intelligence, ed. C. Coello, Dhaenens, L. Jourdan, 272: 167–187. Berlin/Heidelberg: Springer.

Özdamar, L. 2011. Planning helicopter logistics in disaster relief. *OR Spektrum* 33(3): 655–672.

Özdamar, L., Demir, O. 2012. A hierarchical cluster and route heuristic for large scale disaster relief logistics planning. *Transportation Research Part E: Logistics and Transportation Review* 48: 591–602.

Özdamar, L., Ekinci, E., Kucukyazici, B. 2004. Emergency logistics planning in natural disasters. *Annals of Operations Research* 129: 217–245.

Özdamar, L., Pedamallu, C. S. 2011. A comparison of two mathematical models for earthquake relief logistics. *International Journal of Logistics Systems and Management* 10(3): 361–373.

Sayyady, F., Eksioglu, S. D. 2008. Optimizing the use of public transit system during no-notice evacuations of urban areas. Working Paper, Mississippi State University, Mississippi.

Sierksma, G., Tijssen, G. A. 1998. Routing helicopters for crew exchanges on offshore locations. *Annals of Operation Research* 76: 261–286.

Solomon, M., Chalifour, A., Desrosiers, J., Boisvert, J. 1992. An application of vehicle routing methodology to large-scale larvicide control programs. *Interfaces* 22: 88–99.

Stapleton, O., Pedraza, A. M., Van Wassenhove, L. N. 2009. Last mile vehicle supply chain in the International Federation of Red Cross and Red Crescent Societies. Faculty and Research Working Paper, INSEAD, France.

Timlin, M. T., Pulleyblank, W. R. 1992. Precedence constrained routing and helicopter scheduling: Heuristic design. *Interfaces* 22: 100–111.

Tzeng, G., Cheng, H., Huang, T. 2007. Multi-objective optimal planning for designing relief delivery systems. *Transportation Research Part E: Logistics and Transportation Review* 43(6): 673–686.

Wright, P. D., Liberatore, M. J., Nydick, R. L. 2006. Survey of operations research models and applications in homeland security. *Interfaces* 36(6): 514–529.

Yan, S., Shih Y. L. 2009. Optimal scheduling of emergency roadway repair and subsequent relief distribution. *Computers & Operations Research* 36(6): 2049–2065.

Yi, W., Özdamar, L. 2007. A dynamic logistics coordination model for evacuation and support in disaster response activities. *European Journal of Operational Research* 179: 1177–1193.

chapter eight

Planning and management of transportation systems for evacuation

P. Brian Wolshon and Vinayak Dixit

Contents

8.1 Introduction

Emergency evacuations are defined as the prompt and rapid movement of people away from a threat or actual occurrence of a hazard. The extent of an evacuation and urgency at which it must be carried out is generally a function of the threat or hazard, including its size and spread; speed and direction of movement; and potential lethality and destructiveness. These factors also influence the behavioral response of evacuees. Past research, observation, and experience have shown that individual decisions of whether or not to evacuate are based predominantly on the level of threat perceived by that individual. Thus, factors like the level of danger posed by a threat, one's proximity to it at present or in the near future, and the responses of others within the social network of the potential evacuee all influence decision making.

From a transportation perspective, the hazard conditions and behavioral responses they elicit are important because they influence evacuation *travel demand* or the amount of people and/or vehicles that are expected to be moving during an event. Perhaps even more important than just the number of people are the temporal and spatial characteristics of this demand generation. Where evacuees leave from, when they leave these locations, where they travel to gain safe shelter, what routes they will select to undertake the travel, and even what modes of transportation they will use to make the trip all govern the ability of transportation networks and assets to adequately serve this demand in a timely manner. Key performance parameters such as travel time and clearance time; the formation and recovery of congestion; travel delay; and so on are all the direct result of how and where evacuees load themselves on a network.

To illustrate this idea, Tamminga et al. (2011) used traffic simulation modeling to show that the total amount of time needed to evacuate an area, also known as *clearance time*, is increased significantly if the time window for evacuees to respond to an evacuation order is reduced. In effect, transportation networks become fully saturated and then overloaded with vehicles, similar to what might be seen in the evacuation of a crowded room or movie theater. If everyone ran to leave at the same time, the aisles and area immediately in front of the exit doors would become congested, and the smooth and orderly movement of people would break down resulting in longer wait times for those further behind. Conversely, if the movement of people could be regulated such that there was no stopping or waiting, the overall time required to clear them would be significantly less. The daunting question then becomes how to regulate the flow in the face of danger and provide adequate capacity when threatening conditions arise.

The following sections of this chapter highlight these and other key concepts related to the planning, operation, and management of transportation systems, assets, and personnel for emergency evacuations. Although written from a broad perspective that includes many key concepts from the field of emergency management, the chapter focuses on a transportation-based view of evacuations. This includes how transportation planners, engineers, and managers view and assess evacuations from the standpoint of demand generation and providing capacity within transportation systems to best serve those needs. Although many aspects are quite similar to routine transportation practices, the analysis of and planning for emergencies and evacuations present many unique challenges. This chapter highlights them as well as emerging, innovative, and effective practices that have been identified through recent research and experience. The purpose of this chapter is not to make readers experts at all aspects of evacuation-related transportation, but rather to acquaint them (especially those with no or limited backgrounds) with the key needs, aspects, policies, and practices of transportation planning and engineering for evacuations.

8.2   Management and direction of evacuations

In terms of hazard response, evacuations are often the last step in a process that begins several minutes, hours, or even days earlier. Thanks to recent advancements in weather forecasting, threats like hurricanes, for example, can give advanced warning times of nearly a week. This amount of lead time gives emergency response and transportation agencies a significant advantage in being able to anticipate the amount of demand that may be created and then prepare and implement responses for these conditions. Unfortunately, however, hurricanes tend to be an exception to the rule of evacuations. The vast majority of events requiring evacuations give little, if any, advanced warning.

Given the wide variation in factors that influence the scale and urgency of evacuations, it is helpful to recognize the role transportation plays in evacuations and how it influences the process. To help in that understanding, the following sections discuss the management of evacuation in the United States and how evacuation practices have evolved based on recent experiences as well as types of conditions for which evacuations may be needed and some background on.

8.2.1   Role of transportation in evacuations

What often comes as a surprise to those unfamiliar with evacuations is that, historically, they have been carried out with little input from transportation agencies and professionals. The daily activities of transportation planners, engineers, control systems managers, maintenance technicians, and others have a primary focus on providing consistent, safe, and efficient transportation to the traveling public. Traditionally, this has meant keeping up with routine needs like reducing roadway congestion, enhancing highway safety, and designing, constructing, and maintaining physical infrastructure. Given these responsibilities and perpetually limited financial resources of transportation agencies, there has, in the past, been little time, money, or, in some cases, interest in taking on the additional responsibilities of emergency transportation management. Similarly, emergency management professionals have not always been aware of the capabilities and resources available from their transportation counterparts.

It has not been until the past decade or so that the planning and management of evacuations has moved away from the exclusive arena of emergency management agencies to involve agencies with a primary transportation focus. Since the highly publicized failings of Hurricanes Floyd, Katrina, and Rita as well as the more recently recognized potential from threats like terrorist attacks and tsunamis, state Departments of Transportation, regional transit authorities, and other providers of transportation services have become key players in the planning and management of evacuations. This involvement has also been formalized through the development of the National Incident Management System, in which the roles and responsibilities of transportation during emergencies is spelled out in Emergency Support Function 1 (FEMA 2008).

8.2.2   Hazards and the need for evacuation

Evacuations occur more frequently in the United States than is often realized. A recent U.S. Nuclear Regulatory Commission (NRC) report showed that, on average, an evacuation involving 1000 or more people occurs once every 2–3 weeks somewhere in the country (Dotson and Jones 2005). While evacuations are most often associated with hurricanes, the need to evacuate can come from many different types of hazardous events that, themselves, can be classified in numerous different ways. One method distinguishes hazards in terms of origin, *natural* or *man-made*. Common natural hazards include hurricanes, floods, earthquakes, wildfires, and tornados. Man-made hazards include technological hazards like biological, chemical, and radiological hazards, for example, anthrax attacks, toxic chemical spills, and nuclear power plant accidental releases as well as conditions like explosions, fires, hazardous material spills, and so on. Man-made conditions may be further subclassified in terms of *intentional* or *unintentional*. Although it is likely that the majority of man-made hazards are unintended consequences of human activity, some are purposefully initiated. Occurrences such as wars and terrorist attacks are both examples of intentionally man-made events that have required evacuations.

The NRC evacuation study statistics, compiled by the Sandia National Laboratories over the 10-year period between 1993 and 2003, suggest that the vast majority of incidents requiring an evacuation tend to be small, localized events. Of the 230 evacuation events, 171 of them or nearly 80% involved 1000 people or less. Although wildfires and floods were the hazards for which an evacuation was most commonly required, hazardous material incidents and transportation-related events also accounted for more than a quarter of the documented evacuations over this period.

No matter their size, evacuations are by nature disruptive events that can have significant economic and political consequences. The costs of hurricane evacuations, for example, can exceed $1 million per mile of coastline from losses in tourism, commerce, and general productivity. Evacuation orders can also be difficult to implement because the movement and development of hazards can change over a short period of time. The scope and breadth of an evacuation must also be proportional to the threat, so evacuation orders need to be sufficiently large to protect people, but not unnecessarily large that they needlessly disrupt the economic activity of a region, resulting in an over- or "shadow" evacuation, or worse, leading to a "cry wolf" perception among the public.

Currently, the United States is somewhat unique among countries in the world in that evacuations are used as a primary protective action during disasters and emergencies. Most countries use limited, if any, highway-based evacuation for hurricanes. One of the reasons for this is that unlike many other countries, the United States has the ability to move large numbers of people to significant distances in a timely and safe manner and then shelter them at locations away from the hazard zone. This is likely to change in the future as developing countries increase their rates of automobile ownership and roadway infrastructure.

Over time, some critics of evacuation have also argued that private transportation availability has led to an overreliance on evacuations. They contend that the need to evacuate for hurricanes could be reduced by strengthening building codes and increasing the availability of locally based shelter facilities that do not require long travel distances. Such planning could also reduce or eliminate the inability of transportation networks to adequately serve the amount and immediacy of demand generated by a large-scale evacuation.

8.2.3   Evacuation types

For planning and management purposes, evacuations associated with these very hazards are often categorized based on the amount of advanced warning time they give. Although some hazards develop and/ or move slowly, giving days or even weeks of advanced warning time, others can occur with no advanced warning. These temporal differences are important because they limit the ability to issue and receive critical evacuation information as well as allow or prohibit time to implement citizen-assisted evacuation plans and capacity adding traffic management measures.

As part of their information on the planning and implementation of transportation measures for emergency events, the United States Department of Transportation (USDOT) defined several types of incidents in terms of the amount of advanced notice they give (Zimmerman et al. 2007). Although the USDOT definitions are not specific in the precise amount of advanced warning time, they broadly characterize them into the general groups of *advanced notice* and *little/no notice*.

An obvious example of an advanced-notice event is a hurricane that can be identified and tracked up to several days in advance. The key factor is the time such hazards afford to assess options and then formulate and/ or implement actions. The USDOT describes advanced-notice events in the following way (Zimmerman et al. 2007).

> With an advance-notice evacuation, information becomes available during the Readiness Phase regarding the incident that has occurred and the factors that may require an evacuation. Decision makers have time to collect the information they need to determine whether an evacuation should be ordered and, if so, the best way to carry it out.

In contrast, a no-notice event leaves no time for preevent activity and is characterized as the following by the USDOT:

> A little- or no-notice incident is one that occurs unexpectedly or with minimal warning. The lack of warning and the quick response time required introduce distinct challenges for evacuating at-risk populations. No-notice incidents do not provide emergency responders sufficient time to prepare for a specific incident. This greatly affects agencies' abilities to pre-activate emergency protocols, pre-position needed assets, and warn and direct the public. No-notice evacuations require a significantly different approach to planning than advance-notice evacuations because they will be based on a set of capabilities and strategies that will likely be more limited in the time and resources available for implementation.

There is also a third, much more loosely defined, type of evacuation event known as *short-notice* events. Opinions range widely on what constitutes a short-notice event, but in general, they are described as lying between no-notice and long-notice events. Some define the amount of advanced notice as short as 30 minutes, whereas others extend it to 24 hours or more. From a transportation emergency response standpoint, short-notice events appear to permit some level of traffic management, but nothing as extensive as the implementation of regional contraflow or full transit-assisted evacuation plans. One example of a short-notice event is a fast-moving wildfire or nuclear power plant emergency in which one to several hours may be available to issue orders and give route and destination guidance to evacuees.

8.2.4   Evacuation orders

Once an evacuation is deemed necessary, the extent and type of evacuation must be determined and then communicated to the affected populations. The type of order and level of urgency is dependent on the characteristics of the hazard and clearance times described earlier. Currently, there is no formalized language or standards that govern the ordering or communication of evacuations in the United States.

Reviews of evacuation policies and practices (NCHRP 2009) show that, nationwide, evacuations are most often classified as *recommended* or *mandatory*. Recommended evacuations are typically used to warn people when a threat to life and property exists or will likely exist in the immediate future. Although people who receive such warnings are not required to evacuate, it is to their advantage to do so. In hurricane scenarios, recommended evacuation orders are targeted toward people most vulnerable to hurricane storm surge and extreme winds, including offshore workers, persons on coastal islands, and other special populations having particularly long lead-time requirements. From a traffic perspective, recommended evacuations are also used as a way to motivate the most threatened people to move first and clear more heavily populated areas before later, more urgent, evacuation orders have the potential to cause congestion and delays along the travel routes. No special traffic control or transportation measures are usually taken during recommended evacuations and people may remain if they so choose.

Past experience has shown that mandatory evacuation orders are considered to be the most serious by potential evacuees. Realizing this, public officials charged with emergency management use this for the most threatening of conditions. During a mandatory evacuation, authorities put maximum emphasis on encouraging evacuation and limiting ingress into threatened areas. Mandatory orders are also made when many evacuation transportation network management plans (like contraflow and transit-assisted evacuation) go into effect.

Although people are "required" to leave under mandatory evacuation orders, such orders are difficult to enforce, and most government agencies lack both the resources and the legal authority to compel threatened individuals to leave. In the past, many people have resisted orders to leave their homes and property by government officials. Under such conditions, emergency management officials acknowledge that if a person wants to stay, the state will not physically remove them even if it is absolutely certain that they would be harmed. In discussions with county law enforcement officials in California, it was found that some deputies were able to encourage mandatory evacuations by compelling parents to release minor children to authorities under child endangerment laws.

Once children were taken into protective custody, parents would make decisions to leave as well.

Prior study has also shown that other, more ambiguous, evacuation terminology is also common in the United States. Words like *voluntary* and *precautionary* are also used in some locations and the terms "recommended" and "precautionary" are often used interchangeably and are not necessarily as clearly defined as the previous two. One agency described its precautionary evacuations as "prevoluntary" and thought of them as a way to get people or entities in need of long preparation times or those in recognized at-risk areas to move toward action. Again, decisions of whether or not to leave are left to individuals and few special transportation arrangements are made.

The definition and terminology of evacuation declarations are important because they impact people's decision of whether or not to leave. Prior research has shown that people who said they heard mandatory evacuation orders are the most likely to evacuate, while recommended evacuation orders are met with less urgency (PBS&J 2000). The type of evacuation order and how it is communicated is also critical to avoid spontaneous or shadow evacuation.

Shadow evacuations occur when people who believe they are at risk evacuate even though they have not been officially advised or recommended to do so (Gunter 2001). Shadow evacuees most often leave because of concern about safety but could also leave for other reasons. Authorities in Florida and Texas feel that one of the reasons for the extreme number of evacuees during Hurricane Floyd, then later in Rita, was the result of shadow evacuations. However, it has been suggested that the overevacuation problem in Houston also resulted from vague and inconsistent instructions provided by the authorities.

8.3   *Evacuation transportation processes*

The last 15 years have seen tremendous advances in the way that evacuations are planned, implemented, managed, and analyzed. Techniques such as contraflow, staged and phased evacuations, cross-state regional coordination, special needs and transit-assisted evacuation, as well as planning and analysis techniques like regional multimodal traffic simulation have come about due in large part to a series of high-profile failures. Now, there is also a recognition of a greater list of hazards for which evacuations are now required to serve as a protective action. This fact coupled with changing population demographics and changing climatological patterns have also resulted in natural hazards like hurricanes that are being forecast to occur both with higher frequency and increasing levels of severity into the future.

The need for, size of, and urgency at which evacuations are carried out can vary widely based on a number of key variables associated with the timing, coverage area, and movement of a threat. These spatiotemporal variables largely dictate, most notably, the extent of the hazardous conditions that are created and the amount of time until they arrive. This can be further complicated by whether the source of the hazard is also known in advance (known as a fixed-site hazard like a nuclear power plant) or whether the hazard location can occur in a random location as is predominantly the case for wildfires, train derailments, highway chemical spills, and so on. In turn, all of these conditions come together to influence the implementation of protective actions and the guidance of evacuees, including how many evacuees need to leave, how far they will need to travel, and the types of traffic control and emergency proactive actions that can be taken to expedite the evacuation process.

Hazards and their effects, however, are just one part of the equation. Two other important considerations in planning and implementation of evacuations are the demographics and behavioral response of the people who will be evacuating as well as their access to transportation resources. Prior research has shown that socioeconomics can play a significant role in both the decision of when to evacuate (if at all) and how this travel will be accomplished. This dictates when and how fast the network will load as well as the need, amount, and type of additional supporting transportation assistance (shared rides, busses, trains, shelter in place, etc.). Transportation resources also go beyond providing busses for assisted evacuations and the personnel to carry them out. The ability to evacuate is also a function of the network configuration and coverage (e.g., where roads are and where they are not) and capacity of the routes within it.

In the United States, much of the knowledge about evacuations, particularly those carried out on a mass scale, has come from experiences associated with hurricanes. However, the conditions necessitating evacuations extend beyond this single threat. They also include a wide spectrum of naturally occurring and man-made occurrences that occur with and without notice and can be as localized as a single building or city block or as large as a multistate region. Although each of these hazards brings with it a unique set of challenges and preparedness action needs, the evacuation process for all of them are actually surprisingly similar. What changes is the scale and timing at which it is implemented.

The following sections discuss the key spatial and temporal considerations that are considered in evacuation planning and analysis as well as how various hazard conditions impact the planning process. The discussion also includes examples of how these various parameters and systems have been applied, adapted, and modified for use in evacuation. Knowledge of these processes and being able to frame them in terms of time and space are also quite valuable in modeling from two perspectives. First, it makes it possible to disaggregate the often enormously complicated and interrelated processes of the evacuation into separate, smaller components that are easier to observe and record. Then, it permits each of these key components to be represented quantitatively as equations or as a distribution of continuous data so that they can be described and analyzed using computer model representations.

8.3.1   Temporal parameters

Among the temporal evacuation variables, one of the most critical is the amount of advanced notice available prior to the onset of hazardous conditions discussed earlier. Advanced warning time is also important because it dictates the amount of notice that response agencies have to implement in control and management measures like contraflow, road closures, and emergency signal timing plans as well as the time to activate assisted-evacuation plans like evacuation bus services, and medical special needs evacuations for the elderly, infirm, and disabled. It can also limit or extend the amount of time that evacuees have for preevacuation mobilization activities, such as picking up children from school, coordinating with mobility-limited friends and relatives, closing homes and businesses, gathering materials and supplies, and so on. Although it may seem that there is little time for these types of time-consuming activities during short-notice hazmat incidents, for example, there are numerous cases where they could still be used. Nuclear power plant emergencies, in particular, are commonly planned assuming that several hours will be available between the time a reactor emergency occurs and the time at which a containment breach begins to pose a risk to persons within the plume exposure area. This could permit, among other activities, police control check points and contraflow lane reversals to be implemented and school and other assisted evacuation plans to be initiated.

Advanced warning time also affects the ability of officials to issue evacuation orders because they must be communicated through various formal (e.g., media releases, reverse 911 calls, etc.) and informal means, including social networks (e.g., friends, family, coworkers, neighborhoods, etc.). This lead time is particularly important for nonresident transient populations who may be within the evacuation zone for work, shopping, and other recreational activities. Research and development work, particularly related to hurricane evacuations, in modeling behavioral responses under the various advanced-notice conditions is available in the literature. This information has been quantified and adapted to create evacuee departure times and response distributions that can be used in simulation models of evacuation. With additional calibration or adjustment, it is expected that even hurricane evacuation behavioral responses could be adapted for use in analyzing short- to no-notice events such as those commonly associated with hazmat incidents.

Once the evacuation is underway, there are numerous other temporal parameters that can be used to evaluate the performance of evacuation processes. From a transportation analysis perspective, these can include evacuation travel time and delay as well as the time needed to implement and/or remove evacuation traffic management and control measures like contraflow, road and bridge closures, and police control points. From an emergency management perspective, key temporal parameters may include onset time and duration of hazard conditions, time to issue evacuation orders, evacuee mobilization time, and so on. Again, these processes have been quantified in prior work for use in planning and evaluating evacuation alternatives.

8.3.2 Spatial parameters

Like the temporal parameters described previously, evacuations also encompass a range of spatial parameters that also influence the manner in which evacuations are planned and carried out. They can also be used to evaluate the effectiveness of evacuation plans and identify areas of need and improvement.

Among the most essential spatial parameters that dictate the size of the evacuation protective action zone is the spatial extent of the hazard. Obviously, the larger the area of threatening conditions, the larger the area that must be evacuated. However, spatial distribution of the resident and transient population within the threatened population are what actually influence the number of people and vehicles that would be in the evacuation. In the past, large storms threatening thousands of square miles have made landfall in sparsely populated areas of south Texas, for example, did not require major evacuations. Alternatively, although the 9/11 terrorist attacks in New York affected several city blocks, it required the evacuation of several million people. Although most hazmat incidents would be expected to be considerably smaller in scale than a hurricane, there are many hazmat hazard scenarios that could affect hundreds of square miles.

Two other hazard parameters that influence the urgency, extent, and direction of the evacuation are the approach direction and movement of threatening conditions. An illustrative example of the effect of moving hazards on an evacuation is wildfires in urban-wildland interface areas. Because of the highly variable development and movement of wildfires, which are themselves a function of weather and fuel conditions, it is not possible to develop specific detailed evacuation plans. In Southern California, for example, emergency preparedness and response agencies find it more effective to work from a general evacuation framework, rather than a plan, to permit greater flexibility to respond to rapidly changing fire conditions. This includes designating the geographic extent of the threat region, amount of available advanced warning time, available routes, and even shelter destinations. With the exception of nuclear power plant facilities that are typically planned to assume a fixed-site emergency, the majority of hazmat incidents would be expected to create evacuation conditions that are more similar to wildfires where the locations of the protective action zone and urgency at which the evacuation need to be conducted is not known in advance and can change rapidly based on wind strength and direction.

Other important spatial parameters that affect evacuation processes are the location and required travel distance to safe shelters, arrangement and access to transportation networks, and the location and frequency downstream bottlenecks. For hazards like nuclear power plant emergencies and hurricanes, shelters are planned well in advance of the emergency and, as such, evacuation travel is all planned to reach them. However, for wildfires, safe-shelter destinations as well as the routes recommended to reach them may change from event to event or even several times within a single event. Similarly, the available road network including intersections, merges, terminal points, and capacity restrictions (e.g., bridges, tunnels, and so on) influence the direction of movement away from the threat.

8.4   Evacuation travel behavior and response

Understanding the evacuation behavior of the population is critical for devising traffic management strategies to safely evacuate people from the path of a major hurricane. Literature on evacuation behavior suggests the common perception of irrational evacuation behavior during hurricanes (possibly due to panic) is not accurate (Quarantelli 1985; Tierney et al. 2006). In fact, people collectively act rationally during evacuation, and their decision to evacuate depends on factors such as direct perception of threat (Mikami and Ikeda 1985) and issuance of evacuation notice (Mikami and Ikeda 1985; Sorenson and Mileti 1988; Fitzpatrick and Mileti 1991). With the premise of rational evacuation behavior, there should be models that are able to explain evacuation behavior based on certain environmental and demographic factors. Baker (1991) found that housing and storm-specific threat factors also affected the evacuation behavior. Hultaker (1983) noted that families tend to make decisions about evacuation collectively and not on an individual basis. In a study of parishes in Southeastern Louisiana, it was found that people whose homes were damaged by an earlier hurricane were more likely to heed the official recommendation to evacuate (Howell and Bonner 2005).

The evacuation behavior studies can be categorized based on the questions they attempt to answer. Drabek (1983) referred to the research problem of trying to understand why some subjects evacuate whereas others do not, as the "Shall we leave?" question. Another critical group of questions categorized by Sorensen (1991) was the "When shall we leave?" question. This question relates to the variations in departure times during a single hurricane or any short-notice disaster for that matter. An important component of these variations, in the context of a short-notice disaster, is the time spent in preparing for the evacuation after making the decision to evacuate. This duration is referred to as mobilization time in this study and relates to "Why do we leave when we do?" question (Sorenson 1991). Understanding these variations can be used to generate empirical data-based traffic loading rates that can in turn be used for evacuation planning. According to Sorenson (1991), the relationship between mobilization time and characteristics of the evacuees is very critical for developing improved evacuation plans. One of the studies attempting to understand mobilization time, sometimes also referred to as "evacuation delay," has found that households with older members and pets have higher mobilization time due to the need for appropriate transportation (Sorenson 1991; Vogt 1991; Heath et al. 2001).

The time it takes to decide to evacuate has been found to be affected by the time of issuance of the "order to evacuate." However, the use of individual household's decision time to evacuate for defining the mobilization time accounts for not only the time at which orders to evacuate were first issued for an area, but also the time the evacuating households received the information. Hence, the effect of these external factors is already incorporated in the mobilization time because it is obtained by subtracting the decision times (affected by *external conditions*) from the departure times. Therefore, the authors are of the opinion that *mobilization time* for an individual household may be better defined using the time the household actually decided to evacuate.

A study conducted by U.S. Army Corps of Engineers (USACE) (2000) showed that the evacuation response rates follow an S-curve and that 10% of all evacuees had left by the time evacuation orders were delivered. The evacuation patterns are different if the evacuation is due to a hurricane that was recently preceded by another hurricane. On the basis of data from Hurricanes Charley and Frances that hit Florida in 2005, Dixit, Pande, Radwan, and Abdel-Aty (2008) found that the number of evacuees that left at the end of the first day was 10% higher during a recent subsequent hurricane as compared to those observed during an earlier one. They found that home ownership, number of individuals in the household, income levels, and level/risk of surge were significant in the USACE model explaining the mobilization times for the households during a subsequent hurricane. Although pets, the elderly, and children in households are known to increase the mobilization times during isolated hurricanes, they were not significant in the model.

Alsnih et al. (2005) summarized the research on evacuation demand, illustrating both a general model of evacuation behavior, as well as the response curves. Fu and Wilmot (2004) developed a sequential logit model to estimate evacuation response. An attractive aspect of this model is that it is not only able to predict when people will leave, but also how many will leave. Later, Fu and Wilmot (2006) suggested a survival analysis–based evacuation response model. Continuing this work, Fu et al. (2007) calibrated an evacuation response curve model for Hurricane Floyd (made landfall in 1999) in South Carolina and used the model to predict evacuation behavior for Hurricane Andrew (made landfall in 1992) in Southeastern Louisiana. In both regions, these were the first hurricanes of the corresponding seasons, and the population in these regions were not affected by recent prior hurricanes. Their study did not find any statistically significant difference between the predicted response curve and the actual response curve for Hurricane Andrew. Also, the model developed by Fu et al. (2007) was shown to transfer reliably across different regions and hurricanes. The sequential logit model (Equations 8.1 and 8.2) proposed by Fu et al. (2007) is recommended to be used for evaluating departure time choice. L_i denotes the probability of number of evacuees leaving, while S_i denotes the probability of evacuees staying till time i. CL_i is the cumulative probability of leaving at time i (Equation 8.3). Where S_1 is equal to 1 and Table 8.1 explains the parameters that were calibrated by Fu (2004).

$$L_i = \frac{e^{0.555\text{Flood}+0.267\text{Mob}+0.008\text{Speed}_i+1.543\text{TOD1}_i+1.721\text{TOD2}_i+1.681\text{dorder1}_i+1.998\text{dorder2}_i+5.247\text{Gamdist}_i-8.18}}{1+e^{0.555\text{Flood}+0.267\text{Mob}+0.008\text{Speed}_i+1.543\text{TOD1}_i+1.721\text{TOD2}_i+1.681\text{dorder1}_i+1.998\text{dorder2}_i+5.247\text{Gamdist}_i-8.18}}S_i \quad (8.1)$$

$$S_i = \prod_{j=2}^{i-1} \frac{1}{1+e^{0.555\text{Flood}+0.267\text{Mob}+0.008\text{Speed}_i+1.543\text{TOD1}_i+1.721\text{TOD2}_i+1.681\text{dorder1}_i+1.998\text{dorder2}_i+5.247\text{Gamdist}_i-8.18}} \quad (8.2)$$

$$CL_i = \sum_{j=1}^{i} L_j \quad (8.3)$$

It is important to note that the model was developed based on six-hour time intervals, and therefore, the resulting response curve should have the same resolution. To highlight the use of this model, an example hurricane is considered in Table 8.2. To further demonstrate the model, the evacuation response curve is developed for the case when the residences in the region are believed to be flooded, as shown in Table 8.3.

The cumulative probabilities of leaving that were calculated in the last columns of Tables 8.2 and 8.3 are plotted in Figure 8.1.

The total probabilities of evacuating increases when the residences are under flood risk. As mentioned earlier, this model not only captures evacuation departure rates, but also captures how many people will evacuate. The probabilities, when multiplied with the total population of the area, will predict the total people who have evacuated in a given time period.

More recently, there have been more advanced statistical models such as Hasan et al. (2011) and Dixit et al. (2012) that have used more sophisticated modeling paradigms, but as of yet have not been able to demonstrate transferability (Fu 2004).

Table 8.1 Dynamic Evacuation Departure Time Model

Evacuation response model		
Covariate	Definition	Coefficient
Intercept	Model constant	−8.18
Flood	1 if the residence is believed very likely to be flooded, 0 otherwise	0.555
Mobile	1 if residence is a mobile home, 0 otherwise	0.267
Speed	Hurricane wind speed (mph)	0.008
TOD(1)	Time of day, 0 for night (from 6 PM to 6 AM) as reference category, 1 for morning (from 6 AM to 12 PM),	1.543
TOD(2)	2 for afternoon (12 to 6 PM). Two dummy variables	1.721
dorder(1)	Evacuation order. 1 for voluntary,	1.681
dorder(2)	2 for mandatory, and 0 for none. Two dummy variables	1.998
Gamdist	Transformation of distance (mi), with gamma distribution. The parameters selected for the model were 8 and 0.6 for shape and scale, respectively	5.247

Source: Fu, H., "Development of Dynamic Travel Demand Models for Hurricane Evacuation," PhD Dissertation, Louisiana State University, 2004. Available at http://etd.lsu.edu/docs/available/etd-04092004-081738/unrestricted/Fu_dis.pdf, accessed on June 4, 2012.

From a modeling perspective, the next question that needs to be answered is that of destination choice of the evacuees. Modali (2005) used a gravity model to predict evacuation destinations and found that the traditional trip purpose stratification needs to be modified based on destination type (hotel, friend, and relative). Chen (2005) used an intervening opportunity model to model evacuation destination choice, because it incorrectly placed the most importance on travel impedance instead of

Table 8.2 Example Data Set with No Flood

Time	Flood	Mobile	Time of day 1	Time of day 2	dorder1	dorder2	Speed	Dist	Gammadist	CL_i
12 a.m.–6 a.m.	0	0	0	0	0	0	105	543	0.193	0.002
6 a.m. –12 p.m.	0	0	1	0	0	0	115	484	0.231	0.013
12 p.m. –6 p.m.	0	0	0	2	1	0	130	445	0.245	0.334
6 p.m. –12 a.m.	0	0	0	0	1	0	145	394	0.245	0.345
12 a.m.–6 a.m.	0	0	0	0	1	0	165	334	0.209	0.356
6 a.m.–12 p.m.	0	0	1	0	0	2	175	273	0.141	0.599
12 p.m.–6 p.m.	0	0	0	2	0	2	160	202	0.056	0.878
6 p.m.–12 a.m.	0	0	0	0	0	2	145	129	0.008	0.884

Table 8.3 Example Data Set with Flood

Time	Flood	Mobile	Time of day 1	Time of day 2	dorder1	dorder2	Speed	Dist	Gammadist	CL_i
12 a.m.–6 a.m.	1	0	0	0	0	0	105	543	0.193	0.003
6 a.m.–12 p.m.	1	0	1	0	0	0	115	484	0.231	0.022
12 p.m.–6 p.m.	1	0	0	2	1	0	130	445	0.245	0.468
6 p.m.–12 a.m.	1	0	0	0	1	0	145	394	0.245	0.484
12 a.m.–6 a.m.	1	0	0	0	1	0	165	334	0.209	0.499
6 a.m.–12 p.m.	1	0	1	0	0	2	175	273	0.141	0.756
12 p.m.–6 p.m.	1	0	0	2	0	2	160	202	0.056	0.951
6 p.m.–12 a.m.	1	0	0	0	0	2	145	129	0.008	0.96

hurricane path or availability of shelter. In this model, the destinations are considered sequentially based on travel time. Though the model performed reasonably, there was scope for significant improvements.

Cheng et al. (2008) use a discrete choice model to predict evacuation destination choice. The model is based on a logit modeling process and assigns a probability for each destination city based on a destination's

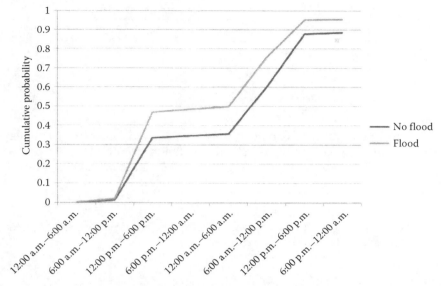

Figure 8.1 Comparison of cumulative probability of departure of evacuees in area under risk of flood and area that is not.

Table 8.4 Logit Model Variables and Their Coefficients

Destination choice model		
Convariate	Definition	Coefficient
DIST	O-D Distance (mi)	−0.004655
POP	Destination city population	1.66E-07
DANGER	Risk indicator (dummy variable)	−0.5171
MSA	Metro area indicator (dummy variable)	1.5562
ETHPCT	White percentage	0.6711

Source: Cheng, G., et al., "A Destination Choice Model for Hurricane Evacuation," Presented at 88th Annual Meeting of the Transportation Research Board, Washington, DC, 2008.

distance from the origin city, whether the destination is likely to be in a hurricane landfall zone, destination population, and destination ethnicity. The model is shown in Table 8.4 and Equation 8.4. The parameters shown in Table 8.4 are from the study conducted by Cheng et al. (2008).

$$PD_i = \frac{e^{-4.655*10^{-2}*DIST_i+1.66*10^{-7}*POP_i-0.5171*DANGER_i+1.5562*MSA_i+0.6711*ETHPCT_i}}{\sum_j e^{-4.655*10^{-2}*DIST_j+1.66*10^{-7}*POP_j-0.5171*DANGER_j+1.5562*MSA_j+0.6711*ETHPCT_j}} \qquad (8.4)$$

To highlight the use of this model, an example hurricane is considered in Table 8.5. The probability is calculated using the coefficients in Table 8.4 and formula shown in Equation 8.4.

Table 8.5 Example Data Set for Destination Choice

	DIST	POP	DANGER	MSA	ETHPCT	PD_i
City 1	80	400,000	1	0	16	0.070
City 2	200	500,000	0	0	15	0.035
City 3	160	600,000	0	1	17	0.780
City 4	280	500,000	0	1	15	0.115

The total number of people leaving at any given time t to a particular destination i is given by N_i^t as in below equation, where PD_i is the probability to choose a destination i and L_t is the probability of leaving at time t.

$$N_i^t = POP * PD_i * L_t$$

This can be then used to develop temporal origin-destination tables that are used in traffic simulation models.

8.5 Evacuation of traffic modeling and simulation

As traffic simulation models have grown in computational speed and analytical detail, it has become easier to apply them for the evaluation of large roadway networks with increasingly larger amounts of traffic. Today, simulation systems like DynaSmart, TRANSIMS, VISSIM, and even CORSIM have the ability to create and track the second-by-second movements of hundreds of thousands of individual vehicles over thousands of miles of roadways for simulation periods that can last for several days. Such massive and intricate models offer the capability to produce detailed data on a large scale and the ability to study and evaluate the system-wide impacts of a nearly infinite set of conditions. Figure 8.2 shows the spectrum of modeling tools available for conducting analysis on evacuation operations and planning.

Because it is recognized that mass evacuations impact transportation systems on regional levels, it is necessary to have tools and techniques to simulate and analyze traffic conditions at these high levels while maintaining the ability to focus on specific areas of concern. Figure 8.3 presents the process to develop and analyze a simulation model to study evacuation operations and plans.

Regional scale evacuation simulation has several key requirements that need to be addressed, including the ability to model vast geographical areas with an enormous number of vehicles and time durations that cover multiple days. In addition, individual vehicular dynamics must be coupled with the ability to calibrate and validate the results, so that the spatial and temporal patterns of traffic observed in the simulation reflect

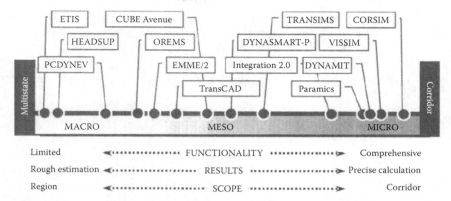

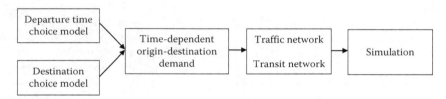

Figure 8.2 Evacuation modeling spectrum. (From Hardy, M. and K. Wunderlich, "Evacuation Management Operations Modeling Assessment: Transportation Modeling Inventory," FHWA Contract No. DTFH61-05-D-00002, U.S. Department of Transportation, Washington, DC, 2007.)

Figure 8.3 Process to develop and analyze evacuation operations and plans.

realistic conditions. Until recently, the ability to achieve any one—let alone all—of these requirements has been difficult, if not impossible.

Early studies to apply traffic simulation for evacuation were limited in their geographical scales and time durations. Evaluation studies conducted with high fidelity microscopic simulation could only be achieved on small networks over time durations of 12 to 20 hours. Several of the earliest focused on the design and traffic operations at contraflow crossovers (Theodoulou 2003; Lim and Wolshon 2005), the evaluation of evacuation routing (Theodoulou 2003; Williams et al. 2007; Dixit, Ramasamy, and Radwan 2008), and the impact of evacuation-level demand on small urban signal networks (Jha et al. 2004; Sbayti and Mahmassani 2006). The simulation of larger regional networks was limited to aggregate-level analyses using macroscopic traffic modeling (KLD 1984; Hobeika and Jamei 1985; Kirschenbaum 1992). From this, only approximate evacuation clearance time and delays could be made, but no understanding of bottlenecks and traffic impacts could be achieved because of the low resolution of the models.

The problems associated with the current state-of-practice for the planning and management of transportation systems during evacuations has been

1. Myopic small-scale modeling of evacuation operations
2. Lack of calibration and validation of simulation models with actual data collected during evacuation
3. Lack of modeling of the movement of low-mobility populations during evacuations

Recently, Chiu et al. (2008) conducted a regional scale simulation study using DynusT to evaluate regional impacts of various evacuation strategies for the Houston-Galveston, Texas metropolitan region during Hurricane Rita. This model represented a significant advance over prior work in that it was among the first to evaluate the traffic impacts of an evacuation at a regional scale.

In fact, several earlier studies have utilized survey data or normal day traffic to calibrate and validate their simulation models. Although this may not be a realistic representation of what actually occurs during a mass evacuation, previous modelers have had little choice due to lack of observed traffic data during evacuations. The calibration and validation of the traffic simulation models are a critical aspect of the modeling process and are needed to ensure that the traffic simulation model replicates the real world with reasonable accuracy.

The Federal Highway Administration's Volume III: Guidelines for Applying Traffic Microsimulation Modeling Software (2004) as well as an in-depth study by Barceló and Casas (2003) identified various statistics for calibrating and validating simulation models. In a recent study conducted by Dixit et al. (2011), the R-squared value of the regression line $y = x$ between the observed and simulated volumes was considered the most suited for evacuation models.

Once calibrated and validated, the simulation model should provide a realistic representation of the evacuation process. This model can then be used to quantitatively and qualitatively examine the spatiotemporal characteristics (speed, volume, density, queuing, and congestion formation/recovery) of the evacuating traffic and evaluate transit evacuation plans.

To visualize the output results of the simulation, color-coded maps of speed were used to graphically represent traffic patterns over space and time. This method was found to be useful to display data in an easy-to-understand manner for analysis. The color maps were particularly valuable for the identification of bottlenecks and understanding the spatiotemporal distribution of speed and volume along various routes. For example, in a study conducted by Dixit et al. (2011), the spatiotemporal graphs were used to identify locational bottlenecks that are indicated within the boxes and dynamics of queue formation, which are represented by the black sloped lines in Figure 8.4.

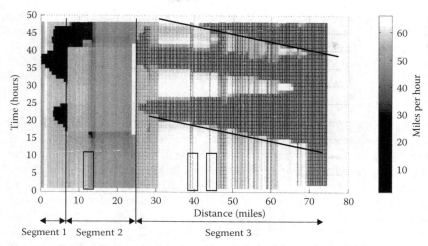

Figure 8.4 Spatiotemporal distribution of speed. (From Dixit, V., et al., *Transport. Res. Record: J. Transport. Res. Board*, 2229, 66–74, 2011.)

After the failure to effectively coordinate the movement of low- and nonmobile populations prior to Hurricane Katrina, there has been an increased interest in developing methods to develop and evaluate transit evacuation plans. The New Orleans transit evacuation plan that was developed, which is also referred to as the New Orleans City-Assisted Evacuation Plan (City of New Orleans 2007), calls for a two-tiered system of buses to first circulate locally within the city to pick up evacuees, then transport them to regional coaches for movement to more remote shelter destinations.

To model transit evacuation plans, it is critical to understand the following:

- The percentage of population expected to utilize this assisted evacuation.
- The temporal demand profile, which is based on the departure time distribution and can be reasonably assumed to have a model the same as those discussed earlier in this chapter.
- A slightly modified response curve and separate bus routes were assumed for tourists seeking transport to the airport.
- The simulation network model can then be used to test scenarios and determine the trips that would be required by bus transit to evacuate all the low-mobility populations to their intended destinations.

A more detailed analysis can be found in Naghawi and Wolshon (2011a,b) as well as Abdelgawad and Abdulhai (2010), who developed a multimodal evacuation plan for cities. Though Naghawi and Wolshon's

(2011a,b) study was for a hurricane evacuation of New Orleans, Abdelgawad and Abdulhai's (2010) study was for a "no-notice" evacuation event in the city of Toronto.

8.6   Conclusion

The history of mass evacuation practice has shown that evacuation traffic management plans have been based largely on lessons learned from failures identified in prior events. This approach has resulted in evacuation plans that have been developed to cater to "failures of the last evacuation." Unfortunately, this does little to address the future and may leave the public vulnerable to unforeseen conditions. The use of simulation modeling to evaluate evacuation plans provides somewhat of a "crystal ball" and holistic view of the entire evacuation operation that can also be used to test any contingency that can be imagined and not experienced in the past. For this reason, simulation can help develop more robust and effective evacuation plans.

The simulation models provide spatiotemporal distribution of traffic, and color-coded maps of speed and volume help to identify bottlenecks. The benefit in using such large models lies in their usefulness to emergency management officials in evaluating evacuation plans and, duration of the contraflow operations, and identifying bottlenecks. The spatiotemporal speed graphs can also be utilized to determine travel times and locations to place fuel containers for vehicles running out of fuel. The insights gained through the regional modeling of evacuation were found to be useful in determining bottlenecks several miles downstream and can be used to develop a more holistic regional evacuation plan. When combined, models and concepts such as those demonstrated here provide a significant resource for policy decisions both large and small. Simulation models could be used to test other policy scenarios, such as providing incentives for people to shelter in certain locations, such as secure hospitals and schools, or to test the feasibility of encouraging evacuees to leave their cars at home and evacuate through transit for the option of designating certain routes as bus-only evacuation corridors. In the future, plans are also being developed to incorporate other modes of traffic such as air and rail to provide an even more robust illustration of evacuation operations.

It is also important to recognize that evacuations vary based on the characteristics of hazards, and therefore it is critical to develop plans that consider various types of hazards and the consequent evacuations. For evacuation plans to be robust, there is a need to develop scenario-based evacuation plans that are flexible and adapt to a hazard type, making the evacuation plans more efficient.

References

Abdelgawad, H. and Abdulhai, B. (2010). "Managing Large-Scale Multimodal Emergency Evacuations," *Journal of Transportation Safety and Security* 2(2): 122–151.

Alsnih, R., Rose, J., and Stopher, P. (2005). "Understanding Household Evacuation Decisions Using a Stated Choice Survey—Case Study of Bush Fires," Presented at the 84th Annual Meeting of the Transportation Research Board, Washington, DC.

Baker, E. J. (1991). "Hurricane Evacuation Behavior," *International Journal of Mass Emergencies and Disasters* 9(2): 287–310.

Barceló, J. and Casas, J. (2003). "Methodological Notes on the Calibration and Validation of Microscopic Traffic Simulation Models," (CD-ROM) 84th Annual Meeting of the Transportation Research Board, Washington, DC.

Chen, B. (2005). "Modeling Destination Choice in Hurricane Evacuation with an Intervening Opportunity Model." Master's Thesis. Department of Civil and Environmental Engineering, Louisiana State University, Baton Rouge, LA.

Cheng, G., Wilmot, C., and Baker, E. (2008). "A Destination Choice Model for Hurricane Evacuation," Presented at 88th Annual Meeting of the Transportation Research Board, Washington, DC.

Chiu, Y., Zheng, H., Villalobos, J. A., Peacock, W., and Henk, R. (2008). "Evaluating Regional Contra-Flow and Phased Evacuation Strategies for Texas Using a Large-Scale Dynamic Traffic Simulation and Assignment Approach," *Journal of Homeland Security and Emergency Management* 5(1), Article 34. ISSN (Online) 1547-7355, DOI: 10.2202/1547-7355.1409, July 2008.

Dixit, V. V., Montz, T., and Wolshon, B. (2011). "Validation Techniques for Region-Level Microscopic Mass Evacuation Traffic Simulations," *Transportation Research Record: Journal of Transportation Research Board* 2229: 66–74.

Dixit, V. V., Pande, A., Radwan, E., and Abdel-Aty, M. (2008). "Understanding the Impact of a Recent Hurricane on Mobilization Time during a Subsequent Hurricane," *Transportation Research Record: Journal of Transportation Research Board* 2041: 49–57.

Dixit, V. V., Ramasamy, S., and Radwan, A. E. (2008). "Assessment of I-4 Contraflow Plans Microscopic vs. Mesoscopic Simulation," *Transportation Research Record: Journal of Transportation Research Board* 2041: 89–97.

Dixit, V. V., Wilmot, C., and Wolshon, B. (2012). "Modeling Risk Attitudes in Evacuation Departure Choice," *Transportation Research Record: Journal of Transportation Research Board* 2312: 159–163.

Dotson, L. J. and Jones, J. (January 2005). "Identification and Analysis of Factors Affeting Emergecny Evacuatons, Main Report," Sandia National Laboratories/U.S. Nuclear Regulatory Commission Report No. NUREG/CR-6864, Vol. 1/SAND2004-5901, Washington, DC, p. 61.

Drabek, T. E. (Fall 1983). "Shall We Leave? A Study of Family Reactions When Disaster Strikes," *Emergency Management Review* 1: 25–29.

Federal Emergency Management Agency. (December 2008). *National Incident Management System*, Department of Homeland Security, Washington, DC. Available at http://www.fema.gov/pdf/emergency/nims/NIMS_core.pdf, accessed on May 7, 2012.

Federal Highway Administration. (2004). "Traffic Analysis Toolbox Volume III: Guidelines for Applying Traffic Microsimulation Modeling Software." Available at http://ops.fhwa.dot.gov/trafficanalysistools/tat_vol3/Vol3_Guidelines.pdf, accessed on November 15, 2010.

Fitzpatrick, C. and Mileti, D. S. (1991). "Motivating Public Evacuation," *International Journal of Mass Emergencies and Disasters* 9(2): 7–18.

Fu, H. (2004). "Development of Dynamic Travel Demand Models for Hurricane Evacuation," PhD Dissertation, Louisiana State University. Available at http://etd.lsu.edu/docs/available/etd-04092004-081738/unrestricted/Fu_dis.pdf, accessed on June 4, 2012.

Fu, H. and Wilmot, C. (2004). "Sequential Logit Dynamic Travel Demand Model for Hurricane Evacuation," *Transportation Research Record: Journal of Transportation Research Board* 1882: 19–26.

Fu, H. and Wilmot, C. (2006). "Survival Analysis Based Dynamic Travel Demand Models for Hurricane Evacuation," 85th Transportation Research Board Annual Meeting CD-ROM, 2006.

Fu, H., Wilmot, C., and Zhang, H. (2007). "Modeling the Hurricane Evacuation Response Curve," 86th Transportation Research Board Annual Meeting CD-ROM, 2007.

Gunter, P. (2001). "Emergency Planning for Nuclear Power Accidents," Reactor Watchdog Project, NIRS, http://www.nirs.org/reactors/emergencyplanning 71301.html, accessed on February 24, 2003.

Hardy, M. and Wunderlich, K. (2007). "Evacuation Management Operations (EMO) Modeling Assessment: Transportation Modeling Inventory." FHWA Contract No. DTFH61-05-D-00002. Washington, DC: U.S. Department of Transportation.

Hasan, S., Ukkusuri, S. V., Gladwin, H., and Murray-Tuite, P. (2011). "A Behavioral Model to Understand Household Level Hurricane Evacuation Decision Making," *ASCE Journal of Transportation Engineering* 137(5): 341–349.

Heath, S. E., Kass, P. H., Beck, A. M., and Glickman, L. T. (2001). "Human and Petrelated Risk Factors for Household Evacuation Failure during a Natural Disaster," *American Journal of Epidemiology* 153(7): 659–665.

Hobeika, A. G. and Jamei, B. (1985). *MASSVAC: A Model for Calculating Evacuation Times under Natural Disasters*, Computer Simulation in Emergency Planning, La Jolla, Society of Computer Simulation.

Howell, S. E. and Bonner, D. E. (2005). "Citizen Hurricane Evacuation Behavior in Southeastern Louisiana: A Twelve Parish Survey." Available at http://poli.uno.edu/unopoll/Summary%20Report%20July%2019%202005%20(2).pdf, accessed on June 4, 2007.

Hultaker, O. (1983). "Family and Disaster," *International Journal of Mass Emergencies and Disasters* 1(1): 7–18.

Jha, M., Moore, K., and Pashaie, B. (2004). "Emergency Evacuation Planning with Microscopic Traffic Simulation," *Transportation Research Record: Journal of Transportation Research Board* 1886: 40–48.

Kirschenbaum, A. (1992). "Warning and Evacuation during a Mass Disaster: A Multivariate Decision-Making Model," *International Journal of Mass Emergencies and Disasters* 10(1): 91–114.

KLD. (1984). *Formulations of the DYNEV and I-DYNEV Traffic Simulation Models Used in ESF*, Federal Emergency Management Agency.

Lim, E. and Wolshon, B. (2005). "Modeling and Performance Assessment of Contraflow Evacuation Termination Points," *Transportation Research Record: Journal of Transportation Research Board* 1922: 118–128.

Mikami, S. and Ikeda, K. (1985). "Human Response to Disasters," *International Journal of Mass Emergencies and Disasters* 3(1): 106–132.

Modali, N. (2005). Modeling Destination Choice and Measuring the Transferability of Hurricane Evacuation Patterns. Master's Thesis. Department of Civil and Environmental Engineering, Louisiana State University, Baton Rouge, LA.

Naghawi, H. and Wolshon, B. (2011a). "Operation of Multimodal Transport Systems during Regional Mass Evacuations," 90th Transportation Research Board Meeting, Washington, DC.

Naghawi, H. and Wolshon, B. (2011b). "Performance of Multi-Modal Evacuation Traffic Networks: A Simulation Based Assessment," 90th Transportation Research Board Meeting, Washington, DC.

National Cooperative Highway Research Program (NCHRP). (2009). *Transportation's Role in Emergency Evacuation and Reentry*, Synthesis of Highway Practice Report No. 392, Transportation Research Board, National Research Council, Washington, DC, p. 142.

PBS&J Inc. (2000). *Hurricane Floyd Assessment - Review of Hurricane Evacuation Studies Utilization and Information Dissemination*, Post, Buckley, Schuh & Jernigan, Inc. Tallahassee, Florida.

Quarantelli, E. L. (1985). "Realities and Mythologies in Disaster Films," *Communications: The European Journal of Communication* 11: 31–44.

Sbayti, H. and Mahmassani, H. S. (2006). "Optimal Scheduling of Evacuation Operations," *Transportation Research Record: Journal of Transportation Research Board* 1964: 238–246.

Sorenson, J. H. (1991). "When Shall We Leave? Factors Affecting the Timing of Evacuation Departures," *International Journal of Mass Emergencies and Disasters* 9(2): 153–165.

Sorenson, J. H. and Mileti, D. S. (1988). "Warning and Evacuation: Answering Some Basic Questions," *Industrial Crisis Quarterly* 2: 195–209.

Tamminga, G., Tu, H., Daamen, W., and Hoogendoorn, S. (2011). "Influence of Departure Time Spans and Corresponding Network Performance on Evacuation Time," *Transportation Research Record: Journal of the Transportation Research Board*, 2234: 89–96.

Theodoulou, G. (2003). Contraflow Evacuation on the Westbound I-10 out of the City of New Orleans, Master's Thesis, Louisiana State University, Baton Rouge, LA.

Tierney, K., Bevc, C., and Kuligowski, E. (2006). "Metaphors Matter: Disaster Myths, Media Frames, and their Consequences in Hurricane Katrina," *ANNALS, The Annals of the American Academy of Political and Social Science* 604(1): 57–81.

U.S. Army Corps of Engineers (2000). Alabama Hurricane Evacuation Study Technical Data Report: Behavioral Analysis. Final Report, 2000, http://www.sam.usace.army.mil/hesdata/Alabama/alabamareportpage.htm. Accessed May 2009.

Vogt B. M. (1991). "Issues in Nursing Home Evacuations," *International Journal of Mass Emergencies and Disasters* 9: 247–265.

Williams, B. M., Tagliaferri, A. P., Meinhold, S. S., Hummer, J. E., and Rouphail, N. M. (2007). "Simulation and Analysis of Freeway Lane Reversal for Coastal Hurricane Evacuation," *ASCE Journal of Urban Planning and Development* 133(1): 61–72.

Zimmerman, C., Brodesky, R., and Karp, J. (November 2007). *Using Highways for No-Notice Evacuations: Routes to Effective Evacuation Planning Primer Series*, United States Department of Transportation Report No. FHWA-HOP-08-003, Washington, DC.

chapter nine

Riverflow prediction for emergency response
Military applications using artificial neural networks

Bernard B. Hsieh and Mark R. Jourdan

Contents

9.1 Introduction

The U.S. Army Corps of Engineers (USACE) Engineering Research and Development Center's (ERDC) Reachback Operations Center provides reachback support to the U.S. major commands and to U.S. personnel deployed worldwide. Requests are submitted not only to support ongoing military operations, but also to assist in humanitarian assistance and disaster recovery. Topics of requests have included hydrology, bridging, airfield expansion, pavements, electrical power, and basecamp designs. Researchers in ERDC and throughout the USACE have solved many of these problems. One of the common topics ERDC had studied is flooding; where does flooding occur, what is the extent of floods, and when will floods happen.

The ability to predict watershed hydrologic conditions and the associated potential for flooding to occur plays a significant role in planning and operational activities. To make highly accurate hydrologic predictions, either physically based or system based, the system parameters and prediction variables are sometimes unavailable or even totally missing. This certainly curtails the capability of prediction particularly for operations where very little time is available to conduct the analysis. In cases where information for a particular watershed may be entirely unavailable, the situation may be resolved by using the similarity concept. The first part of this chapter demonstrates procedures for searching the best match watershed (unsupervised artificial neural networks [ANNs]) for a target watershed from a large knowledge base and determining the reliability of "transplant" watershed information such as hydrologic and climatic parameters. Supervised ANN is then used as a tool to predict the riverflow for this target watershed. The degree of similarity is based on inter- and intra-relationships among many geologic, soil, hydrologic, and climatic factors. Various methods have been employed to analyze the similarities between two objects.

When performing an operational riverflow prediction of a military site, rapid response and high accuracy are required. One such method for delivering answers is a system-based approach, such as using ANNs (Figure 9.1). The ANN model is a mathematical or computational model

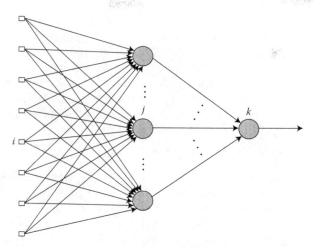

Figure 9.1 Fully connected feedforward network with one hidden layer and output layer. *i*, *j*, and *k* are integer counting indices for the respective layers in the network.

formulated as a network of simple units, each having local memory. A "signal" is transmitted through the network layers, and its value is increased or decreased according to its relevance to the final output. Patterns of signal weights and parameter biases are analyzed to find the optimal pattern that best fits the input parameters and results. Once the network has been trained (calibrated) to simulate the best response to the input data, the configuration of the network is fixed and a validation process is conducted to evaluate the performance of the ANNs as a predictive tool. This research project employed a commercial ANN software tool, NeuroSolution (2003), to perform the computations. Three algorithms, namely multilayer feed forward neural networks (MLPs), Jordan and Elman recurrent neural networks (JERs), and time-lagged recurrent neural networks (TLRNs), were used to test the prediction system. The detailed theoretical development for the algorithms can be found in Haykin (1994) and Principe et al. (2000).

The most widely used methodologies for water level (stage) forecasting use either a conceptual structure with different levels of physical information, a stochastic structure, or a combination of both. These approaches, which became widespread in the 1990s, started in the 1960s. Since 2000, new types of data-driven models, based on artificial intelligence and soft computing techniques, have been more frequently applied. Currently, ANNs are one of the most widely used techniques in the forecasting field (Hsu et al. 1995; Thimuralaiah and Deo 2000). Most applications based on these models consider the discharge as the forecasting variable (Hsieh and Bartos 2000; Imrie et al. 2000; Dawson et al. 2002; Kisi 2008), primarily because of historical contiguity with the classes of conceptual- and physical-based rainfall-runoff models. Such an approach requires the knowledge of the rating curve in the cross section of interest to parameterize the model. However, the knowledge of the stage is required within the framework of a flood warning system, and thus, the rating curve has to be used to transform the forecasted flows into stages.

9.2   Basic concepts of ANNs

Learning or adaptation in which a desired response can be used by the system to guide the learning process is called supervised learning. Unsupervised learning, on the other hand, is learning in which the system parameters are adapted using only the information of the input and constrained by prespecified internal rules. Scientific and engineering communities have reported ANNs' theoretical development and applications for several decades, particularly for supervised ANNs. Here, a brief description of unsupervised ANNs and the concept of the integration of unsupervised–supervised ANNs are discussed.

9.2.1   Supervised and unsupervised ANNs

The vast majority of ANN solutions have been trained with supervision. In this mode, the actual output of a neural network is compared to the desired output. The network then adjusts weights, which are usually randomly set to begin with, so that the next iteration, or cycle, will produce a closer match between the desired and the actual output. The learning method tries to minimize the current errors of all processing elements. This global error reduction is created over time by continuously modifying the input weights until acceptable network accuracy is reached.

With supervised learning, the ANN must be trained before it becomes useful. Training consists of presenting input and output data to the network. These data are often referred to as the training set. That is, for each input set provided to the system, the corresponding desired output set is provided as well. In most applications, actual data must be used. This training phase can consume a lot of time. In prototype systems with inadequate processing power, learning can take weeks. This training is considered complete when the neural network reaches a user-defined performance level. This level signifies that the network has achieved the desired statistical accuracy as it produces the required outputs for a given sequence of inputs. When no further learning is necessary, the weights are typically frozen for the application. Some network types allow continual training at a much slower rate while in operation. This helps a network to adapt to gradually changing conditions.

Under unsupervised training, the networks learn from their own classification of the training data without external help. It is assumed that class membership is broadly defined by the input patterns sharing common features, and that the network will be able to identify those features across the range of input patterns. Unsupervised learning is the great promise of the future as it offers the possibility for computers to learn on their own in a true robotic sense someday. Currently, this learning method is limited to networks known as self-organizing maps. These kinds of networks are currently not in widespread use and are basically an academic novelty. Yet, they have shown they can provide a solution in a few instances, proving that their promise is not groundless. For example, they have been proven to be more effective than many algorithmic techniques for numerical aerodynamic flow calculations.

9.2.2   An unsupervised ANN (SOFM)

The unsupervised self-organizing feather map (SOFM) is trained without teacher signals (Figure 9.2) (Kohonen 1990). SOFM is a special kind of neural network that can be used for clustering tasks. Only one map node (winner) at a time is activated corresponding to each input. The location of the responses in the array tends to become ordered in the learning process as if some meaningful nonlinear coordinate system for the different input features were being created over the network. This illustrates an important and attractive feature of SOFM applications, in that a multidimensional input ensemble is mapped into a (one or) two-dimensional space, preserving the topological structure as much as possible. Boogaard et al. (1998) applied the SOFM to hydrological and ecological data sets.

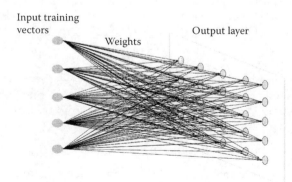

Figure 9.2 A SOFM structure with five input vectors and 5 × 5 projecting lattice.

9.2.3   Visualization of an SOFM

Visualization techniques to depict the data structure of the feature space in the form of clustering of neurons in the 2-D SOFM have been developed (Ultsch et al. 1995). This visualization typically uses grayscale to illustrate the distance between connection weights. The light shading typically represents a small distance and the dark shading represents a large distance.

This type of visualization is useful as long as relatively clear cluster boundaries exist or the granularity of the distance differences is large. When the cluster boundaries get fuzzy or the granularity of the distances become too small to represent with the grayscale, it becomes increasingly difficult to identify fuzzy cluster landscapes. Moreover, because all distance values are normalized, only relative (qualitative) analysis is allowed. Subsequently, this "grayscale distance map" cannot be used to compare different SOFM mapping results.

NeuroDimensions (2003) developed a visual version for the Kohonen topological feature maps to check the performance of an SOFM. Three basic windows used for evaluating the clustering are quantization metric, united distance, and frequency.

9.2.3.1   Quantization metric

The quantization metric produces the average quantization error, which measures the goodness of fit of a clustering algorithm. It is the average distance between each input and the winning process element (PE). If the quantization error is large, then the winning PE is not a good representation of the input. If it is small, then the input is very close to the winning PE. The quantization error is best for comparing the clustering capabilities between multiple trainings of the same SOFM on the same point.

9.2.3.2   Unified distance

The unified distance is the distance between PE clustering centers. The weights from the input to each PE cluster center off the SOFM. Inside a cluster of inputs, SOFM PEs will be close to each other.

9.2.3.3   Frequency

Typically, the number of SOFM PEs is much larger than the number of clusters expected. This allows multiple PEs to capture one logical cluster. The SOFM map is a group of PEs representing a single cluster of the input.

9.2.4   Geographic information system (GIS) linked to ANNs

GIS data often include satellite and other remote sensing imagery. Imagery analysis involves either supervised or unsupervised classifications. Unsupervised classifications of imagery include analyzing the color or black and white pixels of the image for the purposes of classifying image objects and entities where tone, texture, and hue are used. Supervised classification of imagery involves referencing the pixels to actual field or site conditions and color balancing of the image for similar classification purposes.

ANNs are increasingly being used to determine spatial patterns. In the area of landscape ecology, the landscape pattern is an important factor enabling classification. Indeed, more recent developments in the area of remote sensing analysis involve ANNs for the analysis of images for the purposes of classifying objects.

9.3   Watershed similarity analysis (a nonlinear clustering search)

9.3.1   Geospatial knowledge base development

Geospatial data of geographic locations and characteristic natural and constructed features were gathered for database development. GIS databases were utilized for this endeavor. Specifically, the U.S. Environmental Protection Agency's Better Assessment Science Integrating Point and Non-point Sources (BASINS) system provided the 300-m US Geological Service's Digital Elevation Model, Land Use/Land Cover, Soils, and watershed gauge locations within the contiguous United States (Figure 9.3). The selected gauge locations had a complete dataset for medium- to moderately large-sized basins, 6 to 7900 km^2.

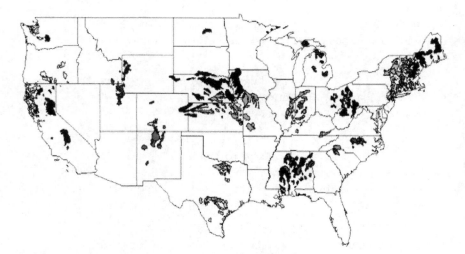

Figure 9.3 Watersheds within the contiguous United States.

Watershed development was conducted with the Environmental Systems Research Institute's ArcGis/ArcView and the Department of Defense's Watershed Modeling System (WMS). From the GIS databases, data were extracted, projected, and shaped into Arc/Info griddled ASCII data as input into the WMS interface where basin delineation and parameter estimations were conducted. Watershed parameters such as drainage area, basin slope, basin length, basin perimeter, and so on were among the variables derived for the ANN's analyses. Watersheds selected were within a 10% margin of error when the areas were compared with recorded drainage areas from BASINS. From these selected watersheds, mean daily flow data for their respective periods of record were compiled for the ANN's verification process. In addition, 30-year mean monthly and annual precipitation as well as temperature data were derived from *PRISM (Parameter-elevation Regressions on Independent Slopes Model)* and presented as GIS coverages. Subsequent GIS analyses produced mean monthly and annual precipitation and temperature data for all selected basins. The final knowledge base had a 1064 watersheds × 70 variables matrix with final relevant parameters listed in Table 9.1.

Table 9.1 Watershed Parameters Considered for Similarity Analysis

Geometric parameters		
Basin area	Basin slope	Basin perimeter
Basin average elevation	Basin shape factor	Basin sinuosity factor
Average overland flow distance	Maximum flow distance	Maximum flow slope
Maximum stream length	Maximum stream slope	Basin length
Centroid stream distance	Centroid stream slope	Centroid to nearest point of MaxFloDist
Land use/land cover		
Residential/industrial	Agricultural land	Rangeland
Forest land	Open water	Wetlands
Exposed rock	Tundra	Glaciers
Soil type parameters		
Well-drained soils (sands and gravel)		Fine textures (silts)
Moderately coarse textures (sandy loam)		Poorly drained soils (clays)
Hydrologic parameters		
Monthly and annual mean	Precipitation	Temperature

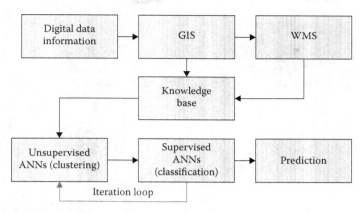

Figure 9.4 System components for similarity analysis.

Figure 9.4 illustrates the system components for a data-driven similarity analysis computational procedure including knowledge base, two components of ANNs (clustering and classification), and prediction (verification).

9.3.2   Demonstration examples

From the knowledge base, the geometric parameters, land use/land cover, soil types, and seasonal and annual mean values of both precipitation and temperature were used to test this calculation procedure. In order to test the reliability of system development, three sizes of watershed were selected to examine the performance.

9.3.2.1   Random selection (watershed 4288000)

In this test, we used a known watershed (gauge number 4288000) to search for another watershed that had similar features. This test included a clustering analysis to identify the similarity between the watersheds, whereas the classification analysis verified the clustering performance. To check the reliability of prediction, time-series hydrographs were used to compare the resulting search pattern. In this procedure, the hydrograph of gauge 4288000 was hidden purposely to check the performance of the system, once the best matched watershed was found.

During the clustering computation, a 5×5 matrix of SOFM was initially selected. Through repeated iterations (usually 200) of examining frequency, unified distance, and quantization of the unsupervised ANN identification, an optimal clustering set to distribute the winner for each watershed was obtained (Figure 9.5). The values in this 5×5 matrix show

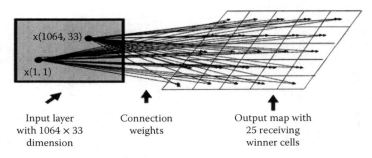

Input layer · Connection · Output map with
with 1064 × 33 · weights · 25 receiving
dimension · winner cells

Figure 9.5 SOFM with 5 × 5 matrix lattice.

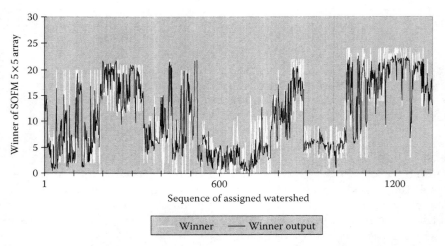

Figure 9.6 Classification verification for the first system iteration with 1064 watersheds.

the most similar watershed within the same group out of the possible 25 groups in this case.

For classification, the problem was trained with MLPs (multilayer perceptron) ANNs. The outcome provided the confidence level of the clustering analysis, which resulted in a successful classification rate of about 91% meeting the target. This result indicated that watershed 4288000 belongs to the group with 103 (group 7) most similar watersheds from the original 1063 possible candidates. This clustering-classification process was repeated until the final target watershed was found. Figures 9.6 and 9.7 show the classification verification during the first and second search iterations, respectively. Note that the size of clustering for this iteration has been reduced to a 3 × 3 matrix.

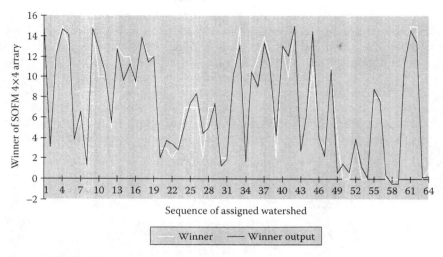

Figure 9.7 Classification verification for the second system iteration with 103 watersheds.

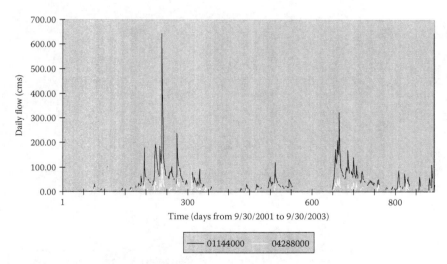

Figure 9.8 Most similar flow (cubic meters per second) (01144000) versus observed flow (04288000).

The final candidate for this search was watershed number 01144000. This implied that the flow patterns from station 01144000 are the most similar to those of station 04288000. Flow hydrograph comparisons between these two stations during the period 1999–2001 are shown in Figure 9.8. Although the flow pattern, particularly the phase, matches very well, the performance of the amplitude representations is dissatisfactory. Therefore, the estimated hydrograph was adjusted by taking the area ratio of station 04288000 and station 01144000 (Figure 9.9).

It is important to tune the clustering group as well as rechecking the performance of the classification process. The identification of the reliability for application also requires data on how well the "transplant" performs. Therefore, a series of combinations including the features of input parameters was adopted. Table 9.2 summarizes the performance sensitivity from input parameters. This indicates that the most important group parameters are hydrologic, geometry, soil type, and land use.

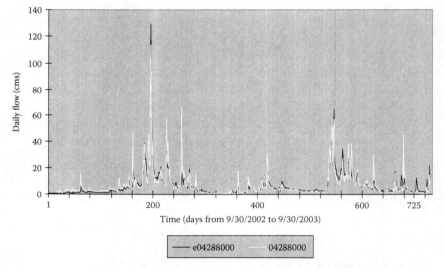

Figure 9.9 Flow (cubic meters per second) estimation (e04288000) versus observed flow (04288000) after basin area ratio adjustment for 33 inputs approach.

Table 9.2 Sensitivity Test Results

Parameters	Candidate watershed	Correlation coefficient	Mean error
All groups	0114000	0.92	0.17
Geometry	4282000	0.82	−7.80
Hydrologic	4288000	0.83	2.99
Land use	2472000	0.10	−28.34
Soil type	1170100	0.67	−4.25

The performance difference between the geometry and hydrologic groups is quite small. The magnitude of hydrographs could not be adjusted by ratios obtained using hydrologic, soil type, and land use groups because they did not contain the basin area factor.

9.3.2.2 Average size (watershed 11427000)
This demonstration example used an average size watershed (856.97 km²) to perform the same search procedure as the first example. Instead of using short-term hydrographs for result comparison, it used a much longer period to compare the daily and monthly flow conditions. Statistical computations checked the degree of similarity between this target watershed and the best candidate watershed.

As the first demonstration example, the clustering-classification iteration process was conducted until the best similar watershed of target watershed from the knowledge base was found. Here we used 3 × 3, 6 × 6, and 3 × 3 clustering sequences to find the best candidate (Figure 9.10).

Watershed number 1144500 was the final candidate from this search process. To examine the similarity, the comparisons of daily and monthly flow for 34 years (1962–1995) between target and best candidate watersheds are shown in Figures 9.11 and 9.12. Although the phase comparison received good results, the amplitude underestimated the results after applying the area ratio factor. This suggests that more watersheds need to be included in the knowledge base, and the area ratio factor may not be the only function for final conversion.

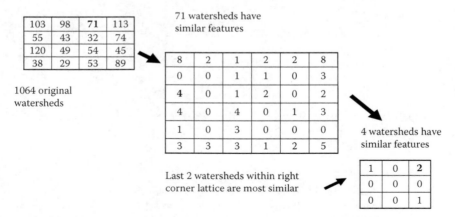

Figure 9.10 Clustering sequences to search for the best similar watershed from a given target watershed.

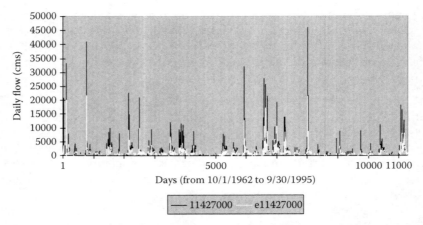

Figure 9.11 Thirty-three-year daily flow estimation (white) from watershed 11445500; $r = 0.883$.

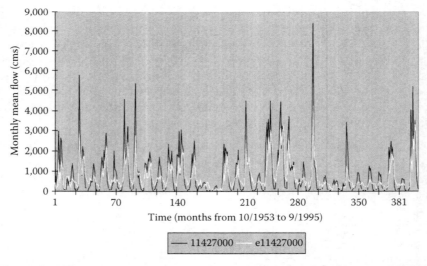

Figure 9.12 Forty-two-year monthly flow estimation (white) from watershed 11445500; $r = 0.882$.

9.3.2.3   Median size (watershed 03153000)

The target watershed 03153000 with a basin area of 419.6 km^2 was used for another test case. It also required three clustering-classification iterations during the final search. The best candidate watershed for the similarity analysis was watershed 03152000, which had a basin area of 1000.2 km^2.

Although 9 years (1967–1975) of daily flow predictions (Figure 9.13) included some overestimates, particularly for the peak flow conditions, the monthly flow prediction (Figure 9.14) agreed well. The frequency analysis (Figure 9.15) shows the frequency distribution comparison for estimate and true monthly flow.

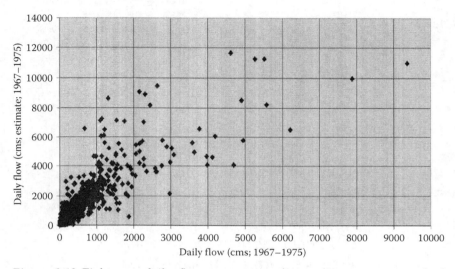

Figure 9.13 Eight-year daily flow comparison for median target watershed 03153000 with $r = 0.896$.

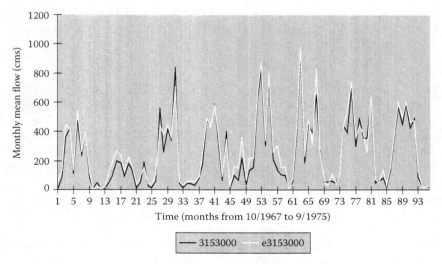

Figure 9.14 Eight-year monthly flow comparison for median target watershed 03153000 with $r = 0.976$.

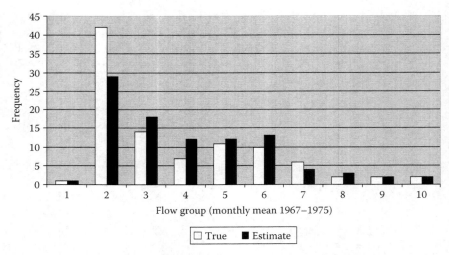

Figure 9.15 Frequency distribution between true (white) and estimate (black) monthly flow for target watershed 03153000.

9.4   An operational riverflow emergency response model (Helmand River, Afghanistan)

The Helmand River Basin (Figure 9.16) is the largest river basin in Afghanistan, stretching for 1150 km. This basin is a desert environment with rivers fed by melting snow from high mountains and infrequent storms. Great fluctuations in stream flow, from flood to drought, can occur annually. The purpose of this study was to identify the best river-flow prediction model, using the upstream Kajakai reservoir gauge and downstream locations, to aid in developing military operation plans. We also developed a daily operational tool for performing predictions using a laptop or desktop computer. River discharge measurements began in Afghanistan in the mid–1940s at a few sites (USGS 2008), but measurements were discontinued soon after the Soviet invasion of Afghanistan in 1979. Discharge for five downstream gauges—Lashkar Gah, Darweshan, Marlakan, Burjak, and Khabgah—with two tributaries, the Musa Qala and Arghandab Rivers, were collected during this period. However, only the Kajakai and Burjak gauges, as well as the two tributaries, had sufficiently complete data. The knowledge base for developing the ANN model included daily discharge measurements from 10/1/1953 to 9/30/1979. Missing data were filled by using ANN algorithms with high correlation relationships.

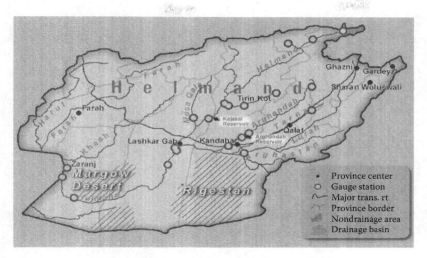

Figure 9.16 Helmand River Basin and its flow gauges.

9.4.1   Knowledge base development for Helmand Riverflow discharge and basic data analysis

To develop a good data-driven model, a knowledge base with minimum uncertainty is a critical factor to make the first successful step. A schematic drawing representing the flow gauges in the Helmand River and two tributaries (Musa Qala River and Arghastan River) were used for building this operational flow prediction model (Figure 9.17). Other than the regular USGS flow gauges, there are two artificial points: "K+M," approximately representing the total flow combined by the Kajakai and Musa Qala gauges and "L+Q," representing the total flow merged by the Lashkargah and Qala-i-Bust gauges. These two control points were used to check how these two tributaries contributed to the downstream flow from historical events because there was only one upstream gauge (Kajakai) currently collecting the measurement.

9.4.2   Knowledge recovery system

About 40% of data from the main stem of Helmand River (Figure 9.18) is missing from the historical records for the eight gauges. Those missing windows must be filled before performing the analysis. The knowledge recovery system (KRS) (Hsieh and Pratt 2001) has been applied to several tidal and riverine ERDC projects. KRS deals with activities that are beyond the capabilities of normal data recovery systems. It covers most

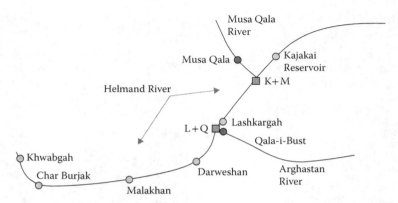

Figure 9.17 A schematic representation of flow gauges for building the Helmend River operational flow prediction model.

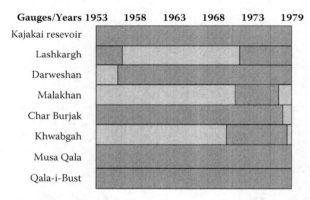

Figure 9.18 Historical record for main stem and tributaries of Helmand River (lighter shade shows missing windows).

of the information that is associated with knowledge-oriented problems and is not limited to numerical problems. Almost every management decision, particularly for a complex and dynamic system, requires the application of a mathematical model and the validation of the model with numerous field measurements. However, instrumentation adjustments and other problems may lead to incomplete data collection or produce abnormal recording curves. Data may also be unavailable at appropriate points in the computational domain when the modeling design work changed. The KRS for missing data was based on the transfer function (activation function) approach (Hsieh and Pratt 2001). The simulated output can generate a recovered data series (missing window) using optimal weights from the best-fit activation function (transfer function). Three types of KRS are defined: self-recovery, neighboring gauge recovery with same parameter, and mixed neighboring and remote gauges recovery with multivariate parameters. It is noted that the knowledge recovery will not always be able to recover the data even with perfect information (no missing values or highly correlated) from neighboring gauges.

9.4.3   Knowledge recovery system for Helmand Riverflow data

To develop the completed knowledge base, the missing window (6/1/1957–9/30/1972) for the Lashkargah flow gauge was first applied by KRS. Instead of using either the Kajakai flow or Darweshan flow to recover this missing window, it was better to use both flow gauges as inputs and the Lashkargah flow as output during the period between 10/1/1972 and 9/30/1979 to build a transfer function (ANN model). Three algorithms—MLP, TLRN, and JERN—were used to compare which one had the best generalization. The TLRN parameters used for the missing windows are listed below.

> System structure: 2 inputs—1 output system
> Training exemplars = 2550; Total iterations = 2500
> Hidden layer = 1
>   Activation function = TanhAxon
>   Learning rate = Momentum; Step size = 1.0; Momentum = 0.70
> Output layer
>   Activation function = TanhAxon
>   Learning rate = Momentum; Step size = 0.1; Momentum = 0.70
>   Memory = GammaAxon function; Depth in samples = 10

This training (TLRN) process was examined by comparing the mean square error (MSE) with a particular Epoch (computational step) for a given maximum desire iteration. The training results were quite good (correlation coefficient [CC] = 0.97). The comparison of training results among three selected algorithms and the measure of statistical parameters—MSE, normalized mean square error (NMSE), mean average error (MAE), minimum absolute error (Min Abs E), maximum absolute error (Max Abs E), and CC (Table 9.3)—indicated the MLP provided a slightly better performance, but the TLRN presented clearer time-delay response from upstream to downstream flow transport phenomenon. The TLRN was selected as the main algorithm for conducting the remaining study. After this training process was complete, the missing window was simulated by using a set of weights for the Malakhan flow gauge (Figure 9.19). The developed KRS procedures were applied to other missing windows with the least uncertainty involved. The black line represents

Table 9.3 Performance Comparison for Three Training
Algorithms (MLP, TLRN, and JERN)

Performance/networks	MLP	JERN	TLRN
MSE	1030	1085	1776
NMSE	0.034	0.036	0.059
MAE	21.56	21.80	28.24
Min Abs E	0.006	0.014	0.051
Max Abs E	474.95	503.74	495.90
Correlation coefficient	0.982	0.981	0.976

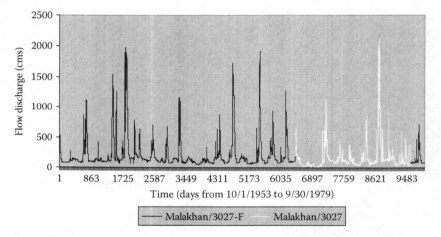

Figure 9.19 Modeled and measured flow for the Malakhan gauge. The white trace
is the measured data and the black trace is the data obtained from the knowledge
recovery system.

the best estimates for missing windows, whereas the white line shows
existing data points. The same procedures were used to simulate the
remaining missing windows for other downstream gauges.

9.4.4 An operational riverflow prediction system for the Helmand River

This task developed an operational downstream riverflow prediction
data-driven model using only the most upper stream Kajakai reservoir
data. Here we present three operational models. The first is for only
the relationship between the Kajakai reservoir flow and the operational
downstream location. The other two also consider the impact of tributary
inflows.

According to the historical record after KRS processing, the best available data cover the riverflow from Kajakai reservoir to five downstream gauges and two tributaries during the period from 10/1/1953 to 9/30/1979 (historical knowledge base). These data are used to perform this operational model development. To test the operational model, the newly collected data below Kajakai from 10/1/2009 to 4/23/2010 is adopted. The historical knowledge base is divided into a training set (10/1/1953–9/30/1973) and a verification set (10/1/1973–9/30/1979). The neural networks architecture is TLR with a 1 × 5 (one input–five outputs) structure. The CCs for training are between 0.95 (Lashkargen) and 0.89 (Char Burjak). The test results are between CC 0.96 for Lashkargen (Figure 9.20) and CC 0.87 for Khwabgah. The operational flow prediction (203 days) was then simulated (Figure 9.21). This simulation only took 20 seconds to complete, once the operational prediction model was trained. The total processing time, including model training, new data transfer, and simulation mode, should be less than around 10 minutes. Once complete, the prediction model does not need to be trained again if the prediction horizon is short (e.g., within a year) or if no significant hydrologic/hydraulic conditions have changed.

As previously noted, two tributaries (Musa Qala River and Arghastan River) merge into the upper main stem of the Helmand River. To examine the impact of the tributary flows on the downstream river flows, the model was adapted to include two inflow data set scenarios. The second scenario dealt with two additional artificial points ("K+M" and "L+Q") and the third scenario only took these two additional points as well as three downstream gauges. Thus, the second scenario had 1 × 7 networks and the third scenario remained as a 1 × 5 structure. Table 9.4 summarizes

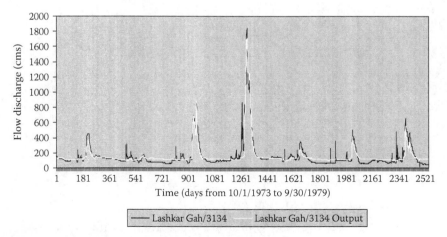

Figure 9.20 Test results for the operational flow prediction model (Lashkargen gauge) from 10/1/1973 to 9/30/1979.

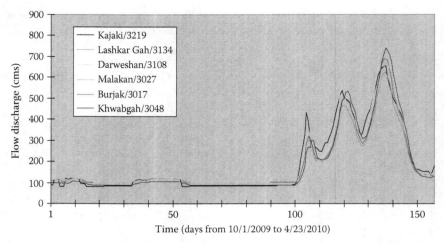

Figure 9.21 Results of operational riverflow prediction model during the period from 10/01/2009 to 04/23/2010 for five downstream gauges when the flow below Kajakai Reservoir is provided.

Table 9.4 Comparison of Correlation Coefficients among Three Prediction Model Scenarios

	Training 1	Test 1	Training 2	Test 2	Training 3	Test 3
K+M	N/A	N/A	0.970	0.972	0.973	0.976
Lashkar Gah	0.949	0.962	0.948	0.960	N/A	N/A
L+Q	N/A	N/A	0.900	0.901	0.888	0.892
Darweshan	0.914	0.876	0.910	0.876	N/A	N/A
Malakhan	0.929	0.870	0.928	0.871	0.931	0.871
Burjak	0.890	0.855	0.887	0.854	0.848	0.848
Khwabgah	0.912	0.870	0.909	0.868	0.908	0.861

the comparison of the CCs among these three scenarios. Except for the Char Burjak gauge, which reduced the CC for training the third scenario runs, no significant variations were found for other conditions. More detailed investigations, such as comparing the overall time history variation as well as other statistical parameters for performance, may better identify more significant correlation impacts.

9.4.5 Prediction improvement analysis for the operational prediction model

After the operational model was established, the most important issues to be addressed were as follows: (1) How can it be improved using the existing and available knowledge base? (2) Where is the best location (apart from existing gauge locations) for obtaining stream measurements other than the gauge below Kajakai Reservoir?

9.4.5.1   Best gauge candidate to be measured

Table 9.5 gives a matrix containing 12 different runs from a historical knowledge base. This table indicates which gauge is used as input (symbol I) and which gauge is represented as output (symbol O). From the design, it consists of one 1×7, five 2×6, one 1×5, and five 2×4 runs. The objective was to see the overall performance from selected statistical parameters (MSE, NMSE, MAE, and CC) and to select the best possible candidate to improve the reliability of the operational model. The idea was based on the concept of the boundary condition that has to be provided from a generalized numerical model development during the model setup process. The prediction reliability would be reduced if the influence of the input signal was dampened due to propagation loss over the long distance traveled to reach the output location. Note that these runs perform the full-scale training (entire length of knowledge base) with the same set of system parameters as the previous section. After the completion of runs, the average statistical parameters are computed from the output series (the last column of Table 9.5). The better performance should be represented by lower MSE, NMSE, and MAE, and higher CC. The overall performance strongly indicated the gauge at Darweshan was the best candidate to be selected as an additional measurement location. This is shown by the high CC in run 9.

9.4.5.2   A two-layered operational flow prediction model

With the above two examinations (best additional measurement gauge and error-correction analysis), a two-layered ANN model was proposed to improve the present capability of the prediction model. The first layer used a 1×1 structure (Kajakai reservoir flow as input and Darweshan gauge flow as output) to simulate the prediction flows at Darweshan. Then it again used Kajakai Reservoir flow as well as the simulated Darweshan gauge flow as inputs to predict the flow at the other four gauges. The ANN structure for the second layer of the prediction model was 2×4. Figure 9.22 shows the training results for the downstream gauge at Lashkar Gah. Table 9.6 shows the improvement in performance from using a 2×4 matrix versus the original 1×5 and 1×1 matrices. Although the performance for the prediction of Darweshan can be neglected, the rest of the gauges gain significant improvement in predictive ability. This test demonstrates how the two-layered operational flow prediction model can generate an increase in predictive capacity.

Table 9.5 Average Corresponding Outputs' Statistical Parameters for 12 Designed Runs

Run	Kajakai	K+M	Lashkar Gah	L+Q	Darweshan	Malakhan	Burjak	Khwabgah	Correlation coefficient
1	1	0	0	0	0	0	0	0	0.916
2	1	0	0	0	1	0	0	0	0.967
3	1	0	1	0	0	0	0	0	0.943
4	1	0	0	0	0	1	0	0	0.957
5	1	1	0	0	0	0	0	0	0.934
6	1	0	0	1	0	0	0	0	0.963
7	1	N/A	0	N/A	0	0	0	0	0.911
8	1	N/A	1	N/A	0	0	0	0	0.944
9	1	N/A	0	N/A	1	0	0	0	0.970
10	1	N/A	0	N/A	0	1	0	0	0.952
11	1	N/A	0	N/A	0	0	1	0	0.964
12	1	N/A	0	N/A	0	0	0	1	0.951

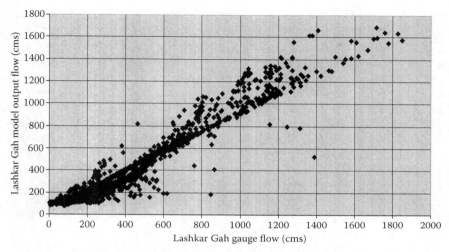

Figure 9.22 Results of test performance for a 1×1–2×4 artificial neural network model (Lashkargen gauge).

Table 9.6 Performance Comparison between Original 1×5 Systems and Alternative 2×4 Approach

Gauges	MSE (1×5)	CC (1×5)	MSE (2×4)	CC (2×4)
Lashkar Gah	3,255	0.949	1,512	0.976
Darweshan	9,964	0.914	10,040	0.913
Malakan	7,487	0.929	927	0.992
Burjak	1,2184	0.890	5,470	0.953
Khwabgah	9,580	0.912	4,329	0.961

9.5   Conclusions

An integration of databases and ANN learning was used to analyze a complex nonlinear watershed similarity for military hydrology applications. Although the unsupervised ANNs, such as SOFM, were used to perform the clustering of watershed characteristics, the supervised ANNs were used to identify the best match candidate watershed for classification analysis. The search procedure required several iterations of the clustering-classification loop.

Three demonstration examples, including random selection, average size, and median size watersheds, were used as the target to search for the best match corresponding candidate. The first example obtained a good correlation coefficient (0.92) for hydrograph prediction (2 years daily flow). The basin area ratio provided a reasonable factor to make the adjustment for hydrograph prediction. The preliminary sensitivity tests indicated that the hydrologic factors were more important to produce a better fitness for transplant than others. In general, monthly hydrograph comparison had better agreement than the daily hydrograph for both average and median size watershed examples. The most significant factor to improve reliability was to include as many watershed patterns in the knowledge base as reasonable. Future research includes developing an automated search procedure for a unique solution.

This study used historical flow records and ANNs to construct an operational riverflow prediction model for the Helmand River, Afghanistan. The tool demonstrated this model has a very short run time as well as good accuracy. However, some significant uncertainties from the historical record show the impact of tributary flow rates on downstream flow rates as well as channel transmission losses—the downstream flow rates are not always higher than the upper stream flow rates. In addition, there is only one upper stream gauge currently being operated. To improve the predictive reliability, the following two external factors need to be considered: adding a measurement location and improving the modeling by a proposed two-layered ANN model. Because the performance of data-driven modeling relies heavily on the quality of the information, it is critical to construct a new knowledge base using more recent hydrologic/hydraulic information beyond the existing historical record. The extrapolation procedures should be extended to include project-level events outside of the range of historic data (e.g., 100-year floods) and a prediction correction analysis.

Acknowledgment

The U.S. Army Corps Engineers Military Hydrology Program, Engineering Research and Development Center funded this work. Permission was granted by the Chief of Engineers to publish this information.

References

Boogaard, H., Ali, Md. S., and Mynett, A. E.: Self-organizing feature maps for the analysis of hydrological and ecological data set, *Proceedings of 3rd International Conference on Hydroinformatics*, Denmark, pp. 733–740 (1998).

Dawson, C. W., Harpham, C., Wilby, R. L., and Chen, Y.: Evaluation of ANNs Technique in the River Yangtze, China, *Hydrology, Earth System Science*, vol. 6, pp. 619–626 (2002).

Haykin, S.: *Neural Networks: A Comprehensive Foundation*, Macmillan, New York(1994).

Hsieh, B., and Bartos, C.: Riverflow/River Stage Prediction for Military Applications Using ANN Modeling, ERDC/CHL Technical Report, TR-00-16 (2000).

Hsieh, B., and Pratt, T.: *Field Data Recovery in Tidal System Using ANNs*, ERDC/CHL, CHEN-IV-38, Vicksburg, MS (2001).

Hsu, K. L., Gupta, H. V., and Srooshian, S.: ANN Modeling of the Rainfall-Runoff Process, *Water Resources Research*, vol. 31(10), pp. 2517–2530 (1995).

Imrie, C. E., Durucan, S., and Korre, A.: River Flow Prediction Using ANNs: Generalizations beyond the Calibration Range, *Journal of Hydrology*, vol. 233, pp. 138–153 (2000).

Kisi O.: River Flow Forecasting and Estimation Using Different Artificial Neural Network Techniques, *Hydrology Research*, vol. 39(1), pp. 27–40 (2008).

Kohonen, T.: The Self-Organizing Map, *Proceedings of the IEEE*, vol. 78, pp. 1464–1480 (1990).

Neurosolutions v5.0: *Developers Level for Window*, NeuroDimensions, Inc., Gainesville, FL (2003).

Principe, J. C., Euliano, N. R., and Lefebvre, W. C.: *Neural and Adaptive Systems: Fundamentals through Simulations*, John Wiley & Sons, New York (2000).

Thimuralaiah, K., and Deo, M. C.: Hydrological Forecasting Using Neural Networks, *Journal of Hydrologic Engineering*, vol. 5(2), pp. 180–189 (2000).

Ultsch, A., Korus, D., and Kleine, T. O., Integration of neural networks and knowledge-based system in medicine, *Proceedings of 5th Conference on Artificial Intelligence in Medicine Europe AIME'95*, Pavia, Italy, pp. 425–426 (1995).

USGS (United States Geological Survey): *Streamflow Characteristics of Streams in the Helmand Basin, Afghanistan*, Fact Sheet 2008–3059 (2008).

section II

Human factors issues

chapter ten

Understanding the influence of the "cry wolf hypothesis" and "false alarm effect" on public response to emergency warnings

Ann Marie Major and L. Erwin Atwood

Contents

10.1 Introduction

Public response to disaster warnings is a critical concern for emergency managers and responders given the potential for human loss of life, economic loss, and social disruption that follows disasters. Preparedness and mitigation programs are based on optimizing public response to disaster warnings to facilitate public compliance with national and local emergency plans and procedures, including evacuation.

This chapter provides examples from a multitude of disasters and applies research from the understanding of public response to disasters to paint a broad picture of the "cry wolf" hypothesis, which is often used interchangeably with the false alarm effect (FAE). In this chapter, the term FAE will be used for both terms.

As of December 31, 2011, the worldwide economic cost of disasters between 2000 and 2011 was more than $1380 billion, a figure reported on March 20, 2012, by Margareta Wahlström, the United Nations Secretary-General's Special Representative for Disaster Risk Reduction (UNISDR 2012a). Worldwide loss of human life during the first decade of this century stood at 1.1 million deaths and affected the lives of 2.7 billion people, 38% of the world's 7 billion population (UNISDR 2012b; U.S. Census Bureau 2012).

The fact that nearly four in 10 people on the planet experienced an extreme weather event between 2000 and 2011 underscores the significance of disaster planning and response. In the United States, 2011 made its mark in history with 12 extreme weather events surpassing 2008's total of eight extreme events. The National Oceanic and Atmospheric Administration (NOAA) initially calculated that the cost of each 2011 event was over $1 billion dollars, with total estimates of damage for the year at $52 billion (Lubchenco 2011).

What becomes an even more problematic situation for emergency planners and responders is that Mother Nature's behavior frequently defies scientific prediction because of the multitude of variables that alter the trajectory of the forecast. The realities of Mother Nature present a formidable challenge for disaster forecasters because, although some events are forecast with keen precision, thus lending credibility to forecasting and reassurance for the public to follow emergency instructions, the likelihood of a "near miss" or "false alarm" far exceeds the probability of an accurate forecast.

This conundrum is a reality of disaster forecasting. The cost attributed to false alarms is estimated at approximately $1 billion annually in the United States (Regnier 2008). For every potential public emergency or disaster, emergency planners and responders and public officials must weigh the costs and benefits of issuing warnings to the public. Decision makers must weigh the risks and benefits of supporting the inexact science of forecasting in the public arena as well as dealing with the response of the news media, which often exacerbates the situation when public warnings need to be issued (Regnier 2008).

The purpose of this chapter is to explore the phenomenon of "false alarms" and the "cry wolf" hypothesis as these constructs relate to disaster warnings and public response to disaster forecasting. Of critical importance is how the public responds to warnings and especially how "false alarms" impact public response to future warnings. The timeliness of the "cry wolf" issue was addressed in an Associated Press story dated April 1, 2012, reporting that five National Weather Service (NWS) offices in Kansas and Missouri are experimenting with ways to better communicate tornado warnings because of concerns arising from the May 22, 2011, Joplin, Missouri, tornado that resulted in 158 deaths (Thomas 2012).

10.2   The problem of false alarms

The prediction of phenomena is at the core of scientific method in the social and natural sciences, yet the science of prediction requires that hypotheses are falsifiable (Popper 2002). The study of public response to disaster warnings brings with it an understanding that false alarms and "near misses" are part of the scientific equation. Huppert and Sparks (2006, p. 1875) argue that "Disastrous outcomes can only increase unless better ways are found to mitigate the effects through improved forecasting and warning, together with more community preparedness and resilience."

The following illustrative example provides a point of departure for discussing the role of false alarms in emergency response. On January 29, 1991, in St. Louis, Missouri, radio listeners were tuned to KSHE disc jockey John Ulett's program; at 7:40 that morning, he played a tape of the Emergency Broadcast System signal followed by the nationally recognizable Don Pardo reporting that the United States was under nuclear attack. The announcement was followed by a few seconds of the Emergency Broadcast System signal. Ulett then announced that station programming would return to music, which it did (Cobo 1991). Shortly after the playing of the tape, the station listeners began calling the station—"some were panicked, others incensed" (McWhorter 1991). Ten minutes later, Ulett apologized to the radio audience for the "false alarm." After an investigation, the Federal Communications Commission, in a letter to the station, wrote (Letter 1991):

> [a]t a time when the United States was at war [the first Gulf War], and the public was in a heightened state of alert and anxiety about the possibility of terrorist violence and other dangers, your January 29, 1991, broadcast obviously had the potential to create widespread panic

(Levine 2000, p. 302)

According to the *St. Louis Post-Dispatch*, John Beck, the general station manager, said that "John [Ulett] wanted to illustrate the erroneous views of those listeners who were in favor of using nuclear weapons in the Persian Gulf, but it was just the wrong approach" (McWhorter 1991, 7A). For this infraction of the rules, the FCC fined KSHE (FM) $25,000 (Puig 1991).

This anecdotal example provides an illustration of a "false alarm" at the extreme because what is of concern in this chapter are false alarms issued by federal, state, and local emergency managers regarding emergency forecasts. However, "false alarms" like the bogus radio announcement of a nuclear attack on the United States are believed to impact public response to emergency disaster warnings and reduce the credibility of future warnings.

The FAE or cry wolf hypothesis has a long history going back as far as Aesop's Fables in Western literature (Perry 1965), with the fable of the shepherd boy who falsely cried wolf. The cry wolf moral appears in Chinese literature around 700 BC with the story of the Zhou Dynasty's King You, known as the emperor who cried wolf, because he ordered the lighting of the warning beacons to summon the feudal lords as a signal of invasion when there were no invaders (Minford and Lau 2000). His purpose was simply to amuse his mistress Baosi, who was said to never take delight in anything except for the lighting of the beacons. After years of false warnings, when the Xirong did invade his city and the warning beacons were lighted, the feudal lords did not respond, thinking it was just another false alarm, and King You was killed in the siege.

From a scientific perspective, a false alarm is the prediction of an undesirable event that fails to occur. The FAE, according to psychologists, holds that every failure of the predicted event to occur reduces the credibility of the alarm, and over time, the public will discount the alarm and not respond appropriately (Breznitz 1984). A second aspect of the FAE introduces an effect of fear caused by the initial prediction. Breznitz (1985) argued that a greater level of fear associated with the initial threat will result in a stronger FAE when the threat is cancelled.

Although the research literature examining public response and behavior in disasters is extensive, support for the FAE exists only in experimental literature and a limited number of field studies. The difficulties social scientists face in studying false alarms in the field are the short time frame surrounding false alarms and the prohibitive financial and personnel requirements for conducting field research following warnings that do not occur. The limited funding available is more likely to be directed to follow up on disasters that do occur.

Further complications arise in studying and understanding false alarms because of what defines a disaster. Chapman (2005) argues that the scientific community in general does not view technological disasters as predictable and has not invested the time in developing prediction systems largely because of a lack of understanding of the complexity of the socio-technological systems that cause disasters. One difference between technological and natural disasters is the complexity of the triggering events, which can be extremely complex in technological disasters because of the human–technology interaction (Chapman 2005). Applebaum (1997) defined socio-technical systems in terms of the intersection of human actions and relationships with technological tools and practices.

Challenges to human decision making in socio-technological systems occur when key players work from outdated models that no longer apply in the situation and interpret information as reliable when it originates from unreliable sources. This then results in a dearth of information within the organization. A key concern that arises in technological disasters is overconfidence or a "false sense of security" because warnings are viewed as a systemic testing device (Reason 1997). Chapman (2005) points to the issue of cognitive factors including "overconfidence," which can be viewed as unrealistic optimism and can lead to a culture contributing to failures.

Another problem arises when an automatic alarm system is considered an inconvenience or a hindrance to a smooth system operation and is turned off. This was the case on Deepwater Horizon, an oil-drilling platform in the Gulf of Mexico, on the night of April 20, 2010. The automatic gas and fire detection alarm had been disengaged at management's direction. Management was reported to have said that an automatic system is less safe than a manual system. Workers on the platform said that the alarm had been switched to manual operation to avoid having false alarms awaken crew members and corporate officials who were sleeping (Barstow et al. 2010; Lin 2010).

A different issue arises when there are calls for more and earlier alarms, especially regarding tornados and hurricanes, while at the same time other emergency managers are calling for fewer alarms to reduce the number of false alarms. The rationale for earlier alarms is that longer lead times will result in fewer deaths and injuries. In their analysis of more than 18,000 tornados covering a period of about 17 years, Simmons and Sutter (2009) found that the decrease in fatalities and injuries owing to longer lead times was equivalent to the increase in fatalities and injuries resulting from the increase in false alarms. Those who argue for fewer alarms argue that fewer will reduce the number of false alarms and decrease the reinforcement of the FAE. In a classic summary of human response to alarms, Janis (1962) argued that a sufficiently high number of false alarms will lead the public to disregard any warning that does not have attached to it additional reasons for compliance.

In theory, a false alarm decreases public attention and response to future alarms because the occurrence of the false alarm reduces the credibility of the warning system, which results in the FAE or the "cry wolf" hypothesis (Baker 1993; Davenport 1993; Levy and Salvadori 1995; Mileti and Fitzpatrick 1993; Mitchell 1993). Warning systems exist for all sorts of disasters including residential, commercial, wildlands, and forest fires; mass casualty incidents (e.g., a train derailment); explosions from terrorist acts or industrial accidents; weather disasters such as hurricanes and tornados; natural disasters such as volcanic eruptions and earthquakes; and technological disasters. In fact, many of these types of disasters overlap. For example, an earthquake can result in an explosion that then results in fire.

Simmons and Sutter (2008), in their study of tornados that occurred between 1986 and 2002, found that a warning, compared with no warning, reduced injuries by 32% but that fatalities were only reduced by 1.0%.

The researchers suggest that the low reduction in fatalities results from the few powerful, atypical tornados that touch down. Just 300 of the more than 18,000 (0.017%) tornados in their analysis produced fatalities, and of those, 15 killed more than 10 persons each (Simmons and Sutter 2008).

Hurricanes offer another problem in that the longer the timespan between the prediction of a hurricane's point of landfall and its occurrence, the more likely the prediction is to be a false alarm (Regnier 2008). Because of their size, hurricanes can extend over hundreds of miles of coastal front and can change course unexpectedly, as was the case with Hurricane Rita in 2005. A hurricane's storm track may follow a more-or-less straight line, wander in a series of sharp turns, or even turn back on itself as it responds to pressures from other systems that steer it. For example, an estimation of where Hurricane Rita would make landfall, made 120 hours (5 days) in advance, was 197 nautical miles in error. An estimation made 12 hours in advance had an error of only 27 nautical miles (Knabb et al. 2006). Thus, the longer the officials delay an evacuation order, the smaller the population area that requires evacuation. Figure 10.1 provides the tracking positions for Hurricane Rita in September 2005.

The problem of hurricane false alarms is not a question of whether there is or will be a hurricane but where the storm is likely to come ashore. This

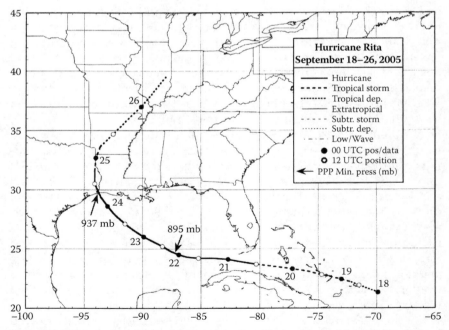

Figure 10.1 Best track positions for Hurricane Rita, September 18–26, 2005. (Courtesy of the National Oceanic and Atmospheric Administration's National Hurricane Center.)

information is critical to the question of evacuation decisions and how long officials can wait before they run out of time and evacuation is no longer an option. A small city needs up to 45 hours for an evacuation (Regnier 2008). For Hurricane Rita, the error in identifying the 48-hour position prediction was 93 nautical miles. That is, Rita would most likely make landfall somewhere within that area. The 36-hour prediction reduced that distance to 76 miles, and the 12-hour prediction error cut the distance to 27 miles. As noted earlier, Rita made landfall near the east edge of the 27-mile landfall line.

10.2.1   Frequency of false alarms

Three out of every four tornado warnings are "false" (Brotgze et al. 2011). If an alarm is issued and there is no report of the predicted twister, as is the case for 75% of tornado warnings in the United States, the alarm is defined as a false alarm. This does not mean that no tornado existed; however, if the tornado was not verified, either by radar or observers in the field, it is classified as a false alarm (Barnes et al. 2007). Thus, it is important to understand that all causes of false alarms are not the fault of the NWS but a combination of NWS prediction accuracy and performance of the distribution systems—mass media and interpersonal communication—over which the NWS has no control. Table 10.1 provides the proportions of false alarms by weather events, where the false alarm ratio (FAR) is calculated by dividing the number of false alarms by the number of disaster predictions (Wilks 2006).

Table 10.1 FARs for 10 Weather Events

FAR	Year	Event
0.75	1992	Tsunamis[a]
0.76	2003	Tornados in the United States
0.32	2003	Hurricanes
0.46	2005	Flash floods
0.31	2005	Winter storms
0.31	2005	High wind warnings
0.48	2005	Severe thunderstorms
0.48	2005	Combined tornadoes and severe thunderstorms
0.43	2003	Tropical cyclone genesis, North Pacific Basin
0.32	2004	Tropical cyclone genesis, North Atlantic Basin

[a] Morrison 2007, 12.

All other events: Barnes et al. 2007, 1–2.

10.2.2   Types of false alarms

An alarm is a double-edged sword. A disaster can occur without warning that results in no warning. In addition, the public is subject to missed alarms, false alarms, near misses, close calls, and direct hits. If a warning is inaccurate and the event does not occur, clearly it is a false alarm. If a warning is accurate and no alarm is given, it is a missed alarm and a disaster may result. To the extent that any warning reinforces a FAE created by previous false alarms, fewer people in the danger zone are expected to heed the next warning. An important finding from field studies that is relevant for emergency managers is that the FAE is not the only likely outcome of a failed prediction. To date, the research indicates that there may be no measurable effect on the beliefs of some of the residents of the threatened area, and there also may be a mobilization effect in which some people believe they must make better preparations for a potential, future disaster. Research shows that all three effects can result from a single false alarm, that is, FAE, the mobilization effect, and no effect (Atwood and Major 1998; Farley 1998).

10.2.3   When false alarms occur

Earthquakes provide no reliable warning of when, or if, they will occur. Observed slippages and tremors in a fault *may* indicate imminent danger, but they do not announce the date, time, or magnitude of the event. In late 1989, an earthquake false alarm with a 50–50 chance for a "tidally triggered," 6.5–7.5 magnitude tremor in the New Madrid Seismic Zone (NMSZ) was predicted for December 3, 1990, plus or minus 48 hours (Davis 1990). Iben Browning, an Arizona businessman with no training in seismology, issued that prediction on December 12, 1989, at the Missouri Governor's Conference on Agriculture in Osage Beach. Although Browning did hold a doctorate in zoology, the news media made an "earthquake expert" out of him by uncritically reporting his false claims to have predicted other earthquakes (Showalter 1993). The NMSZ prediction was not Browning's first experience with earthquake forecasting.

Browning's false alarm for a December 3, 1990, earthquake was reinforced by an unpredicted 4.6 magnitude earthquake on September 26, 1990 (Farley 1993, 273). The news media provided extensive coverage of Browning's prediction and between September 1 and December 3, 1990, the local newspaper printed 147 news stories about the prediction; 46% were front-page stories (Gao 1991). In none of these stories was there an explanation that the ambiguous "50–50" chance provided *no* information about the likelihood of the earthquake appearing on schedule. An estimate by two geologists (Johnson and Nava 1985) of the likelihood of a 6.3 magnitude quake in the next 50 years placed the likelihood of a damaging quake as between 86% and 97%, not quite as precise as the estimate of the businessman from Arizona but far more realistic (Atwood and Major 1998). None of the news media stories pointed out that geoscientists are still unable to predict earthquakes.

10.2.4 No warning and a missed alarm

If there is no warning followed by a missed warning or the warning arrives too late, a double disaster can result as was the case on March 27–28, 1964. At 5:36 p.m. (Alaska Daylight Time), on Good Friday, March 27, 1964, a 9.2 magnitude earthquake, with the epicenter about 20 km north of Prince William Sound on Alaska's south coast, generated a tsunami. The earthquake lasted 4 minutes and caused damage estimated between $300 million and $400 million (1964 U.S. dollars). The tsunami caused 11 deaths and $8 million in damage 3000 miles away in Crescent City, California (Christensen 2002). Known as "Tsunami City," Crescent City has been hit by 34 tsunamis since 1931 including the one generated by the deadly 9.2 magnitude Great East Japan Earthquake of March 11, 2011. The seismic wave struck Crescent City about 10 hours after the earthquake with a maximum wave height of about 2.5 meters (about 8.1 feet); 16 fishing boats were sunk in the harbor (Nishijima 2011).

In 1964, the first three of the four seismic waves that struck Crescent City, though modest in comparison with the fourth wave, damaged or destroyed businesses along the waterfront and damaged docks and boats. On March 28, the fourth and most destructive wave struck Crescent City about 1:45 a.m. and inundated the city (Anderson 1969). The California Disaster Office (CDO) issued two alarms based on information received from disaster officials in Alaska. The first alarm was issued at 11:08 p.m. and the second at 11:50 p.m., 10 minutes before the tsunami was predicted to reach the Crescent City area at 12:00 a.m. local time. (Anderson 1969).

Crescent City disaster officials did not respond to the first warning (Anderson 1969), and when the second CDO warning arrived, it probably appeared to local disaster officials to be too late for an effective local warning. Instead of reaching the Crescent City area at midnight, as predicted, the deadly fourth wave did not arrive for 1 hour and 45 minutes (1:45 a.m., Saturday, March 28) after its estimated arrival time. The harbor was destroyed, the city was under water, and 11 people were killed.

10.2.5 Near miss and a direct hit

Near misses by hurricanes come in various measures of what is "near." Two pairs of hurricanes are used here to illustrate the problem and public response. In 1996, Hurricanes Bertha and Fran were predicted to make landfall in South Carolina: Hurricane Bertha on July 12, 1996, and

Hurricane Fran on September 5, 1996. Both storms struck the U.S. East Coast, not in South Carolina where they were initially predicted to strike, but farther north in North Carolina just above the border between the two states (Dow and Cutter 1998). In South Carolina, 41% of the residents evacuated for Bertha, a Category 1 hurricane, and 59% evacuated for Fran, a weak Category 3 storm (Dow and Cutter 1998). According to the Saffir–Simpson Hurricane Scale, a Category 1 storm yields winds up to 74 to 95 miles/h (119 km/h). A Category 3 storm can yield 111 to 130 miles/h or 178 to 208 km/h (Schneider 2012). The final paths of the two storms crossed as they made landfall on or near the southern edge of Cape Fear (Dow and Cutter 1998, Fig. 1, 241). An estimated 750,000 people evacuated for the storms, and for many, if not most, of the evacuees, it was a false alarm. A survey of evacuees following their return home found that the respondents said they would evacuate again if advised to do so. Overall, then there was no substantial FAE (Dow and Cutter 1998).

10.2.6 A too-close call and a direct hit

Hurricane Rita certainly was not a near miss on Galveston, and it may have been too close to Galveston to be a "close call" as the eye wall tracked along the northern shore of the island (Knabb et al. 2006), but it does provide a clear track, so to speak, to problems associated with evacuations that are not well planned and controlled. Hurricane Rita made landfall on the Texas coast about 85 miles (straight line estimate) north-northeast of Galveston and caused one of the largest evacuations in United States history (Knabb et al. 2006). Knabb et al. (2006) report that more than two million Texans tried to evacuate. An unknown number of evacuees were trapped in miles-long lines of stalled traffic, and 120 people died before Hurricane Rita made landfall (Casselman and Campoy 2008). One news story reported the experience of two people who had tried to follow the evacuation directive. A young woman who reported sitting in traffic for 16 hours trying to leave Galveston ahead of Rita was quoted as saying "I'd rather drown than wait in traffic." A young man who had tried to evacuate gave a clear FAE belief when he said "I've heard about so many hurricane warnings before and nothing ever happened." He did not evacuate for Hurricane Ike in September, 2008, and rode out the direct hit in his Galveston apartment (Casselman and Campoy 2008).

There is a myth that hurricanes turn away at the last minute, and this myth is both supported and refuted by the experiences of Galveston's residents. Hurricane Allen in 1980 was a Category 5 hurricane when it encountered a dry air mass in the western Gulf of Mexico and turned south before reaching Galveston and made landfall north of Brownsville, Texas, as a Category 3 hurricane. Hurricane Ike made no such concession to the estimated 90% of Galveston's residents who did not follow the

Table 10.2 Damage Losses for Selected Disasters

Year	Highest Category	Estimated Losses (U.S. Dollars)[a]	Event
1964	9.2 m	300 to 400 million	Alaska "Good Friday" earthquake[b]
1964	n.a.	8.0 million	Crescent City, CA tsunami[c]
1992	5	125.0 billion	Hurricane Andrew
2005	5	200.0 billion	Hurricane Katrina[d]
2005	3	12.037 billion	Hurricane Rita[e]
2006	1	270.0 million	Hurricane Bertha
2006	3	1.6 billion	Hurricane Fran[c]
2008	4	4.3 billion	Hurricane Gustav[c]
2008	5	24.9 billion	Hurricane Ike[c]
2011	5.8	200–300 million	East Coast earthquake[f]

[a] Not adjusted for inflation.
[b] Alaska Earthquake Information Center.
[c] Knabb, Brown, and Rhone (2006).
[d] Adeola, F. O. (2009).
[e] Knabb, Brown, and Rhone (2006).
[f] Washington Post, August 24, 2011.

official evacuation order and made a direct hit on the island. Hurricane Rita turned northwest shortly before reaching Galveston and made landfall near the Texas–Louisiana border, but the "last minute" was a bit late, and Rita moved along the north shore of the island as a Category 3 hurricane before turning north-northeast. Table 10.2 presents the monetary damages (in U.S. dollars) for 10 selected disasters.

10.3   Public response and the FAE

From experimental research, what is known about the FAE is that people who experience the highest level of fear before the warning's cancellation are assumed to be the least likely group to attend to a future warning; they are the most likely to downgrade the credibility of the warning, and they are the least likely to take protective actions to save their lives and property (Breznitz 1984, 1985). A psychological explanation for these findings is that respondents experience a "loss of face" because of an initial belief in the warning and a belief that the preparations they had undertaken were a waste of time because the warning did not materialize (Breznitz 1984).

Drabek (1986) pointed out that the experimental findings where Breznitz (1984) confirmed a false alarm or "cry wolf" effect did not take into account the role of social interaction and news media coverage of the warning. Survey research does, in fact, provide some evidence of a FAE;

however, it is certainly not far-reaching in large measure, perhaps as a result of the complexity of interacting factors. In a panel mail survey of residents living in Marked Tree and Wynne, Arkansas, and New Madrid and East Prairie, Missouri, Showalter (1993) found no decline in residents' concern about future earthquake predictions in view of a false alarm. This might be because residents of the NMSZ region had experienced earthquakes, given that they resided in an active seismic zone.

Public response following two earthquake false alarms in California provides limited support for an FAE. On April 21, 1976, Dr. James Whitcomb of the California Institute of Technology's Seismology Laboratory issued a prediction that a 5.5 to 6.5 Richter magnitude earthquake would occur before April 1977 in the Los Angeles area. Whitcomb made public that his prediction was a hypothesis as part of his efforts to study predictions; however, in December 1976, he withdrew the prediction. In that same month, on December 2, 1976, a geophysicist with a doctorate named Henry Minturn issued a prediction for an earthquake on December 20 in the Los Angeles area. Minturn was not affiliated with any scientific organization, although he claimed previous success in earthquake forecasting. United States Geological Survey scientists refuted his methods. December 20, 1976, came and went and no earthquake occurred.

In a longitudinal analysis of the withdrawn prediction and the prediction that did not happen, Turner et al. (1986) found that approximately one-third (27%) of their Los Angeles respondents recalled that a warning had been retracted, and 4 out of 10 (43%) recalled that an earthquake that had been predicted had, in fact, not occurred. Although these data provide limited indirect evidence for an FAE, it is important to note that nearly three-fourths (72.7%) of their respondents did not recall the Whitcomb prediction 2 years after it was issued, and the majority (57%) of Los Angeles respondents did not recall the Minturn prediction.

A panel study in southern Illinois and southeastern Missouri communities provided insight into what happens when a disaster prediction fails (Atwood and Major 1998). Iben Browning's 50–50 chance of at least a 6.5 Richter magnitude earthquake for December 2–3, 1990, plus or minus 48 hours, failed to materialize. This study found support for the "false alarm effect" or "cry wolf hypothesis," with 45.6% of the pre- and post-prediction panel respondents reporting a decrease in their perceptions of the importance of future earthquakes (Table 10.3).

However, nearly 1 in 5 (16.8%) respondents reported what Mileti and Fitzpatrick (1993) described as a mobilization effect, where the respondents attribute greater importance to future warnings based on their experience with the false alarm. Nearly 4 in 10 (37.4%) of the panel respondents reported no change in the level of importance that they attributed to future earthquake predictions because of the false alarm. What is important is that nearly half of the respondents in this panel study remained

Table 10.3 Percentages of Respondents Characterized
by the Mobilization Effect, No Change, and FAE
Following a False Alarm

Problem Importance	%	n
Mobilization effect	16.9	44
No change	37.4	98
FAE	45.8	120
Total	100.0	262

Note: These outcomes are from a 1990 survey conducted in Cape
Girardeau, Missouri, by Ann M. Major and L. Erwin Atwood
following the failure of a predicted earthquake to materialize at
the appointed time.

consistent in their perceptions about future earthquake warnings. That
is, the false alarm had no impact on their views about future warnings.

Respondents who reported no change in their views about how important they felt the earthquake problem was in the New Madrid Seismic
Zone were more likely to have thought about the earthquake prediction
than were respondents who were characterized by a mobilization effect.
The nearly 40% (37.4) of respondents who were consistent in their beliefs
reported greater confidence in news stories about the earthquake prediction than did respondents who were more likely to report greater attention
to future warnings (i.e., the mobilization effect). The "no change" respondents reported that preparations and mitigation were very important and
reported greater confidence in information from other people and experts
(Atwood and Major 1998).

What this field study demonstrates is that respondents' reactions to
a warning differ. Indeed, as Breznitz (1984, 1985) argued from his experimental data, there does appear to be a "false alarm effect" or "cry wolf
hypothesis." However, survey research from panel studies points to a
mobilization effect, where respondents are more likely to pay attention to
future warnings as well as a "no change" effect where respondents maintain the same interest in attending and responding to false alarms following a warning.

10.4   Discussion

A key concern about public response to disaster warnings is what the
public learns from emergency warnings whether those warnings transpire or whether they are false alarms. In one of the few field studies that
explores the "cry wolf hypothesis," Atwood and Major (1998) provide
confirmation of the "cry wolf hypothesis." Their study examined whether
their respondents' perceptions of the emergency warning decreased after
a false earthquake prediction.

In their study, the majority of respondents were characterized by the "cry wolf hypothesis" because the mean for the majority of respondents who believed in the earthquake prediction following the false alarm decreased from 3.57 to 2.29 ($t = 13.95$, $p < .001$). The data also make clear that those respondents who did not think that the emergency warning was important maintained their belief that there was no concern to worry about; however, those respondents still reported undertaking mitigation behaviors. Nearly 1 in 5 respondents (16.8%) reported being more likely to believe a future warning. Disaster warning messages that refer in some way to the public's previous experience with disaster events will be more likely to gain the attention of the public and will be more likely to initiate public preparations for a future disaster.

Once emergency management officials decide to issue a disaster warning, the mass media play a key role in disseminating the warning to the public and establishing a chain of events from the initial warning to public response. If a disaster warning fails to materialize, then the research indicates that the public will continue to attend and respond to future warnings as long as emergency managers provide a coherent explanation of why the disaster warning did not transpire (Mileti and Fitzpatrick 1993).

However, post-cancellation explanations disseminated through the news media are not guaranteed to reach the same public who heard the initial warning and cancellation, and there is no guarantee that the public will accept the official explanation for a warning cancellation or false alarm (Baker 1993; Ledingham and Masel Walters 1985; Lindell and Perry 1993; Tonn et al. 1990). In the chain of events that follows a prediction, amplified news media coverage of the warnings leads to public interest and discussion and information seeking about the warning (Mileti and Fitzpatrick 1993; Turner et al. 1986).

Communication problems result from different agencies having different policies, procedures, and equipment, which contribute to message delays and interpretations. The mass media are seen as helpful in disseminating warnings, but after the event, the media rarely report information that would help survivors put their lives back together. Instead, reporters focus on "human interest" stories that highlight past events and hold breathless interviews with survivors that focus on the emotional aspects of the events.

Research has examined the differences in terms of demographics in disaster preparedness and gender. For example, Morrow and Enarson (1996) found that women of minority backgrounds and those from lower socioeconomic status do not have the power and resources to respond adequately to disaster warnings. Despite their lack of resources, women often assume the role of caregivers and have an interest in risk reduction because of their caregiver roles (Enarson 2001; Major 1999). In their study of Hurricane Andrew, Morrow and Enarson (1996) reported that women take on the responsibility of preparing family members by gathering supplies and undertaking protective actions as long as those protective actions are affordable.

Post-disaster analyses make clear that those members of the public with the least financial resources suffer the greatest losses because they live in mobile homes or dwellings that have not been constructed to withstand damage when disasters strike. They are less likely to own a car and are unable to evacuate from an area following a warning because they have no means of transportation nor do they have the money to establish themselves in temporary housing in another city. Finally, they often lack the resources to purchase and store additional food and water.

The "false alarm effect" is a real-world outcome that has at least three characteristics. The FAE may reduce the believability of the warning system, that is, the FAE. It may increase the believability of a warning system, that is, the mobilization effect. The FAE may have no effect on the credibility of the warning system. The five most important reasons people act or do not act on a warning are the following: (1) the amount and quality of the information they have regarding the threat; (2) the extent to which they believe they are at risk from the threat; (3) their ability to afford preparations and evacuation; (4) the strength of their belief that they must remain in their homes to protect their belongings and property; and (5) their previous experience in taking shelter in their own homes during previous disasters and the belief that they can repeat the experience with no ill effects.

Human–technological interaction adds additional factors to the equation in terms of the role of decision making and the wisdom of those decisions to turn off the automatic alarm system to eliminate false alarms, when those very false alarms may have prevented a disaster that resulted in a substantial loss of life and the destruction of billions of dollars in property. When the Deepwater Horizon platform sank and the fires had been extinguished, the world's attention turned to a disaster of unknown proportions, because oil had been escaping from the well and polluting a large area of the Gulf of Mexico and the Gulf Coast. Perhaps the most important lesson to be learned is that false alarms are a blessing and a bane for remaining vigilant in the face of disasters and that the public, researchers, and emergency managers and responders need recurrent reminders that false alarms are essential to disaster prediction, preparation, and mitigation.

The research reported here and the outcomes of many other studies support the false alarm and mobilization hypotheses, and it appears that over the past dozen years, sufficient evidence has accumulated to remove these two variables from the hypothesis category and treat them as permanent members of the sets of variables used in the assessment of natural and man-made hazards. It is no longer a question of whether they function in hazard/disaster situations but of *how* they impact other variables directly and indirectly and are themselves influenced by other variables in the equation. For many years, at least in political communication campaign research, the "no-apparent-effect" respondents tended to be ignored by researchers because "nothing happened."

Would there have been fewer casualties in Joplin, Missouri, if there had been no sirens for the city's residents to ignore the afternoon of Sunday, May 22, 2011, as the EF-5 tornado that killed 158 people in just a few minutes approached the city? The critics of warning sirens and the frequency of their use seem to believe so.

Respondents who reported they were less likely to pay attention to a future earthquake warning (FAE) also were more likely to have believed the prediction than were those who reported a mobilization effect. In addition, respondents characterized by a mobilization effect, that is, those who indicated they would pay more attention to a future warning, initially were less likely to believe the prediction. Breznitz (1985) noted this outcome and commented that where one who has been "fooled" or "made the fool," there is likely to be the temptation to downgrade the importance of the situation as a matter of "saving face." There also is the likelihood that some people at risk from disaster—flash floods, hurricanes, tornados, blizzards—because they are convinced they are in danger are more likely to be concerned about future disasters. To date, there does not appear to be much evidence that a personal belief will insure that the believer will be "bulletproof" when the critical moment arrives, but to do so has so far been a poor bet. So have attempts to "educate" the public to solve the problem. Unfortunately, education appears to be effective only if the target of the educational attempt wants to be educated.

The "false alarm effect" and the "cry wolf hypothesis" are, indeed, effects of warning systems, and those effects need to be taken into consideration when communicating with the public about disaster preparation. In addition, a substantial proportion of the public is neither more nor less likely to believe future warnings. However, these "no apparent" effect respondents, who tend to be ignored in favor of those who are characterized by the FAE, are much more likely to undertake preparations for the event and are more likely to communicate with members of their social networks. The "no apparent" effect respondents may be a critical public to target with information, given that they are undertaking preparations perhaps because they understand that false alarms are a fact of life in the face of potential disasters. Warning officials may take the sting out of false alarms by praising community performance, for in the final analysis, it must be admitted that *no one knows* if any warning will become a false alarm until the danger has passed. After all, false alarms are a critical part of the emergency warning system.

References

Adeola, F. O. 2009. Katrina cataclysm: Does duration of residency and prior experience affect impacts, evacuation, and adaptation behavior among survivors. *Environment and Behavior* 41(4): 459–489.

Anderson, W. A. 1969. Disaster warning and communication processes in two communities. *Journal of Communication* 19: 92–104.

Applebaum, S. H. 1997. Socio-technical systems theory: An intervention strategy for organizational development. *Management Decision* 35: 452–463.

Atwood, L. E. 1991. Illusions of media power: The third-person effect. *Journalism & Mass Communication Quarterly* 71: 269–281.

Atwood, L. E., and A. M. Major. 1998. Exploring the "cry wolf" hypothesis. *International Journal of Mass Emergencies and Disasters* 16: 279–302.

Baker, E. J. 1993. Empirical studies of public response to tornado and hurricane warnings in the United States. In *Prediction and perception of natural hazards*, eds. J. Nemec, J. M. Nigg, and F. Siccardi, 65–73. Dordrecht, The Netherlands: Kluwer Academic Publishers.

Barnes, L. R., E. C. Gruntfest, M. L. Hayden, et al. 2007. False alarms and close calls: A conceptual model of warning accuracy. *Weather and Forecasting* 22: 1140–1147.

Barstow, D., D. Rohde, and S. Saul. December 25, 2010. Deepwater Horizon's final hours. *The New York Times* A1.

Breznitz, S. 1984. *Cry wolf: The psychology of false alarms.* Hillsdale, NJ: Lawrence Erlbaum Associates.

Breznitz, S. 1985. False alarms: Their effect on fear and adjustment. *Issues in Mental Health Nursing* 7: 335–348.

Brotgze, J., S. Erickson, and H. Brooks. 2011. A 5-yr climatology of tornado false alarms. *Weather and Forecasting* 26: 534–544.

Casselman, B., and A. Campoy. September 10, 2008. Many weren't scared enough to flee. *The Wall Street Journal* A3.

Chapman, J. 2005. Predicting technological disasters: Mission impossible? *Disaster Prevention and Management* 4: 343–352.

Christensen, D. 2002. *The great Alaska earthquake of 1964.* Alaska Earthquake Information Center (AEIC), University of Alaska at Fairbanks. Available at: http://www.aeic.alaska.edu/quakes/Alaska_1964_earthquake.html. (Accessed March 30, 2012).

Cobo, L. 1991. False radio broadcast evokes FCC investigation. February 4, 1991. *Broadcasting* 120(5): 29.

Davenport, A. G. 1993. The impact of structural damage due to hurricanes and the prospects for disaster reduction. In *Prediction and perception of natural hazards*, eds. J. Nemec, J. M Migg, and F. Siccardi, 12–31. Dordrecht, The Netherlands: Kluwer Academic Publishers.

Davis, J. F. 1990. *The ad hoc working group on the December 2–3, 1990, earthquake prediction: Evaluation of the December 2–3, 1990, New Madrid Seismic Zone prediction.* Sacramento, CA: Report of the State of California, Division of Mines and Geology.

Dow, K. and S. Cutter. 1998. Crying wolf: Repeat responses to hurricane evacuation orders. *Coastal Management* 26: 237–252.

Drabeck, T. E. 1986. *Human systems responses to disaster. An inventory of sociological findings.* New York, NY: Springer-Verlag.

Enarson, E. 2001. What women do: Gendered labor in the Red River Valley flood *Environmental Hazards* 3: 1–18.

Farley, J. E. 1993. Public, media, and institutional responses to the Iben Browning earthquake prediction. *International Journal of Mass Emergencies and Disasters* 11(3): 271–277.

Farley, J. E. 1998. Down but not out: Earthquake awareness and preparedness trends in the St. Louis metropolitan area, 1990–1997. *International Journal of Mass Emergencies and Disasters* 16(3): 303–319.

Gao, D. 1991. Content analysis of earthquake coverage in two local newspapers: *Southern Illinoisan and Southeast Missourian.* Carbondale, IL: Unpublished manuscript, School of Journalism, Southern Illinois University.

Huppert, H. E., and S. J. Sparks. 2006. Extreme natural hazards: Population growth, globalization, and environmental change. *Philosophical Transactions of the Royal Society A: Mathematical, Physical & Engineering Sciences* 364: 1875–1888.

Janis, I. L. 1962. Psychological effects of warnings. In *Man and society in disaster,* eds. G. Baker and D. Chapman, 55–91. New York, NY: Basic Books.

Johnson, A. C., and S. J. Nava. 1985. Recurrence rates and probability estimates for the New Madrid Seismic Zone. *Journal of Geophysical Research* 90: 6737–6753.

Knabb, R. D., D. P. Brown, and J. Rhone. 2006. *Tropical cyclone report: Hurricane Rita,* March 18–26, 2005, National Hurricane Center. Miami, FL: NOAA National Weather Service.

Ledingham, J. A., and L. Masel-Walters. 1985. Written on the wind: The media and hurricane Alicia. *Newspaper Research Journal* 6: 50–58.

Letter. 1991. 6 F.C.C.R. 2289, 2289, 69 Rad. Reg.2d (P & F) 155.

Levine, J. 2000. The FCC and radio hoaxes: The turning point: Genesis of modern regulation. *Federal Communication Law Journal* 52: 273–320.

Levy, M., and M. Salvadori. 1995. *Why the earth quakes.* New York: W. W. Norton and Company.

Lin, R.-G. II. July 23, 2010. Alarms, detectors disabled so top rig officials could sleep. *Los Angeles Times.* Available at: http://latimes.com/print/2010/jul/23/news/laoil-spill-disabled-alarms-20100723. (Accessed March 1, 2012).

Lindell, M. K., and R. W. Perry. 1993. Risk area residents' changing perceptions of volcano hazard at Mt. St. Helens. In *Prediction and perception of natural hazards,* eds. J. Nemec, J. M Nigg, and F. Siccardi, 159–166. Dordrecht, The Netherlands: Kluwer Academic Publishers.

Lubchenco, J. December 7, 2011. Predicting and managing extreme events. A speech delivered to the American Geophysical Union, San Francisco, CA: National Oceanic and Atmospheric Administration. Available at: http://www.noaanews.noaa.gov/stories2011/20111207_speech_agu.html.

Major, A. M. 1999. Gender differences in risk and communication behavior: Responses to the New Madrid earthquake prediction. *International Journal of Mass Emergencies and Disasters* 17: 313–338.

McWhorter, D. 1991. KSHE Suspends DJ Over Phony Warning. *St. Louis Post-Dispatch* 7A.

Mileti, D. S. and C. Fitzpatrick. 1993. *The great earthquake experiment: Risk communication and public action.* Boulder, CO: Westview Press.

Minford, J. and J. S. M. Lau. 2000. *Classical Chinese literature: An anthology of translations,* Volume 1. New York: Columbia University Press/The Chinese University of Hong Kong, 44.

Mitchell, J. 1993. Natural hazard prediction and responses in very large cities. In *Prediction and Perception of Natural Hazards,* eds. J. Nemec, J. Nigg, and F. Siccardi, 29–37. Dordrecht, The Netherlands: Kluwer Academic Publishers.

Morrison, W. A. 2007. *Tsunamis; monitoring, detecting, and early warning systems*. Congressional Research Service Washington, D. C. (updated May 10, 2007).

Morrow, B. H. and E. Enarson. 1996. Hurricane Andrew through women's eyes: Issues and recommendations. *International Journal of Mass Emergencies and Disasters* 14: 5–22.

Nishijima, T. 2011. U. S. town learned from tsunami. *Daily Yomiuri Online*. Available at: http://.www.yomiuri.co.jp/dy/national/T120315004995.html.

Perry, B. E., trans. 1965. *Babrius and Phaedrus*. Cambridge: Harvard University Press.

Popper, K. R. 2002. *The logic of scientific discovery*. London, UK: Routledge.

Puig, C. May 20, 1991. How FCC deals with hoaxes: Radio: The federal agency is mulling the KROQ case. It can admonish or levy a fine, but hasn't used its power to suspend a license in more than a decade. *The Los Angeles Times*. Available at: http://articles.latimes.com/1991-05-20/entertainment/ca-1596_1_radio-hoaxes/2. (Accessed February 17, 2012).

Reason, J. T. 1997. *Managing the risks of organizational accidents*. Surrey, UK: Ashgate Publishing.

Regnier, E. 2008. Public evacuation decisions and hurricane track uncertainty. *Management Science* 54(1): 16–28.

Schneider, B. 2012. *Extreme weather: A guide to surviving flash floods, tornadoes, hurricanes, heat waves, snowstorms, tsunamis and other natural disasters*. New York, NY: Palgrave MacMillan.

Showalter, P. 1993. Prognostication of doom: An earthquake prediction's effect on four small communities. *International Journal of Mass Emergencies and Disasters* 11: 279–292.

Simmons, K., and D. Sutter. 2008. Tornado warnings, lead times, and tornado casualties: An empirical investigation, *Weather and Forecasting* 23: 246–258.

Simmons, K. M., and D. Sutter. 2009. False alarms, tornado warnings, and tornado casualties. *Weather, Climate, and Society* 1: 38–53.

Thomas, J. L. March 31, 2012. New warnings will pull no punches for strongest storms. *The Kansas City Star*. Available at: http://www.kansascity.com/2012/03/31/3527480/warnings-will-pull-no-punches.html.

Tonn, B. E., C. B. Travis, R. T. Goeltz, et al. 1990. Knowledge-based representations of risk beliefs. *Risk Analysis* 10: 169–184.

Turner, R. H., N. Joanne, and P. D. Heller. 1986. *Waiting for disaster: Earthquake watch in California*. Berkeley and Los Angeles: University of California Press.

U.S. Census Bureau. 2012. Work population clock. Available at: http://www.census.gov/main/www/popclock.html. (Accessed March 25, 2012).

United Nations International Strategy for Disaster Reduction (UNISDR). March 20, 2012a. Disaster losses top one trillion dollars as donors underfund risk reduction. Press Release PDF. Available at: http://www.unisdr.org/archive/25831. (Accessed March 25, 2012).

United Nations International Strategy for Disaster Reduction (UNISDR). March 20, 2012b. The economic and human impact of disasters in the last 12 years. Infographic PDF. Available at: http://www.unisdr.org/archive/25831. (Accessed March 25, 2012).

Wilks, D.S. 2006. *Statistical methods in the atmospheric sciences*. 2nd Edition. Burlington, MA: Academic Press.

chapter eleven

Accessible emergency management

A human factors engineering approach

Tonya L. Smith-Jackson, Woodrow W. Winchester,
Ali Haskins Lisle, William H. Holbach,
Hal Brackett, and Tom Martin

Contents

11.1 Introduction

Emergency preparedness (before disasters) and community resilience (after disasters) are foundational to the safety and security of communities. From a human factors perspective, management structures, procedures, and tools to support preparedness and resilience are conceptualized as subsystems of larger and more complex sociotechnical systems. Sociotechnical systems that underpin disaster events comprise technologies, infrastructure, physical environments, emergency personnel (e.g., first responders, command, and control), and people. These systems must be managed effectively and inclusively to reduce hazards and risks to the public. A working group of experts in emergency management defined emergency management as the function charged with designing and implementing the framework to provide services to communities to "reduce vulnerability to hazards and cope with disasters" (Blanchard 2007, p. 4). What is key in this definition is the concept of "communities," which in most instances reflects a diversity of individuals with differing resources, capabilities, and needs. Building a framework around the complexity inherent to any community is a major challenge to emergency management personnel.

Communities are, in fact, likely to consist of levels of diversity that are not easily realized without thorough analysis using census and other demographic data, as well as methods providing more accuracy such as face-to-face counts of homeless citizens; citizens estranged from partners, spouses, or families; and citizens living in shelters. Most importantly, citizens' capabilities to prepare for, survive, and recover from disasters are strongly influenced by personal factors that are dynamic and difficult to count or understand using cross-sectional methods like demographic surveys. At any point in time, citizens' capabilities and vulnerabilities change and may do so in significant ways. For example, citizens are constantly aging, experiencing accidents, or living with diseases or syndromes. As a consequence, communities are always in flux in terms of the number of individuals who are experiencing challenges that introduce sensory, perceptual, mobility, and cognitive impairments.

Disaster management preparedness comes from what is referred to as "the hazard cycle" (Tierney et al. 2001), which characterizes disasters using four time-sensitive stages: mitigation, preparedness, response, and recovery, though the disaster itself is time delimited. Emergency preparedness refers to all actions taken before the disaster by responders and those directly affected, which enable proactive engagement of social units when the disaster occurs. To develop an effective accessible community emergency management plan, preparation is vital. All citizens should outline an emergency preparedness plan and practice executing it. It is also important to note that during emergency situations, the usual methods of support and assistance may not be available or may be limited. In planning for successful emergency response, the system must comprise both the emergency response system's optimization and preparedness by the community. However, community members' capabilities are constantly changing and will influence the effectiveness and efficiency of emergency management and response.

However, preparedness is still not at the desired level considering the disasters experienced in the last decade. During Hurricane Katrina, 31.6% of the residents of New Orleans (Metaire-Kanner region) had disabilities and 16.6% required the use of special equipment during the hurricane (McClure et al. 2011). In 2007, organizations such as the Department of Homeland Security and the Department of Education found emergency preparedness plans for most states to be inadequate for people with disabilities. Kailes (2005) and Rooney and White (2007) suggested that the shortcomings in accessibility of emergency management systems were attributable to agencies' lack of partnerships with disability advocacy groups or other stakeholders such as senior citizens groups.

The prevalence of individuals with disabilities demonstrates the need to identify managerial frameworks in emergency management, which will provide accessible services to these most vulnerable members of the community. Yet, ensuring individuals with disabilities receive equitable services and support before, during, or after disasters has received little empirical research attention. Efforts to enhance accessibility have rested on educating individuals with disabilities to prepare themselves for disasters, have information in hand to respond during a disaster, and to be aware of services available to support resilience. But few efforts have focused on designing the emergency management framework, so the system responds to the needs of the individual rather than resting the burden of safety and security on individuals' awareness and ability to make effective use of the available systems. Limitations in the use of outreach and education, rather than designing accessible emergency management systems, were indicated by Fox et al. (2007). These researchers found that 56% of people with disabilities did not know whom to contact about emergency plans, while 61% had not developed emergency plans. Thirty-two percent of individuals with disabilities who were employed did not have plans to evacuate their own workplaces. There is clearly a gap in what individuals with disabilities need and the existing mechanisms in place for emergency management targeting these groups. The need for accessible emergency management is significant and will require the integrated, coordinated, and focused efforts of emergency management personnel, researchers, and community stakeholders.

11.2   Gross statistics on individuals with disabilities

The U.S. Census defines disability using an ecological or contextual description, which recognizes the interaction between the person and the physical environment (U.S. Census Bureau 2009). This contextualized definition further demonstrates the dynamics of the label "disability" and the need to avoid the idea that simply counting individuals with disabilities in one cross section in time will always provide the information necessary to design systems to meet their needs. It is clear that an individual with quadriplegia or an elderly person in a wheelchair have mobility challenges that require specialized services during an evacuation. But a high school football player who has sustained a broken leg and a concussion or a woman who has decided to give birth at home will also require similar specialized services for an evacuation. These realistic examples illustrate the importance of broadening and situating conceptualizations of disability in meaningful contexts that recognize the dynamic and transient nature of populations with disabilities. With a broader perceptive, we will be able to engage with individuals who identify as having a disability to elicit the needs and capabilities that define requirements for the design of accessible emergency management systems. When emergency management systems are effective and beneficial to individuals both with and without disabilities, we have "universal design." Universal design as a practice works to positively appropriate what is learned from users with disabilities to benefit the broadest range of users (Center for Universal Design 1997).

Unlike gender and ethnicity, the number of individuals with disabilities is difficult to estimate. Between 2001 and 2005, 30% of the U.S. population reported having some type of disability (Altman and Bernstein 2008). The U.S. Census Bureau estimates about 51.2 million people with disabilities, nearly 1 in every 5 persons (2000), with almost half of those actively employed. With the increasing aging population in the United States, the graying of America is reflected in the disability data from the 2010 census data. However, it is more helpful when designing accessible systems to understand communities in terms of functioning within context. For example, estimations based on integrations of census data, surveys, and medical and support service usage provide more useful information for emergency managers. Steinmetz's (2002) research on functioning within context revealed the following important estimations for the U.S. population:

- 10.7 million individuals aged 6 and above need assistance with activities of daily living or instrumental activities of daily living.
- 11.8 million individuals aged 15 and above use mobility aids, including wheelchairs, canes, crutches, and walkers.
- 9.6 million individuals aged 15 and above are unable to see or have difficulty hearing.
- 7.9 million individuals reported cognitive, mental, or emotional difficulties that impair daily functioning.
- Compared to people without disabilities, people with disabilities are more likely to live alone or with nonrelatives.

The subsequent sections illustrate how collection and use of empirical data aid design of accessible emergency management. Multiple methods are reported to demonstrate the relative ease of conducting user-centered empirical research to address the issue of accessibility in emergency management.

11.3   A human factors approach

The authors conducted a 1-year study, funded by the Virginia Tech Center for Community Security and Resilience, to use a human factors approach to conduct a needs analysis and identify requirements for accessible emergency management systems. Although the phrase "human factors approach" is broad, we used it to describe a combination of mixed methods that are traditional and reflective of the interdisciplinary nature of human factors. Additionally, we focused on accessibility and usability as targeted goals. Accessibility was defined according to 28 CFR Part 36 (1990) as referring "to a site, facility, work environment, service, or program that is easy to approach, enter, operate, participate in, or use safely and with dignity by a person with a disability." Usability was defined using ISO 9241-11 (International Standardization Organization 1998) as "the extent to which a product [or system] can be used by specified users to achieve specified goals with effectiveness, efficiency, and satisfaction in a specified context of use." Both definitions apply to emergency management systems.

Upon review of the definition of emergency management as a framework to reduce vulnerability, the application of accessibility to emergency management may be demonstrated by the development and implementation of an emergency management system that is easy to use or participate in safely and with dignity by those with disabilities. Likewise, usability would be achieved by an emergency management system that can manage various tasks imposed by disasters (i.e., evacuations, sheltering) in a manner that is efficient and consistent with its performance standards, usually expressed in measurements of timeliness of response, number of fatalities or injuries, or number of individuals returned to their homes after having evacuated due to a disaster. Both constructs, when viewed as system attributes, can be used to advance the knowledge domain and application of design principles in emergency management systems.

11.4   Method

The study was conducted in three phases as follows:

1. *Phase 1:* Focus groups to more specifically define the problem space and identify user requirements
2. *Phase 2:* Development of a conceptual design of an accessible and usable emergency management system
3. *Phase 3:* Formative evaluation of conceptual designs using heuristic guidelines

11.4.1   Phase 1 (focus groups)

Participants: Nine participants representing community members and emergency management personnel were recruited for this phase. Four participants were male and five were female; six reported having a disability (auditory, learning, visual, mobility) and four were emergency management personnel.

Questionnaires: We developed a questionnaire eliciting ratings of confidence in emergency management and perceived level of preparedness. Table 11.1 summarizes the questionnaire and scale design.

A semistructured focus group format was used to acquire additional information related to accessibility and usability. Participants were given

Table 11.1 Questionnaire Items and Scaling

Item/question (tag name in brackets)	Anchors
1. If a regional emergency such as a tornado, ice storm, or flood happened today, how confident are you that the emergency management system could meet your needs (such as evacuation, food, supplies, other assistance). **[ConfidenceinEMS]**	1 (not at all confident) ……. 10 (very confident)
2. Rate your overall opinion of the extent to which the emergency alert and communication systems are designed to be effective for people with disabilities. **[CommunicationAlertEffectiveness]**	1 (not at all effective) ……. 10 (very effective)
3. Rate your level of preparedness for an emergency that would require you to be evacuated from your home. **[EvacuationPreparedness]**	1 (not at all prepared) ……. 10 (very prepared)
4. Rate your level of preparedness for an emergency that would require you to remain in your home and survive for 1 week. **[HomePreparedness]**	1 (not at all prepared) ……. 10 (very prepared)
5. Rate your level of preparedness for an emergency that would require you to live in a shelter for 1 week. **[ShelterPreparedness]**	1 (not at all prepared) ……. 10 (very prepared)
6. Rate your level of confidence in the emergency management services to: Effectively evacuate you during an emergency. **[EvacuationConfidence]**	1 (not at all confident) ……. 10 (very confident)
7. Rate your level of confidence in the emergency management services to: Effectively meet your needs if you are sheltered in your home during an emergency. **[HomeShelterNeeds]**	1 (not at all confident) ……. 10 (very confident)
8. Rate your level of confidence in the emergency management services to: Effectively meet your needs if you are sheltered in a center or other site during an emergency. **[CenterShelterNeeds]**	1 (not at all confident) ……. 10 (very confident)

a disaster scenario to provide a context for discussion of the topics. Topics discussed in the focus group were as follows:

1. Specific problems encountered by individuals with disabilities before, during, and after disasters
2. Steps to take to ensure people with disabilities receive the information and support needed before, during, and after a disaster
3. Greatest fears regarding emergency-related issues such as evacuation, sheltering in-place, or being vulnerable to disaster hazards
4. Requirements for a quality emergency management system that is inclusive and accessible to all

Procedure: Participants were divided into two groups for the focus group session. Group 1 comprised those participants who worked in emergency management, and Group 2 comprised those participants with disabilities or who were caregivers for individuals with disabilities. These two groups were not mutually exclusive. For example, two participants who were in Group 1 had a disability. Each focus group had one moderator and one technician. The moderator used a semistructured focus group script. The focus groups lasted for a total of 1.5 hours. Each question of the script was read, and each participant's opinion was collected before continuing to the next question. Also, some questions were expanded using probes to gather more pertinent information. The focus groups were audio recorded, and transcripts were developed from the audio recordings. The technicians also hand-recorded notes during the focus group sessions.

Quantitative Results: SAS 9.2™ software was used to analyze the quantitative data. Nine questionnaires were collected, one of which was incomplete. Thus, the sample size for the descriptive and inferential statistics varied between $n = 8$ and $n = 9$ participants. A Shapiro–Wilk (SW) normality check was conducted on the questionnaire ratings (Table 11.1). SW values ranged from 0.79 to 0.93. All items except item 5, *"Rate your level of preparedness for an emergency that would require you to live in a shelter for one week,"* fit a Gaussian distribution. Therefore, parametric inferential statistics were conducted on the data set. A Cronbach's alpha reliability check was also conducted on the questionnaire items, and the coefficient alpha demonstrated an acceptable level of reliability at $r_{alpha} = 0.82$ (per Nunnally [1978] criteria). The sample size was not large enough to conduct analyses of ratings by disability type. However, we were able to conduct a valid statistical test of the differences in ratings between the emergency personnel (those with emergency experience as an employee or volunteer) and the nonemergency personnel (private citizens with disabilities). An independent samples *t*-test was conducted, comparing the means on each of the questionnaire items (using four in each group because of one incomplete questionnaire). The Satterthwaite adjustment equation was used to compare the means for one of the items (evacuation preparedness) as a folded-F test indicated nonequivalent variances. At an alpha level of 0.05 for significance, emergency management personnel consistently expressed greater confidence in all issues. In particular, the emergency personnel group reported significantly higher confidence in the extent to which the system could meet the emergency needs of individuals with disabilities (t (6) = 5.00, p < 0.01). Similarly, emergency personnel were more confident in the accessibility of emergency communication and alerting (t (6) = 2.89, p < 0.05) compared to private citizens with disabilities. Figure 11.1 illustrates the mean ratings between both groups.

Qualitative Results: Table 11.2 summarizes the focus group discussions. We highlight that there was shared concern about the need to shelter in-place since participants perceived evacuation as a last resort and would place individuals with disabilities at greater risk of harm during transport. The accessibility of emergency notification systems was also a shared concern, and the discussions centered on the need for more inclusive types of language, accessible communication to individuals with visual and hearing impairments (beyond what is already available), and easy-to-understand communications for individuals with cognitive disabilities (graphical systems rather than auditory or text-based).

System Requirements: The design team reviewed the content and the developed system requirements based on the results of Phase 1, literature, and existing standards and guidelines. System requirements are

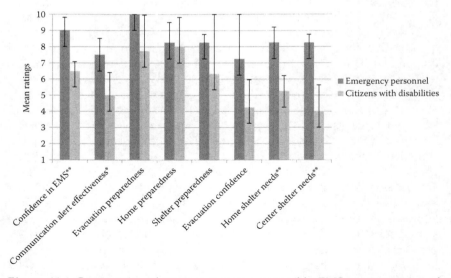

Figure 11.1 Comparison of mean ratings on accessible EMS questionnaire of emergency personnel and citizens with disabilities. * Significant at p < 0.05; ** significant at p < 0.01.

Table 11.2 Content Summary of Focus Groups Notes

Item/code	EMS	Citizens with disabilities
Specific problems encountered by individuals with disabilities that you believe have been overlooked in national discussions and that require attention now.	• Biggest gaps are knowing where people are. They need help but they do not identify themselves. • Online directories are not updated and not current. • Not enough firefighters to go around and check on everyone. Citizens must check. • Databases with people's names and disabilities are ineffective because of the transient nature of the population. Is obsolete in 2 years. • Getting to shelters is a huge problem. People cannot travel, especially in an ice storm or disaster event that impedes movement of traffic. • People with disabilities need generators, but some cannot afford them. • Need to learn to make more efficient use of the EMS personnel's time during disasters. Some things should be left up to citizens who also serve as first responders. • EMS capacity is limited and can only be stretched so far.	• Shelter foods are not healthy so this causes additional medical problems, i.e., too much sodium • Shelters are not accessible: beds are too low and no back support for those with a mobility challenge. • Cell phone service goes out during emergencies so there is no communication backup to find relatives and friends to help. • Red Cross has a communication, notification system but you have to take the time to learn it beforehand. • Need networks in place that include churches and other organizations. *(Continued)*

Table 11.2 (Continued) Content Summary of Focus Groups Notes

Item/code	EMS	Citizens with disabilities
EMS question: What steps do you think should have been done to ensure people with disabilities received the information and support they needed? *People with disabilities question:* What steps have you taken to use the tools provided by the county such as emergency evacuation plans that are online, for instance?	• The three areas in the region have an extremely close knit relationship but we need to integrate better. • After each event, need to have debriefing meetings with individuals with disabilities to find out what worked and what did not work. OEM should hold these meetings. • Need early warning systems. • Blast alerts and verbal announcements are not used frequently enough. • What bogs down the system are those who are not prepared but who could have been prepared. We want to focus on those who cannot prepare because of difficult circumstances, but many CAN prepare but did not do it. • Not prepared for exotic languages; translations such as Urdu. • Emergency kits and generators need to be provided to citizens. Many are not aware of how to get them and, because of changing/transient population of our area, these messages should be sent out constantly.	• Without electricity, cannot do anything. • Transportation to shelters is difficult. • Why not use street lights as beacons. • Not familiar with a registry, so have not used it. • Did not know how to notify county of special needs. • TV and cell phones are the main sources for communication and information. • Text alerts work well. • Reverse 911 • No eye–hand coordination, so cannot write down audio recorded messages with essential information.

What can be done to design a better emergency management system, if you could design it yourself? What technologies do you think should be used to enhance the effectiveness of emergency management? This includes people, ways of doing things, and the technologies that are used.

- Prioritizing or triaging people can be difficult.
- People with disabilities must determine if it is safer to remain in the home/shelter in place or evacuate. If there is a call to evacuate, what will happen if they end up sitting in traffic?
- There are too many people and not enough infrastructure.
- Build better communications. Need interoperability.
- Black Hawk County, Iowa is a good example.
- Video 911 may not work. Dispatchers don't want to see certain things because that is not what they agreed to do.
- Text messaging system might help.
- Verbal warnings are important and still need to be used.
- There will never be a silver bullet.
- Additional resources like the National Guard were helpful.

- Fear of being stranded without a way to communicate.
- Can't use the Metro because it stops working in many disaster scenarios.
- Evacuation is a major concern because of limited mobility.
- Alert pendants to seek information, but some will resist.
- TV alerts must be shown in text and spoken.
- Need more support networks.
- County has limited special needs equipment.
- 72 hours is too much time to wait for help.
- Need more capacity for shelters.
- Generators are vital.
- What about visitors?
- Need contingency plans for when 911 fails.
- Tell public what recourses are available; organizations need to help advertise.
- Need to address privacy issues.

(Continued)

Table 11.2 (Continued) Content Summary of Focus Groups Notes

Item/code	EMS	Citizens with disabilities
What are YOUR requirements for a quality emergency management system that is inclusive of and accessible to all?	• Goal is to be inclusive. • Have to improve on what we have. • Hard decisions need to be made regarding where to put the investments and where to draw the line. Who is going to put money into developing an accessible system? • System needs to provide timely and accurate information. • People with disabilities best know their capabilities at the time of an emergency, and these capabilities are dynamic. This needs to be recognized in some way. • OK1 system in Oklahoma should be reviewed. • Businesses need to do orientations for incoming employees to give them information about emergency preparedness similar to the benefits orientation sessions they get already. • Systems need to account for different types of sheltering possibilities: in cars, in malls, in schools, at work. • We need to use an all-hazards approach and the system needs to address all possibilities: ice, floods, fire, quakes, etc.	• Should make people aware of where to go and what to do. • Clear, quick instructions to ensure language and literacy are not barriers. • Intelligent, readily available. • Read "When the unexpected happens"

descriptive features that must be included in any emergency management system to ensure accessible, effective, and efficient management of emergencies for individuals with disabilities. These requirements were summarized using a method developed by Smith-Jackson et al. (2003) and were applied to the iterative design of the conceptual models.

The system requirements were as follows:

- System must be compliant with existing regulations and harmonization principles (i.e., International Standardization Organization, European Telecommunications Standards Institute, American National Standards Institute, Federal Emergency Management Agency).
- System should align with Global Initiative for Inclusive Information and Communication (G3ict) (United Nations Global Alliance 2006).
- System must support just-in-time registration and updating.
- System must be designed assuming that at least 50% of individuals with disabilities will experience significant mobility and communication challenges during a disaster.
- System must support data sharing across agencies, counties, and regions using secure, web-enabled software.
- System should utilize mobile platforms such as smart phones.
- System should include postdisaster follow-up and updating capability.
- System implementation must encourage emergency preparedness among all citizens.
- System must be sustainable and integrated into current budget allocations for emergency preparedness.
- System should utilize technologies to identify and locate persons with disabilities, including wearable identifiers and radar, radio frequency identification, global positioning system, and geographic information systems.
- System must use alerts that are customized for accessibility by audio, teletypewriter/telecommunications device for the deaf (TTY/TTD), and language codes.
- System must accurately identify and locate all land and mobile lines.
- System must be able to notify or alert all land and mobile lines.
- System must be able to conduct just-in-time assessments of special needs for assistance during emergencies.
- System must provide rapid, accurate, and up-to-date information to emergency dispatch and first responders.
- System must be sufficiently compatible with existing command, control, communications (C^3) structure.

- System prioritization/triage scheme should incorporate levels of dependency on medical devices such as oxygen and other in-home life-support systems.
- System must support personnel and citizen decision making regarding evacuation or sheltering in-place.
- System must be affordable to citizenry and local and regional emergency planning and management agencies.

11.4.2 Phase 2 (conceptual models)

System requirements were used to design conceptual models of different components of emergency management systems that could be reviewed and formatively evaluated by experts and representative participants. Although most of the conceptual models cannot be shown here for proprietary reasons, some of the key features are discussed.

The initial analyses revealed a strong need for an emergency management system that would support triaging of emergency response and recovery. Mobile social media, such as Twitter, could help meet this need. If participants are registered, information can be collected in a knowledge base or repository. This repository could be populated using geographic tagging, emergency calls for alerts and information gathering, and geographic information systems to assist with prioritizing and locating citizens with disabilities. The system would also utilize FEMA's commercial mobile alert system (CMAS), which, as of 2012, will require all mobile subscribers to support functionality, allowing users to register to receive mobile alerts from offices of emergency management in all geographic areas.

Reverse 911 is also a component that could be effectively integrated into any accessible emergency management system. The reverse 911 system cannot be modified by local units, but developers of the system should be alerted to the need for expanded functionality and integration with other systems. A modification or system overlay may offer a way to enhance FEMA's CMAS compliance functions to ensure that mobile service providers not only allow phones to be registered to receive emergency alerts, but also to allow the mobile system data to be stored in a knowledge base so that individuals with disabilities can receive customized information in the form of TTY/TTD, haptic tones, and other special need-based alerting formats. However, this expanded functionality goes beyond the compliance regulations. Reverse 911 can also be enhanced by providing more information to public safety personnel regarding special needs of those they assist (disability, language, etc.). When any event occurs, the database could be updated with the accessibility needs of individuals derived from the most recent emergency event. After each event, people with disabilities could provide feedback to improve the system, resulting in iterative, community-based continuous improvement. Feedback can occur as postevent inputs from citizens and analysis of postevent information by third parties such as universities. The postevent information can be used to update the registry and update response protocols as necessary.

Figures 11.2 and 11.3 are examples of a user interface (UI) that would present information as a dashboard that a dispatcher may use to gather information about the emergency. Emergency personnel could use the same UI to assess emergency situations and plan accordingly. The dashboard would display mobile phone call volumes. Emergency personnel could click a zone to view relevant data for that specific location. Rather than a dispatcher having to populate this emergency information, a voice recognition program could be created that would accumulate words that it recognizes from the various calls that come in to 911. The age, language, needs, and any other demographics that are important would be provided by this system and activated by touch. At present, the design team did not consider this interface to be useful to grassroots organizations, but these organizations could provide emergency personnel with updates on the types of information that may be critical in addition to what is identified as important by emergency personnel.

Additionally, a dispatcher could click on or touch the upper left hand call volume map to receive live video/pictures or recently updated

Figure 11.2 Dashboard (shown to human factors evaluators only). Initial display of user interface during emergency. After user selects a geographical area using click and drag to make a box around the areas of interest, more information is provided.

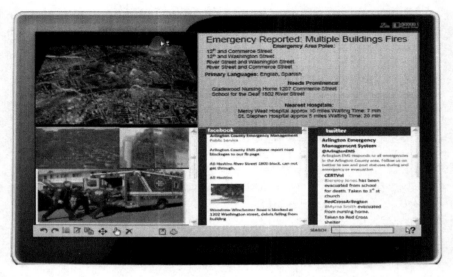

Figure 11.3 Expanded dashboard user interface (shown to human factors evaluators only).

pictures of the current status of a specific location. A traffic camera could provide the traffic on the road nearest the emergency, a picture of a building on fire, and an EMT crew loading an ambulance. These pictures could be provided by

- Traffic cameras
- Building security cameras
- Pictures taken and uploaded by EMTs/police/firemen/EMSs

Facebook newsfeeds and Twitter updates could also be integrated with this system. Newsfeeds could show streets that are blocked by debris or intentionally barricaded for safety reasons.

The dashboard shown in Figure 11.4 illustrates a similar concept shown in Figures 11.2 and 11.3, but was only shown to the emergency management and outreach experts. Figure 11.4 shows a less detailed concept to elicit unbiased opinions from these experts. Thus, Figures 11.2 and 11.3 were shown to the human factors evaluators and Figures 11.2 and 11.4 were shown to the emergency management and outreach/citizen evaluators.

11.4.3   Phase 3 (formative evaluation using heuristic guidelines)

The goal of Phase 3 was to conduct a formative evaluation using heuristic guidelines using the conceptual models.

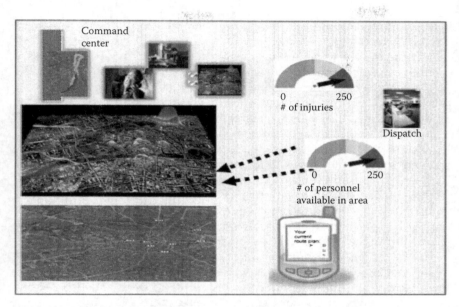

Figure 11.4 Dashboard layout shown to emergency management personnel and community outreach experts.

Participants: Six external evaluators were recruited with the following areas of expertise:

- Doctoral researcher, accessibility, human factors (1 expert)
- Graduate student, usability, computer science (1 expert)
- Office of emergency management personnel (3 experts)
- Community outreach coordinator (Urban League, American Red Cross, United Way) (1 expert)

Questionnaires: Questionnaires were designed for each expert group. The human factors evaluation experts were provided with a set of guidelines (ISO 9241-11 and Usability and Section 508 Rehabilitation Act) along with the conceptual models in one questionnaire that was administered via e-mail. A 6-point Likert-type rating scale was used, with "1" representing various negative anchors (i.e., not at all) and "6" representing various positive anchors (i.e., very well). The emergency management experts received a questionnaire with the conceptual models and a set of rating and open-ended questions. A 6-point Likert-type rating scale was used, with "1" representing various negative anchors (i.e., not at all accurate) to "6" (very accurate). Both questionnaires included spaces for responses to open-ended question to elicit additional comments from the evaluators.

The community outreach expert was administered a semistructured interview by phone.

Procedure: The two human factors experts were provided with a list of heuristics reflecting accessible and usable systems (Section 508 and ISO 9241-11) and requirements identified from Phase 1. These experts conducted a walk-through of the conceptual models for a cell phone alert and a UI using the heuristic list and provided ratings and comments. Data were collected from the human factors expert via e-mail. The second evaluation was conducted using a focus group (emergency management experts) and an interview (community outreach expert). The community outreach expert was given the conceptual models and provided feedback using a list of questions to support an unstructured interview (Figures 11.2 and 11.4). Two note-takers recorded responses in writing. The community outreach expert was interviewed via phone call, and the interviewer took notes during the interview.

Data Analysis: Quantitative data were analyzed using SAS 9.2™ software. General content analyses were conducted manually by identifying the main themes provided by responses to open-ended questions designed to elicit additional comments from experts.

Results: Ratings by the human factors experts of each heuristic guideline for the dashboard conceptual design (Figures 11.2 and 11.3) are summarized in Table 11.3.

Specific issues raised by the expert evaluators are itemized as follows:

- Intuitive mapping needs to be improved.
- Locations should be given first, followed by the type of event.
- Display size is very important for efficiency and usability of the interface. There is too much information to convey, so display size should be revisited.
- Information regarding who is on scene and who has been notified should be provided.
- Include traffic information for the nearest hospital and the "wait time."
- Twitter and Facebook do not require the 27% of the screen.
- A new pop-up could be used instead to post to Facebook/Twitter feeds. These should disappear after submission.
- Do not give a location without context.
- Implement a way for the control room to communicate with emergency personnel in the field.
- It would be useful for the emergency personnel in the field to have a shared screen/selected information pushed to them.
- Revise notification system. Rethink what information needs to be statically displayed.
- What is the purpose of the pictures/video feeds in the left corner?

Table 11.3 Means and Standard Deviations for Ratings of Accessibility by Experts

Cell phone alert accessibility questions	Mean and standard deviation
To what extent does the cell phone alert provide simple and natural dialogue? (1 = not at all/6 = very well)	4.00, 1.41
Is the cell phone alert designed to speak the user's language? (1 = not at all/6 = very well)	4.00, 1.41
Does the cell phone alert provide consistent dialogue? (1 = not at all/6 = very well)	5.00, 0
To what extent does the cell phone alert eliminate error-prone conditions? (1 = not at all/6 = very well)	3.50, 0.71
Is the cell phone alert designed to provide clear error messages? (1 = not at all/6 = very well)	4.50, 0.71
To what extent does the cell phone alert minimize the user's memory load by making objects, actions, and options visible? (1 = not at all/6 = very well)	5.00, 0
Is the cell phone alert designed to eliminate dialogues that do not contain information that is irrelevant or rarely needed? (1 = not at all/6 = very well)	4.00, 1.41
Is the phone application usable with regard to those users with visual impairments? (1 = not at all/6 = very usable)	4.00, 1.41
Is the cell phone alert usable with regard to those users with motor or tactile impairments? (1 = not at all/6 = very usable)	4.00, 1.41
Is the cell phone alert usable with regard to those users with auditory impairments? (1 = not at all/6 = very usable)	4.00, 1.41
To what extent do you believe a citizen *not* in distress would use this texting interface? (1 = not at all likely/6 = highly likely)	4.00, 1.41
To what extent do you believe a citizen in distress would use this texting interface? (1 = not at all likely/6 = highly likely)	4.00, 1.41

Note: Min = 1, Max = 6.

- Assuming this tool will be used by multiple people not necessarily colocated, a notes section should be implemented where emergency personnel and dispatchers can share textile notes about emergencies.
- Color coding system needs to be more clear.
- Users might have difficulty handling and understanding so much information in emergency situations.
- Users may not trust all information from social networking.

The qualitative results of the formative evaluations of the conceptual designs provided by the emergency management personnel and the outreach coordinator were richer and more meaningful than the Likert-type ratings. The emergency management personnel were of the opinion that the designs targeted geographical areas and focused on locating individuals, but were not necessarily accessible or geared toward individuals with disabilities. They also suggested the addition of a way to present the types of buildings rather than simply the location of a building. For example, one scenario discussed involved the need to know whether the location was an assisted living facility or a large house/private residence. Tapestry mapping was suggested as a design feature that would provide sufficient details. Emergency management personnel suggested that the data collection for after-action review and assessment could be collected before individuals leave shelters or even while they are in shelters. This information could then be used for just-in-time updates and registry database updates.

The semistructured interview format was used to elicit information from the community outreach expert. Based on the outreach coordinator's 20 years of experience, emergency management personnel were believed to hold incorrect assumptions about the resources and capabilities of many in the community. In addition, the technologies on which the conceptual models were integrated, such as smart phones, are often not accessible to low-income individuals with and without disabilities. For instance, access to computers and smart phones, especially for low-income and senior citizens, was low, yet many of the alert systems were developed based on these assumptions (i.e., cell phone access). Even cell phones may not be accessible because of problems with financial stability and credit to sustain a service subscription.

Also, the outreach coordinator observed community groups, churches, synagogues, and other community-based organizations to be underutilized, especially when emergency management services require extensions of their capabilities. Many advisory boards for community-based organizations are willing to provide inputs and feedback to emergency services, but need to be approached. The potential for strong and effective partnerships might address accessibility, usability, and equity challenges more effectively.

11.5   Conclusions

Although this project was a small research effort, the general outcomes yielded important results and identified areas requiring further work as we continue to develop accessible emergency management systems. Collaboration with community-based organizations will be essential to design effective and accessible systems. Based on the evaluations, several design areas require further attention, such as the following:

1. Developing a more feasible postevent feedback system and identifying mechanisms to fund and sustain such a system.
2. Improving the ability of the system to eliminate error-prone conditions
3. Simplifying interface language, dialogue, and error messages.
4. Improving activation of the interface for individuals with motor and tactile challenges
5. Avoiding information overload
6. Increasing the uniqueness and salience of alerts
7. Enhancing the system to improve effectiveness to assist first responders to locate individuals with disabilities

Of particular interest is the use of a variety of communication tools and technologies to alert communities, assist during a disaster, and address issues related to resilience (e.g., returning home). Overall, participants provided a number of suggestions for new ways to use TTY, social networking, television, radio, and websites to inform the public quickly and accurately. Also, issues of privacy were discussed, yet participants were willing to make tradeoffs if the information were used and protected within a database or registry. This is important in light of the current controversies related to privacy and surveillance by cameras and drones.

It is likely that efforts on a larger scale to empirically identify the needs of individuals with disabilities in the context of disasters would help to validate preliminary findings from this effort but would also continue to expand our understanding of how to explore the complexities of managing sociotechnical systems such as emergency management in a manner that minimizes vulnerability for the broadest range of community members.

In summary, we evaluated a proof of concept of several concepts associated with accessible emergency management. In general, the results will assist us to develop actual prototypes based on the evaluations and requirements gained from this project.

Acknowledgments

This project was funded by the Virginia Tech Center for Community Security and Resilience (CCSR). Thanks to Dr. John Harrald, Director of the CCSR, for his support. Additionally, we thank Jack Brown, Jr., Director of the Arlington County Office of Emergency Management, for his exceptional support and input on this project. Our utmost gratitude is extended to the participants who gave of their time to engage in dialogue and interactions to improve the accessibility of emergency management. We also express our gratitude to additional team members Sheldon Fuller and Francis Quek. The Industrial and Systems Engineering Senior Design team put great effort into a portion of this project: Eric King, Matthew Milby, Angad Singh, and Nigel Wray. Finally, we are grateful for the information provided by two of the partners on this project, Anne-Rivers Forcke and Pawan Khera.

References

28 CFR Part 36. (1990). Americans with Disabilities Act. Online: http://www.ada .gov/archive/nprm98.htm. Retrieved on August 1, 2011.

Altman, B. and Bernstein, A. (2008). *Disability and health in the United States, 2001–2005.* Hyattsville, MD: National Center for Health Statistics. Online: http://www.cdc.gov/nchs/data/misc/disability2001–2005.pdf. Retrieved March 1, 2012.

Blanchard, W. (2007). *Principles of emergency management supplement.* Online: http://www.iaem.com/publications/documents/PrinciplesofEmergency Management.pdf. Retrieved on August 1, 2011.

Center for Universal Design. (1997). *The principles of universal design.* North Carolina State University. Online: http://www.design.ncsu.edu/cud/about_ud/ udprinciplestext.htm. Retrieved on March 5, 2010.

Fox, M., White, G., Rooney, C., and Rowland, J. (2007). Disaster preparedness and response for persons with mobility impairments. *Journal of Disability Policy Studies, 17*(4), 12.

International Standardization Organization. (1998). *ISO 9241–11, Guidance on usability.*

Kailes, J. (2005). Disaster Services and "Special Needs": Terms of Art or Meaning-less Terms. Nobody Left Behind: Disaster Preparedness for Persons with Mobility Impairments. Online: http://www2.ku.edu/~rrtcpbs/findings/ pdfs/SpecialsNeeds.pdf. Retrieved on July 31, 2010.

Nunnally, J. C. (1978). *Psychometric theory.* New York, NY: McGraw-Gill.

Rooney, C. and White, G. (2007). Narrative analysis of a disaster preparedness and emergency response survey from persons with mobility impairments. *Journal of Disability Policy Studies, 17,* 206–215.

Smith-Jackson, T. L., Nussbaum, M. A., and Mooney, A. M. (2003). Accessible cell phone design: Development and application of a needs analysis framework. *Disability and Rehabilitation, 25,* 549–560.

Steinmetz, E. (2002). Household economic studies current population reports (P70-107). *Americans with Disabilities: 2002.* United States Bureau of the Census. Online: http://www.census.gov/hhes/www/disability/sipp/ disab02/awd02.html. Retrieved on July 31, 2011.

Tierney, K., Lindell, M., and Perry, R. (2001). *Facing the unexpected: Disaster prepared-ness and response in the United States.* Washington, DC: Joseph Henry Press.

United States Bureau of the Census. (2002). Disability status, employment, and annual earnings: Individuals 21 to 64 years old. Table 5.

chapter twelve

Framework for preparing for the unprepared

Kera Z. Watkins, Anuradha Venkateswaran,
and Katrina Simon-Agolory

Contents

12.1   Overview of the call to action
in preparedness

Disasters of many forms have historically been rare events, but in recent years unusual events have triggered a number of unexpected disasters. Although not all events have resulted in disasters, the increased potential impacts cannot be ignored. It only takes one unexpected event to change the course of life.

Take recent weather events as an example. According to the National Oceanic and Atmospheric Administration, the record-breaking warmth of March 2012 generated more than twice the normal number of tornadoes in the United States. The tornadoes caused a loss of 40 lives and more than US$1 billion in damages. During that month, tornado-related fatalities reached two-thirds of the annual average. Is there something more that could have been done to prepare?

One can get a sense about how prepared the nation is by polling the individuals and groups that compose it. According to a study conducted by the Council for Excellence in Government in 2006, 45% of respondents admitted to not ever thinking about disaster preparedness (Citizen Corps 2007). According to a study conducted just 1 year later by the American Health Public Health Association, 87% of the respondents did not meet the minimum standard for preparedness (Citizen Corps 2007). Using reactive response in lieu of a proactive preparedness approach could cost four times as much, according to Magsino (2009). If 87% of the nation was prepared for the tornado outbreak, then around $250 million may have been spared (Magsino 2009).

Preparedness does not seem to receive serious ongoing forethought until after a disaster occurs. The authors of this chapter conjecture that some remain ambivalent because they have not been given enough reason to think differently. This reactive human nature can be detrimental to resilience in terms of loss of lives and property damages. The solution needs to bring preparedness to the forefront in a seamless way, making it an ongoing part of human life. This chapter suggests a framework to incorporate preparedness into the national fabric using a combination of instant gratification, competition, and communication. Specifically, the framework includes social networks and online multimedia solutions. The rest of this chapter is divided into the following sections: Current issues, Terms and definitions, Community disaster preparedness through social networking, Overall disaster preparedness system, Metrics, and A working example: Wilberforce, Ohio.

12.2   Current issues

For each emergency responder, there are over 1200 people to serve (Bureau of Labor Statistics 2012). It is not reasonable to expect emergency personnel and experts to sufficiently serve a growing population of over 300 million people (according to the 2010 U.S. Census) during a crisis. Therefore, the people need to be empowered to serve themselves and one another.

There are plenty of websites that allow a person or family to get prepared (BDPN-Network 2011; California Volunteers 2012; Get Emergency Prepared 2012; Pavlica 2012; Ready 2012; State of Florida 2012). However, most of these plans have the following issues:

1. *Limited engagement of surrounding community.* Even though a number of resources exist to get prepared, many resources are directed at individual households. However, households can be impacted by their surrounding vicinity. For example, if a neighbor's child typically worked on school projects with their child, then parents would have to ration their emergency resources accordingly if an earthquake suddenly hit. If that household did not take their neighbors into account, their emergency resources to survive during a period of time could be adversely affected. Determining isolated preparedness without considering the surrounding vicinity can have deceiving results.
2. *Disaster preparedness information overload.* Education is a necessary part of preparedness. However, many plans have a problem with information overload. This can cause the plan to become ineffective. If people become too inundated with information, they could get overwhelmed and end up giving up. People need guidance to stick to the most critical information provided in a short and concise manner.
3. *Lack of communication.* Despite efforts to remedy the problem, often there remains a lack of communication among various groups of emergency personnel, as well as communication gaps between emergency personnel and the public. It is important to have familiar

communication settings among various groups prior to disasters so that communications and response systems can be effective once a disaster occurs.

4. *Limited engagement in disaster preparedness.* The current literature expresses an ongoing problem of engaging people because disasters are infrequent and seem to sporadically affect different parts of the country. It is imperative to provide immediate incentives through instant gratification to help engage the public.

5. *Limited resources.* In major disasters, local and state resources get overwhelmed fairly quickly (Hooke and Rogers 2005). Governments tend to rely on community and organizational leaders to prepare regions and less on the general populace. This is most likely due to necessity given limited resources. A process needs to be defined to help maximize current resources without overwhelming them. An incremental and prioritized process may be appropriate based on the season.

12.2.1   Limited engagement of surrounding vicinity

There seems to be conflicting views about the importance of avoiding isolationism in preparedness. For example, Ecevit and Kasapoglu (2002) believe that a lack of education is more detrimental than isolationism (when considering surviving a particular earthquake in Turkey). McIvor et al. (2009) suggest that preparedness is not an isolationist activity. Nevertheless, isolationism is not just a geographical phenomenon. There are a number of ways to define isolation among people such as language, as well as location (Solecki 1992). In certain disasters, the homogeneity of race and culture seems to isolate a group from educational resources (Ecevit and Kasapoglu 2002). Other types of isolationism phenomena include a lack of resources and inadequate preparedness education, which can create their own cyclic isolationism problems.

Regardless of the degree of importance to reduce isolationism in disaster preparedness, communities ought to band together during a disaster (King 2000). Communities are more effective when people work together than alone (Patterson et al. 2010).

12.2.2   Disaster preparedness information overload

According to Watkins et al. (2011), people may get inundated with too many resources at once. They can get overwhelmed by the plethora of information. Also, it is not always clear where they should begin with the materials given. When this happens, information overload is likely to de-incentivize people to respond properly during a crisis. Instead, information and incentives need to be introduced according to the specific needs of the particular audience to be effective. In this chapter, an incremental process of three preparedness activities at once is suggested to help people realize success in stages while building on their previous success.

12.2.3   Lack of communication

Communication can be provided in a number of formats such as the following: text messages, phone, online voice, and video (in the form of direct communication or scenarios). It would be advantageous to provide a forum that provides mixed forms of communication to cater to the communication needs of various groups, as well as provide a single viewable online setting for all forms of communication. Short text messages could relay quick warning or emergency messages. Users could interface with the website using mobile devices or computers to listen to pertinent messages before or after a disaster. Prior to a disaster, users could also have the option of leaving voice messages through their mobile devices. All forms of communication could be used for educational purposes—providing quick or in-depth preparedness information, as well as developing preparedness and disaster scenarios for users to react to.

A limitation caused by scenario-based simulations is that, in general, they do not account for communication barriers between various emergency groups (Desouza 2005; Fortier and Volk 2006). Text-based communication also has limitations. Twitter is primarily recognized as a text service, whereas Facebook provides many types of visual and textual media. Even though Twitter has its limitations in terms of popularity as compared to Facebook as shown in Section 12.3.6, it is fairly more lightweight, less complex, and therefore more likely to be available in peak times of a disaster. Once a disaster strikes, it is important to switch to a more lightweight yet familiar version of the communications to reduce the likelihood of a system failure. It is important to have various types of communication working in concert as apparent options to meet the communication needs of various groups, as well as provide a robust system for communication prior to and following a disaster.

12.2.4   Limited engagement in disaster preparedness

It is commonly accepted that preparedness is imperative for a quicker recovery (Krock 2011). The ongoing challenge is to get people to actively engage in preparedness. Disasters have been historically rare occurrences (Krock 2011). Even though the impacts from an actual disaster can be potentially devastating, individuals may perceive that they have little chance of being affected.

Swan (2001) suggests that to engage communities, one must determine a common social identity that offers incentives for actions they take to get prepared. The concept of building a social identity is not new. Watkins et al. (2009) describe developing a common computing identity using a combination of technical seminars, experiential learning, academic success, mentorship, and research to help underrepresented students define a computing identity through the STARS Alliance. The STARS Alliance, initiated at Wilberforce University, has had measured success in retention and graduation rates (http://www.starsalliance.org). This chapter adopts some elements from the concept of social identity within the realm of a social network.

Freeman and Freeman (2010) report that geographical locations in Australia were used to identify their communities. However, they have noticed minimal success with communities actually setting up their social networks. If the theories presented by Swan and Watkins et al. are correct, then at least one of the following may be true:

1. Geographical proximity is not an adequate social identifier among its communities.
2. The incentives (if any) offered by social networks have not yet been realized.
3. It is considered too difficult or time consuming for communities to run a social network.

The work in this chapter strives to consider various factors for incentives to help drive the social network for community preparedness before a disaster and hopefully community resilience after a disaster occurs.

12.2.5   Limited resources

Hecker (2002) mentions the issue with state and local resources being insufficient to handle highly impactful disasters such as terrorist threats. Cost-effective solutions are needed for state and local communities. For example, Lawler et al. (2007) suggest an IT infrastructure for providing disaster-tolerant computing. This solution may be applicable to large companies. However, its cost is most likely beyond the reach of budgets for state governments or smaller communities, let alone families or individuals. This chapter strives to provide solutions at a sensible cost and alternatives as incentives for individuals, families, localities, and beyond.

12.2.6   Discussion

"Talking about Disasters: Guide for Standard Messages" was designed by a coalition of 18 federal agencies and nongovernmental organizations

(NDEC 2004). This guide suggests that people who begin to prepare for just one hazard can expect to be prepared for almost half of all known hazards that can lead to disasters. Individuals, families, and communities should prepare to purchase a bare minimum of 20% of all suggested preparedness items to even begin reducing their risks of impact.

12.3   Community disaster preparedness through social networking

12.3.1   Terms and definitions

This section includes definitions and equations that are used in this chapter. A neighborhood within the community is shown as a square with length $2r$ (Figure 12.1a) within a corresponding circle of radius, r (Figure 12.1b).

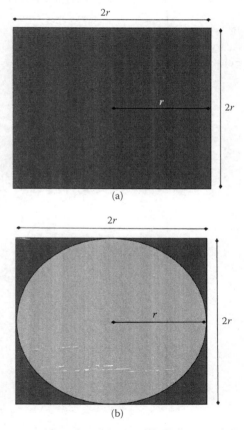

Figure 12.1 (a) Square with radius r units. (b) Square and corresponding circle with radius r units.

Definition 12.1 (Neighbor): Let a neighbor be defined as a household or entity connected through adjacency, proximity, or other commonality to at least one other household or entity.

Definition 12.2 (Neighborhood): Let a neighborhood be defined as two or more connected neighbors.

Definition 12.3 (Community or "Neighborhood of Neighborhoods"): Let a community be defined as one or more connected neighborhoods. Note that those connected neighborhoods may overlap.

Axiom 12.1 (Normal Neighborhood): A normal neighborhood (Figure 12.2) is a circular neighborhood with radius r. It has four sections that appear in one neighborhood (also called single subsections) and four sections that can each be shared with other overlapping neighborhoods within a community (also called dual subsections). For example in Figure 12.3, Neighborhood 4 is represented by N4.Sx and N4.Db where N4.Sx denotes the xth single subsection of the fourth neighborhood and N4.Db denotes the bth dual subsection of the fourth neighborhood.

Axiom 12.2 (Normal Community): A normal community is a square area with a length of $4r$ and is illustrated in Figures 12.2 and 12.3.

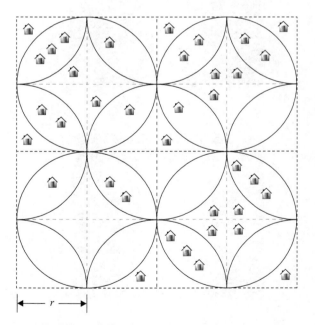

Figure 12.2 Example of normal community (single subsections—maximum one household and dual subsections—maximum three households).

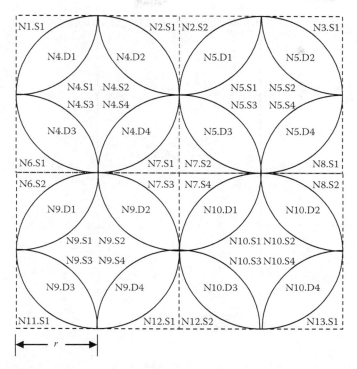

Figure 12.3 Normal community with single and dual subsections.

It is comprised of full and partial normal neighborhoods. A normal community is comprised of a total of eight overlapping neighborhoods: five normal neighborhoods, four halves of normal neighborhoods, and four additional normal neighborhood quarters. Note that dual subsections can be denoted by an overlapping neighborhood. For example, N4.D3 could also be expressed as N6.D2 because that subsection is a part of both Neighborhoods 4 and 6. For simplicity, just one possible label is chosen.

Axiom 12.3 (Connected Community): A community consisting of all adjacent or overlapping neighborhoods is a connected community.

Corollary 12.1 (Radius of a Normal Neighborhood): Suppose that a normal community is comprised of 100 to 150 household adults (Backstorm et al. 2006; Leskovec et al. 2008). In $\bar{r}_{neighborhood}$, $\bar{r}$ represents the radius for the normal neighborhood needed to reach between averages of 100 to 150 household adults within the community. In $\bar{r}_{neighborhood}$, the term $\bar{r}$ can then be estimated from the U.S. Census Bureau (2010) in Equation 12.3 U. S. Census Bureau, (2010).

$$\overline{\text{Household adults}} = \frac{\text{Total people in households - Total children in households}}{\text{Total number of households}}$$

(12.1)

$$\overline{\text{Neighbors}} = \frac{\text{People in occupied housing}}{\text{Total housing units}} \times \overline{\text{Maximum neighbors}} \quad (12.2)$$

$$\overline{r}_{\text{miles}} = \frac{1}{4}\sqrt{\frac{\overline{\text{Number of household adults}}}{\frac{\text{Household adults}}{\text{Square mile}}}} \quad (12.3)$$

$$\overline{\text{Dual subsection}} = \frac{1}{16}\sum_{i=1}^{16} \text{\# Dual subsection households}_i \quad (12.4)$$

$$\overline{\text{Single subsection}} = \frac{1}{32}\sum_{j=1}^{32} \text{\# Single subsection households}_j \quad (12.5)$$

$$\overline{\text{Normal neighborhood}} = 4 \times \overline{\text{Single subsection}} + 4 \times \overline{\text{Dual subsection}} \quad (12.6)$$

$$\overline{\text{Normal community}} = 32 \times \overline{\text{Single subsection}} + 16 \times \overline{\text{Dual subsection}} \quad (12.7)$$

Lemma 12.1

$$\text{Area of single subsection of a normal community} = \frac{1}{4}(4r^2 - \pi r^2)$$

Proof: Using a square with length $2r$ and its corresponding circle for a neighborhood within the community shown in Figure 12.1b, the square has an area of $(2r)^2 = 4r^2$ and its corresponding circle has an area of πr^2. The sections outside the circle but within the square make up the four single subsections within the square. The area of those four single subsections is $4r^2 - \pi r^2$. The area of just one of the single subsections is $\frac{1}{4}(4r^2 - \pi r^2)$.

Lemma 12.2

$$\text{Area of dual subsection of a normal community} = \frac{1}{2}(\pi r^2 - 2r^2)$$

Proof: Using a square with length r and Lemma 12.1 within the community shown in Figure 12.3, the dual area is the square less the area of two single subsections, which is,

$$r^2 - 2 \times \frac{1}{4}(4r^2 - \pi r^2) = r^2 - (2r^2 - \frac{\pi}{2}r^2) = \frac{\pi}{2}r^2 - r^2 = \frac{1}{2}(\pi r^2 - 2r^2).$$

Lemma 12.3

$$\text{Area of normal community} = 16r^2$$

Proof: Using Figure 12.3, a community is a square of length $4r$. The area is thus $16r^2$.

Lemma 12.4

$$\text{Fraction of normal community for a single subsection} = \frac{4 - \pi}{64}$$

Proof: The fraction of the area of a community for a single subsection is $\dfrac{\frac{1}{4}(4r^2 - \pi r^2)}{16r^2} = \dfrac{4 - \pi}{64}$.

Lemma 12.5

$$\text{Fraction of normal community for a dual subsection} = \frac{\pi - 2}{32}$$

Proof: The fraction of the area of a community for a dual subsection is $\dfrac{\frac{1}{2}(\pi r^2 - 2r^2)}{16r^2} = \dfrac{\pi - 2}{32}$.

Lemma 12.6

$$\overline{\text{Subsection adults}} \approx \text{Community adults}$$
$$\times \text{fraction of community for subsection}$$

Discussion: Here, adults refer to household adults. Community and subsection refer to a normal community. There are on average between 100 and 150 household adults in a community. The area of each subsection covers a fraction of the community Subsection adults can be estimated using a fraction of community area and the estimated number of adults in the community.

$$\overline{\text{Dual subsection adults}} \approx \text{Community adults} \times \frac{\pi - 2}{32} \text{ and}$$

$$\overline{\text{Single subsection adults}} \approx \text{Community adults} \times \frac{4 - \pi}{64}$$

Lemma 12.7

$$\overline{\text{Neighborhood adults}} \approx \text{Community adults} \times \left[4 * \frac{\pi - 2}{32} + 4 * \frac{4 - \pi}{64} \right]$$

Discussion: Here, adults refer to household adults. Neighborhood and community refer to the confines of a normal community. There are on average between 100 and 150 household adults in a community. The estimated number of adults within a neighborhood is based on the fraction that the neighborhood takes within the entire community.

12.3.2 Defining the community

There is no question that households within neighborhoods need to be considered as a part of a community. Here, it is assumed that households are determined by noninstitutionalized adults residing within households. However, the concepts of a "neighborhood" and a "community" beg the following questions:

1. How many households are needed to create an effective neighborhood?
2. How many households and neighborhoods are needed to create an effective community?
3. What types of groups need to be included to create an effective community and how many?

According to Backstrom et al. (2006), an effective community should minimally consist of three connected people or households. Leskovec et al. (2008) suggest a maximum of 100 households within a community, whereas Backstrom et al. (2006) suggest a maximum of 150 households. The authors of this chapter prefer to adopt the strategy of reaching out to 150 adults and hope that around 100 of them (which is close to 70%) will become engaged in one or two neighborhoods within a larger community.

Augustine and Siddiqui (2011), along with Hooke and Rogers (2005), support extending the definition of the community to include external groups such as emergency and health groups, and even governments. Hooke and Rogers believe that more diverse organizations and group entities should be generally included in the extended community for preparedness efforts. This suggestion increases the complexity of community collaboration, making this endeavor problematic. Backstrom et al. (2006) warn that unless a mutual interest in preparedness can be clearly defined and identified among the various organizations, sharing and dissemination efforts among groups within the extended community could fail. Because finding the mutual interests among organizations can be nontrivial, the authors of this chapter have limited the extended community to include at most one group or organization along with a number of households. However, the authors do not distinguish between a group and a household (i.e., a household could represent either an actual household or an external entity). Allowing organizations to collaborate within communities helps to at least develop a mutual interest between households and organizations (Hooke and Rogers 2005; Augustine and Siddiqui 2011).

Neighborhoods should have on average 8 to 20 household adults using Lemma 12.3. Each single subsection should have between one and two household adults using Lemma 12.6. In this case, single subsection disaster preparedness plans should include at most one neighbor. Each dual subsection should have between three and five household adults using Lemma 12.6. In this case, dual subsection disaster preparedness plans should consider at most four neighbors. The authors suggest preparing for at least one neighbor regardless of the subsection type.

12.3.3 Why social networking?

Some important lessons can be learned on effective disaster preparedness from game theory, especially multiplayer games (MPGs). There is a theme that remains consistent across most MPGs: satisfaction comes with the accomplishment of some task(s) (Cornett 2004; Rouse III 2005; Tychsen et al. 2008). That accomplishment may be realized in the form of a collection of a new relic, points, annihilation of the bad guys, or finding some new clue. Once one task is successfully accomplished, the desire to have another success sends the person on an additional mission to achieve even more success. Determining the psychological need to repeat success is beyond the scope of this chapter, although there are researches that address this phenomenon (Lawler et al. 2007; Tychsen et al. 2008; McIvor et al. 2009, Sullivan and Hakkinen 2011).

Some social networks have helped redefine the environment of MPGs. Facebook includes social interactions linked with instant gratification (Kavanaugh et al. 2011; Pikimal 2011). For example, people can socially compete in games, and share pictures and videos. This instant gratification largely makes visually rich Facebook a successful model across many groups such as low-income groups, racial minorities, the elderly, as well as majority groups. Rouse (2005) also identified a number of socioeconomic factors for providing a usable and flourishing social network. Some of these factors include the following: challenges, socialization and inter-action, emotional experience, exploration, incremental task accomplish-ments, realism, and game autonomy. The authors conjecture that some of the successful social aspects found in Facebook can also be found in one of the most successful video games—The Sims. So what is it about Facebook, The Sims, YouTube, and other social forums that make them so successful? Some discussion to help answer this question follows.

Rouse helps identify those social elements that need to be considered in providing a usable, flourishing preparedness system that is inclusive of many people with varying socioeconomic factors. These elements are as follows: socialization and interaction, challenge, emotional experience, exploration, virtual mundane activities, incremental task accomplish-ments, realism, and game autonomy and controlled behavior. These ele-ments are expounded on below.

12.3.3.1 Socialization and interaction

Disaster preparedness may begin with an individual household, but it extends much further. For example, a neighbor's lack of preparedness could affect the preparedness potential of a particular household. Once a disaster strikes, unprepared or underprepared neighbors could seek refuge from an unsuspecting nearby household. This could put a strain on resources affect-ing the preparedness level of the unsuspecting household. It is therefore important to consider neighbors within certain proximities in household preparedness efforts. The authors of this chapter suggest that a household preparedness plan includes neighbors within the same single or dual sub-section as seen in Figure 12.3. Some may even actively resist the assistance. Needless to say, it is a feat to include neighbors with preparedness efforts. It will take interaction before disasters strike as Takazawa (2010) suggests.

12.3.3.2 Challenge

Watkins and Barnes (2010) support evidence that competition tends to help motivate users. Steps are combined with the qualitative concept of win–lose for challenges to engage various people as illustrated in Figures 12.4 and 12.5. It is suggested that a community challenge occur once adults from at least three different neighborhoods within the same community have begun participating and have completed at least one system challenge or quiz.

12.3.3.3 Emotional experience

Rouse (2005) claims that people who play games (and probably those who frequent social networking sites) are seeking some type of emotional experience. According to Rouse, people look for a range of emotions and

STEP 1. HOUSEHOLD COMPLETES A PREPAREDNESS CHALLENGE.

STEP 2. HOUSEHOLD CHALLENGES NEIGHBORS TO COMPLETE THE PREPAREDNESS CHALLENGE.

STEP 3. HOUSEHOLD NEIGHBORS ARE INVITED TO REGISTER (IF THEY HAVE NOT ALREADY) AND/OR COMPLETE THE PREPAREDNESS CHALLENGE.

STEP 4. REGISTERED PARTICIPANTS ARE SHOWN THEIR SCORES AS WELL AS ESTIMATED RISK REDUCTION SAVINGS AT THE COMPLETION OF A TASK.

STEP 5. INTERACTIVE (COINED "FOUR LAYER") PREPAREDNESS VERIFICATION VIA PICTURES FOR PRODUCTS ACQUIRED.

Figure 12.4 Community challenges.

STEP 1. ALL COMPETING NEIGHBORHOODS WITHIN THE COMMUNITY WILL RECEIVE 10,000 POINTS FOR THEIR NEIGHBORHOOD JUST FOR COMPLETING A CHALLENGE.

STEP 2. THE VICTORIOUS NEIGHBORHOOD WILL GAIN AN ADDITIONAL 20,000 POINTS FOR WINNING THE CHALLENGE.

STEP 3. EARLY BIRD BONUSES WILL BE POSSIBLE FOR THOSE WHO SUCCESSFULLY COMPLETE A CHALLENGE WITHIN A SMALL WINDOW OF TIME (E.G. 1-2 DAYS).

Figure 12.5 Neighbor activity challenges.

not necessarily just the "happy" feelings. This is important as the social content is implemented for a disaster preparedness system. A number of authors tend to agree with Rouse and suggest using negative reinforcement through fear to help get people engaged (Leskovec et al. 2008; Takazawa 2010; Augustine and Siddiqui 2011). Because people seek varying emotions and have traditionally lacked preparedness as a whole, a system skewed toward eliciting fear may be the most useful as suggested by Augustine and Siddiqui (2011).

12.3.3.4   *Exploration*

From previous research and the authors' experiences, expecting users to read a manual to learn how to use an online system should be a last resort (Tierney et al. 2001; Shimada and Kitajima 2006; Magsino 2009). Active observation through guided exploration (via Sims-like activities) can be crucial for inclusively educating people within various socioeconomic demographics.

Area exploration activities on the website include challenges so that the user may learn about the environment. To keep the challenges within the purview of various audiences, suggestions (in the form of highlighted artifacts or obvious markings) can be used to give hints and/or answers. As a step toward positive reinforcement, some questions may be repeated to help coach the user in learning preparedness within a virtual environment and to help the user to continue being engaged in the preparedness system. As an added feature based on "The Sims" paradigm, the user could have the option to observe the characters in the environment as they act and react to performing tasks. This would be useful for some users who may need further guidance.

Tests could be given sporadically throughout the game scenarios, such as before and after simulated disaster strikes. This would be used to test the user's educational awareness of disaster preparedness. Each time something positive is achieved on the preparedness front, the player is rewarded with points in proportion to the level of achievement. There would also be a fourth layer where the user would need proof of actual implementation using pictures. A discussion of "photographic verification proof" is in Section 12.5.3.

12.3.3.5 Virtual mundane activities

Beyond exploration, familiarity and repetition help with engaging people into a new system. The Sims video game in particular exploits this facet of human behavior (Rouse III 2005). Familiarity and repetition can be used to increase an individual's comfort levels with the online system. This premise is supported by the concept of Ethnocomputing, which associates computing with familiar activities (Eglash et al. 2006). The concept relates familiar aspects of ethnic culture. For example, creating a cornrow hairstyle requires repetition. Using the African American Culturally Situated Design tool software allows a person to repeatedly create ethnically familiar cornrow twists, while introducing that person to more unfamiliar mathematical and computing concepts like dilation and iteration.

Mundane activities are familiar, because they are ordinary and repeated—many times on a daily basis. One example would be eating a meal. Because this activity is universally familiar, it could be a perfect candidate for introducing virtual nuances that might deviate from real life.

The authors of this chapter believe that engaging in familiar, realistic settings in a virtual environment may help elicit confidence in the use of the system for more unfamiliar preparedness activities, such as building a virtual GO kit. A GO kit is a preparedness kit suggested by FEMA, which includes at least 3 days worth of emergency food, water, and supplies (Ready 2012).

12.3.3.6 Incremental task accomplishments

In many cases, users require small task achievements to keep a system successful (Bond and Porter 1984; Rouse III 2005). It would be reassuring to the users to monitor their progress within a reasonable amount of time as they use the disaster preparedness website. While logged in, they can visually see their own accomplishments in the form of virtual badges, levels for experience, and trophies for time engaged on the system and percentage of areas that have been explored within the system, along with the results from challenges. Figure 12.6 shows the tasks that can be completed. The authors suggest that users see the progress of other users as allies and

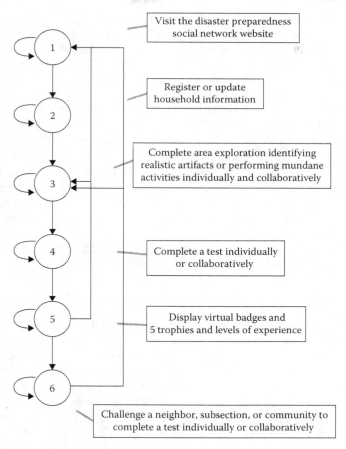

Figure 12.6 Completing incremental tasks.

competitors from within their own subsections, neighborhoods, and communities on summarized daily, weekly, and monthly bases.

12.3.3.7 Realism

Very few would argue that virtual environments are different than real life. However, successful online games have taught that gamers expect some semblance of realism with "make believe" elements sprinkled in fairly sparsely (Rouse III 2005). The Sims game has taught that realism does not necessarily have to equate to actual life. On the contrary, it provides scenarios that could occur given resources within certain physical limitations (e.g., gravity, walls) and time constraints. For example, a Sims character must eat within a certain amount of time. Even the supernatural elements that can appear in the game are constrained by more realistic physical and time bounds. For example, a ghost is not known to float into outer space. It is constrained by gravity to some extent. Also, time does not tend to stop for ghosts. Realism in virtual environments should include mundane activities to help engage people using the world created along with other elements that they would expect. With respect to disasters, the environment would have to preserve the fact that disasters strike when people are going about their normal daily routines.

12.3.3.8   Game autonomy and controlled behavior

One aspect of The Sims game is the concept of game autonomy (Rouse III 2005). Autonomy in the game occurs when characters are left to make (rule-based) decisions and conduct activities without explicit direction from the user. Autonomy could serve as a visual educational tool to illustrate how prepared a community may actually be when a disaster occurs versus how prepared that same community thinks it was. It can also show steps for a community to get prepared—including steps not to follow. Preparing an entire community takes organizational time and effort. It is reasonable to expect that game autonomy may also help users feel more comfortable to become preparedness organizers in their own communities.

12.3.4   Negative community-based reinforcement

According to Kershaw (2005), the notion of preparedness has been assumed for the entire community. Is it possible that scientific advances and the technological revolution have produced a nation with a false sense of security and preparedness? After all, the Citizen Corps (2007) suggests that 54% of people believe that the federal government will be able to support and/or save them during a terrorist attack. It would be prudent to educate the public about the potential consequences of not getting prepared. This concept of negative reinforcement for a community seems to be a well-traveled road. In fact, there is quite a bit to learn from epidemiology, which seems to add science to negative reinforcement. For example, past research suggests engaging the public emotionally with the need for preparedness by explaining how a disaster could possibly make them feel (Augustine 2011). The more the individuals believe that they will be personally affected, the more likely they will be to take action before a disaster strikes (Leskovec et al. 2008).

12.3.5   Historical commemorations

A number of authors would agree that if disasters had not already occurred, then there would be no incentives to get people to prepare in the present or future (Hallgren et al. 1991; Takazawa 2010; Liu 2011). Although it is generally true that disasters have been relatively rare within localized areas, events have occurred more regularly on the global scale. For example, no consistent increase in tornado activity can be noticed in just Ohio (see Figure 12.7). However, when the entirety of the United States is

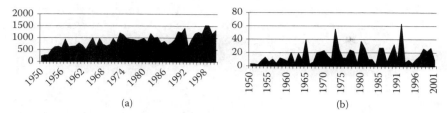

Figure 12.7 The United States versus Ohio annual tornadic events. (a) Tornadoes in the USA (b) Tornadoes in Ohio.

considered (see Figure 12.7), an upward trend can be detected. Hallgren et al. (1991) have claimed that there is a 6-month window after a major disaster to capitalize on postdisaster/predisaster efforts due to alarm and heightened awareness. One can expect local and/or state governments to respond with policy changes regarding concerns of community and professional organizations during this relatively small time frame.

Liu (2011) suggests that commemoration of past disasters, especially those involving technological and social hazards, has been especially useful for reminding people about the importance of getting prepared. Depending on the frequency of events within a 6-month window (the same amount of time that disaster awareness is brought to the forefront after a disaster), on a localized versus a global scale, it may be useful to include major disasters from the past that could affect people around the world today. Figure 12.7 shows an example of tornadoes occurring in Ohio versus on a grander scale in the United States of America. Both of these graphs have one thing in common: tornadoes can be consistently expected on an annual basis. A combination of videos changing at least once every 6 months as aids to commemoration could be used as part of a multimedia campaign to educate the public (Takazawa 2010).

12.3.6   Special considerations: vulnerable populations

Vulnerable populations, namely the elderly, racial minorities, economically disadvantaged, children, those with special needs (physical, communicative, or mental limitations), and visitors to an area are most likely to be negatively affected in a disaster (Tierney et al. 2001; Hooke and Rogers 2005; Shimada and Kitajima 2006; Orcutt et al. 2011; Sullivan and Hakkinen 2011; Nepal et al. 2012). The means of communication must be considered carefully to address the needs of these vulnerable populations. Tierney et al. (2001) suggest that visual imagery is better than written communication for disaster information because visual illustrations offer the means to bypass language and literacy barriers and would reach a wider cross section of people. Furthermore, there is evidence that visual imagery conveys messages more effectively than written text even in nondisaster-related contexts

(Leskovec et al. 2008). For example, shoppers prefer videos in their buying habits, and patrons are more likely to provide video recommendations to friends over books. In fact of all products, books—a popular source of written text—are least recommended to others (Leskovec et al. 2008).

Racial minorities tend to prefer less textual information than Caucasians (Raman et al. 2006; White et al. 2008). Although a wiki and other text-based media have been shown to be effective for certain populations, they may have limited success among vulnerable populations (Raman et al. 2006; White et al. 2008). Visually appealing elements should be social in nature. Foth (2006) found that animated interactions within inner-city neighborhoods were useful. Foth's findings may be useful in other contexts for various vulnerable groups, because a number of inner-city residents tend to have lower incomes.

Twitter is a social media outlet that resembles a specialized blog. It relies on text to send out information and has been used to send out emergency information during a disaster (Starbird and Palen 2011). Although Twitter is a successful medium, it does not appear to be as popular as its media-rich competitor, Facebook. Facebook includes visual elements like pictures and videos, alongside text. It had 40% more active users than Twitter in December 2011 according to Quantcast (2012).

According to Pikimal (2011), five times more Caucasians are on Facebook than Twitter. This would suggest that even within the majority community, there is something different about Facebook that appeals more to the culture of the majority. Both Twitter and Facebook allow textual updates. They both allow still pictures of their respective users. However, Facebook has greater features than its opponent. Also, Twitter does not seem to be well known for multimedia content. This would suggest that focusing on a system that caters to those with special needs would not exclude majority groups.

All preparedness materials should be designed with special needs in mind (Shimada and Kitajima 2006). Customized preparedness materials should include various communication forms: written, sign language, narration, and visual aids (maps, interactive computer graphics, and pictures). An enduring preparedness system would need its users to be engaged on an ongoing basis. It is clear that multimedia that can offer socially appealing media along with text is necessary throughout the life of a useful preparedness system. The system proposed in this framework includes that critical illustrative communication piece in the form of videos, games, text messages, and area maps.

12.4   Overall disaster preparedness system

Augustine and Siddiqui (2011) believe that "resilient social networks" should become the foundation for preparedness systems. The question remains, what should and should not be a part of those social networks to create effective and efficient systems? This chapter expands on this sentiment for disaster preparedness systems. The following elements have been identified as necessary elements for an inclusive social network that considers various needs and hones in on those essential items. Figure 12.8a and b show the general elements that would be complementary in an already existing social network for a particular household. Note that the "Daily Tips" include both disaster-related tips, as well as relevant social tips such as top news stories. Figure 12.9 shows an overview of the framework to help engage people before a disaster strikes.

12.4.1 System elements

12.4.1.1 Loss estimation
On the basis of the premise that every dollar used in mitigation saves $4, it is estimated that each dollar used in mitigation also saves a fraction of a life or a disabling injury (Magsino 2009). Tables 12.1 and 12.2 show the economic costs per death or disabling injury based on 2009 information from the National Survey Council (2012). The average economic cost

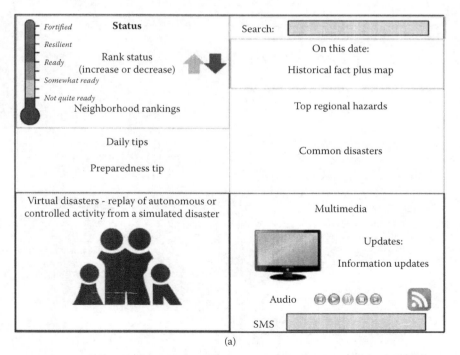

(a)

Figure 12.8 (a) Household general website page 1. (b) Household general website page 2.

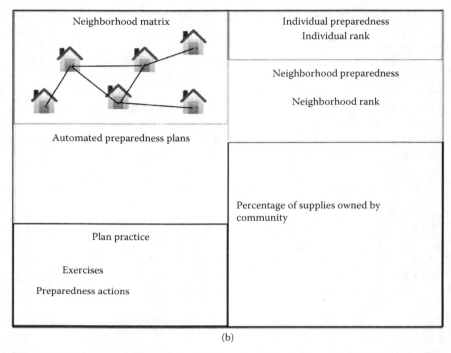

(b)

Figure 12.8 (*Continued*) (a) Household general website page 1. (b) Household general website page 2.

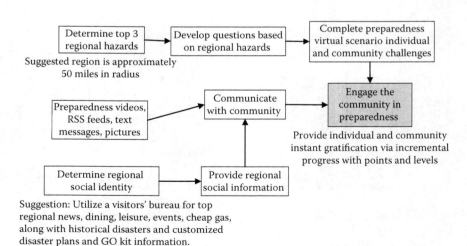

Figure 12.9 Framework flow for online success for disaster preparedness.

per death is $1,130,000. The average economic cost per disabling injury is $27,830. This means that to save someone from death, it is estimated to cost $1,130,000/4 = $282,500. It would cost a little over $6,900 to save a person from a disabling injury. This information would be combined with losses from damages to get an overall cost estimation for losses.

12.4.1.2 Communications

Communications are available in the following forms: videos, text messages, games, and Really Simple Syndication (RSS) feeds from a number of emergency preparedness sources. Table 12.3 shows the RSS feed types and the source(s) for information.

12.4.1.3 Historical commemoration

The historical commemoration feature extends the suggestion by Liu to mention relevant disasters—especially technological and social disasters—at least on a seasonal basis. Terrorist incidents should be included as relevant disasters. Because the main four seasons tend ` last for 3 months each, this fits within the 6-month window of opportunity mentioned by Hallgren et al. (1991). During the winter months, commemorations could include winter storm disasters, such as blizzards and highway technological disasters (e.g., fuel truck accident on an interstate). During the spring and summer seasons, commemorations could include transportation disaster

Table 12.1 Estimated 2009 State Economic Costs per Death

Economic cost per motor vehicle death	$1,290,000
Economic cost per home injury death	$1,050,000
Economic cost per public injury death	$1,050,000

Table 12.2 Estimated 2009 State Economic Costs per Injury

Economic cost per motor vehicle disabling injury	$68,100
Economic cost per home disabling injury	$6,800
Economic cost per public disabling injury	$8,600

Table 12.3 RSS Feed Source Information

RSS feed type	Source(s)
General emergency preparedness education	Crisis and emergency risk communication, disability.gov emergency preparedness feed, NFPA public education
Disaster preparedness tips	Disability.gov emergency preparedness feed
Risk mitigation awareness	FEMA mitigation best practices
Emergency messages	CDC emergency and preparedness response
Disaster occurrences	GDAC RSS information, RSOE EDIS emergency disaster information

information, as well as natural disasters like flooding and severe storms. Fall seasons could include transportation and pipeline disaster information and previous terrorist events. An example of fall commemoration snapshots from the September 11, 2001 terrorist attacks can be seen in Figure 12.10.

12.4.2   Architecture

The suggested preparedness system is generally Web based. The authors suggest having the system run on a Linux machine, such as Fedora, to keep costs low. This includes an open-source Web server (e.g., Apache), an open-source database (e.g., MySQL), and PHP. The social network is based in part on open-source versions of The Sims to partially mimic real-life scenarios. It contains tips transformed from RSS feeds and videos from a number of agencies and nonprofit organizations. The Virtual Disasters video is a replay of disasters encountered within an interactive simulation, while going about daily activities within a virtual realm. Time is accelerated during virtual play, and (seasonally) relevant disasters occur at random during the Virtual Disasters simulation as seen in Figures 12.8a and b.

Questions, tips, RSS, and video feeds are gathered primarily from federal online sources, certain nonprofit organizations, and institutions of higher

FEMA news
photo/Andrea
Booher

FEMA news
photo/Larry
Lerner

FEMA news
photo/Bri
Rodriguez

FEMA news
photo/Andrea
Booher

Figure 12.10 Example: Fall commemoration of the 2001 World Trade Center attacks.

learning. Maps are accessed using Google maps. The SMS text messaging feature is available to all users unless a certain load threshold has been reached. SMS sending would then be temporarily disabled for all nonemergency personnel users. All users would still be able to view new SMS messages from registered emergency personnel, as well as any existing SMS messages.

12.5   Metrics

12.5.1   Cost-benefit analysis

It is important that communities have the bigger picture when determining just how prepared they will be. According to a 2006 survey conducted by the Council of Excellence in Government, 62% of citizens reported that it cost too much to get prepared. If people knew the costs of not getting prepared, those perceptions could shift.

Preparedness up-front costs include overhead costs that the public is generally able to perceive. This includes overall home or renter's insurance premiums, obtaining items for an emergency preparedness kit, and so on. Even though consumers generally perceive overall costs, hidden costs are usually subsumed within insurance premiums across the nation. Preparedness hidden costs include amortized costs from past events. As an example in the state of Ohio, hidden costs include information about where a home is located (Ohio Insurance 2012). Homes that are in frequently flood-damaged areas could increase that area's insurance premiums. Home insurance premiums increase an average of 10%–16% every year in Ohio (Ohio Insurance 2012). By law, many home owners may be required to maintain home insurance regardless of their willingness to get prepared. This becomes a standard expense that fluctuates over time due to variables unseen by the general consumer. Engaging a community in preparedness may lower some hidden costs in favor of those residing within the community according to the Insurance Information Institute (2004).

Preparedness maintenance costs include those costs to continue to be prepared. Maintenance costs include costs to replace damaged, depleted, perishable, or expired items within a GO kit or preparedness plan. FEMA recommends replacing emergency water twice a year (FEMA 2012). The United States Department of Agriculture recommends replacing canned goods after an average of 2 years (United States Department of Agriculture 2012). Other nonfoodstuffs have longer shelf lives. The authors have estimated that one household should expect to spend an estimated $60 a year (or $5 a month) to maintain their preparedness items.

12.5.2   Social network analysis: effectiveness and efficiency

Social network analysis is needed for determining the impacts of using the preparedness system. Effectiveness is defined in terms of the rate of new users, the frequency of usage across members and nonmembers, the percentage of neighborhoods and communities represented within a particular area, and the percentage of repeat users. Other nontraditional elements for effectiveness include instant gratification measures, multimedia communications, realistic localized scenarios, relatable historical commemorations, evaluations using answers from disaster preparedness questions before and after interactively using scenarios, and knowledge of the type of social content (disaster versus nondisaster related) users frequent. Efficiency is defined in terms of affordability measures for smaller communities (including costs to create and maintain the system) and geographical spread (geographical distance between users).

12.5.3   Preparedness verification

The algorithm in Figure 12.11 is used to verify "four layer" activities, which consist of providing pictured evidence of preparedness items and/ or steps in the "real" world. This verification step is used to provide a more decentralized approach to verification using feedback from system users, as well as to provide a fun way to engage with neighbors.

12.6   A working example: Wilberforce, Ohio

Zip code 45384 is located in Wilberforce, Ohio—a rural area approximately 20 miles outside Dayton, Ohio. 45384 has local subdivisions, farmland, and

1) **FIRST LAYER OF VERIFICATION.** Check database to make sure picture does not already exist. If it does, the picture is rejected.
2) **SECOND LAYER OF VERIFICATION.** Check IP address to make sure submitter is in the area claiming residence unless visiting another location. If the IP address for a presumed resident is outside the area of residence, the picture is rejected.
3) **THIRD LAYER OF VERIFICATION.** Have the submitter identify two pictured facts about the area (e.g., popular regional landmark, subdivision sign, street sign, area code, or state flag). If failed after two attempts, picture is rejected.
4) **FOURTH LAYER OF VERIFICATION.** Ask one to two neighbors to identify the submitted picture and vouch that they saw the actual item(s) including time frame. They would then identify three pictured facts about the area, where at least one newer picture is used as a test for determining regional familiarity. The more neighbors who identify newer pictured facts, the more those facts can be used for second layers of verification. This is inspired by captcha/recaptcha methodology, which both verifies that a particular submitter is in fact a person and validates newer information using majority rule of people within a community.

Figure 12.11 Algorithm for verifying preparedness (four layer) activities.

local universities. Data can be gathered from the 2010 Census. For example, from Table G001: Geographic Identifiers from the Geographic Header product type, the area is 3370895 square meters (or 1.301 square miles). Using Table DP-1: Profile of General Population and Housing Characteristics, there are a total of 2001 people in zip code 45384—936 of which are adults aged 18 and more (U.S. Census Bureau 2010). Approximately 92% of the people of one race are African Americans. Using the same table, there are a total of 336 people in households, 83 of whom are children. There are 253 adults in households. Using Equation 12.1, the average number of household adults per square mile is $253/1.301 = 194$. To scale between 100 and 150 people in the community, use Equation 12.3. The radius for a neighborhood is suggested to be between 0.17 and 0.2 miles.

Figure 12.12 shows a portion of a normal community for zip code 45384 starting at the center. Wilberforce University is a part of Neighborhoods 3, 5, 7, 8, and 10. Assuming that all buildings are equivalent to households,

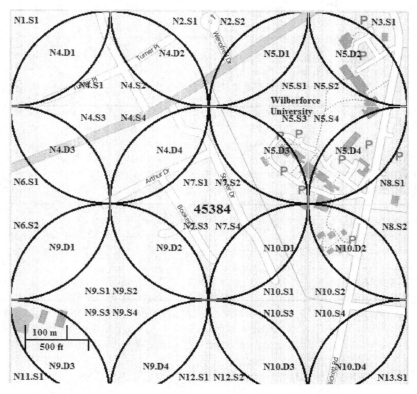

Figure 12.12 Snippet of a normal community for zip code 45384 in Wilberforce, Ohio © OpenStreetMap contributors, CC BY-SA (From *OpenStreetMap*, available at http://www.openstreetmap.org, 2012.

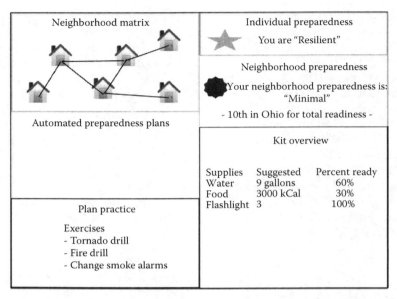

Figure 12.12 (*Continued*) Snippet of a normal community for zip code 45384 in Wilberforce, Ohio © OpenStreetMap contributors, CC BY-SA (From *OpenStreetMap*, available at http://www.openstreetmap.org, 2012.

this map suggests that two buildings in N5.D3 be considered together in preparedness planning, four buildings in N5.D4, five buildings in N10.D2, and three in N5.D2. The building in N3.S1 should not consider other households in preparedness efforts because it is within that subsection. (This building is the old President's House and is approximately 500 ft away from the nearest building.) The authors suggest that households (or in this case buildings) that touch the boundaries or span across dual and single subsections be considered as a part of a single subsection or alone without neighboring buildings because these buildings can be rather large. Also, buildings spanning between single subsections should be considered a part of the lower numbered subsection. The building that is a part of N8.S1 and N8.S2 should be considered with N8.S1 using this convention. The building in N10.D2 touching the N10.S2 boundary should be considered separate from D10.D2. This is a reasonable consideration when considering the proximity of the buildings. Figure 12.13 shows example snapshots of the system for a particular household (in this case, a building).

12.7   Conclusion

Getting prepared is a necessity, but either it is not realized or not understood by the general populace. This chapter provides a framework and a strategy for convincing people to get prepared before a disaster strikes. It

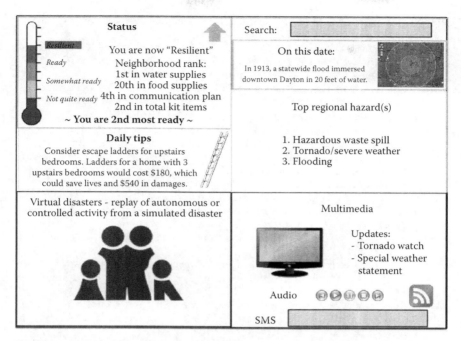

Figure 12.13 Example of household website page 1 for zip code 45384 in Wilberforce, OH and example of household website page 2 for zip code 45384 in Wilberforce, OH.

takes into account everyday social activities that have made social games and websites popular, along with reminders that help people visualize and remember problems from the past. Incremental prototypes for this system are currently underway. Various parts of the system will be tested by the public to garner usability, engagement potential, and future updates.

Acknowledgment

This work is supported by the National Nuclear Security Administration under Grant No. DE-FG52-06NA27597.

References

Augustine, L. A. and Siddiqui, S. National Research Council of the National Academies (Eds). 2011. *How Communities Can Use Risk Assessment Results: Making Ends Meet: A Summary of the June 3, 2010 Workshop of the Disasters Roundtable*, National Academies Press, Washington, DC.

Backstrom, L., Huttenlocher, D., Kleinberg, J., and Lan, X. 2006. Group Formation in Large Social Networks: Membership, Growth, and Evolution, *ACM Proceedings of the 12th ACM SIGKDD International Conference on Knowledge Discovery and Data Mining*, Philadelphia, PA, 20–23 August, pp. 44–54.

BDPN-Network. 2011. Berkeley Disaster Prep Neighborhood Network. http://www.bdpnnetwork.org Last accessed April 2013.

Bond, F. E. and Porter, R. L. 1984. Top Down Architecture for Commercial SATCOM Survivability, *IEEE Military Communications Conference (MILCOM)*, Los Angeles, CA, 21–24 October, pp. 2–6.

Bureau of Labor Statistics, U.S. Department of Labor. 2012. EMTs and Paramedics, in *Occupational Outlook Handbook*. http://www.bls.gov/ooh/Healthcare/EMTs-and-paramedics.htm#tab-1 Last accessed April 2013.

California Volunteers. 2012. We Prepare. Customize Your Family Plan. http://www.californiavolunteers.org/familyplan/plan.html Last accessed April 2013.

Cornett, S. 2004. The Usability of Massively Multiplayer Online Roleplaying Games: Designing for New Users, *ACM Proceedings of the SIGCHI Conference on Human Factors in Computing Systems* 6(1): 703–710.

Citizen Corps 2007. Update on Citizen Preparedness Research, *Citizen Corps*. Available at http://www.citizencorps.gov/downloads/pdf/ready/citizen_prep_review_issue_5.pdf (last accessed: March 2012).

Desouza, K. C. 2005. Scenario Management: From Reactivity to Proactivity, *IEEE IT Professional* 7(5): 42–48.

Ecevit, M. and Kasapoglu, A. 2002. Demographic and Psychosocial Features and Their Effects on the Survivors of the 1999 Earthquake in Turkey, *Social Behavior and Personality* 30(2): 195–202.

Eglash, R., Bennett, A., O'Donnelll, C., Jennings, S. L., and Cintorino, M. 2006. Culturally Situated Design Tools: Ethnocomputing from Field Site to Classroom. *Anthropology and Education* 108: 347–362.

Federal Emergency Management Association. 2012. Maintaining Your Kit. Available at http://www.ready.gov/maintaining-your-kit (last accessed: April 2012).

Fortier, S. C. and Volk, J. H. 2006. Defining Requirements for ad hoc Coalition Systems during Disasters, *IEEE International Conference on Computational Cybernetics (ICCC)*, Tallinn, Estonia, August 20-22, 20–22 August, pp. 1–6.

Foth, M. 2006. Facilitating Social Networking in Inner-City Neighborhoods, *IEEE Computer* 39, 44–50.

Freeman, M. and Freeman, A. 2010. Bonding over Bushfires: Social Networks in Action, *IEEE International Symposium on Technology and Society*, Wollongong, Australia, 7–9 June, pp. 419–426.

Get Emergency Prepared. 2012. Neighborhood Emergency Preparedness—The Vital Next Step. Available at http://www.getemergencyprepared.com/neighborhood.html (last accessed: April 2012).

Hallgren, R. E., Bradley, T., Cluff, L. S., et al.; National Research Council of the National Academies (Eds). 1991. *A Safer Future: Reducing the Impacts of Natural Disasters*, National Academies Press, Washington, DC.

Hecker, J. Z. 2002. Combating Terrorism: Key Aspects of a National Strategy to Enhance State and Local Preparedness. *Before the Subcommittee on Government Efficiency, Financial Management, and Intergovernmental Relations, Committee on Government Reform, House of Representatives* 1:1-23.

Hooke, W. H. and Rogers, P. G.; National Research Council of the National Academies (Eds). 2005. *Public Health Risks of Disasters: Communication, Infrastructure, and Preparedness—Workshop Summary*, National Academies Press, Washington, DC.

Insurance Information Institute. 2004. *12 Ways to Lower Your Homeowners Insurance Costs*. Available at http://www.iii.org/assets/docs/pdf/12Ways.pdf. Last accessed April 2013.

Kavanaugh, A., Fox, E. A., Sheetz, S., et al. 2011. Social Media Use by Government: From the Routine to the Critical, *ACM Proceedings of the 12th Annual International Digital Government Research Conference: Digital Government Innovation in Challenging Times*, College Park, Maryland, 12–15, June, pp. 121–130.

Kershaw, P. J. 2005. National Research Council of the National Academies (Ed.) *Creating a Disaster Resilient America: Grand Challenges in Science and Technology: Summary of a Workshop*. National Academies Press.

King, D. 2000. TREND REPORT: You're on Your Own: Community Vulnerability and the Need for Awareness and Education for Predictable Natural Disasters, *Need for Awareness and Education* 8: 223–228.

Krock, R. E. 2011. Lack of Emergency Recovery Planning Is a Disaster Waiting to Happen, *IEEE Communication Magazine: Network Disaster Recovery*1: 48–51.

Lawler, C. M., Harpter, M. A., and Thornton, M. A. 2007. Components and Analysis of Disaster Tolerant Computing, *IEEE International Conference on Performance, Computing, and Communications Conference (IPCCC)*, New Orleans, Louisiana, 11–13 April, pp. 380–386.

Leskovec, J., Lang, K. J., Dasgupta, A., and Mahoney, M. W. 2008. Statistical Properties of Community Structure in Large Social and Information Networks, *ACM Proceedings of the 17th International Conference on World Wide Web*, Beijing, China, April 21–25, pp. 695–704.

Liu, S. B. 2011. Digital Commemoration: Surveying the Social Media Revival of Historical Crises, *ACM CHI EA '11 Proceedings of the 2011 Annual Conference Extended Abstracts on Human Factors in Computing Systems*, Vancouver, BC, Canada, 7-12 May, pp. 947–952.

Magsino, S. L.; National Academy of Sciences (Ed.). 2009. *Applications of Social Network Analysis for Building Community Disaster Resilience: Workshop Summary*, National Academies Press, Washington, DC.

McIvor, D., Paton, D., and Johnston, D. 2009. Modeling Community Preparation for Natural Hazards: Understanding Hazard Cognitions, *Journal of Pacific Rim Psychology* 3: 39–46.

National Disaster Education Coalition. 2004. *Talking About Disaster: Guide for Standard Messages*, National Disaster Education Coalition. Available at http://www.disastereducation.org/library/public_2004/Talking_About_Disaster_Guide.pdf (last accessed: March 2012).

National Survey Council. 2012. "Estimating the Costs of Unintentional Injuries," http://www.nsc.org/news_resources/injury_and_death_statistics/Pages/EstimatingtheCostsofUnintentionalInjuries.aspx. (Last Accessed: April 2012.)

Ohio Insurance. 2012. Ohio Insurance Premiums and How They Are Determined. Available at http://www.insurance.ohio.gov/Newsroom/Pages/Howratesaredetermined.aspx (last accessed: April 2012).

OpenStreetMap. 2012. *OpenStreetMap*. Available at http://www.openstreetmap.org (last accessed: May 2012).

Orcutt, J. A., Grabowski, M. R., Atwater, B. F., et al.; National Research Council of the National Academies (Eds.). 2011. *Tsunami Warning and Preparedness: An Assessment of the U.S. Tsunami Program and the Nation's Preparedness Efforts*, The National Academies Press, Washington, DC.

Patterson, O., Weil, F., and Patel, K. 2010. The Role of Community in Disaster Response: Conceptual Models, *Population Research and Policy Review* 29: 127–141.

Pavlica, E. 2012. What Happens Now? Available at http://www.whathappensnow.com (last accessed: April 2012).

Pikimal. 2011. Top 10 Best Social Networks. Available at http://social-network.pikimal.com (last accessed: March 2012).

Quantcast. 2012. Quantcast Measure. Available at http://www.quantcast.com (last accessed: March 2012).

Raman, M., Ryan, T., and Olfman, L. 2006. Knowledge Management System for Emergency Preparedness: An Action Research Study, *IEEE Proceedings of the 39th Annual Hawaii International Conference on System Sciences (HICSS)* 2: 37b–46b.

Rouse III, R. 2005. *Game Design: Theory and Practice*, 2nd edition. Worldware Publishing, Inc., Plano, TX.

Shimada, H. and Kitajima, M. 2006. SMMAPS: Scenario-Based Multimedia Manual Authoring and Presentation System and Its Application to a Disaster Evacuation Manual for Special Needs, *ACM Chi Extended Abstracts on Human Factors in Computing Systems*, pp. 1313–1318.

Solecki, W. D. 1992. Rural Places and the Circumstances Acute Chemical Disasters, *Journal of Rural Studies* 8: 1–13.

Starbird, K. and Palen, L. 2011. "Voluntweeters": Self-Organizing by Digital Volunteers in Times of Crisis, *ACM CHI '11 Proceedings of the 2011 annual conference on Human factors in computing systems*, Vancouver, BC, Canada, May 7–12, pp. 1071–1080.

State of Florida. 2012. Disaster Planning for Families. Available at http://www.floridadisaster.org/family (last accessed: March 2012).

Sullivan, H. T. and Hakkinen, M. T. 2011. Preparedness and Warning Systems for Populations with Special Needs: Ensuring Everyone Gets the Message (and Knows What to Do), *Geotechnical and Geological Engineering* 29: 225–236.

Swan, J. 2001. Knowledge Management in Action: Integrating Knowledge across Communities, *IEEE Proceedings of the 34th Annual Hawaii International Conference on System Sciences*, Maui, Hawaii, January 3–6, pp. 1–9.

Takazawa, A. 2010. YouTube Space as the Propagative Source for Social Power: An Experimental Study on the Social Meaning of Disaster, *ACM Proceedings of the 73rd ASIS&T Annual Meeting on Navigating Streams in an Information Ecosystem, American Society for Information Science*, 22–27 October, pp. 47.

Tierney, K. J., Lindell, M. K., and Perry, R. W. Ac, N. (Eds.) 2001. Facing the Unexpected: Disaster Preparedness and Response in the United States, John Henry Press, Washington, DC.

Tychsen, A., Hitchens, M., and Brolund, T. 2008. Character Play: The Use of Game Characters in Multi-player Role-playing Games across Platforms. *Computers in Entertainment (CIE)-Theoretical and Practical Computer Applications in Entertainment* 6(2), Article No. 22: 1–24.

U.S. Census Bureau. 2010. *Profile of General Population and Housing Characteristics: 2010*. Available at http://factfinder2.census.gov (last accessed: April 2012).

United States Department of Agriculture. 2012. Food Safety. Available at http://www.usda.gov/wps/portal/usda/usdahome?parentnav=FAQS_BYTOPIC&FAQ_NAVIGATION_ID=FOOD_FQ&FAQ_NAVIGATION_TYPE=FAQS_BYTOPIC&contentid=faqdetail-24.xml&edeployment_action=retrievecontent (last accessed: April 2012).

Watkins, K. Z. and Barnes, T. 2010. "Competitive and Agile Software Engineering Education," Proceedings of the IEEE SoutheastCon 2010: Energizing Our Future, Charlotte, North Carolina, 111-114.

Watkins, K. Z., Barnes, T., and Thomas, N. 2009. Developing Computing Identity as A Model For Prioritizing Dynamic K-12 Computing Curricular Standards, *ACM Journal of Computing Sciences in Colleges* 24: 125–131.

Watkins, K. Z., Simon-Agolory, K., Venkateswaran, A., and Nam, D. 2011. GET A PLAN! Automatically Generating Disaster Preparedness Plans Using WILBER, *Proceedings of the Eighth International Conference on Information Systems for Crisis Response and Management (ISCRAM)*, Lisbon, Portugal, May 8-11, pp. 1–5.

White, C., Plotnick, L., Addams-Moring, R., Turoff, M., and Hiltz, S. R. 2008. Leveraging a Wiki to Enhance Virtual Collaboration in the Emergency Domain, *Proceedings of the 41st IEEE Annual Hawaii International Conference on System Sciences*, Waikoloa, Hawaii, January 7–10, pp. 322–331.

chapter thirteen

Lest we forget

A critical analysis of bioterrorist incidents, national exercises, and U.S. prevention, response, and recovery strategies*

Tasha L. Pravecek

Contents

* The material contained in this chapter was prepared as an account of work sponsored by an agency of the U.S. government. Neither the U.S. government nor any agent thereof, nor any of their employees, makes any warranty, express or implied, or assumes any legal liability or responsibility for the accuracy, completeness, or usefulness of any information, apparatus, product, or process disclosed, or represents that its use would not infringe upon privately owned rights. Reference herein to any specific commercial product, process, or service by trade name, trademark, manufacturer, or otherwise does not necessarily constitute or imply its endorsement, recommendation, or favoring by the U.S. government or any agency thereof. The views and opinions of the author expressed herein do not necessarily state or reflect those of the U.S. government or any agency thereof.

13.1   Introduction

A large-scale biological weapons attack by a terrorist organization is probably not a matter of "if" but a matter of "when." Terrorist groups in the past and present, for example Al Qaeda, have expressed interest or demonstrated steps in planning for a biological weapons attack. The *National Strategy for Countering Biological Threats*, published in November 2009, reports, "in 2001, while engaging the Taliban in Afghanistan, coalition forces came into possession of a significant body of evidence that Al Qaeda was seeking to develop the capability to conduct biological weapons attacks. Although Al Qaeda has lost many of the resources it had compiled prior to September 2001, it is prudent to assume its intent to pursue biological weapons still exists" (Department of Homeland Security 2006).

To facilitate this desire to use biological weapons, "how to" instructions are readily available. One book currently available for sale on Amazon.com is *Silent Death* written by Uncle Fester (Fester 1997). In this book, the author details the means of production and dissemination of various plant poisons, botulism, and ricin toxin. Many other individuals and groups have published books and websites detailing how to effectively develop biological weapons, contributing to the ease of proliferation.

Current U.S. deterrence and prevention strategies are incapable of completely negating the multiple means a terrorist has to obtain and disseminate biological agents to create a catastrophic U.S. incident. However, a thorough analysis and assessment of previous successful or partially successful biological attacks and national bioterror exercises may provide valuable insights to where most effective deterrence and prevention strategies should be focused to prevent or mitigate a highly successful, large-scale biological terrorist attack, or respond and recover should an attack occurs.

This chapter reviews and analyzes biological incidents that have occurred over the past 30 years and National Exercises that involve simulated bioterrorist events. Next, a brief summary of the current U.S. deterrence and prevention strategies and policies to prevent a biological weapons attack is provided. In addition, U.S. strategies and current programs are examined to determine the focus that will achieve the best effect in light of lessons learned from the recent biological incidents and exercises detailed. Finally, the chapter provides some recommendations to bolster future biological weapons deterrence strategies.

13.2   Recent history of terrorist biological attacks and noteworthy incidents

The proliferation and use of biological weapons are significant concerns to the United States. Although biological terrorists have not attacked since World War II, three incidents in the United States and one attack on U.S. interests overseas demonstrate attacks are not only possible, but can also easily occur (Rodi 2009). In addition, examination of these attacks and the response to them can provide indication of potential areas of improvement in deterrence strategy. It is important to note the similar shortcomings experienced in these events, such as the need for biological agent forensics, clear public communication, and training of first responders and healthcare workers.

13.2.1   Rajneesh cult

"For the first time ever, all of Mid-Columbia's 125 beds were filled; some patients had to be kept in the corridors" (Miller et al. 2002). This sounds like the scene from a Hollywood film. In reality, this mass casualty event occurred in The Dalles, OR, during September and October 1984. By the end of the attack, 45 people were hospitalized and over 750 fell ill suffering from salmonella poisoning. This attack was the act of a religious cult residing in the United States in an attempt to win control of the county government by preventing noncult voters from participating in the November 1984 elections.

In the summer of 1983, two Wasco county commissioners conducted a mandatory inspection of the Rajneesh ranch prior to its annual summer festival. The Rajneeshees gave the two commissioners water laced with *Salmonella typhimurium*, while they waited for their car's mysteriously flat tire to be repaired. Eight hours later, both men became ill, and one was hospitalized. Both suspected the Rajneeshees had put something in the water, but neither had evidence, thus no charges were filed (Carus 2001).

Approximately 1 year later on September 9, 1984, the famed salmonella attack of The Dalles began. This attack was suspected to be a practice run for the primary attack planned in November with the goal to affect participation by local voters in the county elections. In 10 restaurants and a grocery store, the cult members sprinkled salmonella over fruits and vegetables, milk, coffee creamers, and blue-cheese dressing. The first reports of gastroenteritis to the Wasco Sherman Public Health Department occurred on September 17, 1984. By September 21, the county was overwhelmed with sick and frightened people (Miller et al. 2002). That same day, the Oregon State Public Health Laboratory in Portland confirmed the culprit was *S. typhimurium*, only 4 days after the initial reported case (Carus 2001).

Two attacks apparently resulted in two waves of illness, September 9–18 and September 19–October 10, 1984. The final reported count was 751 cases. However, the actual number of victims was likely higher due to the number of out-of-state travelers who consumed contaminated food products (Carus 2001).

Despite the apparent success of the covert attack, the Rajneesh cult abandoned the planned attack in November. A year later, the salmonella illnesses were discovered to be intentional.

Where our prevention and response actions failed. State public health officials and the Centers for Disease Control (CDC) investigated the salmonella outbreaks in The Dalles. They measured salad-bar temperatures, inspected food-handling procedures, and tested cows, raw milk, water, vegetables, and food distributors and found no contamination and no common source for the food (Miller et al. 2002). Despite the apparent randomness, both inspection agencies concluded there was no evidence of deliberate contamination. Instead, they blamed the food handlers at the 10 restaurants impacted by the bacterium. The most senior state epidemiologist went so far as to conclude the contamination "could have occurred where food handlers failed to wash their hands adequately after bowel movements and then touched raw foods" (Miller et al. 2002). In retrospect, this simultaneous coincidence in 10 restaurants seems implausible and clearly demonstrates the lack of forensic detection capabilities or awareness.

To prevent future accidental or intentional outbreaks of salmonella contamination or to induce change to prevent such outbreaks, it is a usual practice to publish an incident report. In this event, a report was not published. The public health officials realized how easily the Rajneeshees spread the disease and did not want to encourage copycats (Miller et al. 2002). A published report would have highlighted clues that a biological agent manufacture capability existed on the ranch, including an incubator and freeze dryer. In addition, the report would have identified the need to register all medical laboratories in the state, regardless of the size.

Finally, and most critically, the report would have underscored the ease with which the cult acquired pathogens. More than 10 years later, Larry Harris's arrest prompted the CDC to take measures through the Select Agent Program to safeguard specific infectious agents and toxins.

13.2.2   Larry Wayne Harris

Larry Harris was a neo-Nazi sympathizer and a trained microbiologist (Stern 2001). In 1995, Larry Harris bought three vials of plague bacteria (*Yersinia pestis*) from the American Type Culture Collection (ATCC) for approximately $300. He pled guilty to wire fraud, the most serious charge possible at the time (Miller et al. 2002). In 1998, he claimed to have military grade anthrax that turned out to be a vaccine with a harmless strain (Stern 2001). His punishment was community service and extension of his probation.

Where our prevention and response actions failed. Today many of our first responders have the capability and training to rapidly analyze certain biological agents on-site. However, in 1995 and 1998, this technology and training was not readily available. Biological terrorism seemed so improbable at the time that one public health official had difficulty persuading a colleague that his call reporting the incident was not a joke (Stern 2001). This incident demonstrates the dire need for training in appropriate HAZMAT response, forensic evidence collection, and the requirement for on-the-spot, immediate biological agent identification.

Due to the ease in which he acquired *Y. pestis*, Larry Wayne Harris was also the motivation behind establishment of the Select Agent Program to control the distribution of 24 infectious agents and 12 toxins. In addition, Larry Harris's arrests and subsequent lax punishments demonstrate the need for more severe laws to deter those who may consider such actions. The laxity in the legal system regarding biological agent possession was further strengthened by the Uniting and Strengthening America by Providing Appropriate Tools Required to Intercept and Obstruct Terrorism Act (USA PATRIOT Act) of 2001. However, these Select Agent Program controls and USA PATRIOT Act legal penalties may not be as global as necessary to prevent acquisition and use of biological agents by groups who desire to harm our troops stationed overseas, as demonstrated by the Aum Shinrikyo and its attack on U.S. Navy installations at Yokohama and the headquarters of the Navy's Seventh Fleet at Yokosuka.

13.2.3   Aum Shinrikyo

On March 20, 1995, the Aum Shinrikyo Japanese religious cult gained world recognition with its sarin gas attack on a Tokyo subway. The Tokyo attack killed 12 and caused over 5000 to seek medical attention (Siddell 1997).

Japanese authorities estimate of these 5000 victims, 73.9% were "worried well," showing no evidence of actual nerve exposure (Kaplan 2001). Prior to this attack, the Aum Shinrikyo tried several unsuccessful attacks on several locations, including U.S. installations overseas, using biological agents.

The cult obtained, developed, and experimented with the dissemination of multiple agents, including anthrax, botulinum, Q-fever, and Ebola. The cult attempted 10 times between 1990 and 1995 to spread botulinum toxin and anthrax in Tokyo and Yokohama (Alibek and Handelman 1999). These attacks included the U.S. Navy installation at Yokohama and the headquarters of the Navy's Seventh Fleet at Yokosuka (Mangold and Goldberg 1999).

Despite possessing the resources and apparent technical knowledge to produce deadly biological weapons, Aum Shinrikyo failed to sicken or kill thousands because the strains of agents used were nonlethal or their dissemination techniques were ineffective (Kaplan and Marshall 1996). Botulinum toxins are extremely difficult to purify, unstable in pure form, and degrade rapidly when exposed to air and sunlight. "These factors led scientists in the U.S. biological weapons program to discard botulinum toxins from their arsenal" (Rosenau 2001). In addition, the cult attempted to isolate the botulinum bacteria from soil and they may not have understood the extreme difficulties in enriching an environmental sample of botulinum bacteria (Rosenau 2001).

The Aum's attacks may have also failed due to inadequacy of dissemination methods and hostile atmospheric conditions. The cult settled on a liquid slurry to disseminate anthrax, versus a more difficult to obtain, yet potentially more stable, powder form. The ineffectiveness in distribution of the slurry may have been related to the material settling to the bottom of the sprayer. In addition, sunlight may have destroyed the viable bacteria or degraded the toxin. Finally, wind conditions may have caused the aerosols to disintegrate or simply not disperse the biological material. Because many of the attacks occurred during Tokyo's warm spring and summer months with strong sunlight, the environment may play some blame in the failure in Aum Shinrikyo's attacks.

Where our prevention failed. In April 1990, cult personnel used a vehicle equipped with a sprayer to disperse botulism toxin near the Yokosuka naval base and the headquarters of the U.S. Navy's Seventh Fleet. According to open source literature, this attack was not detected. Although it is likely the toxin degraded before contact and had no effect on the intended target, it is not published whether the naval installations maintained standoff detectors for perimeter monitoring of low levels of airborne toxins, chemicals, or biological agents that may have impacted overseas U.S. personnel. Nevertheless, this event highlights the need for perimeter monitoring, even in seemingly peaceful locations.

The experience of the Aum Shinrikyo cult demonstrates, despite excellent technical expertise, suitable laboratory equipment, and seemingly limitless funding, large-scale biological weapon dissemination is not a simple task. However, dissemination of biological agents inside a closed space targeting smaller numbers of people can be achieved by even one well-trained individual, as was demonstrated by the 2001 anthrax letter attacks.

13.2.4 2001 anthrax letters

In the initial attack on September 18, 2001, letters containing *Bacillus anthracis*, the bacterium causing anthrax, are believed to have been mailed to four New York City addresses at *ABC News, CBS News, NBC News*, and the *New York Post*, as well as to the *National Enquirer* at American Media in Boca Raton, FL. Only the letters sent to the *New York Post* and Tom Brokaw (NBC) were recovered. In the second attack on October 9, 2001, two additional letters were sent to Senators Tom Daschle and Patrick Leahy in Washington, DC. The Daschle and Leahy letters contained a more potent form of anthrax. The anthrax mailings contaminated the Hamilton, NJ, postal facility and the Brentwood postal facility in Washington, DC, the latter resulting in suspected contamination of many government offices. The anthrax letters caused 22 people to contract the disease and five deaths (Council on Foreign Relations 2006).

Where our prevention and response actions failed. As the nation reacted to this event, people timidly retrieved mail from mailboxes and frequently discarded it unopened or called fire departments and other emergency response teams fearing what may be inside. Hundreds of "suspicious letters" were painstakingly analyzed and found to be harmless. First responders and others spent countless hours and resources responding to the hysteria following these attacks, demonstrating the severe lack of transparent communication of the potential danger and sensitivity of the American people to terrorist action. The extensive and long-term consequence management activities that were undertaken in the months to years following the initial incident demonstrate both adequate and inadequate response, detection, and decontamination capabilities.

Although hospital personnel were able to rapidly distribute treatment and prophylaxis and cleanup crews were able to successfully decontaminate buildings, it could be argued future response to a similar incident should realize dramatic improvement. The financial costs were very high, one FBI report estimating damages over $1 billion. The Brentwood facility alone cost $130 million and took 26 months to clean up, and many other government buildings required decontamination. It was over 3 years after the attack before the American Media building began decontamination activities (Lengel 2005). Despite these shortcomings, the anthrax letter incident spurred additional emergency response training and the passage of the Project Bioshield Act to supply vaccine and drug treatment to protect against future bioterrorist actions.

13.3 National Exercise program and lessons learned

Exercises serve as a training venue, a test of local capabilities, and an opportunity to examine preparedness, status of appropriate resources, and abilities of personnel. National Exercises engage government officials and agencies and test strategic level changes in our national emergency response structure and capability.

Three "Top Official" or "TOPOFF" National Exercises included a simulated release of a biological agent. Now known as National Level Exercises, these exercises engage participation from all levels of government.

The participants and observers of these National Exercises recorded lessons learned to enable everyone, including the top officials, the emergency management workers, and the hospital employees, to gain information and improve future emergency response and consequence-management activities.

13.3.1 TOPOFF 1: May 2000, Denver, CO

TOPOFF 1 was the first National Exercise that involved a response to a simulated chemical attack in Portsmouth, NH, and a simulated biological attack in Denver, CO. This exercise was conducted by the Department of Justice, the Department of State, and the Federal Emergency Management Agency (note: the Department of Homeland Security [DHS] did not exist at this time) (U.S. Department of Homeland Security 2011b).

The exercise event began with a simulated aerosol release of *Y. pestis*, or plague bacteria, on the city of Denver, CO, on May 17, 2000. Exercise play began on May 20, as 500 sick people inundated local hospitals. Players in this exercise included the state and county health agencies, the CDC, the Office of Emergency Preparedness, elements of the Public Health Service, and three hospitals in the Denver area. By the end of the exercise on day 4, hospitals are understaffed, there are insufficient antibiotics and beds for the demand, and there are an estimated 3700 cases of pneumonic plague with 950 deaths (Inglesby et al. 2001).

13.3.2 TOPOFF 2: May 2003, Chicago, IL

TOPOFF 2 was the first major exercise led by the newly formed DHS following the events of September 11, 2001 and the anthrax attacks that followed. This exercise was the first opportunity for DHS to exercise the Homeland Security Advisory System and included a radiological dispersal device attack on Seattle and a plague release in Chicago (U.S. Department of Homeland Security 2011b).

As part of the exercise scenario, in May 2003, Chicago hospitals reported an increase in common illness. During the exercise, 64 hospitals in Illinois participated, the largest mass casualty exercise ever undertaken. The exercise progressed for 3 days. In the end, exhausted and overextended staff at hospitals and the State Department of Health demonstrated the limits on adequate functioning and ability to provide services over an extended period of time (U.S. Department of Homeland Security 2003).

13.3.3 TOPOFF 3: April 2005, NJ

TOPOFF 3 was the first test of the National Response Plan (NRP) and National Incident Management System (NIMS). This exercise included a chemical attack in Connecticut and a biological attack in New Jersey (U.S. Department of Homeland Security 2011b).

TOPOFF 3 was a totally scripted exercise in which the participants were alerted to the biological agent to be employed more than a week before the exercise began. Some criticize that this resulted in a "laser-like" reaction and response and lacked a sense of true chaos (New Jersey Universities Consortium for Homeland Security Research 2006). Nevertheless, in the scenario, it was stated that a vehicle traveling through Union and Middlesex Counties released aerosolized pneumonic plague, and a few days later, hospitals were inundated with the sick. The primary test of this exercise was the operability of the Points of Dispensing (PODs) in three New Jersey counties: Middlesex, Monmouth, and Union. The principle purpose of a POD was to dispense medications, educational materials, and to support the "worried well" and those who may actually have been infected.

13.4 Points of focus to prevent, respond to, and recover from biological attack

Resources and time that can be focused on prevention, response, and recovery from a biological attack are not infinite. Through reflection on lessons learned from past incidents and exercises and consideration of the processes required to achieve a successful biological attack, the points of focus to deter a biological attack with the most "bang for our buck" can be elucidated.

Operational prevention, response, and recovery strategies should include both long-term and short-term goals. The long-term goals include the ultimate objective—international prevention of aggressive biological agent use. The short-term goals include the "stop-gap" measures to reduce vulnerabilities and make American society more resilient, while steps are taken toward the long-term goal. Both long-term and short-term goals require national resolve, planning, resources, and time to achieve their objectives.

There are multiple steps in the process to conduct a successful biological weapons attack. Figure 13.1 displays a notional and logical progression from the formation of a terrorist organization through the attack, including postevent effects or consequences. Between each step of this progression lies an arrow leading to the next step. At these arrows, efforts should be focused to prevent a terrorist from reaching the next step (Kaplan 2001). At a number of places in this progression, strategies to interrupt and prevent biological weapon acquisition and use by adversaries can be focused. Figure 13.1 also displays a sampling of suggested points of focus.

For instance, to manufacture or cultivate and weaponize a biological agent, a terrorist organization must seek subject matter experts who have the knowledge and ability to select not only an appropriate facility to conduct operations, but also to choose a country that will accept and turn a blind eye to, or be oblivious to, a covert biological weapon production activity. From the case studies, it is apparent that the Rajneesh cult found an effective subject matter expert who was able to successfully culture and disseminate salmonella. Conversely, the Aum Shinrikyo cult

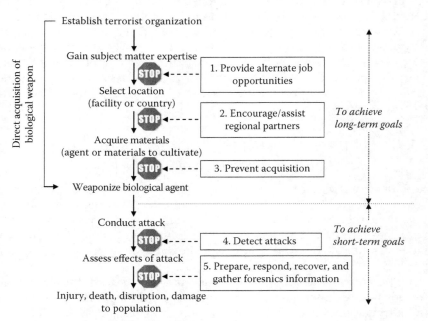

Figure 13.1 Notional progression of biological attack and points of focus.

was not able to attract an expert with the essential scientific training to achieve the ends. An effective U.S. prevention strategy might focus on preventing subject matter experts from being hired by the highest terrorist bidder by offering those experts alternate, more profitable, employment opportunities.

Next, this subject matter expert must know where and how to obtain the necessary supplies, materials, protective equipment, vaccines, antibiotics, and biological agents. In all of the case studies examined in this chapter, the principal biological agent engineer was able to easily acquire the equipment and supplies required. A preventive strategy might focus on identifying who is purchasing these materials and supplies, enact tighter international controls, and thereby help prevent terrorists from obtaining the capability to manufacture biological agents.

It is not enough to have an agent to have an effective attack. Most biological agents are sensitive to environmental conditions and mistreatment, as demonstrated by the failures of the Aum Shinrikyo cult. To ensure viability and adequate dispersal, attention to the weaponization of the biological agent must be considered by the terrorist-hired expert. Again, a counterterrorist strategy could target hiring these biological weapons experts for better uses while, at the same time, providing constant close supervision of all work done in national laboratories.

The arrow running along the left side of Figure 13.1 shows a "shortcut" to bypass the need for an expert and enable the terrorist organization to obtain a fully weaponized, ready-to-install, biological warfare agent. For example, Larry Harris obtained *Y. pestis* and the Rajneeshees obtained a few other pathogens from the ATCC by means of a simple fax and phone call. Regardless of whether a subject matter expert is employed through the entire weapons development continuum or if an agent is acquired directly, preventive measures can be developed to detect and/or prevent acquisition of these agents by terrorist groups.

If a terrorist group is able to successfully acquire and culture a biological weapon, and disperse it with effective viability, the consequences of such an event can be dramatically reduced by early detection and quick responses. For instance, if the Aum Shinrikyo dispersal of an agent near naval facilities in Japan had been detected, even though nonviable, this would have provided a wake-up call, and subsequent Aum Shinrikyo experimentation and potential attacks perhaps could have been prevented. In addition, if sensors had been placed in the Brentwood mail distribution facility and had detected low levels of anthrax spores in October 2001, the subsequent exposure to hundreds of people and the contamination of other buildings may have been prevented. Thus, early detection can mitigate or eliminate devastation due to an attack and may also assist emergency responders engaged in consequence management activities by letting them know early what biological agent they are dealing with.

After the attack, there are two concerns. First, responders like fire-fighters, police, and special response teams will be involved in the initial recovery and potentially long-term consequence-management activities. Second, the terrorist organization must assess whether goals for the attack have been achieved. In both concerns, U.S. activities to prepare, respond, and recover may act as a deterrent to terrorist organizations considering a biological weapons attack on the United States. If the attack is ineffective due to the superior preparedness and resilience of the people to withstand such an attack, then the terrorist's goals will not be met. Yet, in all of the case studies presented (Aum Shinrikyo, Rajneesh, and Larry Wayne Harris) and the TOPOFF 1, 2, and 3 exercises, planning, response, and recovery lacked the robustness needed to be an effective prevention strategy.

Finally, effective forensic techniques can be especially beneficial if they could rapidly determine the perpetrator of an attack (attribution). Again, the recent case studies of biological incidents, especially the anthrax letters attacks, demonstrate the current absence and need for the development of adequate forensic techniques and capabilities concerning biological agents.

The suggested points of focus include methods to achieve both long-term and short-term goals. The first three points of focus on Figure 13.1 are activities that should be accomplished at the national level through international agreements and global strategies. Action in these three points of focus will move us toward achieving our long-term biological weapon prevention strategies. The second two points of focus must be guided by national strategy, but include actions the states and local communities must undertake to achieve the short-term goals.

13.5   Evaluation of U.S. policy

Since the attacks at the World Trade Center in 1993, Khobar Towers in 1996, U.S Embassies in East Africa in 1998, the USS Cole in 2000, and the Pentagon and World Trade Centers in 2001, the United States has progressively reshaped national security and national military strategies. The most current *National Security Strategy* (NSS) was published in March 2006. Supporting this strategy are the *National Strategy for Homeland Security* (October 2007), *National Strategy to Combat Weapons of Mass Destruction* (U.S. Federal Government 2002), *National Strategy for Combating Terrorism* (September 2006), and the *National Defense Strategy* (June 2008).

The *National Defense Strategy* (June 2008) describes the military support to accomplish the plans set forth in the various national strategies. The military, too, has developed strategies to complement the overarching defense strategy of 2008. The *National Military Strategy* (2004), the *National Military Strategy to Combat Weapons of Mass Destruction* (February 13, 2006), and the *National Military Strategic Plan for the War on Terrorism*

(February 1, 2006) detail the methods the military will use to enable the national security deterrence strategy.

Specifically related to bioterrorism, President George W. Bush signed the *Biodefense for the 21st Century* Presidential Directive (Homeland Security Presidential Directive [HSPD]-10) in 2004 (U.S. Department of Homeland Security 2002). "President Bush has made strengthening the nation's defenses against biological weapons a critical national priority from the outset of the administration—investing over $10 billion since 2001. While significant progress has been made to protect America, President Bush instructed federal departments and agencies to review their efforts and find new and better ways to secure America from bio-attacks" (U.S. Department of Homeland Security 2004a). Most recently, President Barack Obama published the *National Strategy for Countering Biological Threats* (November 2009), reaffirming the concern the government still holds toward this threat. Each of these strategies provides a strategic plan for the United States to protect and defend against the nation's enemies and attacks they may perpetrate against it.

13.5.1 Provide alternate job opportunities

"Anecdotal reports persist of former Soviet scientists, especially those in Central Asia and the Caucasus, being approached by officials from pro-liferant states. Further, a 2003 survey of Russian scientists with weapons expertise found that 20% of respondents would consider working in North Korea, Syria, Iran or Iraq for a year or more" (U.S. Department of State 2002a). In each of the four incidents described earlier, the lead subject matter expert was critical in the development of an attack plan, determining the biological agent to be used, obtaining the necessary supplies, and manufacturing the organism for dissemination. If the subject matter experts were enticed to take peaceful positions of employment, it would make it more difficult for terrorists to use biological agents as weapons.

In 2006, the U.S. Department of State and the U.S. Agency for International Development published their fiscal year summary, which addressed the challenge of redirecting former weapons of mass destruction (WMD) scientists to more peaceful programs. The report highlighted activities in the Office of Cooperative Threat Reduction of the U.S. Department of State regarding Iraqi and Libyan scientists through what is currently called the "Iraq Scientist Engagement Program" and the "Libya Scientist Engagement Program" (U.S. Department of State and U.S. Agency for International Development 2005). These programs enable the redirection of former WMD scientists to civilian activities through the enhancement of scientific and economic development (U.S. Department of State 2002b; U.S. Department of State and U.S. Agency for International Development 2005).

Currently, the Nonproliferation of WMD Expertise, which consists of the Science Centers program, the Bio-Chem Redirection program, and the Bio Industry Initiative, aims to redirect former WMD scientists to reduce or eliminate this aspect of the bioterrorist continuum. The Science Centers program supports two international science and technology centers (ISTCs): the ISTC in Moscow and the Science and Technology Center in Ukraine (Science and Technology Center of Ukraine 2010). Both centers provide peaceful and sustainable employment opportunities to Russian and Commonwealth of Independent States scientists possessing WMD knowledge and skills.

The *National Strategy for Countering Biological Threats* recognizes the efforts to redirect former WMD scientists through cooperative international partnerships, reaffirming the level of importance at which the current administration holds these programs. However, to attempt to control or hire all of the individuals with appropriate knowledge to conduct a biological weapons attack would be impossible. It is unclear whether the United States really knows how many scientists and related technical people from former Iraqi WMD programs it should be concerned about; much more nebulous are the numbers of WMD-related scientists from Russia and the other former Soviet states (Roston 2004). Thus, although this is an attractive means to halt or prevent an attack, it is not the most likely or efficient means.

13.5.2   Encourage/assist regional partners

The 2010 *NSS* states that "the United States of America will continue to underwrite global security—through our commitments to allies, partners, and institutions … and our determination to … prevent the proliferation of the world's most dangerous weapons" (U.S. Federal Government 2010). According to unclassified sources, approximately 20 countries have the capability to manufacture biological weapons, including North Korea, China, Russia, Iran, Syria, Libya, India, Pakistan, and the United States (Central intelligence Agency 2001). Because most equipment, technology, and materials for biological agent production are dual use, for peace and war purposes, it is difficult to distinguish between offensive weapons research and development and more peaceful intentions. Biological weapon proliferation is a global issue. Thus, it is vital for the United States to engage with the international community to prevent terrorist activities, including those that involve biological agents.

A keystone to this involvement centers on the *Biological and Toxin Weapons Convention*. Engagement in this convention is highlighted in the 2002 *National Strategy to Combat WMD* (Bush 2002). The 1972 *Convention on the Prohibition of the Development, Production and Stockpiling of Bacteriological (Biological) and Toxin Weapons and on their Destruction*, also known as the *Biological and Toxin Weapons Convention*, went into effect on March 1975. It included 103 cosigning nations, and 140 nations have now signed and ratified the convention. However, it contains no provisions for verification or enforcement (Siddell 1997). The United States has many programs countering proliferation of WMD, in general. However, countering the proliferation of biological warfare agents is problematic.

The 2002 *National Strategy for Combating WMD* also highlights the need to enhance traditional nonproliferation measures, such as diplomacy, arms control, multilateral agreements, threat reduction assistance, and export controls to reduce the threat of attack with WMD (Bush 2002). In addition, the 2006 *National Military Strategy for Combating WMD* details "Strategic Enablers," including deterrence, intelligence, partnership capacity, and strategic communication support, which enhance the effectiveness of military capabilities for combating WMD (U.S. Department of Defense 2006). "The United States alone cannot eliminate this threat, nor can any other single institution or sector. Defeating the threat will take a concerted, collaborative and integrated international approach involving allied governments; law enforcement; the military; and the academic, medical, and scientific communities" (U.S. Department of Defense 2006).

Encouraging or assisting regional partners may help to deter adversary biological agent production. However, it will not completely solve problems presented by the terrorist use of biological agents. Although it is a challenge to entice a country to give up or never obtain a biological weapon capability, it is even more difficult to identify and prevent a terrorist group from developing the same capability on a smaller scale. The Rajneesh cult, Larry Harris, and likely the perpetrator of the anthrax letters did not use U.S. regional partners to develop their biological weapons. Thus, a focus on regional partners and engagement to develop stronger nonproliferation relationships is an important deterrent in a strategic policy, but may not be the most effective at deterring a small-scale terrorist use of biological agents in the United States. However, it is possible to prevent some terrorist groups from acquiring materials and expertise to develop biological weapons through international programs to eliminate national WMD expertise, materials, and arms.

13.5.3   Prevent acquisition

After Larry Harris's arrest, the CDC took measures to safeguard 24 infectious agents and 12 toxins that pose a significant risk to human health. Prior to this incident, the Rajneesh cult obtained the agents on which it experimented, and Iraq obtained some of its lethal strains of anthrax, tularemia, and Venezuelan equine encephalitis from the ATCC. Shippers and receivers of these identified agents must now register with the CDC (Stern 2001). The safeguarding measures, known as the Select Agent Program, became law on June 12, 2002, when President Bush signed the Public Health Security and Bioterrorism Preparedness Response Act of 2002

(Select Agent Regulations 2011). The Select Agent Program requires registration of facilities including government agencies, universities, research institutions, and commercial entities that possess biological agents or toxins deemed a threat to public, animal, or plant health (Centers for Disease Control and Prevention 2008). Although this program makes agents more difficult to obtain, some may still seek them and risk more severe penalties than in the past.

When Larry Harris was arrested, the most severe punishment for his possession and expressed desire to use a biological agent as a weapon was wire fraud. The USA PATRIOT Act of 2001 was the first law to put restrictions on persons who possess select agents and provides criminal penalties for possessing such agents not justified for peaceful purposes. In addition, violation of the Public Health Security and Bioterrorism Preparedness Response Act can result in civil fines of $250,000 for individual or $500,000 for an entity and imprisonment of up to 5 years.

Further legal actions are addressed in the 2007 *National Strategy for Homeland Security*, the Intelligence Reform and Terrorism Prevention Act of 2004 and the Protect America Act of 2007 that promote security and implement portions of both the 9/11 Commission and the WMD Commission recommendations (Homeland Security Council 2007). Finally, the Military Commissions Act of 2006 allows captured terrorists to be tried for war crimes. The more severe penalties may prevent an individual or group from considering these agents as a weapon.

The Select Agent Program has taken the necessary steps to eliminate the ease with which the Rajneesh cult, Iraq, and Larry Harris obtained their biological agents, and the USA PATRIOT Act, and subsequent acts, have strengthened punishment against those who attempt to obtain biological materials for adverse acts. These two types of policy must be adopted by other countries to prevent and deter easy acquisition worldwide. However, if a terrorist group intends to use biological agents, many can be easily obtained and cultivated from natural sources. The growth and development of biological agents require specialized equipment and supplies and thus provide another indicator of an active biological program.

The Aum Shinrikyo cult used a U.S. company to obtain and ship equipment and supplies. It was also assumed the cult had a chemical program based on the numbers of atropine injectors the cult had ordered. An integrated U.S. and international intelligence network that gathers data and tracks certain equipment and supply purchases may provide an early indication of the intentions of a terrorist group or cult. For instance, if a group or individual purchases a large amount of antibiotics or vaccines, intelligence efforts should have the capability to easily track and report these transactions for further examination.

The nation's nerve center for information sharing and domestic incident management is the Homeland Security Operations Center (HSOC).

The HSOC collects and fuses information to deter, detect, and prevent terrorist acts. The HSOC is divided into two halves: Intelligence and Law Enforcement. "The 'Intelligence Side' focuses on pieces of highly classified intelligence and how the information contributes to the current threat picture for any given area. The 'Law Enforcement Side' is dedicated to tracking the different enforcement activities across the country that may have a terrorist nexus" (U.S. Department of Homeland Security 2004b). The HSOC provides real-time situational awareness and monitoring, coordinates incident-response activities, and issues advisories and bulletins concerning threats to the United States (U.S. Department of Homeland Security 2004b).

Preventing acquisition and integrated intelligence are important foci of deterrence. Although some critical steps have been taken in the United States to prevent acquisition of agents, more action should be taken worldwide. In addition, more focus and funding should be provided to the intelligence agencies to enable them to better collect, integrate, and interpret information to obtain a clear picture of biological weapon development and the intent of terrorist groups and others.

13.5.4   Detect attacks

Detection of a bioterrorist attack should occur at the strategic and tactical levels. At the strategic level, detection of a terrorist organization's plan and developing capability to use biological agents is a key to prevent such an event. At the tactical level, detection of the agent release is vital to rapid and effective response and consequence management activities.

On the strategic level, Argonne National Laboratory (U.S. Department of Energy) has developed two computer-based capabilities to aid in the identification of terrorist organizations and prediction of future actions. The *Joint Threat Anticipation Center (JTAC)* anticipates long-term threats to U.S. national security by integrating social science and technology. *JTAC* conducts research in areas of terrorist strategy and tactics, failed states, sociocultural process and precursors to terrorism, and language studies (U.S. Department of Energy).

Complementary to the *JTAC* is *NetBreaker*. *NetBreaker* uses dynamic social network analysis and agent-based modeling with social network formation rules to find and model terrorist networks. *Netbreaker*'s simulation determines what a terrorist group can do, how it interacts, and the probable threats from its network (U.S. Department of Energy).

On the tactical level, detection of the release of an agent is critical to a rapid response to negate the harm done by a bioterrorist's action. Aum Shinrikyo attempted biological attacks on Tokyo a number of times, yet these went undetected. If only one of these unsuccessful attacks had been detected, then appropriate antibiotics could have been administered (although in this case, they were not necessary) and actions could have been taken to prevent a future successful attack. HSPD-21, *Public Health and Medical Preparedness*, states, "The United States must develop a nationwide, robust and integrated biosurveillance capability, with connections to international disease surveillance systems, in order to provide early warning and ongoing characterization of disease outbreaks in near real-time" (Bush, Management of Domestic Incidents 2003). To accomplish this edict, BioWatch is a DHS program, assisted by CDC and the Environmental Protection Agency (EPA), which performs 24 × 7 environmental surveillance using the existing EPA and Department of Energy air quality monitoring systems. Air samples are tested as an early warning indicator of biological attacks (Centers for Disease Control and Prevention, Agency for Toxic Substances and Disease Registry, Department of Health and Human Services 2004). The BioWatch system has been successfully operating in more than 30 urban centers since early 2003 (U.S. Department of Homeland Security 2004).

The *Biological Warning and Incident Characterization* (*BWIC*) is a support system integral to the BioWatch program. *BWIC* integrates a number of diverse computer modeling programs and provides a common view of an event to emergency responders and critical agencies involved. *BWIC* includes geographic information system (GIS) maps, air dispersion models, population information, epidemiological tools, subway and facility models, and links to public health surveillance information.

The BioWatch system is an example of "detect to treat" defensive system. People have already been exposed to an agent when the system provides an alarm and triggers a response. A better capability is a "detect to warn" alarm system that would enable people to take shelter to avoid exposure. Although many organizations conduct research in the area of biological detection, more attention should be focused on an accurate and rapid detection. If an enemy cannot infect the desired target, there is no point to a biological attack—an effective deterrence-by-denial system.

However, if an enemy is successful in releasing an agent, enhanced situational awareness may be improved by another system under development at Argonne National Laboratory. The *Integrated Chemical, Biological, Radiological, Nuclear and Explosives* (*ICBRNE*) Detection System is supported by the DHS's Science and Technology Directorate. This integrates CBRNE sensor systems and supports regional capabilities to enable information sharing, enhance CBRNE detection, and improve situational awareness. The first phase of the program extends coverage to Chicago, Seattle, New York, Los Angeles, and Boston (Cirillo 2009).

Some bioterrorist events will not be detected through capabilities such as BioWatch, *BWIC*, and *ICBRNE*, but instead may be discovered through epidemiological surveillance activities following people's exposure to a biological agent. To this end, the U.S. military utilizes the Electronic System for the Early Notification of Community-based Epidemics (ESSENCE). ESSENCE is a near-real-time global monitoring system to detect infectious disease outbreaks. ESSENCE monitors outpatient and pharmacy data on over 9.2 million military beneficiaries and reports alerts to local and state public health officials, the World Health Organization (WHO), and the CDC BioSense System.

Also in accordance with HSPD-21, *Public Health and Medical Preparedness*, the CDC BioSense System was established as an epidemiologic surveillance system for human health. This system allows for two-way information flow among federal, state, and local government public and clinical healthcare providers (Bush 2003a). The CDC BioSense System tracks patients' health complaints and symptoms to identify trends that may indicate an increase in disease rates, indicating a bioterrorist event. Emergency response personnel can then use the BioSense System to detect, track, and respond more rapidly to disease outbreaks and enhance emergency response and consequent management activities (Centers for Disease Control and Prevention 2012). However, this BioSense System has not yet been tested in a National Exercise to determine whether it is capable of rapidly providing necessary information to decision makers. In addition, no common database exists that can be shared within and across states to aid in response actions, drug distribution, and public awareness.

Nevertheless, detection of an agent release by healthcare workers can be improved *now* with a few simple actions. Healthcare workers and first responders should be immediately trained on the epidemiological clues that indicate a bioterrorist event is unfolding or has occurred. From the Anthrax Letters and Rajneesh incidents, there were clear signs, epidemiological clues, which healthcare workers should have recognized to raise their suspicion of a potential attack. Below is a list of events that give strong indication of a bioterrorist event. This list was taken from selected aspects of the American College of Physicians—American Society of Internal Medicine (ACP/ASIM) Guide to Bioterrorism Identification (American College of Physicians 2002), Public Health Reports' Epidemiologic Clues to Bioterrorism (Treadwell et al. 2003), Chapter 3 of Epidemiology of Biowarfare and Bioterrorism (Dembek et al. 2007), and the California Hospital Bioterrorism Response Planning Guide. Parenthesis indicates the National Exercise or real-world event that demonstrates the usefulness of that epidemiological clue.

1. Large epidemic, with large numbers of casualties (*all National Exercises and Rajneesh*)
2. Large epidemic in a discrete population, or point source outbreak (*all TOPOFF exercises*), or discrete population like people who ate at a specific restaurant (*Rajneesh*) or who live, work, or recreate in a common geographical area

3. Single case of disease caused by an uncommon agent without adequate epidemiologic explanation (*anthrax letters, first victim*)
4. Multiple epidemics (*Aum Shinrikyo if they had been successful*)
5. A disease that is outside its normal transmission season or is impossible to transmit naturally in the absence of its normal vector (*anthrax letters, first victim; plague in TOPOFF exercises*)
6. Uncommon disease, such as anthrax, pneumonic plague, or smallpox (*all events and National Exercises*)
7. A rapid increase in the number of previously healthy persons with similar symptoms (*Rajneesh, anthrax letters, and all National Exercises*)
8. Direct evidence such as when a terrorist group announces an attack has occurred (*Rajneesh, Harris*)

Healthcare workers should be provided with examples of what these epidemiological clues would look like for each of the major bioterrorist agent threats we should expect. For instance, use of salmonella, anthrax, and plague appeared in both the National Exercise scenarios and in actual incidents.

13.5.5   *Prepare, respond, recover, and gather forensic information*

We may deter or dissuade terrorists from using biological agents if convinced they cannot achieve their goals. Thus, robust preparations to ensure a rapid and focused response may prevent any future need to use such a capability. With this in mind, the DHS was tasked by both the 2006 *National Strategy to Combating WMD* and HSPD-5 to develop a plan to prepare, respond, and recover from a WMD attack (Figure 13.2). Specifically, HSPD-5 directs the former Homeland Security Council to develop and administer a NRP. "This plan shall integrate Federal Government domestic prevention, preparedness, response, and recovery plans into one all-discipline, all-hazards plan" (Bush 2003a).

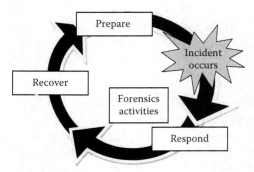

Figure 13.2 Phases of effective response actions.

The NRP was published in 2004 and revised in 2006. In January 2008, the DHS published the *National Response Framework* (NRF) that supersedes the corresponding sections of the previously published NRPs (U.S. Department of Homeland Security 2009). Chapter 3 of the NRF details specific response actions to respond to incidents, which includes three phases of effective response: prepare, respond, and recover.

13.5.5.1 *Prepare*
The NRF describes the following six essential activities encompassing preparedness: plan, organize, train, equip, exercise, evaluate, and improve (U.S. Department of Homeland Security 2009). These activities are displayed in the "Preparedness Cycle," Figure 13.3.

13.5.5.2 *Plan*
The review of the TOPOFF exercises commented that inadequate plans existed regarding distribution of limited medications. In TOPOFF 3, Points of Distribution (PODs) were established; however, there were incomplete logistical or manpower plans for distribution of the medication. Planning "includes the collection and analysis of intelligence and information, as well as the development of policies, plans, procedures, mutual aid and assistance agreements, strategies, and other arrangements to perform missions and tasks. Planning also improves effectiveness by clearly defining required capabilities, shortening the time required to gain control of an incident and facilitating the rapid exchange of information about a situation" (U.S. Department of Homeland Security 2009). The Integrated Planning System is the national system used to develop interagency and governmental plans. However, local emergency personnel must engage and develop plans that incorporate specific capabilities and resources to respond to a biological incident.

Figure 13.3 The preparedness cycle. (U.S. Department of Homeland Security, *National Response Framework*, January 2009. Retrieved from http://www.fema .gov/pdf/emergency/nrf/nrf-core.pdf.)

Plans and the action of planning can be extremely complex under-takings. Thus, Argonne National Laboratory has developed a number of computer-based products to assist with the planning process. The first is the *Synchronization Matrix Planning Process*, a systems-based and problem-solving approach to emergency planning used to integrate emergency response plans across jurisdictions (U.S. Department of Energy 2007). The *Matrix* is an interactive planning tool that allows the user to see a broad view of a response, visualizing the interactions that occur as the event progresses over time. The *Matrix* allows emergency managers to plan and practice interactions among agencies resulting in a more effective and coordinated emergency response.

The second Argonne program to aid in the planning process is the *Special Population Planner* or *SPP*. The *SPP* uses GIS-based software to aid in mapping communities, facilities, and households where special-needs populations reside. Then the registry can be integrated in emergency response planning models to facilitate assistance of special needs individuals. The *SPP* is in use in six Alabama counties, enhancing emergency preparedness for 4500 persons with special needs.

Planners should also consider the effects on the environment and potential alternate routes of biological agent distribution, such as drinking water distribution systems. Argonne's *Threat Ensemble Vulnerability Assessment (TEVA)* simulates threats to drinking-water distribution infrastructure to analyze vulnerabilities, measure public health impacts, and aid in the design of threat mitigation and emergency response strategies (U.S. Department of Energy 2008b).

Finally, Argonne's *Fort Future* is a virtual installation that provides information to aid in the analysis of deployment requirements and the impact of disruptive events, such as a biological attack (U.S. Department of Energy 2008a). *Fort Future* integrates DTRA's Hazard Prediction and Analysis Capability models to determine the contamination levels infiltrating buildings, thus enabling informed decision making by command-ers. *Fort Future* was developed for military applications; however, it could be modified to model a local town or city to aid emergency managers in planning.

13.5.5.3 *Organize*

"Organizing to execute response activities includes developing an overall organizational structure, strengthening leadership at each level, and assembling well-qualified teams of paid and volunteer staff for essential response and recovery tasks. The NIMS provides standard command and management structures that apply to response. This common system enables responders from different jurisdictions and disciplines to work together to respond to incidents" (U.S. Department of Homeland Security 2008). The concept of organizing is especially important when a prolonged situation causes healthcare workers and first responders to be overextended. By considering the expected organization in a plan, potential shortages can be determined and plans modified accordingly.

13.5.5.4   Train

In 1995, when public health officials and fire department personnel responded to Larry Harris's *Y. pestis*, few had trained or been educated regarding any biological agents. "Building essential response capabilities nationwide requires a systematic program to train individual teams and organizations – to include governmental, nongovernmental, private-sector, and voluntary organizations – to meet a common baseline of performance and certification standards" (U.S. Department of Homeland Security 2009). Since September 11, 2001, almost $3 billion in federal bioterrorism preparedness funding has been funneled to states (Harrell 2004). In 2002, the Department of Health and Human Services announced more than $1 billion in federal bioterrorism preparedness grants (Harrell 2004). These grants and funding opportunities, in combination with local and National Exercises, have provided increased training and improved the skills of first responders and other hospital personnel.

13.5.5.5   Equip

After the covert attack conducted by Rajneesh cult members, Oregon hospitals were overwhelmed with sick and frightened people. A future terrorist may be even more successful in generating a mass casualty attack that taxes emergency responders, hospitals, and stockpiles of vaccines and antibiotics. "Effective preparedness requires jurisdictions to identify and have strategies to obtain and deploy major equipment, supplies, facilities and systems in sufficient quantities to perform assigned missions and tasks" (U.S. Department of Homeland Security 2009).

In 2004, President Bush signed legislation called Project BioShield as a new bioterrorism countermeasure. BioShield committed $5.593 billion over 10 years (Garamone 2004). It is a comprehensive effort to develop, stockpile, and make available drugs and vaccines to protect against biological and chemical weapons attacks (Gottron 2003). In short, BioShield provides incentives to pharmaceutical makers and biotechnology companies for the development of medicines and vaccines to treat people exposed to bioterrorist agents. It should strengthen research and development and enhance the ability to counter bioterrorism.

In addition to vaccines and medical countermeasures, first responders also need protective gear, detection equipment, and decontamination equipment in a quantity extensive enough for a predicted attack. These equipment requirements are called for in the 2007 *National Strategy for Homeland Security* (Homeland Security Council 2007). In addition,

HSPD-8, *National Preparedness*, requires caches of equipment be maintained at such levels to meet the national preparedness goal (Bush 2003b). Although it is of great benefit to possess these resources, they are of little use if personnel are not trained and experienced in their use. Exercises offer excellent opportunities to gain proficiency with novel biohazard response equipment.

13.5.5.6   Exercise

The 2007 *National Strategy for Homeland Security* identified the establishment of the National Exercise Program to increase preparedness to respond to the consequences of terrorist attacks (Homeland Security Council 2007). "Exercises provide opportunities to test plans and improve proficiency in a risk-free environment. Exercises assess and validate proficiency levels. They also clarify and familiarize personnel with roles and responsibilities. Well-designed exercises improve interagency coordination and communications, highlight capability gaps and identify opportunities for improvement" (U.S. Department of Homeland Security 2009).

The National Exercises clearly show the capability gaps that exist, especially concerning the distribution of medical countermeasures. To assist Point of Dispensing operations in the future, the *Community Vaccination and Mass Dispensing Model* (*CVMDM*) developed by Argonne may prove useful in the execution of exercises to verify maturing plans. *EpiPOD* or *CVMDM* helps local public health agencies develop and test mass vaccination and prophylaxis dispensing plans. *EpiPOD* is configured for response to a pandemic influenza outbreak; however, it can be customized for other infectious diseases. *EpiPOD* is consistent with the NIMS and Incident Command System (NIMS/ICS) standards (U.S. Department of Energy).

13.5.5.7   Evaluate and improve

"Evaluation and continual process improvement are cornerstones of effective preparedness. On concluding an exercise, jurisdictions should evaluate performance against relevant capability objectives, identify deficits and institute corrective action plans. Improvement planning should develop specific recommendations for changes in practice, timelines for implementation and assignments for completion" (U.S. Department of Homeland Security 2009). This aspect of the preparedness cycle is lacking at the national level in the United States. Although there have been three National Exercises specifically dealing with biological scenarios, there is little publicly available literature to indicate the exercise reviews completed were incorporated in future planning. This shortfall is also clearly shown by the following consistently similar list of lessons learned from each of the National Exercises: lack of a treatment distribution plan, insufficient public communication, failure to adequately collect needed information, and workers overwhelmed by the deluge of patients and problems.

13.5.6   Respond

An adequate response involves the execution of emergency plans. An effective response should save lives, protect property, and protect the environment. The NRF states that "Four key actions typically occur in support of a response: (1) gain and maintain situational awareness; (2) activate and deploy key resources and capabilities; (3) effectively coordinate response actions; then, as the situation permits, (4) demobilize" (U.S. Department of Homeland Security 2009). These response actions are shown in Figure 13.4.

13.5.6.1   Situational awareness

Situational awareness is "Providing the right information at the right time" (U.S. Department of Homeland Security 2009). As stated previously, HSPD-21 directed the establishment of the CDC BioSense System as an epidemiologic surveillance system for human health. This is one of the many systems that enhance the common operating picture needed by officials directing response and recovery operations. However, as the National Exercises demonstrated, further development and training with the tools that enhance situational awareness is needed.

13.5.6.2   Deploy resources

In each real-world incident and National Exercise described earlier, a "first responder" began initial actions to save lives and prevent further damage. First responders are defined in HSPD-8, *National Preparedness*, as "those individuals who in the early stages of an incident are responsible for the protection and preservation of life, property, evidence and the environment ... as well as emergency management, public health, clinical

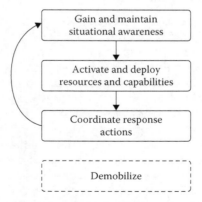

Figure 13.4 The response process. (U.S. Department of Homeland Security, *National Response Framework*, January 2009. Retrieved from http://www.fema .gov/pdf/emergency/nrf/nrf-core.pdf.)

care, public works, and other skilled support personnel (such as equipment operators) that provide immediate support services during prevention, response and recovery operations" (U.S. Department of Homeland Security 2009). During an incident response, first responders "assess the situation, identify and prioritize requirements, and activate available resources and capabilities to save lives, protect property and the environment, and meet basic human needs" (U.S. Department of Homeland Security 2009). These actions are critical to ensuring success in a bioterrorist response.

However, the National Exercises elucidated a shortfall in our hospital and POD staffing, which has also been noted in HSPD-21, *Public Health and Medical Preparedness*:

> *Mass Casualty Care*: The structure and operating principles of our day-to-day public health and medical systems cannot meet the needs created by a catastrophic health event. Collectively, our Nation must develop a disaster medical capability that can immediately re-orient and coordinate existing resources within all sectors to satisfy the needs of the population during a catastrophic health event. Mass casualty care response must be (1) rapid, (2) flexible, (3) scalable, (4) sustainable, (5) exhaustive (drawing on all national resources), (6) comprehensive (addressing needs from acute to chronic care and including mental health and special needs populations), (7) integrated and coordinated, and (8) appropriate (delivering the correct treatment in the most ethical manner with available capabilities) (U.S. Department of Homeland Security 2009).

To affect this flexible and scalable response, the military is tasked with a major role in support to civil authorities. The 2008 *National Defense Strategy* requires the military to maintain the capacity to support civil authorities in times of national emergencies, such as a large-scale bioterrorist event. "The Department will continue to maintain consequence management capabilities and plan for their use to support government agencies" (U.S. Department of Defense 2008).

An additional aid to bioterrorism response is the Strategic National Stockpile (SNS), formerly known as the National Pharmaceutical Stockpile, which provides a resupply of large quantities of essential medical materiel to states and communities during an emergency. The SNS is managed jointly by the DHS and Health and Human Services. The SNS is a repository of antibiotics, chemical antidotes, antitoxins, life-support medications, IVs, airway management supplies, and medical/surgical items. The first line of support from the SNS arrives within 12 hours as "Push Packages."

The Push Packages are strategically prepositioned in various locations in the United States to meet the 12-hour window (Centers for Disease Control and Prevention 2003). During TOPOFF 1 and 2, resources from the SNS were required. However, officials were unfamiliar with the administrative and logistical aspects of acquiring this support. Thus, deployment of these resources was delayed, reinforcing the need for further intense, hands-on training, experience, and coordination with this SNS resource.

13.5.6.3   Coordinate

Specific response actions will be based on the incident, agent, priorities, and resources available. The NRF describes coordination of emergency functions, actions, support, resources, and capabilities and information (U.S. Department of Homeland Security 2009, pp. 36–37). Coordination depends primarily on complete situational awareness. As discussed previously, the National Exercises reiterate the need for a common database, with accurate information to enhance the flow of information.

13.5.6.4   Recover

Recovery occurs, at times, simultaneous to response actions. "Once immediate lifesaving activities are complete, the focus shifts to assisting individuals, households, critical infrastructure and businesses in meeting basic needs and returning to self-sufficiency" (U.S. Department of Homeland Security 2009, p. 45). The National Exercises ceased prior to recovery-type actions. However, the anthrax letters incidents and subsequent decontamination activities speak to the cost in time and resources for years following the bioterrorist incident. In a future, wide-scale incident, issues such as burial of contaminated remains; decontamination of hospitals, homes, and public meeting places; and dealing with and mitigating economic losses will be only a few of the major recovery concerns to be addressed.

13.6   Gather forensic information

The anthrax letters incidents draw attention to the gaping hole in our forensic detection capability. If the perpetrator of the letter attack desired, he could have periodically sent other anthrax-contaminated letters in the mail to random recipients during the 7 years it took investigators to gather enough evidence to point a finger. Imagine the impact to the mail system as the death toll continued to rise over the 7 years. As noted in the 2009 *National Strategy for Countering Biological Threats*, "The primary objectives of any investigation into the alleged use, intended use or development of a biological weapon are to prevent casualties, protect the public health and attribute the activity to its perpetrator" (U.S. Federal Government 2006). The strategy calls for establishing a national-level research and development strategy for microbial forensics. In addition, the strategy reaffirms the need to maintain the National Biological Forensics Analysis Center as the lead federal facility for forensic analysis of biological material (U.S. Federal Government 2006, p. 17).

The National Biological Forensics Analysis Center is part of the National Biodefense Analysis and Countermeasure Center (NBACC) (U.S. Department of Homeland Security 2011a). NBACC is located on the new National Interagency Biodefense Campus at Fort Detrick, MD. The mission of the NBACC is to "support national security, law enforcement and medical communities by improving our understanding of potential bioterrorism pathogens that may be weaponized, transported and disseminated against U.S. targets for the purpose of improving our protection of human health and agriculture against biological terrorism, and sustaining homeland security through knowledge of the threat, prevention of surprise and attribution of use" (Korch 2004). The NBACC is expected to employ a staff of 150 with an annual operations budget of $50 million (U.S. Department of Homeland Security 2011a).

To develop tools to aid biological agent forensics, scientists at Argonne National Laboratory, with Loyola University, have developed the use of a proteomic biochip for identifying biomarkers or signatures indicative of specific growth conditions for *B. anthracis*. By growing anthrax under different conditions and observing changes in the protein and sugar content of the spore coat, a sort of fingerprint can be developed. This fingerprint, or detailed signature, may provide investigators of an attack clues to determine how an agent was produced, what equipment was employed, and the level of technical expertise of the operator. Thus, this information may aid investigators in a rapid identification of the perpetrator of an attack.

All of the programs and initiatives described here are a solid foundation toward building a credible prevention and response to a biological weapon attack. If the response to an attack is efficient and recovery happens quickly, then the attack may not have the consequences desired by the terrorists. If they realize this, they may choose not to strike with biological weapons. They may be deterred. In building a more resilient people, we will continue to improve our ability to defend against and deter such a biological attack.

References

Alibek, K., and Handelman, S. (1999). *Biohazard*. New York: Random House.

American College of Physicians. (2002). *ACP/ASIM Guide to Bioterrorism Identification*. Retrieved from www.acponline.org/clinical_information/resources/bioterrorism/bio_pocketguide.pdf.

Bush, G. W. (2002, December 11). National Strategy to Combat WMD. *Homeland Security Presidential Directive (HSPD)-4*. Washington, DC.

Bush, G. W. (2003a, February 28). Management of Domestic Incidents. *Homeland Security Presidential Directive (HSPD)-5*. Washington, DC.

Bush, G. W. (2003b, December 17). National Preparedness. *Homeland Security Presidential Directive (HSPD)-8*. Washington, DC.

Carus, W. S. (2001). The Rajneeshees (1984). In J. B. Tucker, *Toxic Terror: Assessing Terrorist Use of Chemical and biological Weapons* (pp. 115–158). Cambridge, MA: MIT Press.

Centers for Disease Control and Prevention. (n.d.). Retrieved from http://www .cdc.gov/biosense/files/BioSenseOverviewWeb_20090421.pdf.

Centers for Disease Control and Prevention. (2003, August 13). *Strategic National Stockpile*. Retrieved from www.bt.cdc.gov/stockpile/index.asp.

Centers for Disease Control and Prevention. (2008, December). *National Select Agent Registry*. Retrieved from www.cdc.gov/od/sap.

Centers for Disease Control and Prevention, Agency for Toxic Substances and Disease Registry, Department of Health and Human Services. (2004, March). *A National Public Health Strategy for Terrorism Preparedness and Response 2003–2008*.

Centers for Disease Control and Prevention. 2012. http://www.cdc.gov/osels/phsipo/ docs/pdf/factsheets/DNDHI_BioSense_12_232372_K_remediated_10_26_2012 .pdf Last accessed April 2013.

Central Intelligence Agency. (2001). The Worldwide Biological Warfare Weapons Threat. Washington, DC: Government Printing Office.

Cirillo, R. R. (2009, August). Director of the Center for Energy, Environment, and Economic Systems Analysis in the Department of information Sciences. *Presentation to Air Force Fellows*.

Council on Foreign Relations. (2006, May 3). *Backgrounders*. Retrieved from http:www.cfr.org/publication/9555/.

Dembek, Z. F., Pavlin, J. A., and Kortepeter, M. G. (2007). Epidemiology of Biowarfare and Bioterrorism. In M. K. Lenhart, D. E. Lounsbury, and J. W. Martin, *Medical Aspects of Biological Warfare* (pp. 39–68). Washington, DC: Office of The Surgeon General, Department of the Army, United States of America and U.S. Army Medical Department Center and School.

Department of Homeland Security. 2006. National Strategy for Combating Terrorism, September 2006, 2.; http://georgewbush-whitehouse.archives .gov/nsc/nsct/2006/. Last accessed April 2013.

Fester, U. (1997). *Silent Death*. Green Bay, Wisconsin: Festering Publications.

Garamone, J. (2004, July 23). *Bush Signs $5.6 Billion BioShield Legislation*. Retrieved from www.defenselink.mil/news/.

Gottron, F. (2003). *Project BioShield*. Washington, DC: Congressional Research Service Report for Congress.

Harrell, J. (2004, May 6). Hospitals Recive Bioterrorism Grants. *The Daily Reporter*.

Homeland Security Council. (2007). *National Strategy for Homeland Security*. Washington, DC: Department of Homeland Security, U.S. Federal Goverment.

Inglesby, T. V., Grossman, R., and O'Toole, T. (2001). A Plague on Your City: Observations from TOPOFF. *Clinical Infectious Diseases, 32*, 436–445.

Kaplan, D. E. (2001). Aum Shinrikyo. In J. B. Tucker, *Toxic Terror: Assessing Terrorist Use of Chemical and Biological Weapons* (pp. 207–226). Cambridge, MA: MIT Press.

Kaplan, D. E., and Marshall, A. (1996). *The Cult at the End of the World: The Terrifying Story of the Aum Doomsday Cult, from the Subways of Tokyo to the Nuclear Arsenals of Russia.* New York, New York: Crown.

Korch, G. (2004, February 9). Leading Edge of Biodefense. Jacksonville Naval Air Station: The National Biodefense Analysis and Countermeasures Center.

Lengel, A. (2005, September 16). Little Progress in FBI Probe of Anthrax Attacks. *The Washington Post.*

Mangold, T., and Goldberg, J. (1999). *Plague Wars.* New York: St. Martin's Press.

May, S. (2003, Fall/Winter). TOPOFF Exercise Offers Lessons for Preparedness. *Northwest Public Health*, 22–23. http://www.nwpublichealth.org/docs/nph/f2003/may_f2003.pdf. Last accessed April 2013.

Miller, J., Engelberg, S., and Broad, W. (2002). *Germ: Biological Weapons and America's Secret War.* New York, New York: Simon and Schuster.

New Jersey Universities Consortium for Homeland Security Research. (2006, May 18). *TOPOFF 3 Comments and Recommendations.* Retrieved from http://dimacs.rutgers.edu/People/Staff/froberts/topoffsubmission5-18-06.pdf.

Rodi, D. J. (2009). *Biological Warfare, Bioterrorism, and Dual Research.* Annual Refresher Training Course IBS1012009. Argonne National Laboratory.

Rosenau, W. (2001). Aum Shinrikyo's Biological Weapons Program: Why Did It Fail? *Studies in Conflict and Terrorism, 24*, 289–301.

Roston, M. (2004, April). Ridirection of WMD Scientists in Iraq and Libya: A status report. *RANSAC*, 1–9. http://carnegieendowment.org/pdf/npp/ransac_iraqlibya_scientists.pdf. Last accessed April 2013.

Science and Technology Center of Ukraine. (2010). Retrieved from http://www.stcu.int/.

Select Agent Regulations. (2011, December 22). *42 CFR 73.* Retrieved from http://www.selectagents.gov/.

Siddell, F. (1997). *Medical Aspects of Chemical and Biological Warfare.* Office of the Surgeon General. Washington, DC: TMM Publications.

Stern, J. (2001). Larry Wayne Harris (1998). In J. B. Tucker, *Toxic Terror: Assessing Terrorist Use of Chemical and Biological Weapons* (pp. 227–246). Cambridge, MA: MIT Press.

Treadwell, T. A., Koo, D., Kuker, K., and Khan, A. S. (2003). Epidemiologic Clues to Bioterrorism. *Public Health Reports, 118*, 92–98.

U.S. Department of Defense. (2006). *National Miltiary Strategy for Combating Weapons of Mass Destruction.* Washington, DC: Chairman of the Joint Chiefs of Staff.

U.S. Department of Defense. (2008). *National Defense Strategy.* Washington, DC: United States Federal Government.

U.S. Department of Energy. (2007). *Synchronization Matrix.* Retrieved from http://www.dis.anl.gov/publications/fact_sheets/SyncMatrix_General_FactSheet_2007May.pdf.

U.S. Department of Energy. (2008a, April). *Fort Future: Virtual Installation Supports Rapidly Changing Needs of Armed Forces in the Field.* Retrieved from http://www.dis.anl.gov/publications/fact_sheets/Fort_Future_FactSheet.pdf.

U.S. Department of Energy. (2008b, April). *New Simulator Helps Analyze, Mitigate Threats to Water Distribution Infrastructure.* Retrieved from http://www.dis.anl.gov/publications/fact_sheets/TEVA_FactSheet.pdf.

U.S. Department of Energy. (2008c). *Software Package Coordinates Response to Biological Threats*. Retrieved from http://www.dis.anl.gov/publications/fact_sheets/BWIC_FactSheet.pdf.

U.S. Department of Homeland Security. (2002, June 12). *Fact Sheet: President Bush Signs Biodefense for the 21st Century*. Retrieved from http://www.fas.org/irp/offdocs/nspd/biodef.html.

U.S. Department of Homeland Security. (2003). *Top Officials (TOPOFF) Exercise Series: TOPOFF2, After Action Summary Report*. Washington, DC: United States Federal Government.

U.S. Department of Homeland Security. (2004a). *Fact Sheet: A Better Prepared America: A Year in Review*. Retrieved from http://whitehouse.gov/news/releases/2004/05/print/20040525-4.html.

U.S. Department of Homeland Security. (2004b). *Fact Sheet: Homaland Security Operations Center (HSOC)*. Retrieved from http://www.dhs.gov/dhspublic.

U.S. Department of Homeland Security. (2008, January). *National Response Framework*. Retrieved from http://www.fema.gov/pdf/emergency/nrf/nrf-core.pdf.

U.S. Department of Homeland Security. (2011a, November 22). *National Biodefense Analysis and Countermeasure Center*. Retrieved from http://www.dhs.gov/files/labs/gc_1166211221830.shtm.

U.S. Department of Homeland Security. (2011b, November 22). *National Exercise Program*. Retrieved from http://www.dhs.gov/files/training/gc_1179350946764.shtm.

U.S. Department of State. (2002a, December). *Nonproliferation of WMD Expertise*. Retrieved from http://www.state.gov/t/isn/c12265.htm.

U.S. Department of State. (2002b, December). *Office of Cooperative Threat Reduction*. Retrieved from http://www.state.gov/t/isn/58381.

U.S. Department of State and U.S. Agency for International Development. (2005). *FY 2006 Performance Summary, Strategic Goal Number 4: Weapons of Mass Destruction*. U.S. Federal Government.

U.S. Federal Government. (2002). *National Strategy to Combat Weapons of Mass Destruction*. Washington, DC: U.S. Federal Government.

U.S. Federal Government. (2006). *National Strategy for Combating Terrorism*. Washington, DC: U.S. Federal Government.

U.S. Federal Government. (2010). *National Security Strategy of the United States of America*. Washington, DC: U.S. Federal Government.

chapter fourteen

Resilience to WMD

Communication and active participation are key

Michelle L. Spencer

Contents

14.1   Introduction

On Sunday afternoon, May 22, 2011, Joplin, Missouri, was hit by an EF-5 tornado as many residents were driving home from high school graduation ceremonies. The tornado, which cut a 13-mile-long swath of destruction, took 158 lives. As severe weather threatened much of southwest Missouri that evening, two women watching events on television less than 20 miles away searched the Internet for storm details and vital information. They were frustrated by the lack of real-time news. Deciding to remedy the situation, 23-year-old Genevieve Williams and her mother, Rebecca, created the "Joplin Tornado Info (JTI)" Facebook page using an iPhone. The page received its first post at 7:36 p.m., less than 2 hours after the tornado touched the ground. The JTI page, as it came to be known, went viral almost immediately, but remained organic, gaining numerous volunteer administrators to help monitor and post vital rescue and recovery information continuously. For nearly 2 months, the site was the primary information source for public and private resources including police, Joplin city officials, and utility company representatives—as well as relief organizations, media outlets, and donors. Almost a year later, JTI had more than 87 million post views from over 20 countries, and it has more than 47,000 followers 18 months after the event (Williams and Williams 2012).

The actions taken by the two women should come as no surprise. When catastrophic events occur, the first inclination is to want to actively participate in the response and recovery. But, often, the vastness of the resources can overwhelm the consumer. The Internet and social media have dramatically increased access to real-time information, which is vital during the lifecycle of a crisis. Several issues are raised by these advances in technology and the impact they have on individuals and communities in times of crisis. This chapter examines the evolution of information and communications through the lens of six case studies, which provide valuable lessons in public resilience. Several questions arise from these cases:

- How has the government's response affected citizens' reactions before, during, and following an event?
- Has the government integrated lessons learned into its policies and programs?
- How does public perception fuel reality?
- How did the government communicate with the public during the event?
- How can the government address the changing role of communications?
- How has social media affected the expectations and actions of society during and after disasters?

This chapter explores the concept of resilience by examining six incidents, which provide insights into the effects of a catastrophic weapon of mass destruction (WMD) attack on a resilient population. These cases were viewed from the perspective of community response to chemical, biological, radiological, and nuclear (CBRN)-related incidents. Actions taken by government agencies and the media during these incidents were considered in terms of their impact, both positive and negative, on affected populations. Although most of these incidents do not specifically involve WMD, they provide insight into how WMD might further complicate the response and recovery. Natural and man-made disasters provide fundamental lessons for resilience and necessary government policy and actions because of their frequency, and at times, their effects are analogous to WMD incidents. We may, therefore, study and extrapolate lessons learned from the numerous natural disasters and crises that occur. The first three cases of this chapter represent foreign cases and the last three domestic, providing a range of threats and their consequences. They include the following:

- 1995 Aum Shinrikyo attack on the Tokyo subway
- 2001 foot-and-mouth disease (FMD) outbreak in the United Kingdom (UK)
- 2003 severe acute respiratory syndrome (SARS) pandemic
- 2001 anthrax attacks
- 2005 Hurricane Katrina
- 2011 Joplin, Missouri tornado

Before delving into the individual cases, we must explore what is meant by resilience. One of the challenges regarding resilience lies in the difficulty of defining it or coming to a common understanding of its relevance to society. Despite this lack of clarity, the term "resilience" has gained currency in the post-9/11 world. As one recent report commissioned by the U.S. Government stated, "Although resilience with respect to hazards and disasters has been part of the research literature for decades … the term first gained currency among national governments in 2005 with the adoption of the Hyogo Framework for Action by the United Nations to ensure that reducing risks to disasters and building resilience to disasters became priorities for governments and local communities …" (National Research Council 2012). The U.S. Government defined resilience in 2011 Presidential Policy Directive 8 as "the ability to adapt to changing conditions and withstand and rapidly recover from disruption due to emergencies." Alternatively, the Community and Regional Resilience Initiative (CARRI) at the Oak Ridge National Laboratory defines resilience in terms of a community's "capability to prepare for, respond to, and recover from significant multihazard threats with minimum damage to public safety and health, the economy, and national security" (Wilbanks 2007). The CARRI after-action report on Hurricane Katrina further clarifies the concept by adding that "enhancing a community's resilience is to improve its capacity to *anticipate* significant multihazard threats, to *reduce* overall the community's vulnerability to hazard events, and to *respond* to and *recover* from significant hazard events when they occur" (CARRI 2008).

In addition to defining resilience, determining the level of resilience present in a community is equally challenging. Most communities are capable of "bouncing back" after a major disaster, such as the tornadoes of 2011 or the numerous floods in the Plains States or hurricanes suffered throughout the coastal United States for decades. Does this make them inherently resilient? Following a disaster, help comes from far and wide. But, given the economic challenges of the modern era, is it possible to assist local communities in being better equipped to help themselves before and after a catastrophic event? We may more accurately define resilience not as the ability to "bounce back," but rather as the ability to "bounce forward" (Manyena et al. 2011; Paton 2006)—integrating the four elements of *anticipation* of threats, *reduction* of vulnerability, and *responding* and *recovering* from disastrous events.

14.2   Measuring resilience

Both physical and psychological preparations are required to make populations—individually and collectively—more resilient. Therefore, resilience must be assessed at the individual and community level, with a consideration of the influences of national, state, and local actions. This begs the question: More than 10 years after 9/11, and 8 since Hurricane Katrina, are we really any better prepared for a catastrophic event? Many experts agree that the question of a WMD event in the United States is not whether, but when one will occur, yet it is questionable whether we have prepared our population for that eventuality. How then can the public be resilient to a threat they do not know exists?

One threat regularly faced by Americans is turbulent weather from floods, hurricanes, and tornadoes. According to the National Oceanic and Atmospheric Administration (NOAA), 2011 was an unusually active and deadly year for tornadoes across the United States, with a total of 1691 tornadoes recorded. Several tornado records were broken in 2011, including the greatest number of tornadoes in a single month (758 in April) and the greatest daily total (200, on April 27) (NOAA National Climatic Data Center 2011). Even though more than 600 people died from these events, a recent survey found that most Americans are *consciously* unprepared for tornadoes and similar disasters. A 2012 Persuadable Research study found that less than half of its survey respondents had "gathered the necessary items such as water, food, first aid, radios, or candles that could see them through, even if only for a few days" (PR Web 2012). The poll queried respondents on their preparedness for a disaster. Thirty-eight percent said they had never thought about it much, 48% stated that they did not have the money to prepare, and 15% claimed that they did not have time. One out of five respondents said they intend to "just wing it" if a disaster strikes.

When asked for their views on the government's response capabilities, 28% were unsure whether the government would be able to adequately respond to a major natural disaster in their area, while another 30% thought the government would be incapable. The numbers are even worse when people were asked about man-made disasters. In this case, 41% were unsure and 35% said the government would not be able to handle it. Even more telling is that 68% of those surveyed believe that the government is not doing enough to educate the public. Two-thirds had never even heard of the government website Ready.gov (PR Web 2012).

The results of this poll raise questions regarding the degree of preparedness and resilience that exists within the United States. In addition, given the scale and pace of societal change within the United States, with many different inputs vying for our attention, there is a real concern over how government can focus public attention to improve our individual and collective ability to be resilient. The six cases below provide insights and direction for future actions at the national, community, and individual level to improve public resilience.

14.3   Foreign cases

14.3.1   Sarin attacks on the Tokyo subway

In the spring of 1995, Japan experienced the world's first major terrorist attack using chemical weapons. On March 20, 1995, Aum Shinrikyo, a little known religious cult in Japan, killed 13 people in a sarin attack on the Tokyo subway system. The incident was the first lethal case of a non-state actor using a CBRN agent against a civilian population.

At 8:00 a.m. that fateful March day, during the busiest time of the Monday morning rush hour, sarin was released on the Tokyo subway. The nerve agent was carried into the metro in plastic bags wrapped in newspaper by five teams working in coordination. The teams placed their deadly packages on separate subway lines that converged at Kasumigaseki station, where the police headquarters, the seat of the Japanese government, and the largest fish market in the city are located. At the appointed time, the individuals boarded their respective trains, set packages on the floor and punctured the bundles with sharpened umbrella tips, releasing sarin onto the subway floor. The nerve agent sarin is most lethal when dispersed through the air, but the less efficient method was chosen at the last minute because of a lack of preparation time. Analysts believe that the goal of the attack was to divert police attention from cult activities and prevent a scheduled police raid. The timing was chosen to maximize police casualties during the early morning shift change at police headquarters.

Sarin is extremely potent, and even low concentrations can be fatal. The chosen delivery method, lack of purity of the agent, and the decision of some Aum operatives not to go through with the attack likely saved thousands of lives. Had the cult been able to take more time to prepare the sarin, it likely would have been far more potent. Many experts cite those reasons for the fact that there were only 13 deaths and 50 serious injuries from such a lethal weapon, in a confined space, which had the potential to do far greater harm.

14.3.1.1   The challenge of WMD

In less than 1 hour after the Tokyo subway attack, over 500 people sought care at St. Luke's Hospital, which is within walking distance of the Kasumigaseki station. As victims arrived at St. Luke's and other local hospitals, doctors tried to assess the symptoms and arrive at a diagnosis. At 9:30 a.m., a physician who had treated victims of a previous Aum attack recognized the symptoms while watching the television coverage. The doctor called St. Luke's to suggest that sarin nerve agent may have been used in the attack. This was the first confirmation medical professionals had received of a chemical attack—and, more specifically, that sarin was the agent used. Japanese public health officials were unable to confirm that sarin was the agent for more than 3 hours. Given the uncertainty of diagnosis and the lack of medical experience in treating sarin casualties, it is not surprising that over 100 staff members of St. Luke's, as well as the majority of emergency personnel who transported the injured, reported symptoms of exposure (Smithson 2000). There was no mechanism for the government to communicate with doctors or hospitals or for an expert within the medical community to reach out to his colleagues via mass communication means. Instead, one doctor called the affected hospitals and provided the information to another individual doctor who passed the information to his fellow medical responders.

14.3.1.2   Role of the media

The intensity of the media coverage during the events limited the ability of the government to use the news outlets to communicate with the public. Given the lack of direction from the government, sensationalism within the Japanese press and other media quickly overcame "need to know" reporting. A new word, "sarinoia," was born and heard repeatedly in media reports as the attack and its perpetrators were analyzed for months. The Japanese government noted its lack of direct communication with the public as a hindrance in the official investigations following the 1995 attacks.

14.3.1.3 Implications for resilience

One of the primary concerns following the subway attacks was the reality that terrorist groups could obtain weapons of mass destruction through both legal and illegal means. Aum Shinrikyo made many mistakes in the creation and formation of its weapon. Had the purity been higher and the delivery method more sophisticated, hundreds, if not thousands, could have died. With the Internet age comes much more accessible and detailed information about chemical and biological weapons. The medium also provides numerous avenues for communication between those with information and those seeking it. Addressing these issues requires different approaches to government-led planning and prevention programs, including understanding that today a government is more likely to face a biological or chemical threat that is significantly more virulent or capable than in the past. The "unthinkable" has become "doable."

It falls to the public health community and emergency planners to provide accurate information to the population, allowing those who are able to make better decisions about what types of symptoms require emergency care. This problem may be more severe in the United States, where millions of people depend on emergency rooms as their primary source of medical care and even more dramatic in less prosperous areas within the United States. With many people seeking care for nonemergent conditions on a daily basis, increased flow to emergency rooms during a CBRN event may be unmanageable. Specific guidance regarding symptoms and at-home care options may better enable possible victims to care for—or, at least, reassure—themselves, rather than reporting to an emergency room (Stone 2007).

Finally, given the 24-hour news cycle and prolific nature of information on the Internet, it is important to accurately inform the public in the event of a WMD attack. While the U.S. Government has countless outlets for information, ensuring easy public access to the best information the government has to offer is vital to an effective CBRN response.

14.3.2 UK foot-and-mouth epidemic

While we do not always expect a naturally occurring event to resemble a WMD incident, biological events, especially, provide relevance and highlight best practices for government-supported resilient communities. The 2001 foot-and-mouth outbreak in Great Britain is a useful example of a naturally occurring biological event that has ramifications for government actions and the involvement of the public in providing solutions.

FMD has a long history in Britain with numerous outbreaks of the disease. On February 19, 2001, pigs and sheep in Brentwood, Essex, were diagnosed with FMD (The Cabinet Office 2002). The most recent outbreak before 2001 happened in 1967 and cost the government approximately £27 million, primarily in compensation payments to farmers for destroyed livestock. By 1968, approximately 400,000 animals were slaughtered to bring the disease under control (BBC 2008). In 2001, over 4 million animals had to be slaughtered for the disease to be eradicated (Department for Environment, Food and Rural Affairs 2008).

The public's response to the foot-and-mouth outbreak varied depending on its position relative to the farming community. The outbreak of FMD had a much more devastating impact on rural communities than on urban areas. By 2001, FMD was believed to be a disease of the past. This feeling was shared by farmers and government entities. One reason was the length of time between outbreaks. In terms of resilience, there was no sense of mastery of this problem because the public had little experience with this type of crisis. Farmers could do little but try to limit the spread of disease on their own farms. Contamination of a farm could come from a bird, wild animal, or simply airborne viruses. Once the contamination occurred, the farmer's livelihood was at the mercy of government policies. The dearth of experience, planning, and control created feelings of frustration and despair in the farming community.

Communication between farmers and government officials was limited to formal announcements. The lack of two-way communication left the public with feelings of fear and outrage. Most did not understand the need for mass slaughter and pushed for vaccination. Much of this turmoil would have been avoided had the government explained its policy, objectives, and public benefit in a real and tangible way.

The 2001 FMD outbreak in the United Kingdom highlights important public resilience issues and problems faced by both citizens and government during the crisis. The lack of timely disease detection limited governmental efforts to contain the outbreak and, to many observers, led to overreaction when solutions were sought. Government planning and preparation did not anticipate the extent of the outbreak. Communication with the public lacked credibility because of widespread distrust of government motives, especially within the farming community. Disease eradication policies may have been more widely accepted had the government included farmers in the decision-making process and involved organizations that the farmers trusted, such as the National Farmers Union in Scotland.

14.3.2.1   Implications for resilience

While the 2001 FMD outbreak was an accidental or natural occurrence, it offers insight into issues that could follow a large-scale chemical or biological attack, including the ability to manage perception, reduce confusion, and utilize established relationships between government and trusted nongovernment organizations, which can act as liaisons with the public. The government must understand in advance who its natural allies are in a given situation, as well as how and when they should be engaged. Had the government engaged trusted, local civic and farming organizations, information could have been passed via known entities, rather than public proclamation. Early engagement and two-way communication will lead to ownership of problems and their solutions.

14.3.3   The SARS epidemic

In November 2002, Guangdong Province in China experienced a fast moving and highly contagious respiratory disease outbreak. By February 2003, this respiratory disease was given the name severe acute respiratory syndrome (SARS) by the World Health Organization (WHO) (Johnson 2003). SARS "is a viral respiratory illness caused by a corona virus, called SARS-associated corona virus (SARS-CoV)" (Centers for Disease Control and Prevention 2003). Its symptoms are similar to pneumonia, including high fever, dry cough, and shortness of breath or breathing problems (University of Toronto Joint Centre for Bioethics 2007), and it has a mortality rate of around 15%. Transmission occurs through contact with infected persons, and the mortality rate heightens as the illness progresses.

On February 1, 2003, an ambulance driver in Guangzhou transported a patient who was known to have an extremely contagious respiratory illness. While in contact with the patient, the driver used all necessary precautions, including triple layer surgical mask and gloves. After transporting the patient, he cleaned and disinfected the ambulance. Three days later, the ambulance driver became ill. This ambulance driver is thought to be the first of more than 1,700 health care workers to become infected with SARS. By June 2003, there were more than 8000 cases of SARS in 32 countries and 514 reported deaths (World Health Organization 2003).

14.3.3.1   Information flow

The outbreak of SARS in China highlights the challenges and benefits of social media during a crisis. In 2003, more than 60 million Chinese citizens had access to the Internet, while 55 million had cell phones with text messaging capability (Chiu 2003). As the outbreak worsened, public anxiety grew and rumors of infections and quarantines of entire cities flourished. According to the Hong Kong newspaper, *Wen Wei Po*, 2.1 million e-mails claimed that SARS cases in China had reached 10,000 (Chiu 2003), while the official government number was 44. As Chinese officials tried and failed to control information about SARS, both the mayor of Shanghai and the health minister were fired, and the government finally admitted to more than 396 cases and 67 deaths (Chiu 2003).

Although official information was sparse, citizens took it upon themselves to share information by sending text messages on their mobile phones, quickly spreading the news of disease outbreaks. In response, telecommunication companies began to offer innovative tools to assist their clients in preventing contagion. One Hong Kong telecommunications carrier introduced a short message service (SMS) tool that showed areas where infections had been confirmed to help citizens avoid those areas. The government quickly saw the capability as an asset and sent out an estimated 6 million SMS messages to quell a rumor Hong Kong that would be quarantined (Jardin 2003).

14.3.3.2   Implications for resilience

Use of push technologies, such as SMS, smart phone applications, e-mail, and OnStar capabilities is vital to sharing time-sensitive information with the public. These innovations are readily available and new technologies are announced daily. Technology is not, however, the silver bullet when it comes to the dissemination of timely and accurate information. However, the multitude of resources that exist within the public sphere—from cell phone and Internet capabilities to home security systems—now have the capacity to provide information and warning to the public (Market Watch 2012). Ready access to reliable information can provide real-time capability to prepare for threats.

14.4   Domestic events

14.4.1   2001 Anthrax attacks

In September 2001, the United States suffered the first major attack on its homeland since Pearl Harbor 60 years earlier. Following the terrorist attacks on September 11, 2001, fear of follow-on attacks, including bioterrorism, loomed on the horizon. While the nation anxiously awaited additional attacks, one man fell violently ill and died in Florida. Other cases followed in New York and later, Washington, DC.

On October 4, 2001, public health officials confirmed the first case of intentional infection by *Bacillus anthracis* in the United States. During the following 7 weeks, the Centers for Disease Control and Prevention (CDC) reported 10 confirmed cases of inhalational anthrax and 12 confirmed or suspected cases of cutaneous anthrax. The outbreaks were concentrated in six locations within the United States: Palm Beach County, FL; New York City; Newark, NJ; Capitol Hill in Washington, DC; the wider Washington, DC area; and Connecticut (U.S. Government Accountability Office 2003). By the end of November 2001, five people had died. The Federal Bureau of Investigation (FBI) stated four letters were found containing anthrax, while the Environmental Protection Agency (EPA) confirmed that over 60 sites had been contaminated with anthrax spores.

14.4.1.1   Public reaction

Public response to the anthrax attacks as a separate incident from those that occurred on September 11, 2001, is difficult to gauge. Only 1 week after the attacks, federal officials warned there were more terror cells ready to attack, and some of the 19 hijackers had shown an interest in crop dusting. The specter of bioterrorism elevated to a realistic threat to public health. Federal authorities cautioned public health facilities to be vigilant for "unusual disease patterns" (Weiss and Nakashima 2001). Health and Human Services (HHS) Secretary Tommy G. Thompson warned Americans to "be on the lookout for mysterious health symptoms." Three weeks before any sign of anthrax was reported, some members of the public responded to government warnings by purchasing the only Food and Drug Administration–approved drug for anthrax, Ciprofloxacin, rapidly exhausting existing stocks. In addition, other members sought vaccines from their health care providers and supplies of increasingly scarce Ciprofloxacin from the Internet.

Although having intentionally raised public awareness after events began to unfold, the government then tried to assure the public that the situation was under control. Initially, it was unclear if the first case was an attack. In their haste, officials gave reassuring statements without facts to support their assertions. The most serious example occurred on October 5, 2001, when HHS Secretary Thompson said the death of the first victim was not terrorism, but an "isolated case"—one that was most likely caused by natural events. By October 21, 2001, events had evolved to the point that President Bush called the attacks an "act of terror," stating that while no link to September 11th attacks had been established, clearly the incidents were intentional and, therefore, terrorism.

Americans were asked to be on alert, and many responded exuberantly. Aside from hoaxes, thousands of reports were called in by vigilant citizens who came across white powder in their daily lives. While it is an overstatement to call this panic, the level of attentiveness was so high that white powder was suspect wherever it was found—from doughnut shops, to pharmacies, to makeup counters, to print shops. Common sense did not always prevail over anxiety. As cases continued to show up in Florida, New York, Washington, and New Jersey, responding to guidance to be on alert and perhaps fearful of becoming victims, new and generally false reports began to overwhelm first responders across the country. The FBI received more than 2500 false reports of anthrax contamination. By October 11, 2001, Florida officials had received more than 1000 often-frantic calls. In Miami, police ran out of disposable protective suits after responding to 19 calls in one day (Kocieniewski 2001). In the Washington, DC, area, Prince George's County hazardous materials teams responded to as many as 50 calls a day, working around the clock for almost 2 weeks. First responder agencies scrambled to create protocols for threats. As one health official later commented, "a single anthrax-laden letter pushed the Washington area emergency response system to the limit." False alarms were raised in all 50 states and were not limited to major metropolitan areas. They occurred from Darien, CN, to Honolulu, HI, and Covington, KY, to San Francisco, CA.

14.4.1.2   Public perception and media coverage

As events unfolded during the fall of 2001, the media played a vital role in informing the public. However, this service cut both ways. While Americans gained vital details about conducting themselves in the face of daily threats, anxiety spiked after each new case was announced. There were many false reports, some of which were due to incorrect information, while others were based on false positives in testing for the bacteria. The level of coverage in print media was intense. Between October 4, 2001, the initial government confirmation of an anthrax case, and November 15, 2001, the *Washington Post* had 852 stories mentioning anthrax. By the end of 2001, the number had risen to over 1200. The public had a strong desire for information, a fact reflected in the level of media coverage. Vigilance was high, but clarity was not. This led individuals to take actions that made sense to them.

In many cases, access to some information prompted individual desire for more. At the height of the crisis, the U.S. Government, as well as local entities, struggled to respond. Most often, the public was directed to overburdened hotlines staffed by everyone from law enforcement officials to public health volunteers. On the Internet, the Federal Emergency Management Agency (FEMA), CDC, United States Postal Service, the Red Cross, and local entities such as Fairfax County, VA, tried to maintain accurate and up-to-date information. Although the effectiveness of the information is unclear, it is certain that better capabilities would have been required had the incidents continued beyond the 8-week duration.

During the anthrax incidents, most of the American public got their information from the media, rather than directly from government sources. Interestingly, 46% of the public felt the general media got the facts correct, a rise of 11% from early September and the highest grade for accuracy since 1992 (Kocieniewski 2001). Fifty-eight percent of Americans believed media coverage of the news on anthrax was accurate, and, when it was not, the public blamed "misleading information by the government" rather than sloppy reporting (Kocieniewski 2001).

14.4.1.3   Implications for resilience

Despite the events of October–November 2001, the vast majority of Americans went on with their lives as "normally" as possible, demonstrating that even in the face of a deadly, widely dispersed attack, citizens responded in a largely resilient manner by following the guidance provided and continuing to carry out their responsibilities. As one health official stated, "the closer people have been to anthrax, the more realistic and reasonable they are" (Fisher 2001; Stone 2007). This statement embodies the concept that a population will be more resilient the more prepared it is to deal with a threat. The examination of the anthrax attacks reveals that despite widespread potential threat, conflicting information from official sources, and inadequate flow of information, the public response was surprisingly unproblematic. Individuals, families, communities, and businesses continued to function.

14.4.2   Hurricane Katrina

Many lessons on community resilience can be learned from the events leading up to and following Hurricane Katrina. It seemed that everything that could possibly go wrong did. American society had not seen that level of social fraying since the 1918 influenza pandemic affected one quarter of the U.S. population. Along the Gulf Coast of the United States for a few days in August 2005, anarchy reigned, people died unnecessarily, and government officials spent much of their efforts blaming one another for all that had gone so horribly wrong.

On August 23, 2005, Katrina started in the Atlantic Ocean as a tropical depression, and by August 29, it had reached the Gulf Coast as a Category 5 hurricane. Katrina killed more than 1500 people and destroyed approximately 300,000 homes. This damage cost the U.S. Government between $125 and $150 billion, exceeding the total costs of Hurricane Andrew and the terror attacks on September 11, 2001, combined. As a result of the storm, 95% of the daily oil output from the Gulf of Mexico ceased, and almost 2 million people were without power (Committee on Homeland Security and Governmental Affairs 2006).

14.4.2.1   Planning and communication in advance of the storm

Although the National Weather Service can offer advice, only state and local officials have the power to order the evacuation of a population. As the storm approached the Gulf Coast, officials in Alabama, Mississippi, and Louisiana faced difficult decisions. Louisiana officials chose to stall, waiting to determine whether the severity of the storm would require total evacuation. Much of the able population departed voluntarily as early as August 27, while others prepared to weather the storm. In repeated television interviews, Louisiana officials declared that the storms were likely to overtop the levees and massive devastation could occur, but no mandatory evacuation order was given for New Orleans until 19 hours before the storm hit the city.

The delayed mandatory evacuation order had fatal consequences. First, the lack of clarity meant citizens doubted the veracity of statements when the mandatory evacuation order was finally given. Second, local and state governments made no provisions for those who had no means to leave the area. The city of New Orleans was aware that approximately 100,000 people would be unable to evacuate in the event of a major hurricane making landfall on the coast of Louisiana, yet no provisions were made to assist or protect them (Committee on Homeland Security and Governmental Affairs 2006).

Thus, as Hurricane Katrina descended on the city of New Orleans, the population turned to its leaders for help. When local leadership failed to give succinct directions, citizens headed for higher ground. As the storm subsided, 60,000 people arrived at the entrances to the Superdome, a "shelter of last resort," which had hastily stockpiled enough supplies for 15,000 people for 3 days. Another 20,000 convened at the Convention Center, a location that was never planned as a shelter and, therefore, had no supplies, governance, or the ability to acquire either.

Additionally, officials did not have contingency plans for a flooded New Orleans, even though they knew that it was virtually inevitable. There were no plans for evacuating the "shelters of last resort" hastily designated in the city, nor were vital resources protected or sent to higher ground so they could be retrieved as the flooding subsided. The lack of planning—and the lack of attention to existing plans—caused the greatest challenge: maintaining order in New Orleans.

14.4.2.2   Communication failure

Efforts by local, state, and federal agencies to coordinate actions during recovery were complicated by the destruction of the communication infrastructure and the lack of compatible equipment for responders. Local authorities undertook little to no advance planning for continuity of communications. After Katrina's landfall, the only method for Mayor Ray Nagin to communicate with the city's emergency managers was for him to walk across the street. In contrast, the efficiency and planning of commercial operations that were able to maintain power and communications offer a useful model for future incidents (Committee on Homeland Security and Governmental Affairs 2006). Since Katrina, FEMA has worked with private companies to better facilitate communication during crises, dedicating bandwidth and other assets to first responders and critically devastated areas. In June 2012, AT&T was the first private firm to receive disaster certification from DHS for its business continuity efforts (Homeland Security News Wire 2012).

14.4.2.3   Implications for resilience

There are many lessons to be learned from Hurricane Katrina, and the U.S. government is still assessing and dealing with the aftermath of Katrina more than 7 years after its disastrous impact and response. The lack of planning and proper execution has to be the greatest, overarching failure for all levels of government in Katrina. The dearth of leadership and the chaos that ensued forced thousands of individuals to make their own choices based on their personal priorities, not those of the broader citizenry. The communication failure throughout the crisis impeded all aspects of efficient response, wasting time and resources. However, numerous tools were borne from the hardships from Katrina's aftermath, many of which have made a dramatic difference in rescue and recovery efforts following earthquakes in Haiti and Japan, as well as flooding in Pakistan.

14.4.3   2011 Joplin tornado

As mentioned in the beginning of this chapter, in May 2011, Joplin, MO, was devastated by a tornado that killed 158 people and injured well over 1000 more. According to the NOAA, there were an estimated 7500 residential dwellings damaged by this storm, with 585 households needing temporary housing, 1308 pets displaced, and more than 500 businesses affected (NWS Central Region Service Assessment 2011).

The Joplin tornado is remarkable for several reasons. Most importantly, it is one of the first events in which responders had streamlined access to vital information. The response to the Joplin tornado was well orchestrated, partly because of the efforts of the JTI Facebook page. The page "became a clearinghouse for information on recovery, how to volunteer, where to donate supplies, media updates, and requests for information about loved ones" (Mazmanian 2012). It provided an amalgamation of information, which had not been seen previously. This remarkable social media page meant that everyone—from survivors to rescue workers to volunteers—knew where to go to get the information they required. The timely access to this information can be essential for saving lives in the aftermath of a disaster. There are many examples of past disasters that did not have the benefit of this tool. Following the 2011 Alabama tornado that inflicted significant damage on Tuscaloosa and Birmingham, volunteers turned out in droves to help in recovery and to clear debris in the days following the storm. Many were trained and registered volunteers. But even those who had registered and knew where they were supposed to go could not easily access the check-in locations, and they eventually took it upon themselves to organize teams to walk grids of destroyed neighborhoods and assist the injured or clear debris.

It is clear that the work the JTI Facebook page accomplished would likely have failed as a government resource. There are several reasons why. First, the speed of incoming information would have overwhelmed government resources. As Google executive Shona Brown testified before a Senate panel in May 2011, the Internet is a more reliable platform than traditional communication networks in times of crisis (Brown 2011). The capacity of the Internet and search engines such as Google could never be replicated by the government.

Many web resources, such as Google Person Finder and Ushahidi (an interactive mapping platform used to assess flooding in Pakistan, post-election violence in Kenya, and earthquake damage in Haiti), were created in the aftermath of Katrina and other crises and then fine-tuned in the wake of the 2010 earthquake in Haiti and 2011 tsunami in Japan. During the earthquakes in Japan and Haiti, Google experienced a spike in searches related to the events during and immediately following the disasters from both public and private sector individuals and organizations. Google responded by continually monitoring and updating the priority status of search terms minute by minute. Google also created Person Finder (PF). PF is an amalgamated database allowing people to either update their own status or search for a loved one across numerous sources that were previously unrelated (e.g., the Red Cross or the UN). PF was initiated in the months following Hurricane Katrina, but it was fully operational within 72 hours after the quake in Haiti and in just 90 minutes following Japan's earthquake. Rather than having to search multiple sites, an individual can use PF to locate a loved one or update his status, allaying the fears of family and friends and avoiding lengthy and costly searches by rescue personnel for people who are safe.

In the aftermath of the Joplin tornado, FEMA refused to post information on the JTI Facebook page. In an interview, JTI founder Genevieve Williams stated, "Some of the professionals in charge of Joplin relief efforts actually thought of Facebook as a place for rumors and disinformation, so [we] knew [we] had to work hard to earn a reputation for credibility. In the early days, everything [we] posted was initialed by the person who wrote it, so [we] could have full transparency and accountability" (Sifry 2012). Craig Fugate, the Administrator of FEMA, is aware of the problems his agency faces. He was not surprised that FEMA refused to collaborate with the JTI Facebook page. He said, "Trying to change cultures in any organization is a challenge. We're getting better" (Mazmanian 2012). Fugate admits that FEMA still considers social media to be a method of broadcasting information, and the agency is not yet fully prepared to utilize citizen-driven resources. In fact, for FEMA, along with most other government entities, social media is a function of public information offices, which have no role in acquiring or assessing information, only distributing it.

This may change if provisions now being considered in the U.S. House of Representatives become law. A Fiscal Year 2013 House bill would require FEMA to deploy social media and have the capability to integrate it into their response. According to Fugate, FEMA has begun this process. During the Joplin tornado, he and his staff monitored social media sites such as Twitter, Facebook, and YouTube to ascertain the severity of the damage. As he said, "That was the first really good information that I was able to see that really started to quantify how bad this was, well before any official reports or requests for assistance came through.... We're looking at things that people are hashtagging" (Mazmanian 2012). However, Fugate cautioned, the @FEMA Twitter handle is not "a 911 system where we can respond" (Mazmanian 2012). Yet, this is not understood by most Americans. A 2010 American Red Cross survey found that 69% of respondents felt that emergency responders "should be monitoring social media sites to quickly send help—and nearly half believe a response agency is probably already responding to any urgent request they might see" (American Red Cross 2011). Seventy-four percent expected help to come in less than an hour after their tweet or Facebook post (American Red Cross 2011).

14.5 Lessons learned

The beginning of this chapter contained a number of questions to consider as one reads the case studies. In each of the six cases, individuals utilized information provided by the government, but when confusing or no direction was given as in the cases of Katrina and Anthrax, citizens took it upon themselves to make informed decisions affecting their fate. Most individuals accepted government information during a crisis with significant skepticism, since in most cases, such as the Anthrax attacks, the origin and extent of the threat was unknown and representatives, in trying to allay fears, overstepped the limits of their knowledge.

In general, a number of studies have found that citizens trust media outlets more than they trust government sources. Thus, in times of crisis, it is vital that government agencies reach out to trusted partners to release accurate information. As was the case with FMD in the United Kingdom, trusted organizations, whether local or issue-specific, must be engaged *before* the crisis so that the relationship is established and the lines of communication are open.

With the 1995 sarin attack on the Tokyo subway, the WMD agent was novel as a chemical agent had not been used against a civilian population in peacetime. Therefore, civilian doctors had never seen it, nor did they have the knowledge to treat it without help. While the U.S. Government plans and prepares for pandemics and attacks of biological agents such as anthrax, how we prepare for unknown threats is not clear. The government needs to define how it will communicate information about the threat and its eventual remediation to the public. The two-way conversation should begin now—before an incident occurs.

The 2003 SARS epidemic was one of the earliest examples of citizens using social media and cell phones to share lifesaving information. Rumors went viral, but officials in Hong Kong quickly realized that they would reach a broader audience by engaging the public through text messaging rather than attempting to quash it. Government officials should consider the monitoring and use of social media and the Internet to address the novel challenges of pandemics or a WMD event. Both will be unfamiliar territory and will require nimble responses.

Before Katrina's landfall, officials had ample opportunity to provide well-defined, concise directions to the public. However, the lack of clarity was the first domino to fall in a long line of failures that led to anarchy and death. All levels of government should continue to glean the lessons of Katrina and consider changes for future events. Foremost is trusted communication with a singular purpose: to protect the lives and well-being of citizens.

Although communication during a crisis has been researched extensively, what has not been considered is how communication can empower recovering communities (Nicholls 2012). The value of communication must be measured not only in terms of information broadcast, but also in terms of information received and integrated into an evolving process. The JTI Facebook page retains over 48,000 followers for more than 18 months after the event because the page now functions as a resource for the Joplin community. The focus remains on Joplin's recovery, but JTI now has a wider community foundation.

14.6   Conclusion

Following the 2009 Black Saturday Bushfires, which killed 173 people and destroyed more than 5000 homes and structures in the state of Victoria, the Australian Government created a *National Strategy for Disaster Resilience*, which explains that "achieving disaster resilience is not solely the domain of emergency management agencies; rather, it is a share[d] responsibility across the whole of society." The U.S. Government has enunciated similar concepts in its 2010 Quadrennial Report on Homeland Security and more recently in a 2012 National Research Council study, which states:

> Disaster resilience is everyone's business and is a shared responsibility among the citizens, the private sector, and the government. Increasing resilience to disasters requires bold decisions and actions that may pit short-term interests against longer-term goals.... Such a path requires a commitment to a new vision that includes shared responsibility for resilience and one that puts resilience in the forefront of many of our public policies that have both direct and indirect effects on enhancing resilience.

While the U.S. Government seems to agree that resilience is in the collective hands of the populace, it has not yet figured out how best to harness the ingenuity of the masses to support this concept. As the aforementioned quote states, it requires "new vision"—the laying aside of old methods of thinking and doing business. Social media and the innovative technologies driven by the Internet provide exciting new tools for disaster response. The challenge for governments is to integrate public participation in the creation and use of these tools into its responses. While the federal government may not be capable of interesting individuals in "preparing" for disasters, social media and the Internet may provide the avenue for citizens to get there on their own.

References

American Red Cross. 2011. "Press Release: More Americans Using Social Media and Technology in Emergencies: New American Red Cross Survey Finds High Expectations on Response Organizations (online)," Washington, DC: American Red Cross. http://www.redcross.org/news/article/More-Americans-are-Using-Mobile-Apps-in-Emergencies (May 24, 2012).

British Broadcasting Company. March 2008. "1967: Foot-And-Mouth Slaughter Rate Soars," *BBC On This Day*, http://news.bbc.co.uk/onthisday/hi/dates/stories/november/21/newsid_3194000/3194490.stm.

Brown, S. (Senior VP, Google.org), "Testimony of Shona L. Brown before the Senate Homeland Security Ad Hoc Subcommittee on Disaster Recovery and Intergovernmental Affairs." (May 5, 2011). Senate Hearing 112–390, *Understanding of the Power of Social Media as a Communication Tool in the Aftermath of Disasters, Washington, DC:* U.S. Government Printing Office. http://www.gpo.gov/fdsys/pkg/CHRG-112shrg67635/html/CHRG-112shrg67635.htm

Cabinet Office, The (UK). July 2002. "Foot and Mouth Disease: Lessons to be Learned Inquiry Report HC888," p. 21.

Centers for Disease Control and Prevention. November 19, 2003. "Basic Information about SARS," http://www.cdc.gov/sars/about/fs-SARS.html.

Chiu, L. May 7, 2003. *Outbreak of Rumors Has China Reeling/Conspiracy Theories Explaining SARS at Epidemic Level*, San Francisco Chronicle, http://www.sfgate.com/health/article/Outbreak-of-rumors-has-China-reeling-Conspiracy-2618397.php#ixzz22ycFF1NP.

Committee on Homeland Security and Governmental Affairs. 2006. *Hurricane Katrina: A Nation Still Unprepared* (Special Report 109–322), Washington, DC: U.S. Senate, p. 133.

Community and Regional Resilience Initiative (CARRI). September 2008. *Community Resilience: Lessons from New Orleans and Hurricane Katrina*, CARRI Research Report 3.

Department for Environment, Food and Rural Affairs (UK.). March 14, 2008. "Animal Health and Welfare: FMD Data Archive," http://footandmouth.csl.gov.uk/secure/fmdstatistics/slaughtmonth.cfm.

Fisher, M., November 1, 2001. "Irrational Fear Is Untreatable," *The Washington Post*, p. B1.

Homeland Security News Wire. June 11, 2012. "AT&T Receives DHS Disaster Preparedness Certification," http://www.homelandsecuritynewswire.com/dr20120611-at-t-receives-dhs-disaster-preparedness-certification (July 12, 2012).

Jardin, X. 2003. "Text Messaging Feeds SARS Rumors," Wired.com, http://www
.wired.com/medtech/health/news/2003/04/58506.

Johnson, J. A. May 23, 2003. "Severe Acute Respiratory Syndrome (SARS): Public
Health Situation and U.S. Response," Congressional Research Service Report for
Congress (RL31937), https://opencrs.com/document/RL31937/2003-05-23/
(November 18, 2003).

Kocieniewski, D. October 11, 2001. "A Nation Challenged: The Jitters: Nervousness
Spreads, Though Illness Doesn't," *The New York Times*, http://www.nytimes.
com/2001/10/11/nyregion/a-nation-challenged-the-jitters-nervousness-
spreads-though-illness-doesn-t.html?src=pm.

Manyena, S. B., O'Brien, G., O'Keefe, P., and Rose, J. 2011. "Disaster Resilience;
A Bounce Back or a Bounce Forward Ability?" *Local Environment*, Vol. 16,
No. 5, pp. 417–424.

Market Watch. March 8, 2012. "Over 14,000 Tornado Real-Time Alerts Sent to 2GIG's
Home Security Panels Last Weekend," http://www.2gig.com/news-a-
events/121-over-14000-tornado-real-time-alerts-sent-to-2gigs-home-security-
panels-last-weekend.

Mazmanian, A. June 3, 2012. "Of Hurricanes and Hashtags: Disaster Relief in the
Social-Media Age," *National Journal*, http://www.nationaljournal.com/tech/
of-hurricanes-and-hashtags-disaster-relief-in-the-social-media-age-20120603.

National Research Council. 2012. *Disaster Resilience: A National Imperative*.
Washington, DC: The National Academies Press.

Nicholls, S. February 2012. "The Resilient Community and Communication Practice,"
The Australian Journal of Emergency Management, Vol. 27. No. 1, http://www
.em.gov.au/Publications/Australianjournalofemergencymanagement/
Pastissues/Pages/AJEM27ONE/The_resilient_community_and_
communication_practice.aspx.

NOAA National Climatic Data Center. December 2011. "State of the Climate:
Tornadoes for Annual 2011," http://www.ncdc.noaa.gov/sotc/tornadoes/
2011/13 (May 13, 2012).

NWS Central Region Service Assessment. May 22, 2011. "Joplin, Missouri, Tornado,"
http://www.nws.noaa.gov/os/assessments/pdfs/Joplin_tornado.pdf.

Paton, D. 2006. "Disaster Resilience: Building Capacity to Co-Exist with Natural
Hazards and Their Consequences," In Paton, D. and Johnson, D. (eds.),
Disaster Resilience: An Integrated Approach, Springfield: Charles C. Thomas
Publishers Ltd.

PR Web. 2012. "Most Americans Unprepared for Disasters According to a February
Persuadable Research Survey," http://www.prweb.com/releases/disaster-
preparedness/market-research-2012/prweb9225008.htm (May 28, 2012).

Sifry, M. L. July 2, 2012. "Five Lessons from Joplin Tornado—Social Media and
Disaster Relief," *Personal Democracy Media*, http://www.govloop.com/profiles/
blogs/five-lessons-from-joplin-tornado-info-about-social-media-and.

Smithson, A. October 2000. "Rethinking the Lessons of Tokyo," *Ataxia: The
Chemical and Biological Terrorism Threat and the U.S. Response*, Washington, DC:
Publication of the Henry L. Stimson Center, p. 19, http://www.stimson.org/
images/uploads/research-pdfs/atxchapter3.pdf.

Stone, F. P. June 2007. *The "Worried Well" Response to CBRN Events: Analysis and
Solutions*, Maxwell AFB, AL: USAF Counterproliferation Center Future
Warfare Series No. 40, p. 29.

University of Toronto Joint Centre for Bioethics. October 10, 2007. "Ethics and SARS: Learning Lessons from the Toronto Experience," www.yorku.ca/igreene/sars.html.

U.S. Government Accountability Office. (2003, October). *Bioterrorism: Public Health Response to Anthrax Incidents of 2001*, (Publication No. GAO-04-152). Washington, DC: U.S. Government Printing Office. http://www.gao.gov/products/GAO-04-152.

Weiss, R. and Nakashima, E. September 22, 2001. "Biological Attack Concerns Spur Warnings: Restoration of Broken Public Health System Is Best Preparation, Experts Say," *The Washington Post*, p. A04.

Wilbanks, T. J. November 1, 2007. "The Research Component of the Community and Regional Resilience Initiative (CARRI)," Presentation at the National Hazards Center, University of Colorado-Boulder.

Williams, R. and Williams, G. March 15, 2012. *The Use of Social Media for Disaster Recovery*, University of Missouri Extension, http://joplintornado.info/joplin_documents/Using%20Social%20Media%20for%20Disasters.pdf (July 2012).

World Health Organization. May 9, 2003. "Cumulative Number of Reported Probable Cases of Severe Acute Respiratory Syndrome (SARS)," http://www.who.int/csr/sars/country/2003_05_09/en/print.html (November 19, 2003).

chapter fifteen

First responders
A biomechanical evaluation of supply distribution

Susan Gaines, Arturo Watlington III, and Pamela McCauley-Bush

Contents

15.1   Introduction

The purpose of this chapter is to evaluate and compare results from laboratory experiments recreating the first-responder task of supply distribution and verify these results using simulations of the task on commercially available software tools in ergonomics and biomechanics research. The project provides statistical data from laboratory research; results from commonly used biomechanical tools including rapid upper limb assessment (RULA) and the National Institute for Occupational Safety and Health (NIOSH) guidelines; consideration of heart rate and ratings of perceived exertion (RPEs) (Borg scale); and data from two biomechanical software Tools, Three-dimensional Static Strength Prediction Program (3DSSPP) and JACK.

The research project was conducted primarily in the Ergonomics Laboratory at the University of Central Florida (UCF), Orlando, Florida. The software tools are resident on computers located in this laboratory and were used to conduct the simulations. Volunteers performed the task in the ergonomics laboratory and physiological measurements were collected. This research project will contribute to a proposed study between UCF and the National Science Foundation, entitled "A Human-Centered Assessment of Physical Tasks of First Responders in High Consequence Disasters," which looks at three of the physical tasks typically associated with first responders in disaster management (i.e., emergency management and response). These three tasks include victim extraction (Figure 15.1), supply distribution (Figure 15.2), and moving the injured (Figure 15.3). This chapter focuses on supply distribution. The results analyze the physical demands of the supply distribution task, which may place subjects at risk of physical injury and possible cumulative trauma disorders (i.e., work-related musculoskeletal disorders [WMSDs]). The results of the experiment and

Figure 15.1 Los Angeles search and rescue pull a woman out of rubble in January 12, 2010, Haiti earthquake. (Courtesy of LA County SAR. 2010. "Haiti Earthquake." LA County SAR, http://edwardrees.wordpress.com/2010/01/21/3-days/ (accessed on June 15, 2011))

Figure 15.2 New Jersey national guard's response to Hurricane Katrina. (Courtesy of New Jersey National Guard. 2005. Guardlife, Volume 31 No.5.)

Figure 15.3 Rescuers carry injured quake victim from collapsed building, Beichuan County, China, in May 2008. (Courtesy of www.nytimes.com.)

simulation analysis can be useful to researchers in assessing risks, developing worker training, selecting appropriate personal protective equipment (PPE), and recommending ergonomic interventions to mitigate risks.

The tasks associated with first responders are as follows:

- Task 1: Victim extraction
  Often victims of disasters become trapped in the rubble. Rescues often require awkward and sometimes dangerous postures to keep the victim from incurring additional injury.
- Task 2: Supply distribution (the focus of this study)
- Task 3: Moving the injured
  Keeping a victim's head and neck stationary sometimes requires an awkward position by the rescuer. In this case, other rescuers should be taking some of the load at the feet and mid-body so that the rescuer does not have to support the entire weight of the victim while keeping the neck stationary.

For this research project, one specific task element for first responders, supply distribution, was identified for evaluation. This task element represents a typical physical task performed by a first responder (i.e., load lifting, carrying weight, or awkward posture). This task was simulated from photographs taken during actual task performance, still photographs taken from videos, and photographs taken during actual disasters. The task was simulated by positioning virtual models in the same postures as workers, including the loads. Variables that were considered included uneven ground in which workers must work, lifting loads, and body and limb postures. Variables that could not be simulated using the software tools included temperature; humidity; physical fatigue; mental stress; and chemical, biological, and environmental hazards.

15.2   History and significance

This human-centered study is the initial step in developing a methodology to categorize and analyze physical tasks performed by first responders in high-consequence disasters from a human factors and biomechanics perspective. The National Science Foundation and the National Research Council have challenged the research community to study hazards and disasters to ultimately provide research that will mitigate results of disasters and aid in preparedness and recovery (National Research Council of the National Academies, 2006). Four key phases of disaster management include preparedness, response, recovery, and mitigation. The tasks analyzed in this study occur in the response phase. The software tools 3DSSPP and JACK allow evaluation of biomechanical risks associated with the physical tasks. For comparative purposes, the most common physical tasks for first responders during disasters can be partitioned into three categories: (1) victim extraction, (2) distribution of supplies (food, water, or temporary housing supplies) and (3) moving the injured. Photographs of emergency workers and volunteers fulfilling these roles were retrieved from past disasters and subject-matter experts. Task activities and related postures and load handling associated with distribution of supplies were simulated in the software environment. Experiments were done in which subjects re-created this task, and those results were analyzed using the statistical tool Microsoft Excel. Other biomechanical tools also were utilized, including RULA and NIOSH calculations.

Specific biomechanical risks have not been widely studied among responders in high-consequence emergency response. The volume of rescue workers in high-consequence disasters is difficult to quantify. Although professionally trained rescuers such as firefighters and the police will provide aid, according to Kano et al. (2005, p.59) "It is (also) recognized that members of the lay public are often the actual 'first responders' in many disaster events." Because of this, injuries incurred as a result of rescue or first-responder tasks are not always well documented. Issues related to the mental stress of witnessing widespread death and devastation have been widely researched with regard to first responders. As a result of the World Trade Center disaster in 2001, first-responder health problems related to pulmonary issues due to ingested dust and particles at the disaster site have also been well documented. PPE is endorsed by the Occupational Safety and Health Administration (OSHA) based on data from previous disasters; however, recommended PPE equipment tends to be in response to environmental, biological, or chemical risks. Therefore, potentially valuable technology and PPE such as lifting aids, which might facilitate lifting tasks, are missing from disaster PPE recommendations.

The frequency of weather-related disasters has increased over the past 10 years. From 1980 to 2009, there have been 96 weather-related disasters in which overall damages reached or exceeded $1 billion per event (National Climatic Data Center 2012). Scientists theorize the increase is related to global warming. Whatever the cause, it is suspected that the frequency with which disaster workers and volunteers will need to provide aid will continue to increase. The lack of training and literature regarding mitigation of risks to first responders performing physical tasks points to the need for more research in this area.

Research has focused on mental health risks such as posttraumatic stress disorder (PTSD), environmental risks such as chemicals, electrical risks due to downed power lines, and biological hazards that include "insect bites/stings, mammal/snake bites, and exposure to molds and other biological contaminants as a result of water damage, and sewage infiltration in low-lying areas" (Stull 2006, p. 18). Despite the lack of research and literature regarding risks and injuries of first responders as a result of the physical tasks performed, back injuries account for 31% of all workers' compensation claims in the United States. This fact alone indicates the need to study rescue worker safety with respect to the physical tasks performed and the subsequent risks incurred by carrying out these tasks.

If physical tasks can be categorized and evaluated for risk utilizing software tools for simulation, researchers can identify those tasks and pinpoint the movements that place rescue workers at greatest risk. Once these tasks are identified, real-time data from disaster sites can be collected and analyzed. These results can be compared and validated by re-creation of the tasks in a laboratory setting and analysis with software simulation tools. Action in the form of enhancements to training and additional PPE recommendations can be taken to reduce the risks to these workers. Ultimately, both victims and responders will benefit from having a healthier workforce that can provide faster and more efficient response, further preserving lives and expediting rescues.

15.3   Literature review

Over the years, there have been many improvements in mitigating the effects of disasters. However, little focus has been given to ergonomics or human-centered analysis of first-responder needs. Existing research has focused on the communications, logistics, and cognitive aspects of emergency response (Scavetta 2010). Existing literature on risk assessment has focused on PPE (Youngstown State University 1997), but information on guidelines for PPE assessment in high-consequence emergency environments is lacking. Literature on human factors and ergonomics in disaster response is almost nonexistent. Lifting guidelines for hospitals and musculoskeletal issues in various manufacturing environments exist (Patient Safety Center of Inquiry, Veterans Health Administration 2001; OSHA 2008b; OSHA 2009), but no literature could be found with respect to upper musculoskeletal requirements, lifting recommendations for first responders, and other physical tasks associated with emergency response in disasters, especially when neither mechanical nor electrical devices are available. An extensive review of the literature to date has not yielded any comprehensive study of physical measurements and human factors in emergency response scenarios. Portions of human factors issues have been extracted from Emergency Service Function documentation (FEMA 1999), OSHA guidelines (Ippolito et al. 2005), NIOSH guidelines, firefighter training (U.S. Fire Administration, FEMA 1996), and the National Earthquake Hazards Reduction Program, to name a few.

The nature of disaster response thus far has been reactive rather than strategized. Disaster response studies discussed by the Social Science Committee (2006) confirm that initial first response is typically improvised and often performed by untrained volunteers, usually survivors.

An article by Bennett (2010, p.26) defines human factors as the "golden thread that runs through integrated … systems." He continues, "Studies have shown that a high percentage of major accidents are attributable in some degree to human failures. This includes 'technical failures' that have a human error root cause." These human factors components include the following:

- The job (what): task, workload, environment, display and controls, procedures
- The individual (who): knowledge, skills aptitude, behavior, risk perception
- The organization (how): leadership, culture, resources, work patterns, communications
- The culture and working environment (where): national, sector, and local workplace cultures; social and community values; country economics; and legislative framework

All of these components can be applied to human-centered disaster management research in defining the tasks of first responders, along with all of the other components resulting in an approach to emergency response that is efficient and effective. Clancy and Holgate (2005) David Clancy, fire operations officer in Melbourne, Australia, proposes that a "safe person" approach to emergency response is preferred over the current "safe place" approach. He calls for future research into safety management to take account of existing knowledge of human factors in designing safety programs with a focus on human cognitive abilities, specifically addressing information capacity, decision making under stress, typical biases by decision makers, factors that compromise situation awareness, and factors influencing risk perception.

15.4   Methodology

Distributing relief supplies is a task often performed by first responders. This study allowed the evaluation of risks associated with this task by evaluating the strength required to perform the task from a cardiovascular standpoint. Evaluation of this task included performing a NIOSH lifting analysis and RPE using the Borg scale. Analysis included comparing the lifting indices (LIs) from the NIOSH equation with OSHA-recommended loads and calculating a RULA score.

15.4.1   Selection of subjects

There were a total of 33 subjects (13 female and 20 male). Subjects for the laboratory study were selected from a population of university students. The subjects were selected not only because of availability but also because professional first responders typically have physical fitness requirements, and tasks such as supply distribution might fall to younger responders. Also, supply distribution may be considered an unskilled task as opposed to moving the injured or victim extraction, and spontaneous volunteers are commonly placed in these types of roles. In addition, volunteers to disaster areas tend to come from a younger demographic of the population and are in college or have received a bachelor's degree. According to U.S. Department of Labor (2012)—February 22—Bureau of Labor Statistics 2012 news release, volunteers aged 16–24 years make up 22.5% of the volunteer population and volunteers aged 25–34 years make up 23.3% of the volunteer population. The supply distribution task, which was duplicated by the subjects in the laboratory, was also used as the basis to create simulations and generate data with respect to loads, balance, strength exertion, posture, and other task performance descriptors.

The photograph in Figure 15.2 was the inspiration for the task simulated in the laboratory by the subjects and in the software. This photograph shows the awkward shoulder and arm angles at which supplies are sometimes lifted and moved. This is a typical first-responder task. For the purposes of the laboratory experiment and the simulation, minor changes were made to the postures to ensure conformity in the way the subjects performed the task. For example, instead of the subject lifting the load and twisting with it in their hands as seen in Figure 15.2, the subjects were asked to lift the load, turn 180°, and walk two full steps with no twisting. This was intentionally done to remove the asymmetric multiplier from the NIOSH calculations. This stipulation was made for two reasons: first, a worker might lift and twist to hand the load to another worker at the beginning of the task when the truck is full, but as the truck begins to empty the worker would have to lift the load, walk a few steps, and then hand the load to another person. The task would be performed without a twisting motion the majority of the time. Second, the task was simplified for the laboratory environment so that the subjects were not placed at excessive risk. If the task is shown to be at risk without an asymmetric multiplier, then the risk would only increase if an asymmetric aspect were added to the performance of the task.

15.4.2   Tools used

This study utilized a laboratory experiment, statistical analysis tools, two commonly used biomechanics tools, NIOSH lifting equation and RULA assessment, and two software tools with which to perform biomechanical analysis. Simulation is a powerful tool that can be utilized to better assess human factors needs and to better prepare communities, responders, and researchers when inevitable disaster events take place. JACK output data was used to analyze risks associated with the task included in the study. This data was also compared with the output generated for the same tasks from another simulation tool, 3DSSPP.

15.4.2.1   3DSSPP biomechanical software from
the University of Michigan

3DSSPP was developed by the Center for Ergonomics at the University of Michigan College of Engineering. This program can be used in analyzing manual material-handling tasks. Ergonomists, engineers, therapists, and researchers use the software to evaluate and design jobs. The program allows users to simulate subject postures and loads and use custom anthropometrics or draw from the installed tables. Output from the software includes spinal compression forces, the percentiles of humans who can perform the task, and data comparisons to NIOSH guidelines, which generate color-coded warnings. The analysis is augmented by graphic illustrations of the positions being studied (University of Michigan 2010). The primary feature of interest in the 3DSSPP software, for the purposes of this study, was the lower back compression forces, the region of the spine most prone to lower back injury.

This software is good for analyzing tasks in which the subject stands in one place to lift or move a load. To utilize the software for this task, where the subject lifts the load, turns, walks, bends, and releases the load, a series of smaller simulations had to be made and analyzed.

15.4.2.2   *Siemens PLM JACK and the Task Simulation Builder*

JACK is a human modeling and simulation tool that allows user input to simulate a task or an environment. JACK utilizes a Task Simulation Builder to enable the use of preprogrammed commands to instruct a human model in a virtual 3D environment (Figure 15.4). The human posturing features predict human postures based on changes to the virtual human's posture with respect to variables including hand force exertions, foot positions, center of gravity, head position, and obstacles. JACK enables human models to be sized to match worker populations, as well as test designs for multiple factors, including injury risk, user comfort, reachability, line of sight, energy expenditure, fatigue limits, and other important human parameters (Siemens 2011).

15.4.2.3   *NIOSH lifting guidelines*

A decade after the first NIOSH lifting guide, NIOSH revised the technique for assessing overexertion hazards of manual lifting. The new document no longer contains two separate weight limits (action limit [AL] and maximum permissible limit [MPL]) but has only one recommended weight limit (RWL). It represents the maximal weight of a load that may be lifted or lowered by about 90% of American industrial workers, male or female, who are physically fit and accustomed to physical labor.

Figure 15.4 Simulation from JACK software. (From Siemens, "JACK," Siemens PLM Software, http://www.plm.automation.siemens.com/en_in/Images/4917_tcm641-4952.pdf, 2011.)

This new equation resembles the 1981 formula for AL but includes new multipliers to reflect asymmetry and the quality of hand–load coupling. The 1991 equation allows as maximum a "load constant" (LC)—permissible under the most favorable circumstances—with a value of 23 kg (51 lb) (Waters et al. 1994):

$$RWL = LC \times HM \times VM \times DM \times AM \times FM \times CM$$

where

LC—Load constant of 23 kg or 51 lb. Each remaining multiplier may assume a value [0, 1].
HM—Horizontal multiplier: *H* is the horizontal distance of the hands from the ankles (the midpoint of the ankles) (Figure 15.5).
VM—Vertical multiplier: *V* is the vertical location (height) of the hands above the floor at the start and end points of the lift.

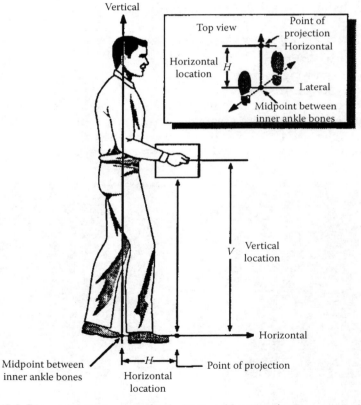

Figure 15.5 Dimensions to measure for the National Institute for Occupational Safety and Health equation. (Courtesy of Waters, et al, 1994, available at www.cdc.gov.)

DM—Distance multiplier: *D* is the vertical travel distance from the start to the end points of the lift.

AM—Asymmetry multiplier: *A* is the angle of asymmetry, that is, the angular displacement of the load from the medial (midsagittal plane), which forces the operator to twist the body. It is measured at the start and end points of the lift.

FM—Frequency multiplier: *F* is the frequency rate of lifting, expressed in lifts per minute.

CM—Coupling multiplier: *C* indicates the quality of coupling between hand and load.

15.4.2.4   Borg scale

The Borg RPE is a way of measuring physical activity intensity levels. Perceived exertion refers to a scale of 1 (minimal exertion) to 10 (strong exertion) for how hard one feels like their body is working. It is based on the physical sensations a person experiences during physical activity, including increased heart rate, increased respiration or breathing rate, increased sweating, and muscle fatigue. Although this is a subjective measure, a person's exertion rating may provide a fairly good estimate of the actual heart rate during physical activity (Borg 1982, p. 380).

15.4.2.5   RULA analysis

RULA is a quick survey method for determining the risk associated with upper-body postures. It focuses on the neck, trunk, and upper limbs. The RULA score indicates the level of intervention required to reduce WMSD risks.

Figure 15.6 is used to calculate the grand score for RULA.

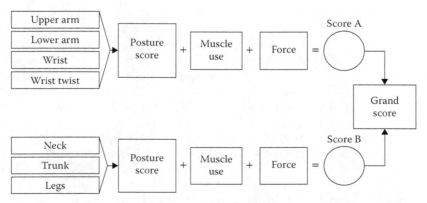

Figure 15.6 Process flow for calculating rapid upper limb assessment score (RULA).

For instance, we add 1 to the score for each of the following actions:

1. Shoulder is raised.
2. Upper arm is abducted.
3. Working across the midline of the body.
4. Working out to the side of the body.
5. Wrist is bent away from the midline more than 15° of radial, ulnar deviation.
6. Neck is twisted.
7. Neck is side-bended.
8. Trunk is twisted.
9. Trunk is side-bended.
10. Legs and feet are well supported and in an evenly balanced posture.
11. Legs and feet are not well supported and/or not in an evenly balanced posture.
12. Mainly static—held for longer than 1 minute.
13. Repeated more than four times per minute.
14. Forces or load score.

RULA total score can be interpreted using the following grading criteria:

Scores	Action
1–2— — — →	Posture is acceptable
3–4— — — →	Further investigation
5–6— — — →	Investigation and changes are required soon
7— — — — →	Investigation and changes are required immediately

15.4.3 Procedure and analysis

Still photographs of rescue workers from a variety of international disasters were used as the basis for the creation of laboratory experiment. The laboratory experiment consisted of a manual material-handling evaluation using people in reasonably good health. This test evaluated the subjects' change in heart rate as well as their subjective rating of exertion while picking up and moving a case of water repeatedly for 10 minutes. Photographs helped researchers to create virtual humans and duplicate postures and estimate loads. Once the tasks were identified and the poses selected for simulation, JACK software was used to create a virtual environment to re-create the task. Figure 15.7 shows an example of a simulation derived from an actual disaster photograph. Figure 15.8 shows samplings of photographs taken during the laboratory experiment (Figure 15.8a and c), and the simulation (Figure 15.8b).

Figure 15.7 New Jersey national guard in Hurricane Katrina and JACK and three-dimensional Static Strength Prediction Program (3DSSPP) animations.

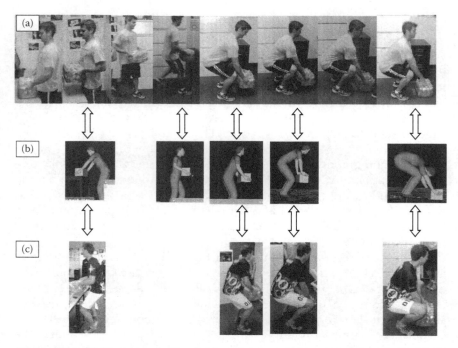

Figure 15.8 (a) Photographs from laboratory experiment—subject 1, (b) JACK simulation of experiment, and (c) photographs from laboratory experiment—subject 2.

15.4.3.1 Laboratory experiment

A survey completed by each subject prior to participating in the experiment ensured the subjects were physically capable of completing the task without health crisis and that the subjects did not have any preexisting health conditions. The subjects' maximum heart rate was determined (220–age) (Mackenzie 1999). If at any time during the experiment the subjects exceeded 85% of the maximum heart rate, they were asked to stop the test immediately. The environment of the study was safe and did not expose the subjects to any adverse environmental conditions or safety issues. The weight of the load was not excessive. The dimensions of the load were reasonable, and adequate coupling (handles) existed.

The subjects picked up a case of water located on a table 30 in. from the floor (76.2 cm). After lifting the water, the subjects turned 180°, walked two steps (54 in. = 137.2 cm), and placed the water on the floor. The subjects performed this task as many times as possible in 10 minutes. At the end of each minute, the subjects' heart rate was measured, they were asked their RPE, and the number of lifts completed was recorded. At the cessation of the activity, the subjects' heart rate was monitored to ensure that they recovered properly.

15.4.3.2 Software simulation procedure

For the purposes of this research, the human posturing techniques were the most useful (Figure 15.11). This feature allowed users to quickly posture the human model while making predictions of the next movements, based on research of actual human movements and mechanics. The postures from the photographs were re-created. An example of the closely simulated posture from the inspiration task in an actual disaster is shown in Figure 15.7.

The photographs in Figure 15.8 were taken during the actual laboratory experiment. JACK draws from a large library of documented human movements and although the postures were not always exactly replicated (Figure 15.8), the simulation is still a relatively accurate depiction of how the task was or may be performed, depending on individual differences in lifting actions.

The JACK software contains a utility tool that allows the user to closely match the distances traveled in the actual experiment. The experiment was set up with tape on the floor to indicate where the subjects should stand when lifting the water initially and how far they should walk before placing the water on the floor (Figure 15.9). This distance was approximately 54 in. (137.16 cm). Figure 15.10 shows the JACK utility that calculates the distance the manikin traveled. Note that JACK calculated the distance at 137.25 cm, with only 0.09 cm difference (0.07%).

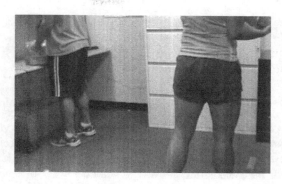

Figure 15.9 Photograph from laboratory experiment showing laboratory setup.

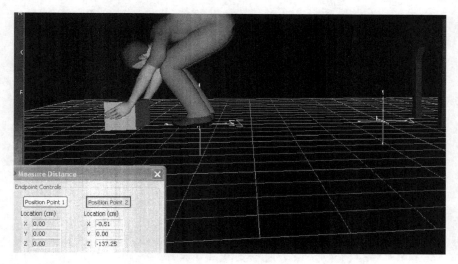

Figure 15.10 Animation from JACK showing closely simulated distance between origin and destination of lift.

The University of Michigan 3DSSPP biomechanical software was used to analyze the poses as well. The weights of the objects were the same as those entered in the JACK software. Hand postures were closely matched as well. The anthropometrics, height and weight, of the virtual figure in JACK were entered into 3DSSPP to keep the variables between the two software packages the same.

The task analysis reports in 3DSSPP predict the percentage of the population who could perform the tasks and were compared with the same percentages generated in JACK. The forces on L4/L5 were also compared.

3DSSPP gives the strength limits for the percentage capable (percentage of the population with sufficient strength) in a graphical format. The green zone is if over 99% of the population can perform the task. The yellow zone is for 25%–99% of the population, and the red zone is if <25% of the population can perform the task. JACK gives a red indicator if <99% of the population can perform the task.

15.4.3.3   Data analysis of results

Both 3DSSPP and JACK utilize the NIOSH lifting guidelines to determine if loads are acceptable. Two metrics were used as the primary tools to compare the software packages: the forces on L4/L5 and the strength capability of the population. The documentation and user guides of the software packages describe the science behind the calculation of these figures.

15.4.3.3.1   Static strength prediction percentage capable.   Whereas JACK bases its static strength prediction percentages on data collected at the University of Michigan, 3DSSPP was actually developed by the University of Michigan and utilizes the same population data to calculate the percentages capable for strength. The 3DSSPP evaluations are based on experimental strength studies done by Clarke (1966); Kumr, Chaffin, and Redfern (1988); and many others (University of Michigan 2010, p.72).

15.4.3.3.2   Low back analysis (forces and moments on L4/L5).   The JACK module that computes the spinal forces at L4/L5 uses the distributed moment histogram (DMH) technique for torso muscle recruitment (Siemens 2011). The Low Back Compression Analysis Tool helps evaluate the spinal forces acting on a virtual human's back, indicating the compression and shear forces at the L4/L5 vertebral disk and how the compression forces compare with NIOSH-recommended and permissible force limits. The results of a low back compression analysis can be used to design or modify manual tasks to minimize the risk of lower back injuries and conform to NIOSH guidelines. The tool can also pinpoint the exact moments of a lift when the compression forces on a worker's L4/L5 vertebral disk exceed NIOSH force limits (Siemens 2011).

The predicted disk compression force from 3DSSPP, on the other hand, can be compared with the NIOSH back compression design limit of 3400 N (University of Michigan 2010). This study focused on L4/L5 compression and moments because as validated by the 3DSSPP manual, "Torso muscle moment arms and muscle orientation data for the L4/L5 level have been studied more extensively than at any other lumbar level" (University of Michigan 2010, p.15).

15.4.4   Results

Both JACK and 3DSSPP were evaluated for their abilities to simulate forces at L4/L5 using a fiftieth percentile male (height 68.9 in. and weight 185 lb). The results were compared with each other as well as the NIOSH

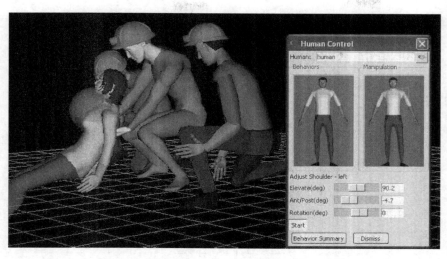

Figure 15.11 The Human Control tab in JACK allows greater manipulation of the shoulder joint.

maximum allowable limit of 3400 N. JACK indicated that the two points of the task that pose the greatest risk are just after the load is grasped and the subject is beginning to arise and right before the load is released to the floor. In this respect, JACK and 3DSSPP agreed with each other.

For this task, JACK appears to be a better simulation tool. JACK provides a fully dynamic simulation of the task, whereas 3DSSPP required a series of static poses to generate similar information. JACK also allows more detailed manipulation of hand postures and gives more flexibility with regard to torso rotations and flexibility. The greater ability to specify the angles of shoulder rotation, elevation, and lifting are also pivotal in this task analysis. See Figure 15.11 for an example of how JACK allows this detailed input. The force on the shoulder was probably one of the greatest, other than on L4/L5, for this task.

3DSSPP provides a clear graphical depiction of joints that are under the greatest strain. It also provides color-coded data regarding whether the forces on L4/L5 are acceptable or hazardous. Figure 15.12 provides an example of the animation of the task and force data.

15.4.4.1   Analysis of forces
Both JACK and 3DSSPP generated very different L4/L5 compression force calculations (Table 15.1). Although different, the forces on L4/L5 for this task fell below the NIOSH-recommended upper limit of 3400 N for both packages at both of the high-risk points of the task. The frequency with which this task may be repeated was not considered and would have

Figure 15.12 3DSSPP simulation for arise from bend.

Table 15.1 Comparison of JACK and 3DSSPP L4/L5 Forces

	JACK	3DSSPP
L4/L5 compression force (N) at arise from bend	3188	632
L4/L5 compression force (N) just before release	2999	925

inevitably generated a fatigue factor if, for example, an entire truckload of supplies at this weight using this posture was unloaded.

15.4.4.2 Analysis of strength capability
Both software packages calculated that >84% of the population (fiftieth percentile male) could perform this task at the weakest point (Table 15.2). This point was determined to be the forces on the hip just before the load is released to the floor. Because females and smaller males may also per- form this task, <84% of the greater population would be capable of per- forming this task. Regardless, either software package would be able to adequately simulate this task. Also, the majority of the population would likely be able to perform this task without incident.

15.4.4.3 Comparison of NIOSH calculations
A review of the data generated by the 33 subjects in the laboratory experiment indicates that the average NIOSH LI was 2.05 at the origin and 3.31 at the destination. For men, the average NIOSH LI at the origin was 2.06 and at the destination was 3.22 (Table 15.3). For women, the average NIOSH LI at the origin was 1.92 and at the destination was 3.08 (Table 15.4). The lower LIs for women may seem counterintuitive because typically men are stronger than women. These unexpected results can be attributed to two other factors in the NIOSH equation

Table 15.2 Comparison of JACK and 3DSSPP Strength Capability Summary

Joint	JACK Arise (% Capable)		3DSSPP Arise (% Capable)	JACK Before release (% Capable)	3DSSPP Before release (% Capable)
Wrist	100	67	99	93	98
Elbow	98		99	100	99
Shoulder	99	99	99	100	99
Torso	95		93	99	93
Hip	97	91	90	98	84
Knee	100		99	99	99
Ankle	99	87	96	99	80

Table 15.3 Male Subject Data from Laboratory Experiment

	Male				
	NIOSH LI (O)	NIOSH LI (D)	HR	RPE	Number of lifts
JK-RK	2.87	5.04	145.20	5.20	12.30
PH-1	1.67	2.5	160.90	4.40	9.90
AY-2	1.98	2.15	124.80	5.40	8.70
GR1-1	2.37	4.19	150.40	5.90	12.50
GR3-1	1.95	3.16	117.40	6.30	7.80
GR5-1			126.80	3.00	8.20
GR5-2			127.60	4.10	10.10
GR6-1	1.46	1.73	100.30	2.70	6.40
GR6-2	1.92	2.49	146.50	3.30	9.60
GR7-1	2.25	3.82	74.40	4.50	11.80
GR7-2	2.21	3.55	117.40	4.50	10.60
GR8-1	1.89	4.43	135.60	3.20	11.00
GR8-2	1.89	2.96	148.40	3.20	10.50
GR9-2	1.94	3.64	165.00	4.50	10.50
GR10-1	1.92		149.70	3.60	13.10
GR10-2	2.84		178.10	4.10	9.90
GR11-1	1.69	2.65	135.60	3.20	11.00
GR12-1	2.85	4.29	156.00	3.60	12.30
GR12-2	1.52	2.58	156.40	3.90	10.60
GR14-2	1.9	2.33	186.70	5.65	9.80
Average	2.06	3.22	140.16	4.21	10.33

Table 15.4 Female Subject Data from Laboratory Experiment

	Female				
	NIOSH LI (O)	NIOSH LI (D)	HR	RPE	Number of lifts
JK-1	1.6	2.8	159	7.1	9.1
KIM-1	0.969	4.09	157.10	6.00	16.60
KIM-2	3.509	5.58	161.10	7.80	13.90
PH-2	1.21	2.43	132.20	4.50	9.90
AY-1	1.72	2.32	116.30	5.70	10.00
AY-2	1.98	2.15	124.80	5.40	8.70
GR1-1	2.39	3.37	180.60	6.40	10.70
GR3-2	1.66	2.37	168.40	6.20	11.20
GR4-2	2.24	3.45	143.90	5.10	9.50
GR6-2	1.92	2.49	146.50	3.30	9.60
GR9-1	2.11	3.71	149.50	3.40	10.10
GR11-2	1.5	2.42	148.40	3.20	10.50
GR14-1	2.14	2.86	148.60	3.10	8.40
Average	1.92	3.08	148.95	5.17	10.63

besides weight of the load. One reason is that the NIOSH index considers how far from the body the load is carried. Women tended to carry the load closer to their body and actually used their bodies to partially support the load. Also, women typically were closer to the destination when lifting and releasing the load, making the horizontal distances less at both the origin and the destination.

JACK indicated that the load weight was greater than that recommended for this task, and most healthy workers would find this job physically stressful. The recommended modifications included bringing the load closer to the worker. This is consistent with the manual results, since, as noted earlier, men typically carried the load farther away from their bodies. Furthermore, JACK indicated that the arise-from-bend motion was one of the two points in the task that represented the most risk. This point in the task also had the highest NIOSH calculation of 1.66. The other high-risk point was just before releasing the load with the fourth highest NIOSH calculation of 1.57. The NIOSH numbers generated by JACK do not indicate quite as severe a risk as the manual calculations. However, all of these LIs indicate that modifications should be made to the task to lower the risk to the workers.

15.4.4.4 *Comparison of RULA results*
RULA was appropriate for this task since it was one that primarily stressed the upper extremities. Manual RULA calculation and the RULA

Table 15.5 JACK RULA Results

Jack Rapid Upper Limb Assessment
30_Pound_Lift_from_Table
  SG
  Job#1,04-16-12
Analysis Summary
Body Group A Posture Rating
  Upper arm: 5
  Lower arm: 3
  Wrist: 2
  Wrist Twist: 1
Total: 6
  Muscle Use: Normal, no extreme use
  Force/Load: <2 kg intermittent load
  Arms: Not supported
Body Group B Posture Rating
  Neck: 5
  Trunk: 1
Total: 7
  Muscle Use: Normal, no extreme use
  Force/Load: <2 kg intermittent load
Legs and Feet Rating
  Seated, Legs and feet well supported. Weight even.
Grand Score: 7
  Action: Investigation and changes are required immediately.

calculation for JACK (Table 15.5) both indicated that the RULA score for this task was 7 (Figure 15.13). A score of 7 on the RULA scale indicates that the task should be investigated and changed immediately.

15.4.4.5 *Statistical analysis of laboratory experiment results*

Data was evaluated for 33 subjects from a laboratory experiment simulating a supply distribution task. Of the 33 subjects, 13 were female and 20 were male. The average number of lifts per minute for the entire subject population was 10.59 lifts per minute. Females averaged 10.63 lifts per minute, whereas males averaged 10.33 lifts per minute. The heart rates did not appear to necessarily correlate with the RPEs or the number of lifts per minute, although there did appear to be a close correlation between RPE and the number of lifts per minute. Females seemed to feel the exertion slightly more than males with an average RPE of 5.17 compared with an average RPE of 4.21 for males.

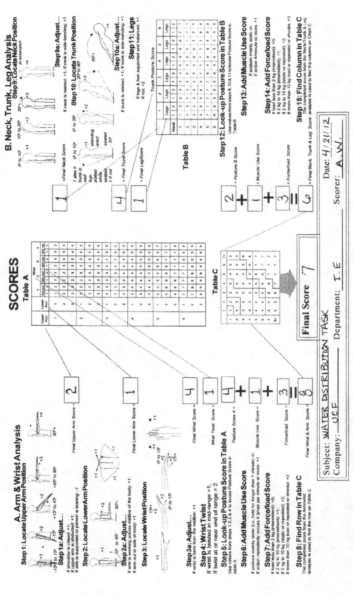

Figure 15.13 RULA assessment form.

15.5   Conclusion

The intention of this project was to evaluate a typical first-responder task, namely, distributing supplies, to determine if a first responder may be compromising his or her health by repeatedly performing this task in a disaster response situation.

Several tools were utilized to assess and compare results to determine if risks exist and to identify specific areas where workers should make mitigating efforts to reduce risk. The traditional tools utilized in this study included calculation of NIOSH LI, RULA assessment, and Borg assessment. Evolving automated tools such as JACK and 3DSSPP also assisted in determining the risks. The researchers found that the results of the two software packages can produce different results that sometimes lead to conflicting conclusions about the safety of a given task. Specifically for this simulation, there were significant differences in the forces on L4/L5. It is noted that some of the variability of the results may be attributed to the input angles and posture manipulation controlled by the user. Furthermore, differences in results may be a factor of the higher degree of freedom of joint movement that is possible by manually manipulating the postures in 3DSSPP, even though those postures may be unnatural and improbable. JACK seems to do a better job of allowing only realistic human contortions. In terms of biomechanics where force calculations are critical, these differences can present conflicting results. Despite the biomechanical conflicts, the two software packages did produce relatively similar results in the ergonomic assessment of risk associated with each task. Both software packages also produced relatively consistent overall risk scores.

In general, this task does appear to place the worker at some level of risk. Although the 30 lb load is well below the OSHA limit of 51 lb, the frequency of repetition of the task and the postures the worker attains can significantly impact the risk factors. A RULA rating of 7 indicates that immediate changes to the task are necessary. Also, the NIOSH calculations consistently indicate that the worker is exposed to a significant amount of risk and that task redesign should occur. Even though the forces on L4/L5 are significantly below the OSHA upper limit of 3400 N, the software still recommended that the worker move the load closer to his or her body. All of these factors combined indicate that this is a task that requires mitigation efforts or redesign.

Using proper lifting procedures, rotating the workforce through different tasks, utilizing PPE, providing tools to aid in lifting and moving supplies (dollies, wheelbarrows, etc.), encouraging frequent breaks, and continually training the workforce in ways to distribute the load (using the torso for partial support of load, getting closer to the origin and destination before lifting or placing, etc.), can significantly reduce risks of WMSDs.

References

Bennett, J. 2010. "Tame the Human Factor" *Industrial Fire Journal*. http://www .hemmingfire.com/news/fullstory.php/aid/858/Human_factors_in_major_ accident_prevention:_the_smoking_gun_.html (accessed on April 17, 2012).

Borg, G. 1982. "Psychophysical Bases of Perceived Exertion." *Medicine and Science in Sports and Exercise*, Vol. 14, No. 5, pp. 377-381.

Clancy, D. 2005. "Can Acceptable Risk Be Defined in Wildland Firefighting?" Proceedings of the Second conference on the Human dimensions of Wildland Fire. http://www.nrs.fs.fed.us/pubs/gtr/gtr-nrs-p-84papers/01clancy-p-84 .pdf (accessed on April 3, 2013).

Clark, H. 1966. *Muscle Strength and Endurance in Man*. Prentice-Hall, Englewood Cliffs, pp.39–51.

FEMA. 1999. "Response & Recovery. Emergency Support Function #6 Mass Care annex." http://www.au.af.mil/au/awc/awcgate/frp/frpesf6.htm (accessed on November 2, 2010).

Hines, E. 2012. "U.S. Army Corps of Engineers Disaster Response Missions, Roles & Readiness." US Army corps of Engineers Building Strong. http://www. trainex.org/osc2012/uploads/517/03%202012%20-%2003%20USACE%20 EM%30Missions_Roles_%20Hines.pdf (accessed on April 3, 2013).

Ippolito, D., Georgiades, G., and Jones, C., OSHA. 2005. "Hazard Exposure and Risk Assessment Matrix for Hurricane Response & Recovery Work: General Recommendations for Working in All Impacted Areas." http://www.osha.gov/ SLTC/etools/hurricane/recommendations.html#gr (accessed on October 24, 2010).

Kano, M., Sigel, J.M., and Bourque, L.B. 2005. "First-Aid Training and Capabilities of the Lay Public: A Potential Alternative Source of Emergency Medical Assistance Following a Natural Disaster." *Disasters*, Vol. 29, No. 1, pp. 58–74.

Kumr, S., Chaffin, D., and Redfern, M. 1988. "Isometric and Isokinetic Back and Arm Lifting Strengths: Device and Measurement." *J Biomechanics*, Vol. 21, No. 1, pp. 35-44.

LA County SAR. 2010. "Haiti Earthquake." LA County SAR, http://edwardrees. wordpress.com/2010/01/21/3-days/ (accessed on June 15, 2011).

Mackenzie, B. 1999. "Maximum Heart Rate." http://www.brianmac.co.uk/ maxhr.htm (accessed April 3, 2013).

National Climatic Data Center. 2012. "Billion Dollar U.S. Weather Disasters." http:// www.ncdc.noaa.gov/oa/reports/billionz.html (accessed on April 3, 2013).

National Research Council of the National Academies. 2006. "Facing Hazards and Disasters: Understanding Human Dimensions." Washington, DC: National Academy Press. http://www.nap.edu/openbook.php?record_ id=11671&page=R1. (accessed April 3, 2013).

New Jersey National Guard. 2005. "Hurricane Katrina Photos" New Jersey National Guard. Guardlife, 31(5). http://www.nj.gov/military/publications/ guardlife/volume31no5/katrina_photos.htm (accessed on April 3, 2013).

OSHA. 2008a. "Ergonomics for the Prevention of Musculoskeletal Disorders." http://www.osha.gov/dsg/guidance/shipyard-guidelines.html (accessed on October 26, 2010).

OSHA. 2009. "Guidelines for Nursing Homes: Ergonomics for the Prevention of Musculoskeletal Disorders." http://www.premierinc.com/safety/topics/back_injury/downloads/Final_OSHA_Guidelines_nursing_homes.pdf (accessed on October 17, 2010).

OSHA. 2008. "Guidelines for Shipyards: Prevention of Musculoskeletal Disorders." http://www.osha.gov/dsg/guidance/shipyard-guidelines.html (accessed on October 11, 2010).

Patient Safety Center of Inquiry, Veterans Health Administration. 2001. "Safe Patient Handling and Movement." http://www.premierinc.com/safety/topics/back_injury/downloads/K_03a_RG_VA_pt_01.pdf (accessed on October 11, 2010).

Scavetta, R. 2010. "Preparing for Medical Response to Disasters." http://africom.mil/Newsroom?Article/7484/preparing-for-medical-response-to-disasters (accessed on April 3, 2013).

Siemens PLM Software. 2011. "JACK." Siemens PLM Software. http://www.plm.automation.siemens.com/en_in/Images/4917_tcm641-4952.pdf (accessed on July 15, 2011).

Siemens PLM Software. 2011. "Low Back Analysis Compression Tool Background." Siemens PLM Software. http://www.intechopen.com/download/pdf/35814 (accessed on July 25, 2011).

Siemens PLM Software. 2011. "Task Analysis Toolkit (Tat) Training Manual." Siemens PLM Software. http://www.plm.automation.siemens.com/en_us/products/tecnomatix/assembly_planning/jack/index.shtml (accessed on July 25, 2011).

Siemens PLM Software. 2011. "Static Strength Prediction Tool Background." Siemens PLM Software. http://www.umich.edu/~ioe/3DSSPP/Manual_606.pdf (accessed on July 26, 2011).

Stull, J. March 2006. "Understanding PPE Selection & Use During Disasters." *Professional Safety*, Vol. 51, No. 3, pp. 18–49.

University of Michigan. 2010. "3D Static Strength Prediction Program." University of Michigan. http://www.engin.umich.edu/dept/ioe/3DSSPP/ (accessed on February 27, 2011).

U.S. Fire Administration, FEMA. 1996. "Fire and Emergency Medical Services Ergonomics: A Guide for Understanding and Implementing an Ergonomics Program in Your Department." http://www.usfa.dhs.gov/downloads/pdf/publications/fa-161.pdf (accessed on October 24, 2010).

U.S. Department of Labor. 2012. "Bureau of Labor Statistics, Volunteering in the United States—2011." http://www.bls.gov/news.release/pdf/volun.pdf (accessed on April 22, 2012).

Waters, T.R., Putz-Anderson, V., and Garg, A. 1994. "Applications Manual for the Revised NIOSH Lifting Equation." http://www.cdc.gov/niosh/docs/94-110/pdfs/94-110.pdf (accessed on April 17, 2012).

Youngstown State University. 1997. "Environmental and Occupational Health and Safety." Back Belts Pros and Cons, In: Youngstown State University. http://cc.ysu.edu/eohs/bulletins/Lifting%20Belts.htm (accessed on June 15, 2011).

chapter sixteen

Dynamics and dangers of therapeutic strategies for organophosphate poisoning
A physiologically based model*

**Gregory G. Seaman, Michael L. Shelley,
Jeffery M. Gearhart, and David A. Smith**

Contents

16.1   Introduction

The physiological effects of nerve agent exposure (principally organophosphates [OPs]) are well documented. There is significant literature on
the treatment of severe nerve agent exposure. The mechanism of action

---

* The material contained in this chapter was prepared as an account of work sponsored
by an agency of the U.S. government. Neither the U.S. government nor any agent thereof,
nor any of their employees, makes any warranty, express or implied, or assumes any
legal liability or responsibility for the accuracy, completeness, or usefulness of any information, apparatus, product, or process disclosed, or represents that its use would not
infringe upon privately owned rights. Reference herein to any specific commercial product, process, or service by trade name, trademark, manufacturer, or otherwise does not
necessarily constitute or imply its endorsement, recommendation, or favoring by the U.S.
government or any agency thereof. The views and opinions of authors expressed herein
do not necessarily state or reflect those of the U.S. government or any agency thereof.

of atropine (administered on the presentation of nerve agent symptoms) is well known and well accepted as the necessary first treatment to save life. However, balancing atropine injection with the severity of symptoms is often precarious, since atropine itself can threaten life after several administrations over a short period after exposure. An oxime (such as 2-PAM Cl) is also considered medically indicated to directly counter the mechanism of the nerve agent, but its efficacy is less certain, and it has been suggested that it could be contraindicated in some scenarios (Szinicz et al., 2007). Since nerve agent attack is now a reality for the civilian population, in counterinsurgency operations, and on the battlefield, clear guidance on the administration of antidotes, particularly for first responders or fellow soldiers on the battlefield, is paramount. Yet, when comparing existing guidelines for military and local health departments, we do not see uniformity as illustrated in Table 16.1.

The uptake, distribution, and metabolism of organophosphates have been modeled using physiologically based pharmacokinetic (PBPK) approaches (Gearhart et al., 1994; Gentry et al., 2002; Timchalk et al., 2002). However, we have not seen the simultaneous modeling of the agent and its antidotes with all of the mechanisms involved nor has there been any exploring of possible undesired effects because of unintended consequences of mechanistic feedback influences within the complex physiological system. This work presents such a model and explores various exposure scenarios along with various dosing regimens of atropine and oximes postexposure combinations.

Table 16.1 Agencies Having Different Treatment Protocols

		CDC[a]	NYDH[b]	USAMRICD[c] (U.S. Army)
Atropine	Initial dose	6 mg	6 mg	6 mg
	Repeat dose	2 mg	Unspecified	2 mg
	Repeat interval	5–10 min	2–5 min	3–5 min
2-Pam Cl (oxime)	Initial dose	1800 mg	1800 mg	1800 mg
	Repeat dose	No instructions	600–1200 mg	No instructions
	Repeat interval	No instructions	Once between 30–60 min and every hour thereafter	No instructions

Source: Cannard, K., *J. Neurol. Sci.* 249, 86–94, 2006.

[a] CDC (Centers for Disease Control) 2008.
[b] NYDH (New York Department of Health) 2008.
[c] USAMRICD (U.S. Army Medical Research Institute—Chemical Defense) 2000.

16.2   Background

16.2.1   Nerve agent and antidote mechanisms

Organophosphates used as pesticides and chemical warfare agents act in the synaptic spaces between serial nerve cells to prevent the natural shutdown of nerve firing, yielding uncontrolled convulsions, secretions, and paralysis. The direct cause of death is often asphyxiation secondary to diaphragm paralysis. Tremors, prolific secretions, and myopic responses of the eyes are examples of nonlethal symptoms that can be observed early in the exposure or shortly after exposure. Stimulating nerve impulses reach the end of a nerve axon and must transfer to the next serial nerve cell to reach the intended point of stimulation. This is accomplished by the release of acetylcholine from the presynaptic nerve into the synaptic space where it can bind to receptors on the postsynaptic nerve membrane and incite continued impulse transmission. This acetylcholine signal to fire is balanced by the presence of acetylcholinesterase, which destroys the acetylcholine to stop the impulse stimulation (see Figure 16.1). An organophosphate binds with acetylcholinesterase to inhibit its balancing function and cause continuous firing of the nerve as demonstrated in Figure 16.2.

On observation of initial symptoms, atropine is administered intramuscularly by first responders, by self-injection or by fellow soldiers on the battlefield. Atropine has the ability to bind with the acetylcholine receptors on the postsynaptic cell to limit acetylcholine binding and partially control the overfiring (Figure 16.3). This inhibits symptoms until the exposure can be comprehensively treated. The use and necessity of atropine as a first treatment is uncontroversial. Repeated doses of atropine can overinhibit nerve firing and cause death owing to cessation of nerve impulse–controlled

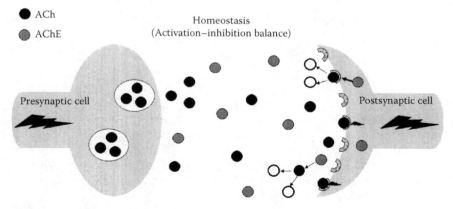

Figure 16.1 The normal action of acetylcholine and acetylcholinesterase in transmitting and balancing nerve impulse transmission.

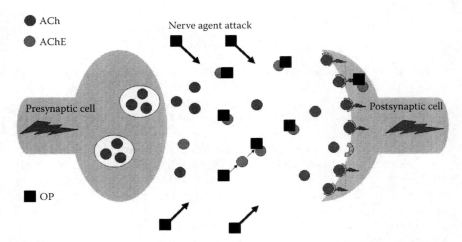

Figure 16.2 The organophosphate nerve agent disrupts the homeostatic balance by inactivating acetylcholinesterase, allowing continuous overfiring of the nerve.

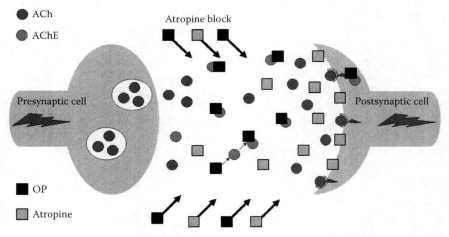

Figure 16.3 Atropine blocks acetylcholine binding by occupying the binding sites on the postsynaptic nerve cell.

life-sustaining functions. The acetylcholinesterase–OP bond is reversible, reactivating previously deactivated enzymes. However, after a period of time, the bond "ages" and becomes permanent. These dynamics are critical to any modeling effort of the simultaneous agent-therapeutic system. With the aging of these bonds, natural restoration of active acetylcholinesterase to homeostatic levels may be too long to prevent serious symptoms or death. Thus, an oxime is typically administered, which acts on the esterase–OP bond to activate the esterase again (Figure 16.4).

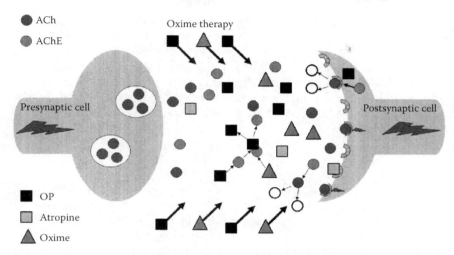

Figure 16.4 An oxime is administered either in the field in accordance with established guidelines or later in a comprehensive medical facility as supervised by a physician. The oxime breaks the esterase–organophosphate bond to reactivate the esterase.

Oxime treatment has become controversial in the literature, and its efficacy has come into question. Szinicz et al. (2007) noted several examples of severe OP human exposure, in which oxime therapy appeared ineffective or, perhaps, harmful. They also noted that oxime's efficacy highly depended on the proper dosing of atropine to sustain life while the oxime was becoming effective. In addition, Eddleston et al. (2009) performed a randomized-control study on 235 patients self-exposed to OP insecticide with WHO-guided oxime treatment versus a saline placebo. They observed a 69% increase in mortality for patients treated with oxime compared to controls. The work presented here suggests why an oxime's effectiveness may be diminished under certain circumstances.

16.2.2 Previous modeling

Gearhart et al. (1994) modeled the uptake, distribution, and elimination of an OP with additional model characteristics and state variables compared to traditional PBPK models to accommodate unique aspects of the pharmacokinetics of an OP, diisopropyl fluorophosphate (DFP). They explicitly modeled the spontaneous and enzyme-mediated hydrolysis of a specific OP (DFP) as well as the bimolecular constants for binding and releasing with common blood and tissue-based esterases (acytalcholinesterase [AChE], butyrylcholinesterase [BChE], and carboxylesterase [CaE]). In their model, the arterial blood and venous blood compartments were explicitly modeled as storage/reaction compartment state variables to account for the significant protective reactions of these esterases with the OP (including the spontaneous reversal of these reactions and the aging of the OP–esterase bond inhibiting bond reversal). Their model results reasonably agreed with *in vivo* data from the literature. They propose this approach as a good model foundation for risk assessment and further exploration of OP-AChE kinetics as well as the imposed action of therapeutic agents.

16.3   Method

A PBPK model is formulated in the style of Gearhart et al. (1994) with the arterial and venous blood compartments explicitly modeled as state variables. Figure 16.5 is a general outline of the model, and Figure 16.6 is a more detailed example of the liver compartment. (In the analysis of this work, the focus was on brain tissue; however, the liver was illustrated here to demonstrate the direct metabolism of OP by liver enzymes as an important removal mechanism from the system.) Two additional general compartments (slowly perfused tissue [SPT] and richly perfused tissue [RPT]) are used to represent the sum of tissues in human physiology and to properly account for the distribution of the entire cardiac output.

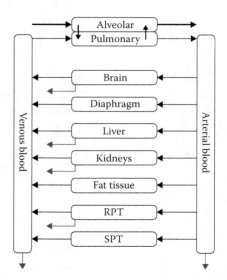

Figure 16.5 General outline of the physiologically based pharmacokinetic model with tissue groups connected by air and blood flows. Arterial and venous blood compartments are explicitly represented because of their role in organophosphate-binding reactions in addition to just flow transport. (Hanging arrows represent disappearance from the system by either exhalation, binding, or metabolic destruction.)

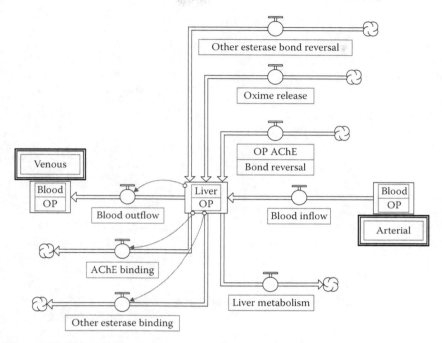

Figure 16.6 Mass balance flow rates of organophosphate around liver tissue. (Binding with esterases is an outflow; reversal of that bond is an inflow. These rates are related to the mass balance of the esterase around the liver.)

Equation 16.1 is the single ordinary differential equation (ODE) in time representing the mass balance of the OP around liver tissue, including its metabolism and its binding and binding reversal with acetylcholinesterase (and other esterases). Each term represents a flow rate in mass per time of the OP as seen in separate flows in Figure 16.6. Each state variable, or "stock," represented by the rectangular icons of the model diagram has its ODE that, all together, presents the series of simultaneous ODEs in time representing the system, which is solved numerically using STELLA version 9.0.2 (STELLA, 2007). Table 16.2 provides the parameter values used in the simulation and their literature source. Model results are produced that demonstrate behaviors in all tissue groups for the agent, the therapeutics, and their interactions for multiple exposure and treatment scenarios. The ratio of the prevailing level of active acetylcholine in synaptic space to the normal homeostasis level is used as a metric for the severity of symptoms and life status of the individual. This metric can be compared with the literature data (Ashani and Pistinner, 2004) on symptoms and survival, given an exposure scenario that is modeled to observe the corresponding value of this ratio.

Table 16.2 Parameter Values

Physiological parameters		
Body weight	60.9 kg	Gearhart et al. (1994)
Cardiac output	302 L/h	Gearhart et al. (1994)
Pulmonary ventilation rate	354 L/h	Gearhart et al. (1994)
Fractional blood flow to tissues		
Arterial	1	Asserted
Brain	0.134	Gearhart et al. (1994)
Diaphragm	0.006	Gearhart et al. (1994)
Richly perfused	0.2	Gearhart et al. (1994)
Fat	0.036	Gearhart et al. (1994)
Slowly perfused	0.1244	Gearhart et al. (1994)
Thigh	0.0066	Gearhart et al. (1994)
Kidney	0.223	Gearhart et al. (1994)
Liver	0.27	Gearhart et al. (1994)
Venous	1	Asserted
Tissue normalization factors (tissue volume per kg body weight)		
Arterial	0.02 L/kg	Gearhart et al. (1994)
Brain	0.0214 L/kg	Gearhart et al. (1994)
Diaphragm	0.003 L/kg	Gearhart et al. (1994)
Richly perfused	0.0343 L/kg	Gearhart et al. (1994)
Fat	0.17 L/kg	Gearhart et al. (1994)
Slowly perfused	0.5238 L/kg	Gearhart et al. (1994)
Thigh	0.0276 L/kg	Gearhart et al. (1994)
Kidney	0.0043 L/kg	Gearhart et al. (1994)
Liver	0.04 L/kg	Gearhart et al. (1994)
Venous	0.057 L/kg	Gearhart et al. (1994)
Organophosphate (DFP)		
Molecular weight	184 mg/mmol	Calculated
Partition coefficients (tissue/blood)		
Brain	0.67	Gearhart et al. (1994)
Diaphragm	0.77	Gearhart et al. (1994)
RPT	0.67	Gearhart et al. (1994)
Fat	17.6	Gearhart et al. (1994)
SPT	0.77	Gearhart et al. (1994)
Thigh	0.77	Gearhart et al. (1994)
Kidney	1.63	Gearhart et al. (1994)
Liver	1.53	Gearhart et al. (1994)
Arterial	1	Asserted

Table 16.2 (Continued) Parameter Values

Partition coefficients (tissue/blood)		
Venous	1	Asserted
Blood/air	12.57	Gearhart et al. (1994)

Metabolic parameters (metabolic hydrolysis [saturable])		
Brain V_{max}	688 mg/h	Gearhart et al. (1994)
Brain K_M	440 mg/L	Gearhart et al. (1994)
RPT V_{max}	560 mg/h	Gearhart et al. (1994)
RPT K_M	51 mg/L	Gearhart et al. (1994)
Kidney V_{max}	5042 mg/h	Gearhart et al. (1994)
Kidney K_M	134 mg/L	Gearhart et al. (1994)
Liver V_{max}	52,474 mg/h	Gearhart et al. (1994)
Liver K_M	237 mg/L	Gearhart et al. (1994)
Venous V_{max}	616 mg/h	Gearhart et al. (1994)
Venous K_M	199 mg/L	Gearhart et al. (1994)
Arterial V_{max}	216 mg/h	Gearhart et al. (1994)
Arterial K_M	199 mg/L	Gearhart et al. (1994)

Oxime (2-PAM Cl)

Molecular weight	132 mg/mmol	Heath et al. (2002)

Partition coefficients (tissue/blood)		
Brain	0.97	Schmitt (2008)
Diaphragm	0.92	Asserted
RPT	0.89	Schmitt (2008)
Fat	0.34	Schmitt (2008)
SPT	0.92	Schmitt (2008)
Thigh	0.92	Asserted
Kidney	0.89	Schmitt (2008)
Liver	0.78	Schmitt (2008)
Venous	1	Asserted
Arterial	1	Asserted
Blood/air	0	Asserted

Metabolic parameters		
Liver V_{max}	15 mg/h/kg BW	Sterner et al. (2010)
Liver K_M	4 mg/L	Sterner et al. (2010)

Kidney partition parameter		
Elimination fraction by glumerular filtration	0.35	Asserted

(Continued)

Table 16.2 (Continued) Parameter Values

Atropine		
Molecular weight	289 mg/mmol	Heath et al. (2002)

Partition coefficients (tissue/blood)

Brain	0.95	Merrill et al. (2010)
Diaphragm	0.92	Asserted
RPT	0.88	Merrill et al. (2010)
Fat	0.36	Merrill et al. (2010)
SPT	0.92	Merrill et al. (2010)
Thigh	0.92	Asserted
Kidney	0.88	Asserted
Liver	0.77	Merrill et al. (2010)
Venous	1	Asserted
Arterial	1	Asserted
Blood/air	0	Asserted

Metabolic parameters

Liver V_{max}	10 mg/h/kg BW	Merrill et al. (2010)
Liver K_M	30 μM	Harrison et al. (2008)

Kidney partition parameter

Elimination partition by glumerular filtration	0.35	Asserted

Acetylcholinesterase		
Molecular weight	75,000 mg/mmol	Levinson and Ellory (1974)

Synthesis rate

Arterial	0.0001 μmol/h	Gentry et al. (2002)
Brain	0.00002 μmol/h	Scaled from Gentry et al. (2002)
Diaphragm	0.000003 μmol/h	Scaled from Gentry et al. (2002)
RPT	0.00003 μmol/h	Scaled from Gentry et al. (2002)
Fat	0.0 μmol/h	Gentry et al. (2002)
SPT	0.0005 μmol/h	Scaled from Gentry et al. (2002)
Thigh	0.00002 μmol/h	Scaled from Gentry et al. (2002)
Kidney	0.000004 μmol/h	Scaled from Gentry et al. (2002)

Table 16.2 (Continued) Parameter Values

Synthesis rate		
Liver	0.00004 μmol/h	Scaled from Gentry et al. (2002)
Venous	0.0001 μmol/h	Gentry et al. (2002)
Initial concentration		
Arterial	0.001212 μM	Gentry et al. (2002)
Brain	0.04928 μM	Gentry et al. (2002)
Diaphragm	0.000909 μM	Gentry et al. (2002)
RPT	0.008314 μM	Gentry et al. (2002)
Fat	0.0 μM	Gentry et al. (2002)
SPT	0.222196 μM	Gentry et al. (2002)
Thigh	0.011708 μM	Gentry et al. (2002)
Kidney	0.000104 μM	Gentry et al. (2002)
Liver	0.002424 μM	Gentry et al. (2002)
Venous	0.003454 μM	Gentry et al. (2002)
Degradation constant (first order)		
Arterial	0.082508251 h^{-1}	Calculated from Gentry et al. (2002)
Brain	0.000405844 h^{-1}	Calculated from Gentry et al. (2002)
Diaphragm	0.00330033 h^{-1}	Calculated from Gentry et al. (2002)
RPT	0.003608371 h^{-1}	Calculated from Gentry et al. (2002)
Fat	0.0 h^{-1}	Asserted
SPT	0.002250266 h^{-1}	Calculated from Gentry et al. (2002)
Thigh	0.001708234 h^{-1}	Calculated from Gentry et al. (2002)
Kidney	0.038461538 h^{-1}	Calculated from Gentry et al. (2002)
Liver	0.01650165 h^{-1}	Calculated from Gentry et al. (2002)
Venous	0.02895194 h^{-1}	Calculated from Gentry et al. (2002)
Butyrylcholinesterase		
Molecular weight	83,000 mg/mmol	Main et al. (1971)

(Continued)

Table 16.2 *(Continued)* Parameter Values

Synthesis rate		
Arterial	0.0001 μmol/h	Gentry et al. (2002)
Brain	0.00002 μmol/h	Scaled from Gentry et al. (2002)
Diaphragm	0.000003 μmol/h	Scaled from Gentry et al. (2002)
RPT	0.00003 μmol/h	Scaled from Gentry et al. (2002)
Fat	0.0 μmol/h	Gentry et al. (2002)
SPT	0.0005 μmol/h	Scaled from Gentry et al. (2002)
Thigh	0.00002 μmol/h	Scaled from Gentry et al. (2002)
Kidney	0.000004 μmol/h	Scaled from Gentry et al. (2002)
Liver	0.00004 μmol/h	Scaled from Gentry et al. (2002)
Venous	0.0001 μmol/h	Gentry et al. (2002)
Initial concentration		
Arterial	0.00606 μM	Gentry et al. (2002)
Brain	0.016859 μM	Gentry et al. (2002)
Diaphragm	0.002 μM	Gentry et al. (2002)
RPT	0.006236 μM	Scaled from Gentry et al. (2002)
Fat	0.0 μM	Gentry et al. (2002)
SPT	0.190454 μM	Gentry et al. (2002)
Thigh	0.010035 μM	Gentry et al. (2002)
Kidney	0.000782 μM	Gentry et al. (2002)
Liver	0.019392 μM	Gentry et al. (2002)
Venous	0.017271 μM	Gentry et al. (2002)
Degradation constant (first order)		
Arterial	$0.01650165 \ h^{-1}$	Calculated from Gentry et al. (2002)
Brain	$0.00118631 \ h^{-1}$	Calculated from Gentry et al. (2002)
Diaphragm	$0.0015 \ h^{-1}$	Calculated from Gentry et al. (2002)
RPT	$0.004810776 \ h^{-1}$	Calculated from Gentry et al. (2002)
Fat	$0.0 \ h^{-1}$	Asserted

Table 16.2 (Continued) Parameter Values

Degradation constant (first order)

SPT	0.002625306 h^{-1}	Calculated from Gentry et al. (2002)
Thigh	0.001993034 h^{-1}	Calculated from Gentry et al. (2002)
Kidney	0.00511509 h^{-1}	Calculated from Gentry et al. (2002)
Liver	0.002062706 h^{-1}	Calculated from Gentry et al. (2002)
Venous	0.005790053 h^{-1}	Calculated from Gentry et al. (2002)

Carboxylesterase

Molecular weight	100,000 mg/mmol	Asserted from Koopmanschop and de Kort (1989)

Synthesis rate

Arterial	0.0001 µmol/h	Gentry et al. (2002)
Brain	0.00002 µmol/h	Scaled from Gentry et al. (2002)
Diaphragm	0.000003 µmol/h	Scaled from Gentry et al. (2002)
RPT	0.00003 µmol/h	Scaled from Gentry et al. (2002)
Fat	0.0 µmol/h	Gentry et al. (2002)
SPT	0.0005 µmol/h	Scaled from Gentry et al. (2002)
Thigh	0.00002 µmol/h	Scaled from Gentry et al. (2002)
Kidney	0.000004 µmol/h	Scaled from Gentry et al. (2002)
Liver	0.00004 µmol/h	Scaled from Gentry et al. (2002)
Venous	0.0001 µmol/h	Gentry et al. (2002)

Initial concentration

Arterial	5.0904 µM	Gentry et al. (2002)
Brain	0.778104 µM	Gentry et al. (2002)
Diaphragm	0.52722 µM	Gentry et al. (2002)
RPT	442.73754 µM	Scaled from Gentry et al. (2002)

(Continued)

Table 16.2 (Continued) Parameter Values

Initial concentration		
Fat	0.0 μM	Gentry et al. (2002)
SPT	73.007244 μM	Gentry et al. (2002)
Thigh	3.846888 μM	Gentry et al. (2002)
Kidney	4.29957 μM	Gentry et al. (2002)
Liver	110.292 μM	Gentry et al. (2002)
Venous	14.50764 μM	Gentry et al. (2002)

Degradation constant (first order)		
Arterial	1.96448×10^{-5} h^{-1}	Calculated from Gentry et al. (2002)
Brain	2.57035×10^{-5} h^{-1}	Calculated from Gentry et al. (2002)
Diaphragm	5.69022×10^{-6} h^{-1}	Calculated from Gentry et al. (2002)
RPT	6.77602×10^{-8} h^{-1}	Calculated from Gentry et al. (2002)
Fat	0.0 h^{-1}	Asserted
SPT	6.848864×10^{-6} h^{-1}	Calculated from Gentry et al. (2002)
Thigh	5.19901×10^{-6} h^{-1}	Calculated from Gentry et al. (2002)
Kidney	9.30326×10^{-7} h^{-1}	Calculated from Gentry et al. (2002)
Liver	3.626674×10^{-7} h^{-1}	Calculated from Gentry et al. (2002)
Venous	6.89292×10^{-6} h^{-1}	Calculated from Gentry et al. (2002)

Acetylcholine		
Molecular weight	146 mg/mmol	Calculated

$$V_L \frac{dC_L}{dt} = Q_L C_A + K_{\text{OP-ChE-R}} C_{\text{OP-ChE-L}} + K_{\text{OP-ChE-OX}} C_{\text{OP-ChE-L}} C_{\text{OX-L}}$$
$$- Q_L \frac{C_L}{P_{\text{OP-L/B}}} - \frac{V_{\text{max-OP-L}} C_{\text{V-OP-L}}}{K_{\text{M-OP-L}} + C_{\text{V-OP-L}}} - K_{\text{OP-ChE}} C_L C_{\text{ChE-L}} \tag{16.1}$$

where Q_L = blood flow to liver; C_A = OP concentration in arterial blood; C_L = OP concentration in liver tissue, $C_{\text{OP-ChE-L}}$ = concentration in liver of a bound OP–cholinesterase complex (separate terms for AChE, BChE, and CaE); $C_{\text{ChE-L}}$ = liver concentration of a cholinesterase (separate terms for AChE,

BChE, and CaE); C_{OX-L} = liver concentration of oxime; C_{V-OP-L} = venous blood OP concentration flowing from liver [$C_L/P_{OP-L/B}$]; $P_{OP-L/B}$ = OP liver/blood partition coefficient; $K_{OP-ChE-R}$ = kinetic rate constant for spontaneous reversal of the OP–esterase bond (separate terms for AChE, BChE, and CaE); K_{OP-ChE} = kinetic rate constant for the binding of OP with the esterase (separate terms for AChE, BChE, and CaE); $K_{OP-ChE-OX}$ = kinetic rate constant for the reversal of the OP–esterase bond by action of oxime (separate terms for AChE, BChE, and CaE); $V_{max-OP-L}$ = Michaelis–Menten maximum velocity of the saturable OP metabolism by the liver; and K_{M-OP-L} = Michaelis–Menten constant for OP metabolism in liver.

16.4   Results and discussion

The model is exercised over conditions of no exposure, mild exposure (0.1 mg/L breathing zone atmosphere for 5 minutes), and severe exposure (1.0 mg/L breathing zone atmosphere for 30 minutes). All plots are given in milligrams of chemical in brain tissue with the single plot of "Brain ACh symptoms" expressed as a unitless ratio of the prevailing level of actively bound ACh on receptors in the postsynaptic membrane to the homeostatic level (values >1 are a metric for the rise in symptoms of nerve agent poisoning, and values <1 are a metric for possible undesired symptoms brought on by the therapeutic agents themselves). First we look at no exposure and consider Figures 16.7 and 16.8; then we look at severe exposure and consider Figures 16.9 through 16.14; and finally, we look at mild exposure considering Figures 16.15 and 16.16. Figure 16.7

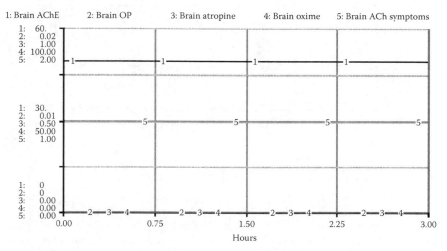

Figure 16.7 Homeostatic levels in brain tissue under conditions of no exposure and no antidotes.

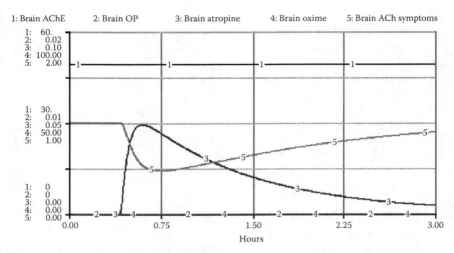

Figure 16.8 IM injection of 6 mg atropine 30 minutes after simulation starts under conditions of no nerve agent exposure.

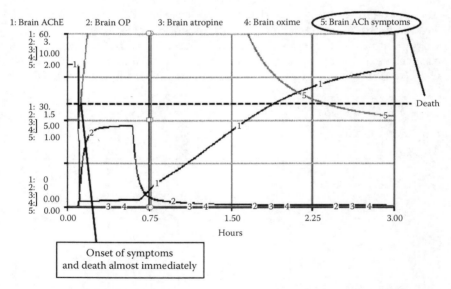

Figure 16.9 Severe 30 minute exposure with no treatment, suggesting quick onset of symptoms and death during exposure. The assessment of clinical symptom onset and death arise from empirical studies of exposure and response (Ashani and Pistinner, 2004), with modeled response of the metric chosen for this study.

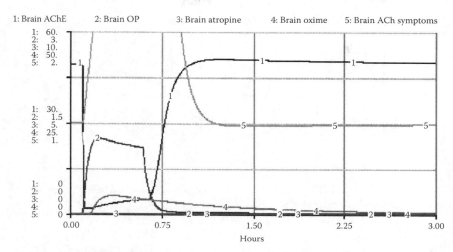

Figure 16.10 Severe nerve agent exposure with a single oxime dose of 1800 mg on onset of symptoms.

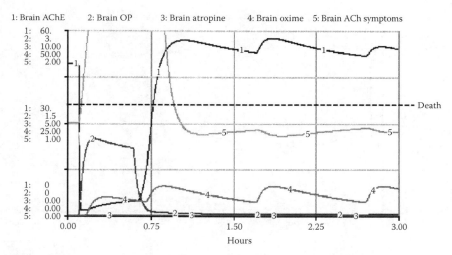

Figure 16.11 Severe nerve agent exposure with a single oxime dose of 1800 mg on onset of symptoms and repeated doses of 600 mg in accordance with current guidelines.

demonstrates the homeostatic level of acetyl cholinesterase and the metric brain ACh symptoms under conditions of no exposure and no antidotes. The amounts of OP, atropine, and oxime remain at their initial levels of zero.

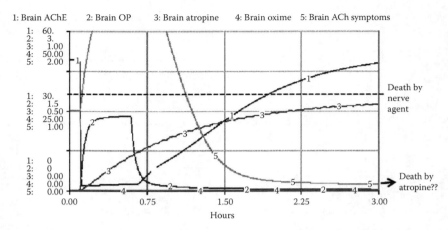

Figure 16.12 Severe nerve agent exposure with a single atropine dose of 6 mg on onset of symptoms and repeated doses of 2 mg in accordance with current guidelines.

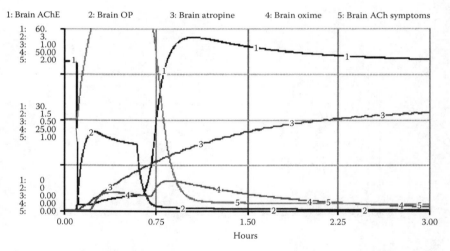

Figure 16.13 Severe nerve agent exposure with a single atropine dose of 6 mg on onset of symptoms and repeated doses of 2 mg combined with an initial 1800 mg dose of oxime and one repeated dose of 600 mg.

To illustrate the effects of antidotes only, apart from countering the effects of an OP attack, Figure 16.8 demonstrates an injection of 6 mg of atropine (as recommended by current guidance) 30 minutes after the simulation begins with no nerve agent exposure. A typical rise in a specific tissue group (brain) with follow-on decay is seen. The presence of the atropine blocks the Ach-binding sites on the postsynaptic membrane,

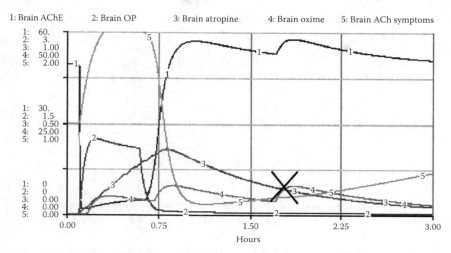

Figure 16.14 Severe nerve agent exposure with a single atropine dose of 6 mg on onset of symptoms and repeated doses of 2 mg until symptoms fall off combined with an initial 1800 mg dose of oxime and two repeated doses of 600 mg.

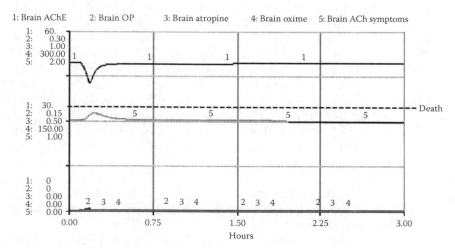

Figure 16.15 Mild nerve agent exposure with no treatment.

decreasing the prevailing actively bound ACh on the postsynaptic nerve and lowering the symptom metric curve. This exhibits the case of nerve impulses not being transmitted as desired, threatening life-critical functions. This illustrates that overuse of atropine in the case of no exposure or mild exposure could be very dangerous and must be employed with careful guidelines. As the atropine naturally decays, bound ACh begins to return to homeostasis levels.

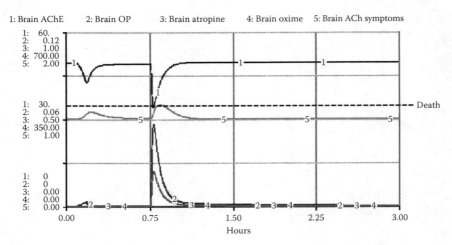

Figure 16.16 Mild nerve agent exposure with a single "prophylactic" dose of 1800 mg oxime while patient is relatively stable.

Figure 16.9 introduces a severe exposure scenario (1.0 mg/L breathing zone exposure for 30 minutes). Note that the intended detrimental response is seen dramatically just a few minutes after beginning of exposure as the AChE levels drop to near zero, offering no inhibition to the nerve impulse firing. Were the patient to survive, the natural decay of OP after cessation of exposure would allow slow recovery; however, in this case, the ACh symptom metric suggests that this individual would likely die before the cessation of exposure and before initial medical response beyond on-site injectors available for self-administration.

Figures 16.10 and 16.11 show the same severe exposure level with oxime treatment only (Figure 16.10 with one initial oxime dose of 1800 mg and Figure 16.11 with repeated doses of 600 mg according to current guidelines). Although the beneficial effect of the antidote is evident as time proceeds, were the patient to survive, it is clear that, under this exposure scenario, oxime treatment alone in accordance with current protocols will not save this patient.

Continuing the exploration of antidote treatment with severe exposure, Figure 16.12 shows the effect of repeated atropine dosing according to guidance. Even with maximum atropine dosing, the exposure appears too severe for survival. By keeping the ACh receptors blocked, this patient might be saved (were we able to control symptoms early on), but the degree of nerve firing eventually drops to lower than desired levels.

Figure 16.13 combines the two antidotes under the same severe exposure scenario. Again, by continuing repeated atropine injections long after the symptoms are under control, survival is endangered by severe understimulation within the nervous system. Figure 16.14 attempts to optimize this protocol by ceasing repeating atropine doses after cessation of symptoms. In this case, we have better results; the third dose of oxime is seen to be unnecessary, and survival may be possible if the initial effect can be controlled under this severe exposure. An exposure of this magnitude appears extremely difficult to overcome.

The case of a mild exposure (0.1 mg/L breathing zone for 5 minutes) minimizes the appearance of symptoms and can create complex treatment scenarios. We start by examining the effect of a mild exposure with no treatment (Figure 16.15). In this case, the subject is asymptomatic for some time, bringing into question whether there was any exposure at all. At this smaller scale, although difficult to see on the graph, it is seen that OP levels do not decay to zero very quickly but, rather, sustain at a nonzero (but small) level for some time and continue to exert the intended effect. This is due to reversal of the OP–esterase bond (which protects the OP from decay) and subsequent rebinding with esterase. This cycle can sustain the OP in a tissue for extended periods of time (the OP decay is significantly hindered, challenging quick and complete recovery in cases of more severe exposure).

In the case of Figure 16.15, some physicians, seeing a nearly completely stable patient with a questionable exposure history, might order an initial dosing of oxime as a prophylactic measure to assure the stable patient does not become symptomatic. In this case, that decision is simulated in Figure 16.16. The oxime appears and then decays in brain tissue as expected. The unexpected and unintended consequence is a resurgence of the OP nerve agent in brain tissue to levels much higher than that experienced during the initial exposure, and the patient (without further atropine treatment) comes very close to death within 5 minutes. This is where we must see "systematically" what is going on physiologically. Much of the OP nerve agent taken up in the original exposure is bound to butyrylcholinesterases and carboxylesterases in the blood compartment. This phenomenon has served as a shield to the other tissues until this point. When administered oximes enter the blood, they break these bonds, releasing esterases and the original OPs (and oxime–OP conjugates also serving as acetylcholinesterase inhibitors) to complete their distribution to tissues. This has major implications for postexposure treatment of OP nerve agent attack.

These results demonstrate that the model is behaving reasonably and as expected for an individual exposed to nerve agent and subsequently treated with conventional antidotes. All results are explained by the actual physiology and pharmacokinetics of the chemicals, lending validity to the model. In addition, the results demonstrate how complex the system is and how ill-considered therapies can do more harm than good. This calls for more in-depth study to promulgate a uniform set of treatment guidelines for both first responders at the scene and physicians at comprehensive care facilities after exposure. Without the knowledge of possible unintended consequences arising from the system when determining appropriate treatment protocols, results can be fatal.

16.5   Conclusions

Nerve agent exposure and the use of conventional antidotes in its treatment are more complex than may initially be assumed. The antidotes themselves (without nerve agent exposure) can bring on deleterious effects. Treatment after exposure can present unintended consequences, which can result in death that could have been avoided. Further work is needed to provide uniform guidelines in the treatment of nerve agent exposure for both first responders in the field and for physicians in more definitive care facilities.

Based on the work presented here, the following conclusions are offered to further thinking and research in this area:

1. Catastrophic incidents of nerve agent release; for example, terrorist attack or military deployment of weapon in the field.
   a.   First responders administer atropine symptomatically.
   b.   Do not administer oximes in the field.
2. Oxime indication is tricky. When atropine is required, both must be properly balanced. When patient is not symptomatic, oxime treatment could be harmful.
3. Accurate and quick data transmission to command medical facility on the level of exposure and specific agent deployed is critical for configuration and protocols used by first responders.

References

Ashani, Y., Pistinner, S. 2004. Estimation of the Upper Limit of Human Butylcholinesterase Dose Required for Protection against Organophosphates Toxicity: A Mathematically Based Toxicokinetic Model. *Toxicol. Sci.* 77, 358–367.

Cannard, K. 2006. The Acute Treatment of Nerve Agent Exposure. *J. Neurol. Sci.* 249, 86–94.

Eddleston, M., Eyer, P., Worek, F., Juszczak, E., Alder, N., Mohamed, F., Senarathna, L., et al. 2009. Pralidoxime in Acute Organophosphorus Insecticide Poisoning: A Randomized Controlled Trial. *PLoS Med.* 6(6), e1000104. doi:10.1371/journal .pmed.1000104.

Gearhart, J.M., Jepsen, G.W., Clewell, H.J., Andersen, M.E., Conolly, R.B. 1994. Physiologically Based Pharmacokinetic Model for the Inhibition of Acetylcholinesterase by Organophosphate Esters. *Environ. Health Perspect.* 102(Suppl 11), 51–60.

Gentry, P.R., Hack, C.E., Haber, L., Maier, A., Clewell, H.J. 2002. An Approach for the Quantitative Consideration of Genetic Polymorphism Data in Chemical Risk Assessment: Examples with Warfarin and Parathion. *Toxicol. Sci.* 70, 120–139.

Harrison, T.R., Fauci, A.S., Braunwald, E., Kasper, D.L., Hauser, S.L., Longu, D.L., Jameson, J.L., Luscalzo, J. (eds.) 2008. *Harrison's Principles of Internal Medicine*, 17th ed. New York: McGraw-Hill.

Heath, A.J., McKeown, R., Balali-Mood, M., Dewan, A., He, F. 2002. *Antidotes for Poisoning by Organophosphorus Pesticides: Monograph on Atropine*. Geneva: IPCS/INTOX; 41 p.

Koopmanschop, A.B., de Kortm C.A.D. 1989. Carboxylesterases of High Molecular Weight in the Hemolymph of Locusta Migratoria. *Cell. Mol. Life Sci.* 45(4), 327–330.

Levinson, S.R., Ellorym, J.C. 1974. The Molecular Form of Acetylcholinesterase as Determined by Irradiation Inactivation. *Biochem. J.* 137, 123–125.

Main, R.A., McKnelly, S.C., Burgess-Miller, S.K. 1971. A Subunit-Sized Butyrylcholinesterase Present in High Concentrations in Pooled Rabbit Serum. *Biochem. J.* 167(2), 367–376.

Merrill, E.A., Robinson, P.J., Ruark, C.D., Gearhart, J.M. 2010. Quantitative Physiological Modeling of an Antimuscarinic Agent. *DTRA Bioscience Review Conference*, May 23–27, 2010. Hunt Valley, MD. Poster Presentation.

New York Department of Health (NYDH). 2008. Chemical Terrorism Preparedness and Response Card: Information card for first responders. Available at: http://www.health.state.ny.us/environmental/emergency/chemical_terrorism/docs/chemical.pdf.

Schmitt, W. 2008. General Approach for the Calculation of Tissue to Plasma Partition Coefficients. *Toxicol In Vitro* 22, 457–467.

STELLA. 2007. Version 9.0.2. ISEE Systems Inc., Lebanon, NH.

Sterner, T.R., Ruark, C.D., Covington, T.R., Hack, C.E., Gearhart, J.M. 2010. Characterizing Uncertainty and Population Variability in a PBPK-PD Mathematical Model for the Oxime HI-6. *Chemical and Biological Defense Science and Technology Conference*, November 15–19, 2010. Orlando, FL. Poster Presentation.

Szinicz, L., Worek, F., Thiermann, H., Kehe, K., Eckert, S., Eyer, P. 2007. Development of Antidotes: Problems and Strategies. *Toxicology* 233, 23–30.

Timchalk, C., Nolan, R.J., Mendrala, A.L., Dittenber, D.A., Brzak, K.A., Mattsson, J.L. 2002. A Physiologically Based Pharmacokinetic and Pharmacodynamic (PBPK/PD) Model for the Organophosphate Insecticide Chlorpyrifos in Rats and Humans. *Toxicol. Sci.* 66, 34–53.

U.S. Army Medical Research Institute of Chemical Defense (USAMRICD). 2000. *Medical Management of Chemical Casualties Handbook*, 3rd ed. Aberdeen Proving Ground, MD: USAMRICD.

U.S. Department of Health and Services, Centers for Disease Control and Prevention (CDC). 2008. Nerve Agents: Report on Medical Management Guidelines for Nerve Agents. Available at: http://www.atsdr.cdc.gov/MHMI/mmg166.pdf. (accessed April 7, 2013)

section III

Managerial models

chapter seventeen

Coordinated project systems approach to emergency response*

Adedeji B. Badiru and LeeAnn Racz

Contents

* The material contained in this chapter was prepared as an account of work sponsored by an agency of the U.S. government. Neither the U.S. government nor any agent thereof, nor any of their employees, makes any warranty, express or implied, or assumes any legal liability or responsibility for the accuracy, completeness, or usefulness of any information, apparatus, product, or process disclosed, or represents that its use would not infringe upon privately owned rights. Reference herein to any specific commercial product, process, or service by trade name, trademark, manufacturer, or otherwise does not necessarily constitute or imply its endorsement, recommendation, or favoring by the U.S. government or any agency thereof. The views and opinions of the author expressed herein do not necessarily state or reflect those of the U.S. government or any agency thereof.

413

17.1   Introduction

Every human endeavor can be modeled as a conventional project, which deserves to be managed with the usual tools and techniques of project management. This chapter presents a concise outline of what needs to be accomplished to manage any emergency response project effectively through all its steps. The usual steps of project management go through project identification, definition, initiation, planning, organizing, resource allocation, scheduling, control, and closing. The steps in the midrange constitute project execution.

According to Hanfling et al. (2012), catastrophic disasters occurring in 2011 in the United States and worldwide have demonstrated that even prepared communities can be overwhelmed. Those disasters range from the tornado in Joplin, Missouri, and the earthquake and tsunami in Japan to the earthquake in New Zealand. Successful disaster response depends on coordination and integration across the full system of the key stakeholder groups: state and local governments, emergency medical services (EMS), public health, emergency management, hospital facilities, and the outpatient sector. Vertical integration among agencies at the federal, state, and local levels is also crucial. At the cornerstone of this coordination and integration is a foundation of ethical obligations and the legal authorities and regulatory environment that allow for shifts in expectations of the best possible response based on the context of the disaster in which that response is being executed.

The contents of the chapter suggest a proactive approach to emergency response. Colonel Cassie Barlow, 88th Air Base Wing commander at Wright Patterson Air Force Base (Dayton, Ohio), commented that "we can't wait for emergencies to happen before we respond." She also advocates closer relationships between business, industry, the military, and the community in executing emergency response programs. Such wide collaborative effort is made possible by using proactive systems-based project management approaches. As a demonstration of integrating human aspects with technology assets in emergency response, in 2012, Greene County (Ohio) initiated a multimillion dollar program to install communication technology hardware and six strategically placed radio towers that would link the county's first responders. The goal is to link the first responders to all of the community emergency agencies in the county as well as providing a linkage to Ohio's Multi-Agency Radio Communication System. All across the nation and in many international locations, efforts are being made to enhance emergency response. The premise of this chapter is that those efforts can yield better and more positive results through the application of project management tools and techniques.

17.2   Occurrence of disasters

It is a fact of life that disasters occur every now and then, regardless of whatever preemptive preparation is made. We can take preemptive and cautionary steps to mitigate the occurrence or gravity of disasters, but they cannot be completely eliminated due to the interplay of human factors and technical imperfection. Catastrophic disasters come in every form and shape ranging from different causes such as the following:

1. *Accidental incidents (e.g., mechanical failure)*
   Random events
   Assignable cause (e.g., design flaw)
   Unassignable cause (e.g., system wear-out)
2. *Natural causes (e.g., weather related)*
3. *Deliberate human acts (e.g., evil acts)*
   Terrorism
   Sabotage
   Vandalism
4. *Inadvertent mismanagement (e.g., error in judgment, improper work process, and incompetence)*
   Operator error
   Carelessness
   Incompetence
   Faulty work process
   Negligence

The risk of disasters cannot be totally eliminated. All that can be done is to mitigate the risk. As a result, it is essential to have a standing guide for emergency response. Conceptual preparedness is one thing, but the reality of execution is another thing. A lack of communication or the occurrence of miscommunication can spell doom for the response effort. The importance of communication and coordination provides the basis for the current exposition of the human systems issues in emergency response. Disasters lead to traumatic incidents that may hamper relief project efforts. Typical results of a disaster include, but are not limited to, the following:

- Infrastructure damage
- Destruction of roads, bridges, and access ways
- Deaths
- Injuries
- Displacement of people
- Asset destruction

- Property damage
- Looting and pillaging
- Lawlessness
- Social unrest

Disasters lead to traumatic incidents that may hamper relief efforts. With the occurrence of disaster outcomes and the accompanying confusion, there will be incidents of communication breakdown, cooperation failure, and lack of coordination. To get everything back in order, a systems-oriented model is needed. The importance of communication and coordination provides the basis for the recommendation to use the Triple C model of project management (Badiru, 2012) in this chapter. The origin of the Triple C model is a successful application to an emergency response at Tinker Air Force Base in 1985 (Badiru et al., 1993).

17.3   Coordinated response to emergency

It is obvious that coordination is the most crucial requirement for effective emergency response. Such coordination is predicated on the human factors and managerial processes in the emergency situation. The foundation for coordination is cooperation, which depends on prior communication. This is why a collaborative structure is needed. On the basis of the diversity of events and issues that can develop during a disaster and the ensuing emergency response, a coordinated plan of action must be put in place and executed on the fly. Prior planning, practice, and reinforcement exercises are essential and must recognize the tools and techniques available from the following topics:

- Project systems analysis
- Systems engineering framework
- Human factors considerations

A coordinated response to emergency will often involve the military, which brings diversity of transportation assets, manpower resources, command and control processes, and well-honed skill sets to bear during an emergency situation. The expedited response of the military can bring efficiency and effectiveness into the overall emergency response. If a coordinated and systematic response program is not in place, citizens will run helter-skelter. As the military response scenario in Figure 17.1 shows, emergency response must be fast, effective, and uncompromised. Mobility, configurability, and modularity are desired elements of coordination in a response effort involving military and civilian entities. Using a systems approach, as suggested by NRC (2007) and Kossiakoff and Sweet (2003), increases the potential for a successful coordinated response.

Figure 17.1 Illustration of military–civilian disaster response scenario.

17.4   Using a project systems framework

Like any other resource-constrained and time-sensitive endeavor, emergency response programs require conventional planning, organization, scheduling, control, and phaseout. These are standard management functions for which project management tools and techniques have proven very effective. Badiru (2012) defines project management as follows:

> Project management is the process of managing, allocating, and timing resources to achieve a specific goal in an efficient and expedient manner.

The above definition suggests a systematic integration of technical, human, and other resources to achieve goals and objectives. This is precisely what is needed in a coordinated emergency response, where a multitude of goals and objectives interplay in a fast-paced environment. In this chapter, emergency response challenges are seen as being amenable to a rigorous project management approach. Far too many response projects are initiated without a careful project management framework. From a project management perspective, there are several questions to consider when executing emergency response. Some pertinent questions include the following:

1. How should emergency response be planned and executed in the presence of time-sensitive relief requirements?
2. What impact will response technology changes have on the emergency response infrastructure and facilities?
3. How agile, adaptive, and resilient is the response infrastructure to shifts in community needs?
4. What resource allocation strategies are in place to ensure sustainability of the response operation?
5. How adaptive is the project plan in the presence of demands and requests from the shareholders, city administrators, and local governments?

Emergency response is very complex with many stakeholders, diverse inputs to the system, constraints on the work processes, and changes in resource availability. As in any project, an emergency response is executed to achieve one or more of three basic objectives:

1. Physical product—for example, configure a rescue tool
2. Service—for example, provide public assistance
3. Result—for example, accomplish a successful rescue

Of importance across these basic objectives is the recognition of the interrelationships between the components of the disaster event touching on the following categories of involvement:

1. People—first responders, victims, and so on
2. Technology—communication hardware, transportation vehicle, medical device, and so on
3. Process—search and rescue processes, policies, procedures, permissions, and so on

17.5   Using a participative and inclusive model

On the basis of the systems view and the human factors issues discussed earlier, it is essential to use a participative and inclusive model as illustrated in Figure 17.2. Everyone has a coordinated role to play, from the federal level to the personal level, where individual actions and reactions are paramount for a successful coordinated emergency response. The role of the federal government is to provide assurance and resources to execute the emergency response. If possible, business and industry must continue to operate to provide basic amenities, goods, and services that

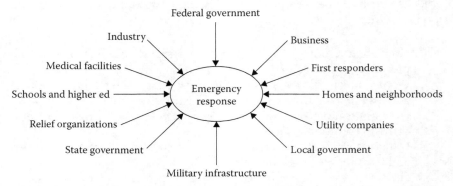

Figure 17.2 Participative and inclusive model for systems-oriented emergency response.

may be needed for the emergency response. This is particularly important for cases where the impact of the emergency situation or disaster may be protracted, stretching over several days or weeks. Medical facilities must remain open and accessible to cater to the needs of victims. First responders must be ready, agile, and adaptive as situations on the ground evolve. Schools and higher education facilities are often used as shelters in emergency situations. As such, they must be robust and resilient. Relief organizations, working in consonance with victims, must be visible but not constitute an impediment to the movement of first responders. Individuals, homes, and neighborhoods must be cooperative in following response instructions, announcements, and alerts. Utility companies must be swift in creating a safe pathway for rescue and evacuation teams to have access to emergency response structures. Finally, local and state governments must cooperate to work within the response decision hierarchy. All units must be involved in a collaborative, inclusive, and participatory systems framework. A single point of being obstinate can prove to be the cog in the wheel of the response system.

17.6   Executing the steps of project management

Figure 17.3 illustrates the steps of a project framework starting with the identification of a need and ending with a phaseout after the need has been met. The steps provide a structured framework for executing an emergency response endeavor ranging from identifying the need for a response, establishing response objectives, rescue planning, personnel organization, resource allocation, task scheduling, progress tracking, process control, and termination of response. Response termination, in this case, does not imply conclusion of the overall effort, but rather the

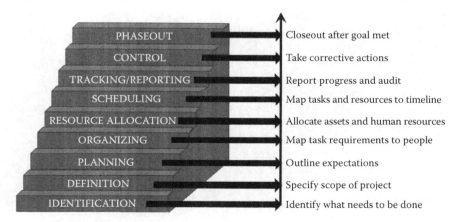

Figure 17.3 Steps of project management applied to emergency response.

completion of a particular phase, which may then serve as the initiating point for other follow-on efforts, such as rebuilding. Further explanations of the elements in the step are provided below.

Identification: Identification is the process of recognizing the need for an emergency response. This recognition may be initiated by the announcement of the occurrence of a disaster.

Definition: Project definition requires an explicit articulation of what exactly should be accomplished by the response effort, identifying goals and strategies to meet them. This step outlines a clear understanding concerning what the final accomplishment is expected to look like. Questions that can be asked during this step that can preempt or mitigate problems later on include the following:

1. Who is in charge of the emergency response program?
2. When will the response be initiated?
3. How will the response contribute to the overall welfare of the community?
4. How will the emergency response affect different groups of people within the community?
5. What organizations within the community will be involved?
6. What will happen after the emergency response is implemented?
7. How do the emergency response objectives fit the overall priority of the community?

Planning: Expectations of the emergency response, quality control techniques, communication hierarchy, technology selection, and transportation modes are established at this step. Planning should not commence until a clear definition of the project objectives in the previous step have been developed. Quality control techniques include expert reviews to detect logical errors and logistical impediments. Stakeholders can benchmark other successful response programs when designing and outlining the structure and activities of the response. Also during this step, methods that will be used to assess the performance of the program should be outlined.

Organization: Organization requires a mapping of tasks and responsibilities to specific individuals involved in the emergency response. Individuals and organizations that will contribute to the response are identified. If possible, a responsibility matrix can be used as an effective tool for organizing projects. Simple codes may be used to outline who has what responsibility. Examples are R (Responsible), I (Inform), S (Support), and C (Consult) while the task tracking codes could be D (Done), O (On Track), and L (Late).

Resource allocation: Resource allocation is a critical part of executing a project effectively. One common complaint about emergency response infrastructure is inadequate allocation of resources (i.e., human assets, technology, and service support). The needs of first responders and victims should dictate the type of technology or combination of technologies used to deliver emergency response services. Without a dependable technology, emergency responders are likely to become frustrated, disappointed, and detached, which can be detrimental to the overall goal. When resource allocation is addressed as an explicit step, the process of communicating resource requirements to decision makers is better structured and more effective.

Task scheduling: Task scheduling is often the most visible part of project management, but it is not the first step, as most people often falsely assume. If the preceding steps of project management are executed properly, they will pave the way for a more effective task scheduling process. This will prevent scheduling conflicts between first responders, rescuers, facility managers, and law enforcement officers.

Tracking: Tracking the response project once it has been launched is important. Tracking entails reporting and auditing the performance of those involved in the operations to ensure that actions are aligned with plans, goals, and standards.

Control: At the control step, corrective actions are taken to determine whether operations meet expectations. Corrective actions must be taken in a timely manner before minor incidents become major events.

Phaseout: A phaseout denotes that a specific goal has been accomplished and efforts can proceed to the next contiguous undertaking in the overall emergency response project.

17.7 Using systems engineering framework

The George Washington University Institute for Crisis, in a guide entitled *Disaster and Risk Management* provides a definition of a systems approach to emergency response as a "management strategy that recognizes that disparate components must be viewed as interrelated components of a single system, and so employs specific methods to achieve and maintain the overarching system." These methods include the use of standardized structure and processes and foundational knowledge and concepts in the conduct of all related activities. This aligns directly with the project management and systems approach theme presented in this chapter as a recommended approach.

Any project is ultimately a component of a larger system. As such, a systems view facilitates a complete look at all factors involved in emergency response. Systems engineering efficiency and effectiveness are of interest across the spectrum of project management for the purpose of improving organizational performance. Emergency responders should be interested in having systems engineering serve as the umbrella for a coordinated approach to emergency response. This will get everyone properly connected with the prevailing organizational goals as well as create collaborative avenues among the responders. Systems application applies across the spectrum of any organization and encompasses the following elements:

- Technological systems (e.g., engineering control systems and mechanical systems)
- Organizational systems (e.g., work process design and operating structures)
- Human systems (e.g., interpersonal relationships and human–machine interfaces)

A systems view of emergency response will make everything work better and projects more likely to succeed. A systems view provides a disciplined process for the design, development, and execution of complex projects both in engineering and nonengineering organizations. One of the major advantages of a systems approach is the win-win benefit for everyone. A systems view also allows full involvement of all stakeholders of a response effort. Every emergency response environment is very complex because of the diversity of factors involved. There are differing human personalities, differing technical requirements, differing expectations, and differing environmental factors. Each specific context and the prevailing circumstances will determine the specific structure of what can and cannot be done in the effort. The best approach for effective project management is to adapt to what each project needs. This requires taking a systems view of the project. Project management represents an excellent platform for the implementation of a systems approach. It integrates various technical and management requirements. Project management requires control techniques, such as operations research, operations management, forecasting, quality control, and simulation to deliver goals. In systems-based project management, it is essential that related techniques be employed in an integrated fashion so as to maximize the total project output. There should not be an overreliance on technology nor should there be an overdependence on human processes. Similarly, there should not be too much emphasis on analytical models to the detriment of common sense human-based decisions. The following is the comprehensive systems engineering definition that illustrates the importance of a systems framework in emergency response:

Systems engineering is the application of engineering to solutions of a multifaceted problem through a systematic collection and integration of parts of the problem with respect to the lifecycle of the problem. It is the branch of engineering concerned with the development, implementation, and use of large or complex systems. It focuses on specific goals of a system considering the specifications, prevailing constraints, expected services, possible behaviors, and structure of the system. It also involves a consideration of the activities required to assure that the system's performance matches the stated goals. Systems engineering addresses the integration of tools, people, and processes required to achieve a cost-effective and timely operation of the system.

Logistics can be defined as the planning and implementation of a complex task, the planning and control of the flow of goods and materials through an organization or manufacturing process, or the planning and organization of the movement of personnel, equipment, and supplies. Just as it is in the corporate world, logistics plays a direct role in emergency response activities. Complex projects represent a hierarchical system of operations. Thus, we can view an emergency response as a collection of interrelated projects all serving a common end goal. Consequently, the following definition is applicable to the diversity of needs in an emergency response situation:

Project systems logistics is the planning, implementation, movement, scheduling, and control of people, equipment, goods, materials, and supplies across the interfacing boundaries of several related projects.

Conventional project management must be modified and expanded to address the unique logistics of emergency response.

17.8 Human factors considerations

Human attitude is a big part of emergency response. Some of the human issues that can impede a smooth emergency response include the following:

- Bureaucracy
- Complacency
- Incompetence

Figure 17.4 Top-down and bottom-up communication interfaces.

- Indifference
- Arrogance
- Inaccessibility
- Personal rigidity
- Obstinate
- Stubbornness
- Administrative inflexibility

Human behavior in emergency situations is subject to the *Cry Wolf Syndrome*, whereby repeated warnings become so persistent that the warnings lose effectiveness, validity, and impact. Citizens may then become complacent and indifferent to warnings by giving emergency announcements a cold shoulder. These adverse behaviors can be mitigated by reemphasizing the structure of top-down and bottom-up communication in emergency response scenarios. Figure 17.4 illustrates the communication hierarchy using a systems framework. This suggests that each interface within the communication structure should be managed to maximize the perception of gravity of what is communicated.

17.9   *Application of the Triple C model*

Managing a project in the wake of a tragedy is complicated and involves various emotional, sentimental, reactive, and chaotic responses. This makes it difficult to clearly pursue the traditional methods of planning and execution. This is the time that a structured communication model is most needed. Unfortunately, practical models are often not readily available. Conventional wisdom and assumptions, in lieu of direct communication, portend failure for a project. At a time of disaster, reactive rhetoric may not coincide with action or reality. Having a guiding model can help put things in proper focus.

Using the basis of communication, cooperation, and coordination presented earlier, we now present a Triple C–based model for disaster rapid response. The Triple C model presented by Badiru (2012) is an

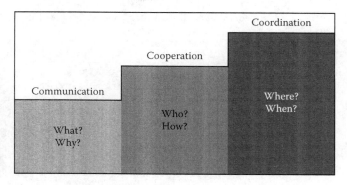

Figure 17.5 Steps of Triple C model for managing emergency response.

effective tool for emergency response project planning. The model states that project management can be enhanced by implementing it within the integrated functions of

- Communication
- Cooperation
- Coordination

Figure 17.5 shows the steps of the Triple C model with the specification of what, why, who, how, where, and when parameters as they relate to managing emergency response.

The model facilitates a systematic approach to project planning, organizing, scheduling, and control. The Triple C model complements the usual military command, control, and communication processes as may be applied to emergency response. The model suggests communication as the first and foremost function in any coordinated endeavor. It highlights what must be done and when. It can also help to identify the resources (personnel, equipment, facilities, etc.) required for each effort. It points out important questions such as the following:

- Does each project participant know what the objective is?
- Does each participant know his or her role in achieving the objective?
- What obstacles may prevent a participant from playing his or her role effectively?

An adaptive version of the Triple C model for disaster response uses integrative and inclusive process so that communication, cooperation, and coordination points are identified throughout the stages of implementing the model. Table 17.1 illustrates the level-to-level interfaces for communication, cooperation, and coordination within the top-down and bottom-up response hierarchy.

Table 17.1 Top-down and Bottom-up Disaster Response
Decision Hierarchies

Top-down response hierarchy	Bottom-up response hierarchy
Federal level	Personal level
State level	Household level
Local level	Neighborhood level
Institutional level	Company level
Community level	Community level
Company level	Institutional level
Neighborhood level	Local level
Household level	State level
Personal level	Federal level

17.9.1   Communication for emergency response

Communication makes working together possible. The communication function of project management involves making all those concerned aware of project requirements and progress. Those who will be affected by the project directly or indirectly, as direct participants or as beneficiaries, should be informed as appropriate regarding the following:

- Scope of the project
- Personnel contribution required
- Expected cost, schedule, performance requirements, and merits of the project
- Project organization and implementation plan
- Potential adverse effects if the project should fail
- Alternatives, if any, for achieving the project goal
- Potential direct and indirect benefits of the project

The communication channel must be kept open throughout the project life cycle. In addition to internal communication, appropriate external sources should also be consulted. The project manager must

- Exude commitment to the project
- Utilize the communication responsibility matrix
- Facilitate multichannel communication interfaces
- Identify internal and external communication needs
- Resolve organizational and communication hierarchies
- Encourage both formal and informal communication links

When clear communication is maintained between management and employees and among peers, many project problems can be averted.

Project communication may be carried out in one or more of the following formats:

- One-to-many
- One-to-one
- Many-to-one
- Written and formal
- Written and informal
- Oral and formal
- Oral and informal
- Nonverbal gestures

Good communication is affected when what is implied is perceived as intended. Effective communications are vital to the success of any project. Despite the awareness that proper communications form the blueprint for project success, many organizations still fail in their communications functions. The study of communication is complex. Factors that influence the effectiveness of communication within a project organization structure include the following:

1. *Personal perception:* Each person perceives events on the basis of personal psychological, social, cultural, and experimental background. As a result, no two people can interpret a given event the same way. The nature of events is not always the critical aspect of a problem situation. Rather, the problem is often the different perceptions of the different people involved.
2. *Psychological profile:* The psychological makeup of each person determines personal reactions to events or words. Thus, individual needs and level of thinking will dictate how a message is interpreted.
3. *Social environment:* Communication problems sometimes arise because people have been conditioned by their prevailing social environment to interpret certain things in unique ways. Vocabulary, idioms, organizational status, social stereotypes, and economic situation are among the social factors that can thwart effective communication.
4. *Cultural background:* Cultural differences are among the most pervasive barriers to project communications, especially in today's multinational organizations. Language and cultural idiosyncrasies often determine how communication is approached and interpreted.
5. *Semantic and syntactic factors:* Semantic and syntactic barriers to communications usually occur in written documents. Semantic factors are those that relate to the intrinsic knowledge of the subject of the communication. Syntactic factors are those that relate to the form in which the communication is presented. The problems created by

these factors become acute in situations where response, feedback, or reaction to the communication cannot be observed.

6. *Organizational structure:* Frequently, the organization structure in which a project is conducted has a direct influence on the flow of information and, consequently, on the effectiveness of communication. Organization hierarchy may determine how different personnel levels perceive a given communication.

7. *Communication media:* The method of transmitting a message may also affect the value ascribed to the message and, consequently, how it is interpreted or used. The common barriers to project communications are

- Inattentiveness
- Lack of organization
- Outstanding grudges
- Preconceived notions
- Ambiguous presentation
- Emotions and sentiments
- Lack of communication feedback
- Sloppy and unprofessional presentation
- Lack of confidence in the communicator
- Lack of confidence by the communicator
- Low credibility of communicator
- Unnecessary technical jargon
- Too many people involved
- Untimely communication
- Arrogance or imposition
- Lack of focus

Some suggestions on improving the effectiveness of communication are presented next. The recommendations may be implemented as appropriate for any of the forms of communications listed earlier. The recommendations are for both the communicator and the audience.

1. Never assume that the integrity of the information sent will be preserved as the information passes through several communication channels. Information is generally filtered, condensed, or expanded by the receivers before relaying it to the next destination. When preparing a communication that needs to pass through several organization structures, one safeguard is to compose the original information in a concise form to minimize the need for recomposition.

2. Give the audience a central role in the discussion. A leading role can help make a person feel a part of the project effort and responsible for the project's success. He or she can then have a more constructive view of project communication.

3. Do homework and think through the intended accomplishment of the communication. This helps eliminate trivial and inconsequential communication efforts.
4. Carefully plan the organization of the ideas embodied in the communication. Use indexing or points of reference whenever possible. Grouping ideas into related chunks of information can be particularly effective. Present the short messages first. Short messages help create focus, maintain interest, and prepare the mind for the longer messages to follow.
5. Highlight why the communication is of interest and how it is intended to be used. Full attention should be given to the content of the message with regard to the prevailing project situation.
6. Elicit the support of those around you by integrating their ideas into the communication. The more people feel they have contributed to the issue, the more expeditious they are in soliciting the cooperation of others. The effect of the multiplicative rule can quickly garner support for the communication purpose.
7. Be responsive to the feelings of others. It takes two to communicate. Anticipate and appreciate the reactions of members of the audience. Recognize their operational circumstances and present your message in a form they can relate to.
8. Accept constructive criticism. Nobody is infallible. Use criticism as a springboard to higher communication performance.
9. Exhibit interest in the issue to arouse the interest of your audience. Avoid delivering your messages as a matter of a routine organizational requirement.
10. Obtain and furnish feedback promptly. Clarify vague points with examples.
11. Communicate at the appropriate time, at the right place, to the right people.
12. Reinforce words with positive action. Never promise what cannot be delivered. Value your credibility.
13. Maintain eye contact in oral communication and read the facial expressions of your audience to obtain real-time feedback.
14. Concentrate on listening as much as speaking. Evaluate both the implicit and explicit meanings of statements.
15. Document communication transactions for future references.
16. Avoid asking questions that can be answered yes or no. Use relevant questions to focus the attention of the audience. Use questions that make people reflect upon their words, such as, "How do you think this will work?" compared to "Do you think this will work?"
17. Avoid patronizing the audience. Respect their judgment and knowledge.

18. Speak and write in a controlled tempo. Avoid emotionally charged voice inflections.
19. Create an atmosphere for formal and informal exchange of ideas.
20. Summarize the objectives of the communication and how they will be achieved.

A communication responsibility matrix can be used to show the linking of sources of communication and targets of communication. Cells within the matrix would indicate the subject of the desired communication. With a communication responsibility matrix, a clear understanding of what needs to be communicated to whom can be developed. Communication in a project environment can take any one of several forms. The specific needs of a project may dictate the most appropriate mode. Three popular computer communication modes are discussed next in the context of communicating data and information for emergency response.

Simplex communication: This is a unidirectional communication arrangement in which one project entity initiates communication to another entity or individual within the project environment. The entity addressed in the communication does not have the mechanism or capability to respond to the communication. An extreme example of this is a one-way, top-down communication from top management to the project personnel. In this case, the personnel have no communication access or input to top management. A budget-related example is a case where top management allocates budget to a project without requesting and reviewing the actual needs of the project. Simplex communication is common in authoritarian organizations.

Half-duplex communication: This is a bidirectional communication arrangement whereby one project entity can communicate with another entity and receive a response within a certain time lag. Both entities can communicate with each other, but not at the same time. An example of half-duplex communication is a project organization that permits communication with top management without a direct meeting. Each communicator must wait for a response from the target of the communication. Request and allocation without a budget meeting is another example of half-duplex data communication in project management.

Full-duplex communication: This involves a communication arrangement that permits a dialogue between the communicating entities. Both individuals and entities can communicate with each other at the same time or face-to-face. As long as there is no clash of words, this appears to be the most receptive communication mode. It allows participative project planning in which each member of the project has an opportunity to contribute to the planning process.

Each member of a response team needs to recognize the nature of the prevailing communication mode in the response environment. An evaluation of who is to communicate with whom about what may help improve the project data/information communication process. A communication matrix may include notations about the desired modes of communication between individuals and groups in the project environment.

17.9.2 *Cooperation for emergency response*

The cooperation of the project personnel must be explicitly elicited. Merely voicing consent for a project is not enough assurance of full cooperation. The participants and beneficiaries of the project must be convinced of the merits of the project. Some of the factors that influence cooperation in a project environment include personnel requirements, resource requirements, budget limitations, past experiences, conflicting priorities, and lack of uniform organizational support. A structured approach to seeking cooperation should clarify the following:

- Cooperative efforts required
- Precedents for future projects
- Implication of lack of cooperation
- Criticality of cooperation to project success
- Organizational impact of cooperation
- Time frame involved in the project
- Rewards of good cooperation

Cooperation is a basic virtue of human interaction. More projects fail due to lack of cooperation and commitment than any other project factors. To secure and retain the cooperation of project participants, you must elicit a positive first reaction to the project. The most positive aspects of a project should be the first items of project communication. For project management, there are different types of cooperation that should be understood.

Functional cooperation: This is cooperation induced by the nature of the functional relationship between two groups. The two groups may be required to perform related functions that can only be accomplished through mutual cooperation.

Social cooperation: This is the type of cooperation affected by the social relationship between two groups. The prevailing social relationship motivates cooperation that may be useful in getting project work done.

Legal cooperation: Legal cooperation is the type of cooperation that is imposed through some authoritative requirement. In this case, the participants may have no choice other than to cooperate.

Administrative cooperation: This is the cooperation brought on by administrative requirements that make it imperative that two groups work together on a common goal.

Associative cooperation: This type of cooperation may also be referred to as collegiality. The level of cooperation is determined by the association that exists between two groups.

Proximity cooperation: Cooperation due to the fact that two groups are geographically close is referred to as proximity cooperation. Being close makes it imperative that the two groups work together.

Dependency cooperation: This is the cooperation caused by the fact that one group depends on another group for some important aspect. Such dependency is usually of a mutual two-way nature. One group depends on the other for one thing, whereas the latter group depends on the former for some other thing.

Imposed cooperation: In this type of cooperation, external agents must be employed to induced cooperation between two groups. This is applicable for cases where the two groups have no natural reason to cooperate. This is where the approaches presented earlier for seeking cooperation became very useful.

Lateral cooperation: Lateral cooperation involves cooperation with peers and immediate associates. Lateral cooperation is often easy to achieve because existing lateral relationships create a conducive environment for project cooperation.

Vertical cooperation: Vertical or hierarchical cooperation refers to cooperation that is implied by the hierarchical structure of the project. For example, subordinates are expected to cooperate with their vertical superiors.

Whichever type of cooperation is available in a project environment, the cooperative forces should be channeled toward achieving project goals. Documentation of the prevailing level of cooperation is useful for winning further support for a project. Clarification of project priorities will facilitate personnel cooperation. Relative priorities of multiple projects should be specified and communicated broadly so that all groups within the organization understand the relative importance of work. Some guidelines for securing cooperation for most projects are the following:

- Establish achievable goals for the project
- Clearly outline the individual commitments required
- Integrate project priorities with existing priorities
- Eliminate the fear of job loss due to industrialization
- Anticipate and eliminate potential sources of conflict
- Use an open-door policy to address project grievances
- Remove skepticism by documenting the merits of the project

Commitment: Cooperation must be supported with commitment. To cooperate is to support the ideas of a project. To commit is to willingly and actively participate in project efforts again and again through the thick and thin of the project. Provision of resources is one way that management can express commitment to a project.

17.10   Triple C + Commitment = Project success

17.10.1   Coordination in emergency response

After the communication and cooperation functions have successfully been initiated, the efforts of the project personnel must be coordinated. Coordination facilitates harmonious organization of project efforts. The construction of a responsibility chart can be very helpful at this stage. A responsibility chart is a matrix consisting of columns of individual or functional departments and rows of required actions. Cells within the matrix are filled with relationship codes that indicate who is responsible for what. The matrix helps avoid neglecting crucial communication requirements and obligations. It can help resolve questions such as

- Who is to do what?
- How long will it take?
- Who is to inform whom of what?
- Whose approval is needed for what?
- Who is responsible for which results?
- What personnel interfaces are required?
- What support is needed from whom and when?

17.10.2   Resolving emergency response conflicts

When implemented as an integrated process, the Triple C model can help avoid conflicts in emergency response. When conflicts do develop, Triple C can help in resolving the conflicts. Several sources of conflicts can exist in large emergency response scenarios. Some of these are discussed below.

Schedule conflict: Conflicts can develop because of improper timing or sequencing of project tasks. This is particularly common in large multiple projects. Procrastination can lead to having too much to do at once, thereby creating a clash of project functions and discord among project team members. Inaccurate estimates of time requirements may lead to infeasible activity schedules. Project coordination can help avoid schedule conflicts.

Cost conflict: Project cost may not be generally acceptable to the clients of a project. This will lead to project conflict. Even if the initial cost of the project is acceptable, a lack of cost control during implementation can lead to conflicts. Poor budget allocation approaches and the lack of a financial feasibility study will cause cost conflicts later on in a project. Communication and coordination can help prevent most of the adverse effects of cost conflicts.

Performance conflict: If clear performance requirements are not established, performance conflicts will develop. Lack of clearly defined performance standards can lead each person to evaluate his or her own performance based on personal value judgments. In order to uniformly evaluate quality of work and monitor project progress, performance standards should be established by using the Triple C approach.

Management conflict: There must be a two-way alliance between the management and the project team. The views of the management should be understood by the team. The views of the team should be appreciated by the management. If this does not happen, management conflicts will develop. A lack of a two-way interaction can lead to strikes and industrial actions, which can be detrimental to project objectives. The Triple C approach can help create a conducive dialog environment between the management and the project team.

Technical conflict: If the technical basis of a project is not sound, technical conflict will develop. New industrial projects are particularly prone to technical conflicts because of their significant dependence on technology. Lack of a comprehensive technical feasibility study will lead to technical conflicts. Performance requirements and systems specifications can be integrated through the Triple C approach to avoid technical conflicts.

Priority conflict: Priority conflicts can develop if project objectives are not defined properly and applied uniformly across a project. Lack of a direct project definition can lead each project member to define his or her own goals that may be in conflict with the intended goal of a project. Lack of consistency of the project mission is another potential source of priority conflicts. Overassignment of responsibilities with no guidelines for relative significance levels can also lead to priority conflicts. Communication can help defuse priority conflict.

Resource conflict: Resource allocation problems are a major source of conflict in project management. Competition for resources, including personnel, tools, hardware, software, and so on, can lead to disruptive clashes among project members. The Triple C approach can help secure resource cooperation.

Power conflict: Project politics lead to a power play that can adversely affect the progress of a project. This can manifest itself in terms of power play between the various groups involved in emergency response programs. Project authority and project power should be clearly delineated. Project authority is the control that a person has by virtue of his or her functional post. Project power relates to the clout and influence that a person can exercise due to connections within the administrative structure. People with popular personalities can often wield a lot of project power in spite of low or nonexistent project authority. The Triple C model can facilitate a positive marriage of project authority and power to the benefit of project goals. This will help define a clear leadership for a project.

Personality conflict: Personality conflict is a common problem in projects involving a large group of people. The larger the project, the larger the size of the management team needed to keep things running. Unfortunately, the larger management team creates an opportunity for personality conflicts. Communication and cooperation can help defuse personality conflicts.

In summary, conflict resolution through Triple C can be achieved by observing the following guidelines:

1. Confront the conflict and identify the underlying causes
2. Be cooperative and receptive to negotiation as a mechanism for resolving conflicts
3. Distinguish between proactive, inactive, and reactive behaviors in a conflict situation
4. Use communication to defuse internal strife and competition
5. Recognize that short-term compromise can lead to long-term gains
6. Use coordination to work toward a unified goal
7. Use communication and cooperation to turn a competitor into a collaborator

The application of the Triple C model is in direct alignment with contingency planning strategies for emergency response as summarized in Table 17.2.

17.11   Coordination of equipment capabilities

Having the right equipment at the right time in the right place is an essential part of having a successful emergency response. Consider the fire truck in Figure 17.6. All elements must be tactically coordinated to cover people-process-equipment interfaces. For Figure 17.6, pertinent questions include the following:

- Can the ladder extend high enough to reach the target firefighting point?
- How far up are we trying to reach?

Table 17.2 Emergency Contingency Planning Compared
to Conventional Planning

Characteristics	Preparedness planning	Contingency planning	Operations planning
When (contingency)	Planning phase	Before the emergency	During emergency
What (scope of plan)	General	Time-frame specific	More specific
Who (involved partners)	Everybody within system	People knowledgeable	People actually involved
How (focus)	All types	Specific/ projected	Actual
How (planning style)	Long term, global	Specific time frame	Actual
What (allocation)	Estimated	Quantified	Precise
Where (planning level)	All levels	Managerial level	Actual/field level
When (time frame)	Annual (1 year)	Specific (but uncertain)	Executed right time, fixed
Who (relationships)	Long term	Developing	Utilizing

- Is the terrain stable enough to accommodate the fire truck?
- What aspects of the ladder height might pose dangers to firefighters?
- Who goes on the ladder first?
- Does the size of the truck adversely affect mobility?

Similar relevant questions should be a priori evaluated for each piece of equipment that is to be used in emergency response. A good strategy to pursue is that we should not be encumbered by limitations of equipment capabilities. The sky should be the limit in our preparations, even though such extreme cases may never occur. Practice and preparation are the essentials of a well-coordinated emergency response.

17.12   Conclusions

Education is an important part of any coordinated emergency response. Using a systems approach will foster collaborative interdisciplinary alliances among schools, higher education academic programs, and community groups through research and development and service programs. Graduate students can plan direct research roles through studies dealing with not only technical systems but also social sciences and behavioral studies. Information technology and cyber infrastructure can also

Figure 17.6 Assessment of equipment capabilities and limitations for emergency response.

impinge upon how a response program is executed. New academic specialization tracks can evolve to address specific issues and unique issues of emergency response. These could include the following:

- Systems Integration, Systems Management, and Resilience Engineering
- Computational Statistics, Simulation, and Risk Analysis to assess emergency-related data sets

- Optimization and Control for Emergency Medical Response
- Organizational Processes and Human Behavioral Tendencies
- Information Technology Research and Product Development for Emergency Response

References

Badiru, Adedeji B. (2012), *Project Management: Systems, Principles, and Applications*, CRC Press/Taylor and Francis, Boca Raton, FL.

Badiru, Adedeji B., B. L. Foote, L. Leemis, A. Ravindran, and L. Williams, Recovering from a Crisis at Tinker Air Force Base, *PM Network*, Vol. 7, No. 2, Feb. 1993, pp. 10–23.

Hanfling, Dan., et al. (2012), *Crisis Standards of Care: A Systems Framework for Catastrophic Disaster Response*, The National Academies Press, Washington, DC. http://www.nap.edu/catalog.php?record_id=13351.

Kossiakoff, Alexander and W. N. Sweet. (2003), *Systems Engineering: Principles and Practice*, Wiley InterScience, Hoboken, NJ.

NRC, National Research Council. (2007), *Improving Disaster Management: The Role of IT in Mitigation, Preparedness, Response, and Recovery*, The National Academy Press, Washington, DC.

chapter eighteen

Overcoming obstacles to integrated response among incongruent organizations

Robert K. Campbell

Contents

> "A rededication to preparedness is perhaps the best
> way to honor the memories of those we lost that day."
> *The 9/11 Commission Report*

18.1 Introduction

In review of the March 11, 2011 earthquake, tsunami, and nuclear power plant release (triple disaster) that struck Japan, there are several notable success stories, namely the earthquake survivability based on building design standards. This remarkable strength is the result of long-term planning in the design and construction sector. There were, however, several systematic failures that were not unique to Japan and have been identified previously in other major disasters such as 9/11 and Katrina. First, most notable was the government's lack of a unified disaster management framework that could enable deployment and integration of different response agencies under a unified operational command. Second, the government's inability to swiftly communicate requests for assistance from the local to the national level inhibited response operations. While individual organizations and those within the organizations are highly trained and capable, the inability to swiftly respond to and integrate into a unified response compounded the tragedy. This chapter focuses on the obstacles and solutions to integrated response operations among incongruent organizations. To examine this topic further, we will explore several case studies including the Japan disaster in 2011 and the response to it.

18.2 Integrated response

18.2.1 Definition of integrated response

Integrated response describes effective response activities among different units or resources that work with interoperability and seamless integration in executing the incident action plan with efficiency and a unified approach. An example includes two different hazardous material response resources such as a local fire department and a civil support team that work together sharing common terminology, training where interfacing on the response, and interoperable communications and procedures. This integrated response team acts efficiently and effectively together in achieving the incident objectives.

Over the past 7 years, the author has managed over 1500 all-hazard exercises, many of which the author designed, controlled, and evaluated. From each of these exercises, the author has collected data through observations and lessons learned from after action reports that indicate several obstacles and challenges to implementing an integrated response (Campbell 2005–2012). There has been one consistent observation throughout most of these exercises: when different response teams attempted to form a response unit or strike team, they were unable to successfully integrate as a team and effectively achieve the incident objectives. Furthermore, the author has conducted root cause analyses to determine why they were unable to integrate effectively. The root causes, as well as solutions to overcoming obstacles, are the main subjects of this chapter.

18.2.2   Obstacles and challenges to integrated response

The following outlines the types of obstacles and challenges to successful integration. Below each type of obstacle are supporting examples from training and exercises conducted around the world. Many of the examples are from hazardous material response teams since those comprise the majority of the sample population in this study.

18.2.2.1   Organizational obstacles among incongruent organizations

"If New York and other major cities are to be prepared for future terrorist attacks, different first responder agencies within each city must be fully coordinated, just as different branches of the U.S. military are. Coordination entails a unified command that comprehensively deploys all dispatched police, fire, and other first responder resources" (9/11 Commission Report, chapter 9). Response organizations within a community may be operated by different management structures that do not agree on priorities, missions, and roles. There are many organizational obstacles that prevent or inhibit different organizations from integrating at the right organizational interface and working together in the integrated planning process to ensure an integrated response. Each of these is discussed in the following with examples from our work around the world.

18.2.2.1.1   Pride in organization supersedes missions.   One organizational barrier to integrated operations is pride in one's organizational mission. Organizations are trained, equipped, and prepared to respond based on their concept of their mission. When missions of similar organizations overlap, there is a potential for conflict over which organization will perform which function. In fact, it is possible that one of the organizations may not be needed or may supplement the other team.

Therefore, it is important to understand the capabilities associated with similar organizations so that they can be effectively integrated if needed. Likewise, there may be a need to integrate the capabilities of each organization in an unconventional manner, such as postponing their support until needed to relieve the initial team.

18.2.2.1.2   Redundant and uncoordinated funding.   Before 9/11, there was a lack of unity of effort among organizations as well as the oversight levels in the executive and legislative branches of government. This issue is also pertinent to response organizations as there are conflicts between federal grant guidance and state limitations as well as an abundance of discipline-based grants that are submitted by specific response organizations rather than communities. For instance, local responders have their own priorities and sources of funding and oversight. In addition, state and federal bodies that fund local response organizations are not well coordinated in how they disburse funds, sometimes emphasizing discipline-specific capabilities that become developed in a stovepipe rather than a community-based capability setting. Furthermore, various grants can inadvertently encourage competition for capabilities among different response organizations. However, the Homeland Security Grant Program (HSGP) encouraged community-based capabilities and is successful in many regions. Ultimately, federal funding lines (i.e., coming from Department of Health and Human Services [HHS], Department of Homeland Security [DHS], Department of Transportation [DoT]) programmed for specific missions and first response organizations has led to duplication of efforts.

Since 9/11, several departments have developed funding for similar purposes and different but sometimes overlapping recipients. Some of these funds/grants are designed to equip, train, or staff organizations such as emergency management, public health and hospital staff, fire and emergency services, and so on. However, if not coordinated within the jurisdiction or region, these funds may lead to duplication of effort. For example, DoT grants that are distributed by the state Hazardous Material Emergency Preparedness coordinator fund local emergency planning committees (LEPCs) and seek to advance their mission of awareness and community planning. This includes industry, fire departments, emergency planners, and the general public. Meanwhile, the DHS HSGP accomplishes similar functions within the same community through approximately 20 different preparedness grants. HSGP grant funds are typically administered by local/regional/state emergency managers. In some cases, the participants in these projects are the same. Additionally, the grants are subject to different regulatory oversight. In 2011, the Government Accounting Office (GAO) found "DHS should also improve the efficacy of the grant application process by mitigating duplication or redundancy within the various preparedness grant programs" (GAO 11-881, 2011)

The problem is that different organizations (i.e., fire, law enforcement, EM, EMS, etc.) may request funds without coordinating with others that have a similar ability to obtain grant funding for similar purposes. When not coordinated, these expenditures may be used in a way that is inconsistent with local plans or a distraction from the region's strategic plan.

18.2.2.1.3 Stovepipes and power struggles.   Due to the size of the government, it is likely that there are redundancies among and within the various departments and agencies. As a former military staff officer, the author observed the complexities of organizations performing similar functions and trying to solve the same problems without realizing that there are many other organizations tackling the same issues. Needless to say, each organization was growing their span of control, overlapping other areas, and spending funds on sometimes the same things while new organizations were emerging. In 2004, the GAO summarized one of its findings as follows: "No single entity has been given the authority and responsibility to integrate and manage departmentwide installation preparedness efforts ... Until organization roles and responsibilities are clarified, and an integrating authority is designated, DOD will be limited in its ability to develop a comprehensive approach, promulgate departmentwide guidance, and effectively coordinate ongoing billion-dollar improvement initiatives at the installation level" (GAO 04-855). Since 2004, some actions have been taken to improve coordination and integration of resources at the installation level. When GAO conducted this audit, they interviewed installation, headquarters, and department level personnel. There is still much work to be done in this area to ensure comprehensive and integrated planning at the installation level can be successful.

The stovepiping that was identified in the GAO 04-855 report has not been fully resolved. While the recommendation to assign an integrator may have occurred, there appears to be some inconsistencies between the service and department level as well as among the end users of such equipment.

18.2.2.1.4 Political interest.   Political interests of different organizations play a factor in the lack of integrated response operations. Competing interests unfortunately take precedence over providing swift aid to those in need. While most government organizations have accountability systems in place, some developing nations rely heavily on nongovernmental organizations (NGOs) such as the UN, Red Cross, Red Crescent, and others. While there are many reputable NGOs capable and interested in providing relief and aid, some are biased or have alternative interests that inhibit an integrated response operation.

In Eastern Afghanistan after the April 2009 earthquake, it took the provincial government about 30 days to respond with assistance to affected communities. Additionally, some NGOs are reluctant to share their capabilities and resources with the government to protect their autonomy and ability to respond in a way that furthers their mission. For example, the Afghan Red Crescent touts its neutrality as an organization that cannot be seen as siding with militants, the government, or foreign militaries. Therefore, they choose efficacious and opportune situations to respond and provide support based on their capabilities and resources. NGOs also derive their funding and resources based on publication of their accomplishments during a response. This independence precludes them from coordinating with the government in a way that optimizes the overall response operations. Therefore, the government often does not know where NGO resources are providing aid to disaster victims. Each organization vies for recognition by being able to meet the needs of disaster victims without coordination or sharing of information and resources. This makes it extremely difficult for the government to know the real-time needs of disaster victims and stage a coordinated response operation that employs the most appropriate organization in the right place based on proximity and capability. Finally, in some cases, disaster response resources are misappropriated by local leaders precluding effective distribution to those in need. This results in a false sense of accomplishment for those leaders that provided the resources to this district or village. Deployment of aid in developing countries like Afghanistan continues to be a major organizational challenge due to political factors.

18.2.2.1.5   Overlapping missions and turf battles.   Overlapping missions of different organizations have the potential to create discordance rather than integration among organizations as they participate in a response. This can be seen with the myriad of volunteer organizations that "show up" to a response and want to help. Depending on the organization's desire to promote itself versus provide aid during the response, there is a potential for conflict. Additionally, when different response teams with similar capabilities are called to a response, the incident commander (IC) and general staff have the challenge of piecing together the capabilities in the most optimal way of achieving the incident objectives. While developing response plans, different organizations may be responsible to donors, a board, or government leaders for executing the mission that they have been funded to accomplish. A lesser mission may result in loss of stature, funding, or perceived value. These factors may result in organizations that vie for missions or compete with other organizations rather than focus on integration to achieve the incident objectives.

18.2.2.1.6 Maintaining currency with new and changing resources. With so many response teams and capabilities, it is challenging for incident commanders to maintain an awareness of all of the available response teams and resources. Since 2005, the author has been involved in bringing together different first response teams and improving their ability to integrate into a unified response team. One of the most frequent comments on course critiques is, "I had no idea that these other capabilities existed," although many of these capabilities have existed for the past decade. Unfortunately, this is evidence that various teams are not effectively training together.

So how much harder is it for incident commanders to keep track of all of the resources outside of their community? Well, the request for assistance process has trained Emergency Operation Centers (EOCs) to request resources, not necessarily specific teams. Regardless of how resources are selected to join a response, it is important for responders to know about the capabilities of other response teams that they are likely to work with so that they can define the interface with those teams and drill response procedures to ensure consistency, interoperability, and synergy. This is one reason we have the National Incident Management System.

18.2.2.1.7 Procedural communications and bureaucracy. Another barrier to integrated responses is bureaucratic communication procedures. This is probably the single-most detrimental factor contributing to the lack of integration among Japan's responders after the triple disaster of 2011. Many of the communities and prefectures were constrained in asking for the right resources by an inefficient communication process. Prefecture plans call for leaders to form a committee and submit a formal, written request for assistance to the government of Japan. In some cases, prefecture leaders were required to present their requests in person in Tokyo. Unfortunately, the triple disaster disabled power and communication systems and in some cases destroyed buildings, preventing leaders from being able to effectively gather the committee together, formulate the proper written request, and submit it to the government of Japan. Had these leaders been able to request assistance directly to the government of Japan, the Japanese military, or specific response teams, then rescue efforts could have begun much sooner.

In the event that a military response is appropriate in Japan, they would only be deployed on the order of the police commander. This requirement for an official request from a regional police commander can in itself lead to a breakdown in capability delivery under the stress of a chaotic disaster situation. The reason why the Hyogo Prefecture (the administrative area where Kobe is situated) took so long in asking the Japan Self-Defense Forces (JSDF) for support in 1995 following the Kobe earthquake was, first, because they spent valuable time in facilitating the official process. The law required that the request to the JSDF should be done with official documentation; yet, the officers in Hyogo prefecture found this problematic as their administrative building had collapsed, and there was a complete breakdown in operational functionality (Takashi 1995).

Second, the lack of a decision-making ability despite disaster management/prevention councils at the national, prefecture, and local levels compounded the bureaucratic communication process. Each ministry or council tends to focus on what they can do for disaster management and response, rather than focusing on what capabilities are needed for disaster response from these ministries. The root cause of this mentality comes from stovepiping, lack of cross-ministry communication and coordination, and protectionist perspectives with respect to capabilities and information within each bureaucratic agency. This point was reinforced by Leo Bosner, a senior director of the U.S. Federal Emergency Management Agency (FEMA), who spent a year in Japan studying its emergency management procedures.

> [Five years after Kobe], I heard this refrain again and again from Japanese emergency planners: excellent work within individual agencies, but still a lack of government-wide disaster response planning. Interagency coordination is still weak, and in most cases the government's disaster response plans are woefully short of detail. Most agencies are unfamiliar with the disaster response plans and capabilities of other agencies, impeding efforts to develop well-coordinated plans. Underlying all of this is the Japanese government system, which does not appear to encourage rapid decision making or interagency coordination.
>
> **(Bosner 2002)**

The U.S. Request for Assistance process is not much better. It takes approximately 72–96 hours for support teams to be tasked and deployed from the federal government after a request from a governor. After Hurricane Katrina, electrical power to parish EOCs was nonexistent making it impossible for requests from parishes to Baton Rouge. Additionally, federal resources were not invited to respond until after it was too late. The federal government, in particular FEMA, took the brunt of criticism whereas the responsibility for response, planning, coordination, and integration of local, state, and federal responders was with the local area leaders.

To overcome the bureaucratic slowdown in mobilizing and delivering response capabilities, more autonomy should be given to local and regional responders capable of participating in the response. For this to be effective, the response must be managed at the local level with an IC who controls, accepts, and rejects response capabilities offered. DoD Directive 3025.15 allows U.S. military commanders to provide immediate response to save lives, prevent human suffering, or mitigate property damage under imminently serious conditions without deployment orders from national headquarters. This example of access enables local incident commanders to coordinate and draw from local resources based on coordination and memoranda of understanding without becoming entangled in bureaucracy. However, there are limits on this authorization, specifically a time limit of 72 hours, after which the normal bureaucratic process must be used to authorize continued support.

Based on observations from recent incidents, such as the Japan triple disaster and the Joplin tornado, one of the most effective integration success stories lies with the faith-based organizations that are prepared to respond and work through networks of local partners. In Japan, faith-based organizations such as the Global Mission Center in Iwaki, Fukushima were well connected and agile enough to coordinate resources, volunteers, and distribution of resources so well that the local government supported this organization in the response and went as far as asking the organization what was needed and how it could help. In many other disasters, the first emergency responders on scene are the local civilian population. In the United States, there are many examples of effective communication in small-scale incidents. However, as the size of the event draws in more bureaucracy, the effectiveness of integrating responders tends to decrease. The lesson learned is that small, agile organizations that can adapt, act autonomously if needed, and collaborate within the community and among other responders will overcome procedural communication barriers and effect an integrated response.

18.2.2.2   Personnel

18.2.2.2.1   Inconsistent standards.   One of the challenges for integrating personnel across organizations during a response is different standards to which responders are trained and/or certified or credentialed. For example, within the DoD, fire department officials have subscribed entirely to the National Fire Protection Association (NFPA) standards for responders and have created DoD certification. This has included personal protective equipment (PPE) standards for fire protection and levels of HAZMAT response. Until recently, other organizations were not permitted to respond without the same certification, training, and PPE even though other organizations were trained and equipped in some cases to higher technical standards (i.e., Occupational Safety and Health Administration [OSHA] HAZWOPER standard and NFPA 1994 for chemical, biological, radiological, and nuclear [CBRN] PPE). While other organizations trained to the DoT HAZMAT response standards and held that all responders needed to comply with this curriculum, the compliance with standards was not reciprocal. This 10+ year disagreement led to confusion and lack of integrated response capabilities on many installations. It also allowed ill-equipped and undertrained responders to engage in operations beyond their capabilities.

In the public sector, civilian emergency responders can look to the Type 3 All-Hazards Incident Management System Qualification Guide (September 2010) for certification and credentialing in key response positions. This qualification guide sets standards for training, experience, and fitness in a performance-based qualification system to ensure that responders to Type 3 incidents are qualified for response. Regulatory agencies like OSHA establish standards for hazardous material response. NGOs such as the NFPA establish consensus standards on how to respond to hazardous material incidents (NFPA Standard 472). While there is a myriad of training courses, there is a lack of consensus standards for technical response–oriented activities making it challenging for different organizations to attain similar standards of performance and implement common operational procedures.

18.2.2.2.2   Variation in knowledge, skills, and abilities.   Another challenge is that personnel from different organizations possess varied levels of knowledge and skills related to their professions. In the United States Air Force (USAF), readiness and emergency management personnel are trained on militarized chemical and biological agents and detectors designed specifically for these agents. The bioenvironmental engineering (BE) technicians are required to complete college-level equivalent training in chemistry and biology, along with understanding of the myriad of chemical hazards that are used in industrial workplaces. This training also includes proper sampling and analysis with advanced field portable instruments to comply with OSHA, Environmental Protection Agency (EPA), American Society for Testing and Materials (ASTM), National Institute for Occupational Safety and Health, and other standard methodologies for compliance sampling and risk analysis. During response operations, both disciplines have been equipped with similar advanced field portable analytical instruments that are capable of detecting, identifying, and quantifying most chemical hazards. However, only the BE technicians are trained to a level that enables accurate, consistent, and reliable interpretation of the results. This has led to a wide variation of personnel abilities when attempting to implement an integrated response capability.

On the contrary, some personnel have recognized and accepted these differences in knowledge and skills and have chosen to find ways to build integrated capabilities that capitalize on the unique attributes that each discipline brings to the IC. This forward-thinking approach has led to successful integration of the teams as demonstrated in USAF CBRN Challenge™* response competitions. During integrated training, teams learn enough about each other's mission, equipment, and procedures so that they can effectively interface with each other while performing their integrated mission. This capability-enhancing training has developed numerous best practices and standards from which these organizations have adopted into their combined operations—simultaneously accomplishing multiple objectives with the complete equipment set. This provides the IC with a multiplicative capability that is beyond which each individual team could present.

18.2.2.3   Policies and plans

18.2.2.3.1   Lack of stakeholder integration. While working with civilian emergency managers and local organizations to develop integrated plans, there is reluctance among some organizations to coordinate and agree on plans. For example, within the medical community, hospitals may be in competition with each other for business and therefore reluctant to share information with others. There are also challenges with cities located within counties to agree upon who will be in charge of an incident. This is compounded when conducting regional planning. Smaller, less populated counties within the region tend to become a lower priority when resources are allocated. In one region, private colleges, public colleges, private health care organizations, public health care organizations, hospitals, nursing homes, county health departments, city emergency managers, and others did not agree on each aspect of the plans the author was developing. Each organization had plans that were not interoperable with those from other organizations and did not agree on key points. However, we were eventually able to overcome these challenges.

18.2.2.3.2   Disagreement on roles and responsibilities. USAF policy defines 42 response missions (Air Force Manual 10-2503, 2011). It further subdivides these missions by primary, secondary, and tertiary organizations responsible for accomplishing these mission essential tasks. In many cases, there are two responsible organizations for the same mission, and in a few cases, there are three responsible organizations

---

* CBRN Challenge™ is a trademark of Alliance Solutions Group, Inc. initiated in 2005 as a competition; it has evolved into a coached training and exercise event that builds CBRN response capabilities.

for the same mission. These are not the types of conflicts that are needed during a response. Rather, these differences should be sorted out during the planning phase. While it may appear that the disagreements are occurring among different disciplines, the primary obstacle is that the different disciplines are not putting the needs of the IC ahead of their own interests. However, in recent years, the USAF has made attempts at the headquarters level to coordinate response planning functions for installations.

18.2.2.4   Unit training and exercises

Some obstacles to integrated response are rooted in training and exercise disparities among different organizations. At the unit responder level, organizations define their training requirements based on certifications, standards (American National Standards Institute, NFPA, mission essential tasks, and regulatory [OSHA, EPA, etc.), and qualification requirements. Unit responders train to these standards that may not directly apply to sister organizations, such as law enforcement's acceptance and implementation of NFPA standards. Another example is the numerous standards set by different organizations for sampling procedures (i.e., Department of Justice, HHS, DHS, ASTM, DoD, etc.). While different training curricula and standards may bring multifaceted approaches and perspectives, thereby adding value, the different training curricula also can create conflict and uncertainty over who is actually trained to the right standards. It is important for responders who are going to work together to define, adopt, and implement consistent standards to which they will train and comply. In many cases, there will be multiple standards that apply due to the various regulatory bodies overseeing various response operations (i.e., OSHA, EPA, DOT, NFPA, etc.), adding to the confusion.

Organizations also train on specific equipment such as detectors that they have purchased. Since other organizations may not use the same detectors or have the capability and knowledge to do so, it is important that partners share basic information about their capabilities with each other and their potential incident commanders that they may work with so that these capabilities can be correctly employed.

Exercises can comprise a wide range of scale, scope, and participation. Tactical drills and functional exercises must be utilized to build up to full-scale exercises to complete the preparedness cycle and ensure proper interfaces among responders exist for the purposes of working together, communicating information up and down the incident command system (ICS), and delivering the right capability within a response. While the application of exercise series is gaining more traction throughout the United States, an overemphasis on evaluation and testing still remains strong. Instead, a strong emphasis on training at the same scale and scope is necessary to gradually build the capabilities that will lead to success during evaluated exercises. Over the course of several hundred, week-long combined training-exercise events, we have found significant capability gains by combining training with exercises using a systematic building block approach. This ensures a successful exercise based on proper training, rehearsals, and integration of first response disciplines interfacing at the right pace to ultimately yield the most value while focusing on more critical issues during a full-scale exercise.

18.2.2.5 Leadership

Leaders (health care directors, mayors, military installation commanders, etc.) are under pressure from various oversight bodies (e.g., The Joint Commission, grant funding requirements, and inspectors) to conduct specific exercises that achieve, in some cases, unrelated, lengthy objectives (e.g., Air Force Instruction 10-2501 requires USAF installations to conduct at least 11 exercises each year), which result in poor quality, unrealistic exercises and truncated training and focus on regulatory-based requirements instead of true capabilities-based response exercises. These exercises often exclude partners and responders in lieu of capturing more objectives with a perceived importance that results in reinforcement of disintegrated response operations. The "box checking" approach to exercises has devalued their importance as a validating tool of the preparedness cycle that ultimately drives capability improvements in preparedness, mitigation, response, and recovery.

18.2.2.6 Equipment and supplies

18.2.2.6.1 Personal protective equipment. An analysis of environmental, health, and safety risks during the 9/11 response highlighted problems with worker exposure to asbestos, heavy metals, silica, and other inhalation hazards from smoke, dust, and debris operations. Few workers wore respiratory protection. Those responders that did wear respirators found replacement cartridges in short supply, and those that were available were often incompatible with available respirators. Today, response organizations are standardizing equipment and PPE to realize savings and maximize interoperability within their communities. In the past 3 years, the USAF has standardized their self-contained breathing apparatuses (SCBAs) across three response organizations. This has led to common testing of suits, respirators, SCBA tank certifications, and interoperability. LEPCs are also beginning to take a closer look at the hazards within their communities and the preparedness of first responders such as fire department PPE and detection systems to address specific hazards. These efforts to standardize equipment sets can overcome a challenge to integration. However, many organizations are not able to sufficiently maintain or replace the PPE due to scarce funding.

18.2.2.6.2 Detection systems. HAZMAT responders, fire departments, and other responders have procured caches of CBRN detection equipment often chasing technology for the next best detector without a strategic approach. In 2009, the global CBRN defense market was $7.9 billion. While many of the detectors are of high quality, little thought has gone into identifying CBRN risks for a local area and then procuring the right detectors to address local threats such as chemicals associated with local industrial activities. Instead, responders have procured equipment based on available (and sometimes, pushed) funding without considering partner capabilities, desired local capabilities, and how their organization meets that need. Uncoordinated procurements have led to duplication of equipment sets among organizations with varied levels of knowledge and abilities to properly employ these detectors during a response. Different equipment can lead to seemingly conflicting interpretation of results impairing integrated response efforts.

Many different organizations that respond together have procured detection systems that perform similar functions but sometimes yield different results. This has led to confusion among responders over which instrument to trust and further raises questions about certification and competence among responders to properly utilize the instrument and properly interpret the results. For example, radiation responders may select different radiation detection instruments that employ different principles of operation. This can lead to unfamiliarity with the other team's system and lead to larger variation in results depending on how the instrument was calibrated and used. Different instruments may have useful technologies that overlap in what they can do, but also complement each other in some ways as well. Depending on the users and their understanding of the technologies, the use of multiple instruments can lead to seemingly conflicting results when the real difference is caused by the user's inability to reconcile differences among technologies. Nonetheless, this has presented an obstacle to effective integration of responders and their tools.

In the U.S. military, the procurement of multiple equipment sets for each responder has led to debates over the proper organizations responsible for employing the instrument based on the overlapping capabilities of detectors. Instead, the goal should be to determine the optimal employment of equipment at the end user's competence level so that instruments and results can be integrated. Based on several HAZMAT response case studies, it is evident that hundreds of thousands of samples will likely be collected and analyzed during a response. It is important for samples collected and data from monitoring to be interpreted properly and withstand the test of validity and comparability. In some cases, organizational protectionism has led to unrealistic assumptions of scalability during a response (i.e., assuming that only one organization should have a specific responsibility during a response). A significant chemical response could require the sampling/monitoring efforts of many responders at the local, state, and federal levels within a community. The focus of the discussion should shift from team responsibilities and team-specific equipment to data quality control, scalable capacity, and data applicability for risk assessment and management.

For example, the USAF has identified and allocated responsibilities for certain detection equipment to specific organizations rather than establishing a framework for the installation to identify and build a set of detection equipment based on its local hazards and risks. In contrast, several HAZMAT response teams that we have worked with in Virginia have identified the level of HAZMAT response capabilities needed in their region based on risk. The fire departments within one region comprising four counties working together to share their different resources and capabilities through mutual aid agreements. Additionally, they defer to state HAZMAT response resources rather than procuring and maintaining the full HAZMAT capability within their small, rural region. This risk-based approach to building capabilities and leveraging state resources creates a defined interface for the different response organizations to work together.

Resource typing has become a popular exercise in communities to ensure that all resources (personnel, equipment, training, and capacity) are mapped to type (based on specific capacity and capability) and kind (team, equipment, etc.). The purpose of resource typing is to make the resource requesting and dispatching process more accurate and efficient. Having been involved in standardizing 15,000 equipment items across the USAF, we have found that there are certainly some benefits to standardization (i.e., common training, familiarity with the same equipment regardless of which unit is involved in the activity, similar procedures for employing the capability, common training for large organizations like the federal government, etc.). However, it is important to ensure the process of standardization does not drive the capability needed. This could lead to a parochial view on the resources available to respond. Rather, the capabilities should be clearly articulated and typed to ensure appropriate utilization. Some communities are facing the challenge of "typing" their resources in a way that enables standardization with other communities due to the broad variation in resource specifications. This has led to the development of new nomenclature despite the best efforts of the region to standardize.

18.3   Solutions

These solutions represent a collection of observed best practices that have been demonstrated within multiple response organizations including the U.S. military, foreign governments, local responders, and federal response teams. While each of these solutions may not be applicable to all responders, they are presented as a menu of successful solutions that may be adopted to address a particular challenge that an organization may be facing. Finally, the solutions presented in the following discussion may be cross-cutting solutions that address a multitude of problems and root causes addressed in the preceding discussion.

18.3.1   Collaborative planning

It is vitally important to build a collaborative, low-threat environment among various organizations from the beginning. It is also important to understand the current political situation and dynamics among the planning team including power struggles, agendas, and other interests early in the process. Effective team building techniques to facilitate collaboration are critical to gaining participation and acceptance from stakeholders early in the process. In the following discussion is an outline of best practices that has been used successfully in building collaborative planning environments. This will help responders integrate before the disaster occurs.

18.3.1.1   Understand the context
Discuss priorities and agendas with the primary stakeholder to ensure that you understand their goals and potentially competing interests from other organizations.

18.3.1.2   Define the process
The process that will be used to develop the plan should include gathering input from all of the stakeholders and adjudicating comments, objections, and conflicts among stakeholders. Communicate this process to everyone at the outset of the project so that everyone understands when they will have opportunities to review the draft plan, provide input, and how their comments will be adjudicated.

18.3.1.3   Do your homework
Before integrating plans, it is important to review all existing plans to know the context, format, and related responsibilities of all the key players. It is also important to read the plans for the county and city as well as the other disciplines in the region to ensure that the integrated plan will

be interoperable with the other plans. Often, the primary challenge is that plans are not interoperable with each other, and there are power struggles over who will be in command, what form the ICS structure will take, and how smaller communities in a region will be represented and addressed during a response.

18.3.1.4   Be humble
Come ready and eager to learn the opinions, positions, and interests of all of the stakeholders. While you may not agree with them or you may have your own biases, it is important to validate everyone's perspective. Whenever there is incorrect or contentious information being discussed, ask probing questions, engage others in the room directly to get their input, and work to address concerns that are at odds with others in room. There is plenty of time to readdress invalid or incorrect information. It does not always need to be resolved in a large group setting.

18.3.1.5   Review lessons learned
Review recent exercise after action reports to capture existing processes for response, lessons learned, and improvement plan items. Discuss these items with stakeholders to ensure the plan will address these issues.

18.3.1.6   Present the plan in phases
Present the plan in phases to gather input from the group. Always send the draft out to the stakeholders with sufficient time for them to review and comment on the plan and then follow up with an in-person working group meeting to discuss comments and gather other feedback. It is important to remember that some comments that are provided may conflict with other perspectives and these should be vetted adequately to build a consensus for the plan.

18.3.1.7   Consider other best practices
Utilize best practices from other communities. There may be some individuals on a committee with perspectives from other communities and they may have valid input on which process will work well.

18.3.1.8   The planning process is more important than the plan
While it may be easier for an individual to write an integrated plan in a vacuum, the planning process and working groups enable the stakeholders to address issues that are both within their discipline and outside of their discipline. These discussions can extend beyond a particular plan and generate substantial improvements to other plans further enabling interoperability within the community.

18.3.1.9   Make the plan useful and actionable

There are many plans that read like narratives, regulations, or a textbook with background information that will not be relevant in an emergency. Rather, a plan is most effective when it is short, objective, and actionable with easy-to-find information. Otherwise, the plan can be useless during an emergency. While the planning process helps educate stakeholders on the processes and creates buy-in, a well-written plan can be utilized whether it is known and understood (in case the hardcopy is destroyed) or the actual plan is read and followed. We have conducted extensive modeling to develop scenarios of varied scale and scope to ensure the plan covers the wide range of incidents that could occur. In the case of pandemic influenza planning, there are multiple guidelines at the international, federal, and state levels. To develop an actionable plan, the author mapped several models and their respective scenarios to these guidelines and created trigger points with associated actions, along with a menu of options that could be employed as the situation dictated. This provided leaders with flexibility to choose from options. Then, capabilities and options available to decision makers were evaluated to ensure that the communities had the right capabilities and determined their capacity to handle this situation. The modeling, trigger points, canned response options, capability and capacity analysis, and useful tools and references made for an easy-to-explain and understandable plan that the stakeholders could visualize and use.

18.3.1.10   Coordinate with potential partners in the response

In 2011, after a tornado swept through Joplin, MO, medical personnel recommended, "Know whom your local, state, and federal response partners are and have an established relationship with them prior to any event" (Joplin, MO Tornado Medical Lessons Learned Workshop). While there have been many strides forward in integrating responders effectively through establishing professional relationships, this remains an incomplete task from the 9/11 Commission Report mainly due to a systemic challenge—turnover in local government, public safety officials, volunteer organizations, and military transfers.

18.3.1.11   Test the plan

After completing a new plan or revision, it is essential to test the plan through seminars and tabletop exercises. The author has found that this is a key step in creating buy-in for different organizations to integrate into a seamless response team. The seminar gives all parties a chance to read and learn about the threat, the responder's capabilities (which should be outlined in the plan), the response structure, key decisions, response actions, and tools. The tabletop exercise gets the responders working together, making decisions (sometimes decisions that could not be made during the planning workshops), and hopefully agreeing on how they will respond. If there are shortcomings or inaccuracies in the plan, then those can be addressed after the tabletop exercise.

Exercise design and planning also can offer opportunities to build collaborations. During the course of several large functional and full-scale exercises where the author reviewed the plans in advance of the exercise design, he was able to engage stakeholders on inconsistencies, inadequate procedures, poor assumptions, and bad procedures that would not work. Due to the expense of the exercises, it is important to resolve these issues as much as possible with the stakeholders before designing and running the exercise. If you can predict the major observations and items that will be on the improvement plan before you design the exercise, then it is not worth running the exercise. In addition, many response teams identify additional problems with their plan during the course of conducting training in preparation for the functional exercises.

At one U.S. military installation's medical group, it was clear that the plan was not going to work during training involving medical teams and a few mock casualties. The teams were tripping over each other, did not have all of the supplies necessary such as IV stands (this was a large clinic with ancillary services, not an inpatient hospital), and the patient flow through the clinic was not logical. This occurred even after several improvements to the plan were made in the 2 months leading up to the exercise. At the end of the training, the medical group commander asked the participants for ideas and feedback. This meeting quickly turned into a white board session that lasted 2 hours with the best ideas coming from the lowest ranking personnel. By the next morning in preparation for the exercise, they had modified team positions and the facility (they hung reshaped coat hangers from the drop ceiling to hang IV bags). After the 100-person mass casualty incident (MCI) exercise, the hot wash was electric with cheers, and the large clinic was ready to handle a MCI. The key to bringing together many on-base response teams and off-base responders was collaborative planning; team-based response training in concepts of operation, equipment, and procedures; collaborative-based functional exercises; more planning; and a full-scale MCI exercise.

Community-based capabilities: Having done an extensive analysis of response capabilities based on current publications, instructions, and observations during 1500+ civilian and military exercises, it is clear that the one-dimensional approach to capability development does not work well. This has led to multiple disciplines claiming responsibility for the same task, multiple equipment sets with different users and different procedures, similar equipment items with different capabilities yielding different results (which appear to be conflicting results and leads to lost confidence in these units by the IC), and unknown capabilities among first response disciplines that should be integrating during a response. One solution to this issue is to take a community-based approach to building capabilities. Military installation commanders are on the receiving end of headquarters staff conducting mass procurements within their stove-pipes without considering the local risks and impact on the installation. This often leaves the installation or community with the maintenance bill for the future and creates obstacles to the integration of response teams. While headquarters staff organizations can help by standardizing equipment sets, procedures, and training, the installation commander who is responsible for the response capabilities on his or her installation should be the one who defines the capabilities based on the myriad of threat, risk, and vulnerability assessments conducted on the installation.

Civilian communities have faced similar challenges where the fire department applies for one grant to fund their equipment and the medical community uses another grant to fund their equipment and so on until there are multiple trailers full of equipment in a community with little opportunity to integrate these resources. Ultimately, the local community becomes responsible for maintaining the redundant equipment and training different responders on different equipment while the cost skyrockets. There should be a prohibition on specific public safety disciplines building internal capabilities without a community-based all-hazards risk and vulnerability assessment, and clear decisions about the risks that the community is willing to take. This would enable a community-based approach to defining the required resources and capacity for those capabilities that the community is willing to develop and maintain. Then, community emergency planners could identify which disciplines will be funded to build and maintain the capabilities. This might include several disciplines working together to procure equipment, train together, and create the right team capability.

For U.S. military installations, first response disciplines should plan together at the installation to define, for example, how many electronic personal radiation dosimeters are required to attain the desired capability for dosimetry. This plan would outline who receives them, who maintains them, who calibrates them, common procedures, standard alarm settings, how they will be worn, and the radiation experts that will interpret the results. Under the current situation, few experts know how many are on the installation, if they are being calibrated and by whom, and whether all disciplines are utilizing the same alarm settings, because they were procured centrally from higher headquarters and sent to individual units rather than the installation.

An effective community-based plan ensures that the capabilities are built based on the location-specific risks. This enables the locality to tailor their capabilities and capacities accordingly. It also helps consolidate resources and integrate responders while eliminating unnecessary redundancies. When integration occurs, responders become more familiar with their counterpart's resources, knowledge, and equipment so that they can train together effectively and standardize their procedures. Ultimately, the community becomes better prepared for response because they are ready to respond together as a knowledgeable, integrated response team.

18.3.2   Integrated training and exercises

"Regular joint training at all levels is, moreover, essential to ensuring close coordination during an actual incident" (9/11 Commission Report, chapter 12). Furthermore, "It is a fair inference, given the differing situations in New York City and Northern Virginia, that the problems in command, control, and communications that occurred at both sites will likely recur in any emergency of similar scale. The task looking forward is to enable first responders to respond in a coordinated manner with the greatest possible awareness of the situation" (9/11 Commission Report, chapter 12).

In 2005, the USAF adopted the CBRN Challenge concept. The concept has evolved from the initial concept of realistic, all-hazards CBRN exercises into an integrated response training and exercise event for fire and emergency services, hazardous material response teams, and emergency management/responders. This event has had a significant impact on community response capabilities by focusing on integrating CBRN tactics, techniques, and procedures (TTPs) with a suite of advanced field portable analytical instruments. First-hand observation of hundreds of training and exercise events comprising different combinations of hazardous material responders in civilian and military settings has provided a perspective of HAZMAT response capabilities in different venues that highlights common obstacles to integration as well as best practices that support integration. Some of the key observations related to integration include the following:

- Each discipline was not familiar with the other discipline's equipment and procedures.
- The various disciplines were not familiar with each other's response capabilities.
- Each discipline used different and sometimes incorrect procedures to calibrate equipment, operate equipment, and interpret results leading to incomparable data.
- Attempting to integrate procedures led to conflicting arguments over who should do what task and which task took precedence. Teams had difficulty recognizing that they could accomplish multiple objectives simultaneously if they worked together.

- The organizations struggled to integrate unit-specific capabilities into the overall response. For example, the risk assessment and management was typically based on only one discipline's input since the teams were not integrating operationally. Since the fire department typically filled the roles of IC and operations section chief, the other teams were frequently left out of the response.

During a recent 2-week Integrate Base Emergency Response Capability Training (IBERCT™*) and exercise event with 60 representatives from three first response teams, each team learned from the others. One of the most frequently cited comments from the fire department participants was that they had no idea about the response capabilities of the other teams. An observer from the fire and emergency services community commented, "This event was invaluable for [all teams] to fully integrate capabilities and operations. This type of training should be mandated on a semi-annual basis at the base level where [all teams] can consolidate their response outside of the base exercise arena." Other observations and data captured from this event included the following:

- Only 2 out of 10 military installations participating in this event held joint training among all HAZMAT response teams on a recurring basis.
- Almost no fire department personnel expressed knowledge of the other first response disciplines at an emergency scene outside of minimal exposure during base exercises. At the end of the IBERCT program, all participants expressed knowledge of each organization's roles and requirements. All fire personnel stated they would take this experience back to the base to champion joint training.
- Terminology use and standards were not consistent among the teams.

These observations and comments from observers and participants are consistent with other similar training and exercise events. These types of training events that focus on familiarizing all public safety disciplines with each other's capabilities, increasing depth of knowledge at the operation and application level, practicing TTPs with detection equipment, and then integrating responders in realistic scenarios with expert coaches/evaluators has had consistent results in building effective integrated capabilities. Unfortunately, it is impossible to measure the lasting effect due to frequent movement and turnover of military personnel. To

---

* IBERCT™ is a trademark of Alliance Solutions Group, Inc. This integrated training and exercise event employs a variety of learning techniques including classroom training, hands-on practice, coached vignettes, and exercises to build the capabilities of integrated response teams.

better address movement and turnover, standardized and integrated initial training must be completed to ensure the new troops going to their first assignment possess a common operational knowledge of capabilities, equipment, and procedures. Second, common procedures and equipment sets must be adopted across the enterprise to enable capability growth in an expeditionary environment. Third, recurring integrated training at the installation, where risk and capabilities are defined, must be conducted outside of the annual exercise program. The annual exercise program has become a "box-checking" exercise instead of a capability-validating exercise that leads to improved capabilities—plans, equipment, and training.

18.3.3 Informed leadership

The 9/11 Commission Report identified several problems and shortfalls to effective integration. Before 9/11, the intelligence community suffered from numerous barriers to unified operation. This is still common today among emergency response organizations; however, fusion centers where different disciplines meet and share information have helped close this gap. The 9/11 Commission noted, "There was a lack of comprehensive coordination between FDNY, NYPD, and Port Authority Police Department personnel climbing above the ground floors in the Twin Towers" (9/11 Commission Report, chapter 9). Each organization was focused on their respective elements of the larger operation.

In the UAE, the government has established integrated operations centers that are composed of various ministries and government agencies that can support effective response operations. These fusion centers are touted for their seamless integration and ability to work together as a team (International Security and National Resiliency Conference 2012). The UAE has a distinct advantage over other countries, namely that the Crown Prince can issue an edict without getting bogged down in bureaucracy. This top-down approach ensures a common operating picture and a solid understanding of capabilities that exist within each ministry.

For western countries, there is a long history and culture of the capabilities existing in the local community. As leaders within the ICS, EOC, or other levels of government get farther away from the actual responders and their operations, there is a tendency to not only become less knowledgeable about the local capabilities but also to attempt to redefine the local response capabilities into something that they can manage, understand, and standardize across the larger area of operation. This tendency loses resolution on specialized local resources and can lead to a mismatch in sourcing resources, or myopia and biases when it comes to tasking resources with which an IC or coordinator is more familiar.

The challenge and solution is for those that may be designated as general staff, ICs, or Emergency Support Functions to become familiar with the local resources and ensure that different public safety disciplines are interoperable so that they will work together seamlessly during a response. Another challenge is that some of these leaders do not have regular control or authority over these units during day-to-day operations. This applies to both the military and among community first responders. However, exercises provide an opportunity to identify cross-functional issues and problems that need to be addressed through the improvement plan. Exercises also afford an opportunity for peer accountability as long as the observations are not diluted through an easy-going self-assessment. Ultimately, strong relationships among the parties of a working group can help address these issues. Political leaders, organizational leaders, and ICS-based leaders have a shared responsibility to facilitate integration of responders. Exercises are an excellent venue to ensure that this integration is occurring.

18.3.4   Strike teams

The number of CBRN response teams at the local, state, regional, and national levels continues to grow. However, teams at the community or military installation level may not fit perfectly into the resource type criteria. However, each community should identify different resources that could (and should) come together during a response to form a strike team. While they may receive different funding and report to different commanders or leaders, a strike team could help resolve many obstacles (political will, turf battles, different procedures, equipment, etc.). The USAF Incident Management System defines a branch within the ICS for each discipline even though there is much overlap in capabilities among the various first response organizations. Trying to divide the response force will not result in an effective unified response team. It will also result in inefficiency. The strike team offers a distinct advantage—common terminology, training, equipment, procedures, and unified purpose before the response occurs.

18.4   Conclusion

We should look to successful models to creating integrated teams from among incongruent organizations. One of the best models is the U.S. Special Operations Forces (SOF), which has succeeded in this mission for over 30 years. U.S. SOF are composed of highly trained, elite specialists from the various military services. Their highly specialized training and joint interoperable TTPs enable them to communicate and operate seamlessly as one unit regardless of which service the members originate. They train together on their common TTPs and capitalize on each service's capabilities and strengths during joint operations. According to

USAF Col. John Jogerst (2002), "Real interoperability requires constant testing and training not only to work out equipment problems, but also— and more importantly—to work out the human problems of command, control, and communications as well as unit tactics." As response teams look to overcome the obstacles associated with achieving a truly integrated response operation, they should examine the solutions provided in this chapter—collaborative planning, community-based capabilities, integrated training and exercises, informed leadership, and strike teams. SOF provide a lasting and tested example of these characteristics that achieve an integrated operation.

References

Bob Campbell, PE. 2005–2012. Alliance Solutions Group, Inc. After Action Reports.

Dr. Abdullatif bin Rashid Al-Zayani/Gulf Cooperation Council (GCC) Secretary-General. Mar 2012. International Security and National Resiliency Conference. Security of GCC Countries.

GAO 04-855. August 2004. *Combating Terrorism. DOD Efforts to Improve Installation Preparedness Can Be Enhanced with Clarified Responsibilities and Comprehensive Planning.* http://www.gao.gov/assets/250/243726.pdf.

GAO 11-881. September 2011. Department of Homeland Security. *Progress Made and Work Remaining in Implementing Homeland Security Missions 10 Years after 9/11.* http://www.gao.gov/assets/330/322889.pdf.

John Jogerst. June 2002. What's so Special about Special Operations? Lessons from the War in Afghanistan, *Aerospace Power Journal* 16, no. 2, 100.

Leo Bosner. 2002. Emergency Preparedness: How Japan and the United States Compare, *Asia Perspectives* 4, no. 2, 17-20 (Mansfield Center for Pacific Affairs).

Lessons Learned Conference. Aug 2011. Medical Response to Joplin Tornado May 22, 2011.

Takashi Nagata. 1995. Comparing Hurricane Katrina to Japan's Kobe Earthquake in 1995: Sharing Policy and Institutional Lessons from Two Large Scale Natural Disasters in the United States and Japan. http://www.hsph.harvard.edu/takemi/files/2012/10/RP239.pdf.

chapter nineteen

Decisions in disaster recovery operations

A game theoretic perspective on organization cooperation*

John B. Coles and Jun Zhuang

Contents

* Originally published in *Journal of Homeland Security and Emergency Management*, 8(1), 2011. Reproduced with permission from De Gruyter Publications (http://www.degruyter.com).

19.1 Introduction

The Indian Ocean tsunami that struck on December 26, 2004, killed over 230,000 people, destroying housing and critical infrastructure everywhere it landed. As a result, one of the largest international relief efforts in modern history was mounted to save life and property and to stabilize the devastated region. Among the successes and failures of the response, an overwhelming need for cultural understanding and sensitivity became apparent to make the response and recovery efforts sustainable. Although sensitivity to cultural issues is challenging in the initial response phase, identifying sustainable methods of aid distribution throughout the recovery phase is critical to local acceptance while minimizing the chance of long-term dependence on outside assistance. A lack of sensitivity to these critical issues could reduce the positive long-lasting changes or recovery in the region due to the fact that aid delivery does not address economic and social issues in a culturally acceptable manner. With the recent disasters in Haiti and Chile in January and February 2010, respectively, the need for a more holistic approach to actors' cooperation has become increasingly clear. Using perspectives from game theory in the problem of cooperative interactions between international and local actors, we discuss the potential for improvement in disaster management and cooperative strategies across the developing world.

Large-scale emergencies, also called disasters, are a global phenomenon. From Indonesia to Haiti, disasters have killed hundreds of thousands of people, destroying local infrastructure and leaving millions of people homeless. In the United States, Hurricanes Katrina and Ike showed the remaining vulnerability of developed countries to natural disasters, despite the large number of actors dedicated to responding to these scenarios at the local, state, and federal levels. Differences in culture and context make it essential for organizations, government, and individuals responding to a disaster to be prepared to work effectively in an unfamiliar environment even within a common national border. Disaster relief and emergency management have played an increasingly significant role in foreign policy for the United States and other developed countries.

Destruction and loss following a disaster tends to attract a variety of organizations offering resources and services to support the redevelopment of the stricken area. These services range from medical care to business guidance. Because of the unique nature of each disaster and cultural differences between different impacted areas, it is critical that any actor entering such a situation approach it with a clear objective, maintaining an open mind as to how it might be accomplished. This chapter provides new insight into the dynamics that may occur when actors enter a new environment, and well-established local actors interact and develop working relationships. By analyzing the problem of actor partnerships in disaster recovery, this chapter provides a new game theory perspective on how to model these relationships and how emergency managers could better utilize their resources during the recovery effort.

19.1.1   Problem definition

In the complex and dynamic environment that follows after a major disaster, it is essential that organizations, agencies, and individuals, collectively called "actors," be able to manage and use their resources to effectively respond. The choice of when, where, how, and with whom these resources should be deployed is a complex problem in emergency management. The actors involved in emergency management choose how to use the available resources to maximize the impact in disaster environments. This problem of optimal resource allocation becomes more complicated when the decision makers act in environments that are unfamiliar to them. The lack of familiarity with an environment could decrease the efficiency of decision makers due to the increased complexity and the potential addition of unidentified factors in the new environment. By providing a decision support framework for emergency managers in the process of developing partnerships, and framing it as a game theory problem, this chapter explores a new methodology to maximize the efficiency of actors involved in disaster recovery.

19.1.2   Objective and structure

Since the time necessary to sufficiently understand an unfamiliar situation may not be available to the actors in an emergency environment, here we discuss a methodology for using interactions between actors to increase efficiency in the final stage of emergency management—the recovery phase. By analyzing the dynamics of relationships that may occur in disaster recovery through the lens of game theory, we provide a new perspective on improving the efficiency of disaster relief operations.

The remainder of the chapter is organized as follows: Section 19.2 introduces game theory terminology and how it could be applied to decision making in the context of disaster recovery operations. Section 19.3 explores emergency management and delves into the nuances of disaster recovery and actor–actor partnerships. Section 19.4 discusses the new framework developed to support emergency managers by integrating game theory and disaster recovery. Section 19.5 provides a discussion of the limitations of this approach and real-world examples to which this research could be applied. Section 19.6 concludes the chapter with an overview of what our work contributes to the body of knowledge.

19.2   Game theory

In the development of a holistic approach to disaster recovery operations and actor relationships, we first review some terminology of game theory used in this chapter as follows:

- *Actor (player):* This term is broadly used to mean an organization, agency, individual, government, or business that is involved in a game with other actors.
- *Benefactor:* An actor that provides resources to another actor.
- *Game:* A framework for interaction between actors.
- *Entering actor:* An organization, agency, individual, government, military, or business that did not normally operate in the affected region prior to the disaster. This includes actors based internationally as well as actors that operate in the same country but are not familiar with the specific nuances of the local area affected by a disaster.
- *Local actor:* An organization, agency, individual, government, military, or business that operated in the affected region prior to a disaster.
- *(Nash) equilibrium:* The balance point in a game where no actor could benefit by changing his/her strategy while the other actors keep their strategies unchanged.
- *Payoff:* The benefit(s) received by each actor at the conclusion of a game.
- *Objective:* The maximization or minimization of certain goals for each actor in a game.
- *Outcome:* The result of a game after it is played, including the payoffs for each actor.
- *Partnership:* A relationship between two actors where goods or information is exchanged and all involved are perceived as equals without one being subject to another. Accountability is mutually given and received.
- *Game with perfect information:* A game where each actor knows the options that the other actors are faced with, and if it is a sequential game, the choices that previous actors in the game have made.
- *Sequential game:* A game where the actors involved make decisions in sequence, and some information regarding the decision made by the first actor may be available to the second.
- *Simultaneous game:* A game where actors make their decisions at the same time. As a direct result, neither party is able to know what the other has decided when making their decision.

Using the terminology defined above, we discuss how game theory could be applied to decision making in the context of disaster recovery operations. Game theory studies the interaction of multiple actors and the outcomes that occur (Camerer, 2003). These interactions can be broken down into specific "games" where a finite set of actors interact for some time, resulting in a payoff for each actor involved. Using mathematics and probability to model these interactions, game theory can help to predict the outcomes of future interaction between actors, and even provide decision support for future circumstances (Rasmusen, 2007). Game theory has been applied to a variety of different circumstances and environments providing, among others, valuable insight on counter-terrorism operations (Hong and Apostolakis, 1993; Hashagen 2002; Zhuang and Bier, 2007).

Games can take many forms where two actors may move sequentially, with actors "playing" one after another, or simultaneously, where both actors choose a strategy prior to approaching negotiations (Shor, 2006). Another element that impacts the games in coalition formation is when varying amounts of information are available to the actors and there is uncertainty as to what degree the information available can be trusted (Kapucu, 2005).

By increasing local participation in the recovery process, efforts to restore what was lost would be motivated by local individuals rather than by external capital and personnel that leave when the recovery phase for a given organization is over (Tolentino, 2007). This is greatly desirable for both the local economy and the society in the long run, as it could improve the livelihood of proactive local individuals, and even may initiate a locally driven movement to mitigate future disasters (Mainville, 2003). The development of relationships within the disaster community is also critical to an effective and efficient recovery (Kapucu, 2008; Hall, 2008). Although resources such as food, machinery, and clothing are necessities in response and recovery efforts, they cannot be effectively used without properly trained and located personnel. Thus, the formation of partnerships allows for maximization of the intersection of the necessary materials and conditions needed to effectively meet local needs.

Given the large number of criteria involved in disaster recovery operations, the problem of formulation related to interactions between actors could be complicated. However, decision and risk analysis methods could help to quantify these comparisons (Cox, 2009). By focusing on the dynamics of actor–actor interactions individually rather than collectively, the system dynamics as a whole could be modeled using game trees (Hong and Apostolakis, 1993; Myerson, 1997). Game theory has proven useful in the analysis of multiple complex dynamics within disaster management processes (Sheremetov et al., 2004). Using game theory, people can examine probabilistic subproblems and identify the total system output for the complex recovery environment (Hashagen, 2002; Marschak and Radner, 1972).

The issue of actor–actor coalition formation is complex due to the unique nature of each separate relationship. Although partnership formation is not a simple problem to begin with, dynamics during emergency response and recovery operations add some additional elements of complexity, which makes the proof of a general coalition theorem for this particular problem elusive. There are four primary factors to be addressed to ensure that the problem of coalition formation is effectively approached by all parties during an operation: (1) the different types of actors that could be involved, (2) the challenge of cross-cultural coalitions, (3) the differences in overarching scenario methodology directing operations, and (4) the different dynamics of a partnership that could fundamentally alter the progression of the game. Although the presence of each of these particular dynamics may be easy to identify in practice, the understanding and adaptation necessary to appropriately respond to these unique dynamics may prove challenging in practice.

19.3   Emergency management

Emergency management is the process of preparing for and responding to any emergency or disaster. In this chapter, we use the definition of "disaster" provided by the National Governors Association of the United States that defines a disaster as any "event that demands substantial crisis response requiring the use of government powers and resources beyond the scope of one line agency or service" (Haddow et al., 2008). The four stages of emergency management are mitigation, preparedness, response, and recovery (Comfort, 1990). Mitigation and preparedness occur prior to the disaster and serve to minimize the impact of a disaster and plan the response and recovery phase ahead of time. Response and recovery are the short- and long-term approach, respectively, where actors assist an affected area following a disaster. Because of time constraints and the significant effect of decisions made regarding lives and property, the development of relationships between actors is a complex problem during all four stages of emergency management. For our discussion of relationships in disasters, we focus on actors entering a disaster scenario, defined here as "entering actors," and those actors that had an established operation prior to the disaster, the "local actors."

Advanced planning assists in the achievement of greater efficiency during disasters, primarily because it decreases the number of unknowns and allows for a focused and cooperative effort by multiple actors. Much of the research on partnerships during emergencies points to greater efficiency being achieved through previously established relationships (Telford and Cosgrave, 2007; Kapucu, 2008). To sustain a high level of preparedness, external actors would ideally sustain partnerships with local actors in as many disaster-prone areas as possible, allowing them to work solely through previously established relationships.

Although the type of actors involved in disaster recovery operations are situation dependent, here we refer to all representatives of these actors as emergency managers for the sake of consistency. Emergency managers are defined here as the individual(s) representing an actor in an emergency situation. These individuals are empowered to make decisions in real time, including decisions about resource allocation and partnership formation. Since both local and external actors have limited resources and a vast number of areas could require such partnerships, it is critical that methods be developed to generate a high level of efficiency in response efforts through local and external partnerships formed in a short time frame (Rowan, 1994; Kapucu, 2006). To perform this task effectively, it is critical that emergency managers be given the proper tools and information to make the best decisions for their organization.

19.3.1 Emergency management and game theory

The application of game theory in dynamic situations should be performed with caution. The fundamentally indeterminate elements that compose a disaster or emergency make it helpful for any decision framework applied to be inherently flexible. The maximization of organizational cooperation is often more realistic when applied during the recovery phase of a disaster. It is difficult to provide emergency managers with information accurate enough to create an optimal, long-term strategy that could be implemented at the outset of an event in the immediate aftermath of a disaster (Hashagen, 2002).

One key driver of cooperation and partnership between actors in disaster response and recovery operations is field interaction between first responders (Wilson and Musick, 1997; Smith and Dowell, 2000). When actors are simultaneously working in the same geographic location, the probability of them cooperating is much higher because of natural convenience and the potential gain of a partnership. Any holistic framework modeling the development of partnerships between actors would incorporate the subjective assessments of field personnel. Such quick judgments made by the field personnel are critical to operational success and actor efficiency in the field. These subjective assessments of other actors provide a referent for emergency managers when developing partnerships.

19.3.2 Actor–actor partnerships and disaster recovery operations

The maximization of organizational cooperation is often unrealistic for large-scale scenarios during the response phase. Because of the necessity of rapid response and the tendency of individual actors to maximize individual rather than collective outcomes, uncoordinated work by multiple actors may not lead to the best collective solution (Hausken, 2002). Hence, a key driver of cooperation is field interaction between first responders on the ground in the regions to which they are actively responding (Wilson and Musick, 1997; Smith and Dowell, 2000). The best achievable minimization of loss in life and property in these areas is often the direct result of coalitions formed out of necessity, the basic human drive to preserve life, and previously formed procedures (Kapucu, 2006). Since there may be little time for stable relationships to form during this time, need-based cooperation facilitates the sharing of resources during the response phase (Bergantinos et al., 2007; Kapucu, 2008). Although there are a variety of rule-based approaches to risk communication, it is critical that general heuristics are developed to aid sustainable development of partnerships formed under suboptimal conditions (Rowan, 1994).

As shown by large-scale disasters such as the Indian Ocean tsunami in 2004 and the recent Haiti earthquake in January 2010, the significance of effective cross-cultural partnerships has been clear. Without such relationships, there may be less effective use of resources due to miscommunication, misunderstood objectives, and a lack of a common operating perspective. The challenges of cross-cultural cooperation may occur even within a community, and responding actors, even within their jurisdictions, should always be aware of the potential need to address local customs or boundaries.

Some of the challenges that may arise during an attempt by multiple actors to coordinate their activities are differences in relief methods, goals, and terminology. These challenges make the common operating perspective more difficult to create and cooperative behavior more challenging to model, since the resulting combination of these three possible differences is difficult to predict. Actors attempting to partner across cultural boundaries would need to be aware of what objectives could be subject to change/interpretation and which are nonnegotiable to effectively determine what partnerships would be productive.

19.4 Application of game theory in disaster recovery

In the case of multi-actor cooperation within a disaster, one of the key challenges faced by responders in disaster recovery is the development of a common operating perspective. Communication is essential for the development of a stable operating perspective between different actors. To that end, identifying the similarities in organizational objectives would provide a useful starting point. When developing a model for actor–actor interactions, it is critical that the techniques to determine the estimated payoff values and different outcomes are standardized across costs, benefits, and objectives. Because of the differences in how actors measure success in disaster recovery operations, it could be challenging to create a common operating perspective to measure efficiency and productivity from an independent perspective. Once a common operating perspective is established, actors could coordinate their decision making to get to the optimal outcome to their collective decision. This process allows the different actors to decide if their common interests are close enough to warrant the creation of a team or partnership (Marschak and Radner, 1972). If additional perspective is added to both actors' operating picture, then there is potential for a shift in perceived payoffs, which could improve the combined outcome for both actors and result in a more stable long-term partnership.

19.4.1   Theoretical basis for application

In our problem formulation, we propose to break the actors involved in disaster relief into two specific subsets: local actors and external actors. As defined earlier, the local actors are those that were present prior to a disaster, while external actors are those that entered the scene after a disaster. By dividing the actors into these two subsets, we then discuss some of the perceived characteristics of these different groups.

Based on the proposed separation of actors, we then attempt to identify whether there are fundamental differences in their objectives. For our analysis of the operations of entering actors, we argue that the primary objective of partnerships with local actors is to maximize operational efficiency and perceived impact. This assertion is based on the definition of an "external actor," where we assume that the actor is new to the disaster zone and is coming in with some constrained set of resources of skills with the goal to apply them optimally. Partnership with local actors is one avenue for the entering agency to expend resources to achieve the greatest perceived outcome. To provide a more general framework, we intentionally avoid defining exactly what desired and perceived outcomes an actor might be seeking.

The other set of actors active following a disaster are the local actors. These are the actors that were active in the disaster zone prior to the causal event/disaster. The objective that we associate with local actors, in their development of partnerships with entering actors, is the maximization of resources acquired that could be applied to the relief goals of the local actors. This assumption is based on interviews conducted by the researchers in Haiti following the earthquake on January 12, 2010. It was found that over 50% of local organizations interviewed wanted free or subsidized products out of the partnerships with an entering actor.

An additional aspect of differences in direction methodology that needs to be considered is the differences in cultural norms and objectives, which may be integral to the central response methodology. Although some of these issues come out in cross-cultural partnership formation, it is also important to mention the effect that the culture of the directing methodology may have even on partnerships between actors with similar cultures. When considering partnership formation, it is critical to ensure that such partnerships increase actors' abilities to achieve their objectives. Furthermore, when these objectives and methods are dynamic because of an unfamiliar direction methodology, it is critical that more caution is exercised in partnership formation by both entering and local actors.

19.4.2   Historical real-world examples

Here, we explore three case studies that provide anecdotes for why this model could be useful in disaster recovery environments.

19.4.2.1   Hurricane Mitch

In the case of Hurricane Mitch in 1998, bean production in Honduras suffered massively with a national cumulative loss of 35% of the bean production for that year (Mainville, 2003). Mainville compares three different markets and the intervention methods used to highlight the importance of working with localized channels while diverting resources from the profit-based commercial market. Some of the systems that normally rely on commercial channels for the bean seed were challenged, but did not suffer extensively since pricing was frozen across the country to assist the relief effort. Because of this partnership between the local and international actors, the amount of money spent on diverting relief supplies to the hardest-hit areas was kept significantly lower by not having to pay for the natural increase in price as a direct result of loss in demand. By partnering with the local community to meet the needs of those within it, the external actors had a better opportunity to maintain an external identity rather than being viewed as a permanent benefactor (Donovan et al., 2006).

As noted by Donovan, the timing and method of food aid injection is critical to encourage development within a region that is recovering from a shock. The distribution process could be performed by local entities, which have previously served a similar purpose, rather than by new, theoretically temporary, mechanisms being put in place. By partnering with local actors in the relief effort, external actors adhering to previously existing distribution mechanisms and techniques in the recovery process would be better able to exit a scenario without the population feeling abandoned. Since the local actor would still be in place after the external actor leaves, the transition challenge could be reduced through a minimization of the entitlement phenomenon. By examining this problem from a game theory perspective, it was necessary for the external actor to alter their utility function to integrate both short- and long-term goals to succeed in assisting their partner without being viewed as a benefactor.

19.4.2.2 Indian Ocean tsunami

After the Indian Ocean tsunami in 2004, an international coalition was created to evaluate the response to the disaster, and the results were summarized in the Tsunami Evaluation Coalition's (TEC) Synthesis Report, authored by Telford and Cosgrave in 2006. The report noted that some of the most effective work was done through the partnership of international nongovernmental organizations with local response efforts. A key conclusion from this report was that such relationships are vital to effective and sustainable response operations. This result was confirmed through later work done by Telford and Cosgrave (2007). The TEC also noted that, "The engagement of international actors with local capacities was most effective and efficient when it was built on sustained prior partnerships with the local actors." However, it was also found that several international actors satisfied their staffing requirements by poaching from local actors. Though it is not known how widespread this practice was, it is clear that there was a significant negative effect on the local actors that had lost the personnel. From the perspective of the developed framework, it is clear that actors that chose to partner with local actors were more effective in an absolute sense than those that poached.

The massive influx of aid to the region following the Indian Ocean tsunami of 2004 not only met the needs of the victims in the region but also overwhelmed the area. One case of poor resource management is discussed in the work of Telford and Cosgrave (2006). Following the devastation wrought on the Indonesian region of Aceh, especially in coastal businesses, thousands of fishing boats were imported by aid organizations. However, when the boats were given to local fishermen, they proved to be unsuitable to the water conditions because they were made of the wrong kind of wood (Telford and Cosgrave, 2006). Massive waste could have been avoided in such situations by partnering with the local people and understanding the needs of the area (Smith and Dowell, 2000). Had the external actors accounted for their lack of local familiarity, they could have avoided being perceived as a benefactor by partnering with local actors. This partnership could have allowed them to change their utility function to incorporate a success of meeting the local conditions as a part of their success metrics.

19.4.2.3 Haitian earthquake

The situation that unfolded following the earthquake in Haiti provides a unique look into the potential devastation that could be wrought on a country's central government. With the U.S. military and other American aid actors serving in positions normally occupied by a country's government, this potential phenomenon is one which actors entering into such complex international scenarios need to be aware of since avenues of official communication and overarching objectives may be unstable or more challenging to identify.

Disaster management operations have advanced greatly in recent years in the developed world with improved equipment and communication as well as standardized organizational structures. In the developing world, the advantages of these new advances have yet to be realized, even when international actors respond to a local disaster as highlighted by the recent disaster in Haiti. The differences in individual and cultural objectives are significant enough between regions that implementing standard practices of a developed nation in developing regions often creates dependencies and becomes ineffective in assisting the local population in sustainable ways. To improve the outcome and perspectives of work, it is important to include a cultural aspect to a game theory utility function. This may allow the actors to recognize when they may accomplish their objectives in name and action by adapting to local cultural norms.

By partnering with local actors during the recovery phase of a disaster, external actors have the opportunity to assist the population in sustainable and culturally appropriate ways. Through the process of adapting organizational objectives to a region's cultural and social norms, aid dispersed may have the potential to reach further, even when measured by the organization's original objectives. The exception to this would be cases in which the organizational objectives violate local norms, and the assistance rendered by the external actors would likely do more harm than good in the long run.

19.5   Discussion

The perspective presented in this chapter provides insight into the problem of actor partnerships to support decision makers in disaster recovery environments. Here, we further develop the suggested framework by addressing some limitations of a game theory–based framework for disaster recovery operations, and then applying the proposed framework to real-life examples of disaster recovery operations.

19.5.1   Approach limitations

While the framework suggested may help solve some of the problems encountered by emergency managers, it is not sufficiently developed to provide a holistic assessment tool for emergency managers. Although we describe some of the problems facing emergency managers during the recovery phase of the disaster, it is clear that this approach is not all-encompassing. The use of game theory in our approach to the problem of actor partnership development and stabilization during emergencies provides additional insight into the deterministic aspects of the relationships developed, but may not completely account for all possible emergent phenomena.

19.5.2   Future work

In future iterations of this framework, it would be interesting to include an analysis of relationships between external actors and funding sources. An analysis of the local–external actor relationship could then be analyzed with greater depth by integrating the additional needs of the external agencies (e.g., demonstration of effective recovery operations to obtain additional/future funding). Although this piece is challenging to quantify, it has been identified as a significant factor in relationships that remain stable over a long recovery period. This factor could be incorporated into the external actors' efficiency calculations, providing insight into how partnerships are tailored to improve public image and develop, augment, and sustain resource flows.

As this approach becomes more accessible to emergency managers and first responders, the integration of actors' perceptions of each other prior to the formation of a relationship should become a larger part of the model. By including these subjective observations in a quantitative framework, we can better account for the subjective nature of such relationships while simultaneously utilizing the value that such observations have when received from well-trained personnel. Though the relationships that form during emergency response and disaster relief are dynamic and complex, it is essential that work continues to overcome the obstacles and provide managers with tools to make organization-wide and tactical decisions.

19.6   Conclusions

The development of a holistic framework for actors interacting across cultural boundaries has the potential to greatly increase the efficiency of those responding to a disaster. The development and provision of support tools could provide guidance to emergency managers during the recovery phase of a disaster, making operations more efficient, productive, and sustainable. Further improvement of this framework, and development of parallel utility functions from disaster data, could provide support mechanisms for emergency managers when considering how best to approach potential partners in future scenarios. In future research, it would be helpful to use probabilistic components in the model to incorporate the variety of different actors that participate in disaster recovery.

Acknowledgments

This research was partially supported by the U.S. Department of Homeland Security through the National Center for Risk and Economic Analysis of Terrorism Events (CREATE) under award number 2010-ST-061-RE0001. However, any opinions, findings, and conclusions or recommendations in this document are those of the authors and do not necessarily reflect views of the U.S. Department of Homeland Security or CREATE. This research was also supported by the National Science Foundation (NSF) under Award #1034730, a University at Buffalo's School of Engineering Senior Scholar Award to John B. Coles, and an NSF Graduate Research Fellowship to John B. Coles. We also thank Dr. Ann Bisantz, Elizabeth Newell, Jodie-Ann Duquesnay, Heather Coles, and all at University at Buffalo for their helpful comments and support.

Bibliography

Alexander, D. 2006. World Disasters Report 2005: Focus on Information in Disasters. *Disasters 30(3)*: 377–379.

Brunner, B. 2007. *The Great White Hurricane.* http://www.infoplease.com/spot/blizzard1.html (accessed October 5, 2010).

Cheshire, L. 1997. *All about Snow.* http://nsidc.org/snow/shovel.html (accessed October 15, 2010).

Christiano, G.J. *Blizzard of 1888; the Impact of this Devastating Storm on New York Transit.* http://www.nycsubway.org/articles/1888-blizzard.html (accessed October 5, 2010).

Comfort, L.K. 2007. Crisis Management in Hindsight: Cognition, Communication, Coordination, and Control. *Public Administration Review 67(Suppl 1)*: 189–197.

Coombs, W.T. 1999. *Ongoing Crisis Communication: Planning, Managing, and Responding.* Thousand Oaks, CA: Sage.

Corbacioglu, S., and N. Kapucu. 2006. Organizational Learning and Self-Adaptation in Dynamic Disaster Environments. *Disasters 30(2)*: 212–233.

DMA. 2000. *Hazard Mitigation Plan.* New York: Suffolk County, New York, 2008.

Fischer, H. W. 1998. *Response to Disasters: Facts Versus Fiction & Its Perpetuation.* New York: University Press of America.

Jennings, N.A., and McC. Lingan. 1888. *New York in the Blizzard.* New York: Rogers & Sherwood.

Kennedy, R.C. 2001. *On This Day.* http://www.nytimes.com/learning/general/onthisday/harp/0303.html (accessed October 30, 2010).

Levy, J.K., K.W. Hipel, and N. Howard. 2009. Advances in Drama Theory for Managing Global Hazards and Disasters. Part 1: Theoretical Foundation. *Group Decision and Negotiation 18(4)*: 303–316.

Lindell, M.K., C. Prater, and R.W. Perry. 2007. *Introduction to Emergency Management.* Hoboken: John Wiley & Sons.

New York City Office of Emergency Management. *NYC Hazards: Winter Storm History.* http://www.nyc.gov/html/oem/html/hazards/winter_history.shtml (accessed October 14, 2010).

Stern, R.A.M., T. Mellins, and D. Fishman. 1999. *New York 1880:* In *Architecture and Urbanism in the Guilded Age.* New York: The Monacelli Press.
Stobo, J.R. 2008.*Organized Labor, Housing Issues, and Politics: Another Look at the 1886 Henry George Mayoral Campaign in New York City.*
Squires, M.F., and J.H. Lawrimore. 2006. *Development of An Operational Northeast Snowfall Impact Scale.* Ashville: NOAA National Climatic Data Center.
The Northeast States Emergency Consortium. *Winter Storms.* http://www.nesec.org/hazards/winter_storms.cfm (accessed October 10, 2010).
Wachtendorf, T., and J.M. Kendra. 2004. *Considering Convergence, Coordination, and Social Capital in Disasters.* Canadian Risk and Hazards Network 1st Annual Symposium, Winnipeg, Manitoba, Canada.
Woods, M., and M.B. Woods. 2008. *Blizzards.* Minneapolis, MN: Lerner Publications.

References

Bergantinos, G., B. Cases-Mendez, M. Fiestras-Janeiro, and J. Vidal-Pugo. 2007. A Solution for Bargaining Problems with a Coalition Structure. *Mathematical Social Sciences 54(1):* 35–58.
Camerer, C.F. 2003. *Behavioral Game Theory: Experiments in Strategic Interactions.* Princeton, NJ: Princeton University Press.
Comfort, L.K. 1990. Turning Conflict into Cooperation: Organizational Designs for Community Response in Disasters. *International Journal of Mental Health 19(1):* 89–108.
Cox L.A. 2009. Game Theory and Risk Analysis. *Risk Analysis 29(8):* 1062–1068.
Donovan, C., M. McGlinchy, J. Staatz, and D. Tschirley. 2006. *Emergency Needs Assessments and the Impact of Food Aid on Local Markets.* Michigan: Michigan State University.
Haddow, G.D., J.A. Bullock, and D.P. Cappola. 2008. *Introduction to Emergency Management.* Oxford: Elsevier.
Hall, R.A. 2008. Civil-Military Cooperation in International Disaster Response: The Japanese Self-Defense Forces' Deployment in Aceh, Indonesia. *Korean Journal of Defense Analysis 20(4):* 383–400.
Hashagen, P. 2002. *Fire Department, City of New York.* Turner Publishing Company.
Hong, Y., and G. Apostolakis. 1993. Conditional Influence Diagrams in Risk Management. *Risk Analysis 13(6):* 625–636.
Hausken, K. 2002. Probabilistic Risk Analysis and Game Theory. *Risk Analysis 22(1):* 17–27.
IRT: *The First Subway.* http://www.nycsubway.org/irtsubway.html (accessed October 30, 2010).
Kapucu, N. 2005. Interorganizational Coordination in Dynamic Context: Networks in Emergency Response Management. *Connections 26(2):* 33–48.
Kapucu, N. 2006. Public-Nonprofit Partnerships for Collective Action in Dynamic Contexts. *Public Administration: An International Quarterly 84(1):* 205–220.
Kapucu, N. 2008. Collaborative Emergency Management: Better Community Organizing, Better Public Preparedness and Response. *Disasters 32(2):* 239–262.
Mainville, D.Y. 2003. Disasters and Development in Agricultural Input Markets: Bean Seed Markets in Honduras After Hurricane Mitch. *Disasters 27(2):* 154–171.

Marschak, J., and R. Radner. 1972. *Economic Theory of Teams.* New Haven and London: Yale University Press.

Myerson, R.B. 1997. *Game Theory: Analysis of Conflict.* Cambridge, Massachusetts: Harvard University Press.

NY Blizzard of 1888: Stories and Memories of the Blizzard. http://www.virtualny. cuny.edu/blizzard/stories/stories_set.html (accessed October 5, 2010).

NY Blizzard of 1888: Towards a Cleaner City. http://www.virtualny.cuny.edu/blizzard/sanitation/san_set.html (accessed October 30, 2010).

Rasmusen, E. 2007. *Games and Information: An Introduction to Game Theory.* Malden, Massachusetts: Wiley-Blackwell.

Rowan, K.E. 1994. Why Rules for Risk Communication Are Not Enough: A Problem-Solving Approach to Risk Communication. *Risk Analysis 14(3):* 365–374.

Sheremetov, L.B., M. Contreras, and C. Valencie. 2004. Intelligent Multi-Agent Support for the Contingency Management System. *Intelligent Computing in the Petroleum Industry 26(1):* 57–71.

Shor, M. 2006. *Sequential Game.* http://www.gametheory.net/dictionary/SequentialGame .html (accessed December 20, 2010).

Smith, W., and J. Dowell. 2000. A Case Study of Co-ordinative Decision-Making in Disaster Management. *Ergonomics 43(8):* 1153–1166.

Telford, J., and J. Cosgrave. 2006. *Joint Evaluation of the International Response to the Indian Ocean Tsunami: Synthesis Report.* London: Overseas Development Institute.

Telford, J., and J. Cosgrave. 2007. The International Humanitarian System and the 2004 Indian Ocean Earthquake and Tsunami. *Disasters 31(1):* 1–28.

Tolentino, A.S. 2007. The Challenges of Tsunami Disaster Response Planning and Management. *International Review for Environmental Strategies 7(1):* 147–154.

Wilson, J., and M. Musick. 1997. Toward and Integrated Theory of Volunteer Work. *American Sociological Review 62(5):* 694–713.

Zhuang, J., and V.M. Bier. 2007. Balancing Terrorism and Natural Disasters— Defensive Strategy with Endogenous Attacker Effort. *Operations Research 55(5):* 976–991.

chapter twenty

Integrating Department of Defense response with nongovernmental organizations during a disaster*

Harley M. Connors, Helen W. Phipps, and Michael J. Surette Jr.

Contents

20.1   Introduction

The events of 9/11 and Hurricane Katrina are clear examples of the challenges associated with executing large-scale response to disasters involving the "whole of government" including the U.S. Department of Defense (DOD) and nongovernmental organizations (NGOs). Emergency responders from the same city, and in some cases the same organization, were unable to communicate and effectively organize overlapping response activities. Differences with communication equipment, unclear roles and responsibilities, and confusing terminology and jargon all contributed to the challenges experienced by responders. The common belief was that the federal government, DOD, and NGOs had standardized procedures that allowed interoperability. However, active collaboration at all levels of response and across all shared missions was not effectively executed.

Typically, incidents begin and end locally and are managed at the lowest possible geographical, organizational, or jurisdictional level. However, there are instances in which successful incident management operations depend on the involvement of multiple jurisdictions, levels of government, functional agencies, and/or emergency responder disciplines. These instances require effective and efficient coordination across this broad spectrum of organizations and activities. This chapter introduces the major stakeholders within federal, DOD, and NGO disaster response and then focuses on the key strategic considerations for their integration at all levels of response.

20.1.1   National Incident Management System

As a direct result of the challenges experienced during the responses to 9/11, the federal government adopted the National Incident Management

System (NIMS). Originally issued in March 2004 by the Department of Homeland Security (DHS), Homeland Security Presidential Directive (HSPD)-5, *Management of Domestic Incidents,* directed the development and administration of NIMS. NIMS forms the basis for interoperability and compatibility that will, in turn, enable a diverse set of public and private organizations to conduct well-integrated and effective emergency management and incident response operations. Emergency management is the coordination and integration of all activities necessary to build, sustain, and improve the capability to prepare for, protect against, respond to, recover from, or mitigate against threatened or actual natural disasters, acts of terrorism, or other man-made disasters. It does this through a core set of concepts, principles, procedures, organizational processes, terminology, and standard requirements applicable to a broad community of NIMS users.

NIMS is based on the premise that a common incident management framework will give emergency management and response personnel a flexible but standardized system for emergency management and incident response activities. NIMS provides a consistent, systematic, and proactive approach to guide departments and agencies at all levels regardless of cause, size, location, or complexity. The complexity and challenges associated with effective organization and execution of responses to incidents from small scale to national or international levels require NIMS to be flexible and able to evolve as technology evolves and lessons are learned from subsequent incidents. NIMS fosters the development of specialized technologies that facilitate emergency management and incident response activities, and allows for the adoption of new approaches that will enable continuous refinement of the system over time.

HSPD-5 requires all federal departments and agencies to adopt NIMS and to use it in their individual incident management programs and activities, as well as in support of all actions taken to assist state, tribal, and local governments. The directive requires state, tribal, and local organizations to adopt NIMS to receive federal assistance through grants, contracts, and other activities. NIMS encourages the compliance of NGOs and the private sector because of the recognized role they have in preparedness and activities to prevent, protect against, respond to, recover from, and mitigate the effects of incidents.

Identifying and recognizing the need to maintain NIMS relevance within the emergency management and homeland security communities, the then Secretary of Homeland Security, Tom Ridge, established the NIMS Integration Center, which is now known as the National Integration Center. The National Integration Center is a division within the FEMA's National Preparedness Directorate. FEMA is the primary government office responsible for the maintenance, management, and sustainment of NIMS. The National Integration Center recommends 14 activities for NGOs that support NIMS implementation (Federal Emergency

Management Agency [FEMA], 2006). These activities closely parallel the implementation activities since 2004. Effective and consistent implementation of NIMS across the board is meant to strengthen national capability to prevent, prepare for, respond to, and recover from any type of incident.

HSPD-5 also required the Secretary of Homeland Security to develop the *National Response Plan*, which was superseded by the *National Response Framework* (NRF). The NRF is a guide to how the nation conducts all-hazards response. The NRF identifies the key principles, as well as the roles and structures that organize national response. In addition, it describes special circumstances where the federal government exercises a larger role, including incidents where federal interests are involved and catastrophic incidents where a state would require significant support. NIMS works hand in hand with the NRF. NIMS provides the template for the management of incidents, whereas the NRF provides the structure and mechanisms for national-level policy for incident management (DHS, 2008).

20.1.2   Federal Emergency Management Agency

FEMA was established in 1979 under President Carter's Executive Order 12148, and appointed responsibility of disaster relief efforts. As an independent agency, FEMA absorbed several other programs and agencies including the Federal Insurance Administration, the National Fire Prevention and Control Administration, the National Weather Service Community Preparedness Program, the Federal Preparedness Agency of the General Services Administration, and the Federal Disaster Assistance Administration. It also assumed the function of overseeing the nation's civil defense, which was previously done by the DOD. In 1995, the United States and FEMA recognized the potential for chemical, biological, radiological, and nuclear attacks and the potential use of weapons of mass destruction (WMD) when the terrorist group Aum Shinrikyo deployed sarin gas in the Tokyo subways. In 1996, FEMA was elevated to a cabinet rank under President Clinton, yet shortly after it was taken out of the presidential cabinet by President Bush and the priorities shifted heavily to natural disaster preparedness.

Since its creation, FEMA has gone through many changes and has been criticized for many "failures." In 2002, the debate of the Homeland Security Act called for FEMA to remain an independent agency. In 2003, FEMA joined 22 other federal agencies and offices to become the DHS. Shortly following Hurricane Katrina in 2005, many critics called for the removal of FEMA from the DHS. The "Final Report of the Select Bipartisan Committee to Investigate the Preparation for and Response to Hurricane Katrina," released February 15, 2006 by the U.S. Government Printing Office, revealed that federal funding to states for "all-hazards" disaster preparedness needs was not awarded unless the local agencies specified the funding was solely for antiterrorism functions. Following the responses to the investigation,

FEMA pushed on and began implementing new programs and policies to strengthen its own organization and aid U.S. citizens in preparing for all threats, both natural and man-made. FEMA is responsible for managing the National Flood Insurance Program and assists small businesses and individuals acquire low-interest loans. In addition to small loans, FEMA provides funds throughout the United States for the training of response personnel.

Currently, FEMA is divided into 10 response regions as per Figure 20.1. Each region maintains several regional offices as well as a region headquarters. This division allows for FEMA officials in each region to focus mainly on the states they are responsible for. This includes monitoring state and local emergency management plans, ensuring that all are consistent with regulations mandated by the government for response, infrastructure, zoning, and insurance.

20.2   Planning for integration

The general role of the DOD, federal, state, and local government is to protect the lives and property of U.S. citizens. No agency should work alone in this role. DOD and NGOs must collaborate at every level to take on three specific roles before, during, and after a disaster has occurred:

1. Sustain life
2. Reduce physical and emotional distress
3. Promote recovery of disaster victims (when assistance is not available from other sources)

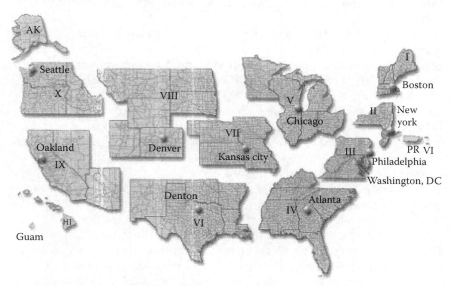

Figure 20.1 FEMA regions (http://www.fema.gov/about/regions/index.shtm).

These services do not solely pertain to human life and services. There are many NGOs that focus on the environment and animals. Greenpeace, Sea Shepherd, the Conservation Fund, and the National Wildlife Federation are a few examples of environmental NGOs that are based in the United States. In the event of a disaster or an emergency, these organizations reach out and provide as many resources as they can to aid in response and recovery. For this reason, the categorization, orientation, operation, and classification of NGOs is key to planning.

By identifying DOD and NGOs early, response plans can be written to include the "whole-of-government" in the overall response package. The DOD and NGOs bring a wealth of knowledge and response acumen with them to disaster relief and recovery efforts. Incident commanders should understand each of their capabilities to effectively use them in times of emergencies.

20.2.1   Integrating DOD into NGO plans

Integration and planning are not only accomplished by the DOD. NGOs also plan for natural and man-made disasters and relief efforts. NGOs bear fiduciary responsibility in this process. NGOs can and should understand the working relationship with the DOD during exercises, tabletop discussions, and open forums. In doing so, NGOs gain insight into the DOD operational response capability and enhance their presence by gaining acknowledgment for participation and capabilities. More importantly, NGOs may gain a better understanding of the DOD mission and responsibilities.

Involvement with inputs into combined gaming exercises would allow civilians, NGOs, and the DOD to learn how the others respond to complex emergencies and how each must modify expectations to accommodate different operating styles. Not only would these exercises bring together different agencies to explore typical problems and solutions, but they would also highlight any dialogue limitations between organizations. What appears clear is that the DOD and NGOs must understand the need for a comprehensive and integrated response to crisis, even if that comprehensive and integrated response necessitates subjugating their charter to the overall good of the mission.

When developing response plans, NGOs must understand that their needs and requirements are part of an integrated overall response to crisis. Depending on the situation, NGOs may have a lead role or a follow-on role, and plans should be tailored accordingly. Understanding working relationships with the DOD and the key points of contact will go a long way in a mutually beneficial and successful response capability.

Contingency response plans are conducted at all levels (local, state, and federal) of government. It is incumbent on the response planners for the NGOs to recognize the need for training and a working relationship with all levels of government in order to have a plan that reflects the best practices in crisis situations. Many states have emergency response and training classes as well and planning seminars for disaster relief efforts. NGOs should contact the state's Division of Emergency Management and review the State Emergency Response and Recovery Plan (or similarly titled by state) to identify classes and/or points of contact for use in their organizational response planning.

20.2.2   Integrating NGOs into DOD plans

In writing emergency response plans, NGOs should be listed with local and national contact information to include the skill sets and capabilities that the NGO has at their disposal. The most prominent NGO is the American Red Cross. The American Red Cross has helped people recover from disaster worldwide since 1881 (American Red Cross, 2012). There are also other large volunteer organizations, such as The Salvation Army, that align themselves with the Voluntary Organizations Active in Disaster group. Although each voluntary organization is a stand-alone group, they readily communicate with each other and exchange ideas, supplies, equipment, and volunteers. The following link includes NGOs that often play a major role in disaster relief with a brief description about the organization: http://www.disastercenter.com/agency.htm. Although the American Red Cross is not a government agency, it has been chartered by the U.S. Congress to "carry on a system of national and international relief, in response to fire, floods and other great national calamities" (The American National Red Cross, 2012). In addition to its support of local authorities for small tragedies, such as a house fire, it is an integral part of the NRF.

Working with and including the DOD and NGOs in local, state, and federal response plans, emergency exercises, and tabletop discussions will allow for clear and consistent response. The more the DOD and NGOs collaborate, the more comfortable the integrated response becomes. To aid in the collaboration, the next sections provide an introduction to categorizations, orientations, operations, and classifications of NGOs.

There are many instances where government agencies work with the private sector and groups as partners during times of emergency to help manage the incident. It is here that the NGOs are a valued asset during relief efforts.

20.2.3   Categorization of NGOs

We live in a world of constant change. While some of these changes are for the better, there are just as many that are for the worse. The possibility and probability of disasters, both natural and man-made, is a constant threat to all. In order to better prepare for these threats, DOD must effectively integrate with its nongovernmental counterparts. This section focuses on defining NGOs and the various types of NGOs within our society.

An NGO is a legally constituted organization created by natural or legal persons that operates independently from any government. The importance of NGOs is recognized and expressed in the NRF. As stated in the NRF, "A key feature of NGOs is their inherent independence and commitment to specific sets of interests and value. These interests and values drive these groups' operational priorities and shape the resources they provide" (DHS, 2008). NGOs support all levels of government effort for disaster response planning and operations. "In some cases, however, NGOs may need and are often provided with direct assistance from some government organizations when preparing and planning the allocation of local emergency management resources and structures" (DHS, 2004). NGOs can be categorized into two types: orientation and operation.

20.2.3.1 NGO orientations

The NGO types by orientation refer to the type of activities the organization takes on as a priority, such as charitable, service, participatory, and empowering. A charity-oriented NGO involves a paternalistic effort with little participation by the "beneficiaries." The activities of such NGOs are directed toward meeting the needs of the less fortunate such as food, water, shelter, and clothing. A service-oriented NGO deals with activities such as the provision of health, family planning, and education services where beneficiaries are expected to volunteer in order to receive support. A participation-oriented NGO often involves and is characterized by self-help projects where local people are directly involved in the implementation of specific projects by contributing all the necessities and resources to complete the project. An empowerment-oriented NGO is one that aims to aid less fortunate communities to develop a clearer understanding of the social, political, and economic factors affecting their lives; to strengthen the way they envision themselves; and to discover the potential power to take control of their lives.

20.2.3.2 NGO operations

The NGO types by level of operation indicate the scale on which they may operate, be it international, national, or local/community. Community-based NGOs arise from personal initiatives such as sporting clubs, recreation clubs, friendly organizations, and religious or educational organizations, to name a few. A variety of community-based organizations are supported by NGOs while others remain completely independent of outside help. Citywide organizations include Rotary Clubs, Knights of Columbus, educational groups, coalitions of business, and community associations. National NGOs include The American National

Red Cross, Young Men's Christian Association, Young Women's Christian Association, and Boys and Girls Clubs, to name a few. International NGOs span across many countries such as the Save the Children Organization and the Ford and Rockefeller Foundations. The activities of an international NGO range from funding local NGO projects to implementing the projects themselves.

20.2.3.3   NGO classifications

Apart from the categorizations, orientations, and operations of NGOs, there are two classifications: operational and campaigning. An operational NGO is the one that seeks to "achieve small scale change directly through projects" (Willitts, 2002). Through the mobilization of financial resources, materials, and volunteers, they create localized programs in the field. By hosting fundraiser events and submitting applications to organizations and the government for grants and contracts, they raise money to support the projects they develop and work on. Often dealing with a wide variety of issues, they are mostly associated with the delivery of services and welfare, emergency relief, and environmental issues.

Alternatively, a campaigning NGO seeks to "achieve large scale change promoted indirectly through influence of the political system" (Willitts, 2002). To be effective and efficient, a campaigning NGO needs a group of professional members who are able to keep supporters informed and motivated. The main activity of a campaigning NGO is the holding of demonstrations.

20.3   Human factors to integration

Regardless of planning, educating, exercising, and training, disasters stress human factors to the limit even if responding with a familiar DOD or NGO team. Factors such as competency, workload, physiological stress, and cultural barriers all must be considered prior to any disaster to ensure the most seamless response possible. The following sections outline the key considerations to ensure efficient integration in the face of diverse human factors and large-scale disasters.

20.3.1   Competency level

When considering competency levels, one must consider formal education and practical experience of personnel at all levels. Table 20.1 lists examples of degrees and their utility in response. Although the DOD and NGOs have hierarchies, each member has its own niche; some are experts in technical and tactical aspects whereas others are experts in management and strategic responses. This makes communication of requirements all the more important. For example, some NGOs require

Table 20.1 Examples of Degrees Useful to Emergency Response

Degree	Utility
Business Management and Accounting	Funds management and execution, budgeting, accounting
Biology, Chemistry, or Radiology	Reachback, technical interpretations, lab response network
Criminal Justice	Evidence collection and preservation, forensics
English	Writing and editing policy and procedures
Journalism	Public affairs, social media integration
Communication	Public affairs, evacuation, warning and reporting
Education	Training (responders and populace)
Foreign Languages and Culture	Translation and integration/relations

Note: This list does not include the traditional response-related degrees (e.g., emergency management, fire science).

a bachelor's degree in a related field prior to employment, whereas the DOD does not. To address the variety of competency requirements, the DOD Fire and Emergency Services Certification Program established certifications recognized by all International Fire Service Accreditation Congress or Pro-Board Accredited Programs except Hazardous Materials Branch Officer Certification Level. Likewise, the U.S. Air Force established a Certified Emergency Manager Program acknowledged by the International Association of Emergency Managers Certified Emergency Manager Program.

In addition to formal education and certifications, the years and sites of practical experience of personnel must be considered. For instance, availability of personnel with experience in specialized facilities and/or teams such as Poison Control Centers would be helpful in emergencies involving contamination.

20.3.2   Workload/burden/rotation

Demanding work schedules and combat environments often lead to physiological and emotional burden. These human factors must be considered prior to aligning resources in disasters. The Occupational Safety and Health Administration (OSHA) and the National Oceanic and Atmospheric Administration offer guidance and resources for developing plans for work rates/loads and training personnel for conditions relevant to various environments. Each DOD Service and NGO may modify these values based on their particular needs and mission requirements.

20.3.3 Cultural, ethnic, and religious background

Cultural, ethnic, and religious aspects of an operating environment play a significant role in successful NGO integration in and outside the United States. The most obvious of the considerations is language. However, language barriers refer to both the operational terminology and the native spoken language of personnel. Diet should also be considered when requesting support. For example, Judaism permits only kosher food, Islam permits only halal foods, and Buddhism and Hinduism generally permit or encourage only vegetarian foods.

20.3.4 Operational terminology and protocols

Operational environments are synonymous to cultures with their own behaviors, traditions, and languages. For this reason, NIMS implemented a universal operational language at the emergency management level. However, not all organizations implemented NIMS, especially at the tactical level. For this reason, the DOD developed JP 1-02, *Dictionary of Military and Associated Terms* and international and/or national standards are referenced to ensure continuity of operations throughout the response environment whenever possible. Table 20.2 lists some common standards used by the DOD and NGOs.

Owing to diverging operational language and signals, the U.S. Air Force developed standardized visual aids in addition to a new campaign supporting community awareness and educational requirements (DODI 6055.17 and AFI 10-2501). The Air Force Emergency Management "Be Ready" Awareness Campaign spans over 32 categories within four major topic areas: basic planning, natural disasters, man-made events, and after a disaster (www.BeReady.af.mil). This campaign was designed with a holistic approach to provide emergency managers at every U.S. Air Force installation a standardized method to disseminate hazard and basic

Table 20.2 Examples of Standards Common to DOD and NGOs

Within the continental U.S.	Outside the continental U.S.
National Fire Protection Association	American Society for Testing and Materials
OSHA	American National Standards Institute
National Institute for Occupational Safety and Health	UNHCR Emergency Standards and Indicators
Environmental Protection Agency	International Organization for Standardization
American Industrial Hygienist Association	

Note: Some standards are only recognized in the U.S. while others are recognized internationally.

awareness information. The program also enhances the ability to raise awareness, provide warning, and guide public behavior with consistent operational language and signals.

20.3.5   Native language barriers

Teaming the DOD with NGOs is of utmost importance when responding in areas that do not speak the native language. Although it is preferable to have translators within each organization, disaster situations often require more translation than there are translators. In permissive environments, this integration is near seamless. However, nonpermissive environments pose multiple concerns including the need to ensure malice is not promoted due to misinterpretation.

20.4   Integration in nonpermissive environments

Disasters come in all types and degrees from natural occurrences to actual or potential terrorist attacks using WMD. The best time to plan or prepare for war is during times of peace. The same can be said for planning and preparing for a natural or man-made disaster. It is during this time frame that federal, state, and local governments and the DOD need to look at all avenues of response capabilities not only from the inside but also from outside NGOs. NGOs' performance in permissive environments and other more secure environments has been comparatively more successful than in less secure environments. For this reason, this section deals with integrating the DOD response with NGOs during man-made or natural disasters in nonpermissive environments.

The U.S. security strategy, policy, and military doctrine recognize NGOs as vital for the United States and the United Nations (U.N.) when conducting relief and reconstruction operations. Recent operations provide keen insights into the effectiveness of NGOs in permissive and nonpermissive environments. There are many publications that provide for the integration of NGOs within the framework of military operations and are the foundation for Joint doctrine:

- Joint Publication 0-2, *Unified Action Armed Forces* (Joint Staff, 2006), states in many locations and ways that unity of effort "requires coordination among government departments … with non-governmental organizations (NGOs), international organizations (IOs), and among nations in any alliance or coalition."
- Joint Publication 3-08, *Interagency Coordination during Joint Operations* (Joint Staff, 2008), states military operations must be synchronized with those of other agencies of the U.S. government, foreign forces, NGOs, and international organizations. It is believed that this

coordination helps forge the link between each agency and other instruments of national power.

- Joint Publication 3-57, *Joint Doctrine for Civil–Military Operations*, identifies Military Public/Civil Affairs personnel as the military organization charged with coordinating and integrating NGOs to best achieve military objectives. They conduct most of this integration with the NGOs, the U.N. agencies, and Department of State agencies such as the United States Agency for International Development (USAID) out of the Civil–Military Operations Center (CMOC).

Over the course of the last decade, military operations (permissive and nonpermissive) have stimulated new Service and Joint doctrine emphasizing the importance of including NGOs in planning to achieve strategic and operational objectives. The latest version of the Quadrennial Defense Review Report (DOD, 2010), reminds readers that stability operations are on par with combat operations and emphasized the use of Provisional Reconstruction Teams and Human Terrain Teams.

Security has also been a problem for NGOs. Adequate security is needed before NGOs and contracting partners in both Iraq and Afghanistan can begin reconstruction and humanitarian work (OCHA, 2001). In Iraq, because of the lack of security, most of the humanitarian community left the country or operated from other countries. Hundreds of civilian foreigners were detained and killed in the last decade in an attempt to drive foreigners out of the country or to gain ransoms including aid workers, journalists, and contractors. The United Nations High Commissioner for Refugees (UNHCR), the U.N.'s preeminent humanitarian NGO coordinator, alluded to their inability to provide adequate physical protection for themselves, the Iraqi population, and the humanitarian communities.

Lessons learned in Iraq and Afghanistan with reference to civil–military relationships continue to confirm that security allowing NGOs and contractors' freedom of movement within an area of responsibility (AOR) for reconstruction and humanitarian work is paramount (Mitchell, 2006; OCHA, 2003). An AOR is the geographical area associated with a combatant command within which a combatant commander has authority to plan and conduct operations (AFDD 2-1.8). It is also considered a predefined area of enemy terrain in which supporting ships are responsible for covering by fire on known targets or targets of opportunity and by observation (JP 1-02). Additionally, NGOs and contractors need neutrality and coordination with the military in order to synchronize efforts.

NGOs have a bleaker report card of performance in nonpermissive theaters of operation such as Iraq and Afghanistan. There are regular reports of contractors and NGOs being targeted by terrorists because they are soft targets. Because of this targeting of NGOs by terrorists in a nonpermissive environment, the NGOs' mission, relief, and recovery efforts are at risk.

The NGOs are stakeholders and vital players in implementing national strategy, especially involving nation building and providing humanitarian aid. When conditions become less stable, the NGOs become less effective. The military must understand and be ready to mitigate, and possibly compensate, for NGO ineffectiveness. The following descriptions illustrate lessons from recent experiences about the performance of NGOs in different environments:

- NGOs may avoid association with the military in order to maintain their neutrality; militaries are often involved in protecting humanitarian relief supplies out of necessity. In lawless environments, warlords and/or armed militants may steal relief supplies or restrict freedom of movement to the indigenous population as well as relief workers.
- In Somalia and Sudan, NGO workers found that over 80% of food supplies were lost due to theft or raids (Byman, 2001). In Angola, Burundi, Chechnya, Rwanda, Sierra Leone, Sudan, Iraq, and other countries, relief workers have been targeted and killed in the process of providing humanitarian relief.
- Coordination between the military and NGOs is essential to provide an effective and synchronized effort to relieve the pain and suffering in the early days of a crisis. NGOs depend on neutrality and innocence for protection. They want the host nation population and the armed parties to see them as harmless and useful. Being associated with the military threatens the NGO's impartial image and may put them in danger.
- In Somalia, fears were realized when World Vision personnel were attacked by militia forces expressing their displeasure with the U.S.-led attempt to capture Mohammad Farah Aideed.

20.5   Budgeting for integration

Despite the desire for impartiality and neutrality, NGOs are warming up to a better working relationship with militaries and governments involved in crisis situations. "Many NGOs report a lessening of respect for neutral parties present in a conflict, a breakdown of spoken and unspoken rules safeguarding helpers" (Byman, 2001). Not only is security important, but NGOs also cooperate with militaries and government entities because they need funds. USAID and the Office of the UNHCR primarily hire NGOs to conduct their humanitarian and reconstruction tasks.

Funding for NGOs by governmental agencies causes the perception of taking sides though it arises out of necessity. Neutrality is compromised when one agency is being paid by one of the opposing stakeholders. Another reason NGOs want to keep closer ties with the DOD includes sharing of information and emergency evacuation. Because of the NGOs' need for neutrality, the DOD–NGO relationship is fragile and requires a balanced approach. Coordination, information sharing, and security arrangements must be discrete and controlled in order to preserve the veil of security and perceived neutrality for the NGOs.

NGOs are a valuable tool in implementing strategy as expressed in all strategic documents and military doctrine. The DOD robustly planned for the increasing use of commercial contractors to fill capability short-falls and improve efficiency under best business practices. Recent history shows that civilians on the battlefield do not always perform as desired, especially in nonpermissive hostile environments. NGOs are the experts at conducting humanitarian and reconstruction tasks, but they are inde-pendent operators, and they require a degree of security. Although they covet neutrality for their own protection, their need for security, funds, and information outweighs the desire for neutrality. The military needs to coordinate and work discretely with NGOs to respect their neutrality when possible. Additionally, the military should coordinate with NGOs through the UNHCR and U.S. Department of State (USAID) when possible.

20.6   Managing integration within and outside the United States

While disasters typically begin locally, the resources of a local community or state can quickly become overwhelmed during a disaster. During larger-scale disasters, the federal government may be requested to assist and will inevita-bly come in contact with NGOs also providing support. For disasters within the continental United States, U.S. Northern Command (USNORTHCOM) is tasked with providing and managing the DOD response.

USNORTHCOM was established October 1, 2002, to provide command and control of DOD homeland defense efforts and to coordinate defense support of civil authorities. Its AOR includes air, land and sea approaches, and encompasses the continental United States, Alaska, Canada, Mexico, and the surrounding water out to approximately 500 nautical miles. It also includes the Gulf of Mexico, the Straits of Florida, and portions of the Caribbean region including the Bahamas, Puerto Rico, and the U.S. Virgin Islands.

USNORTHCOM's civil support mission includes domestic disaster relief operations that occur during fires, hurricanes, floods, and earth-quakes. Support also includes counter-drug operations and WMD event management. In providing civil support, USNORTHCOM generally oper-ates through established Joint Task Forces subordinate to the command. An emergency must exceed the capabilities of local, state, and federal agencies before USNORTHCOM becomes involved. In most cases, sup-port will be limited, localized, and specific. When the scope of the disaster is reduced to the point that the Primary Agency can again assume full control and management without military assistance, USNORTHCOM will exit, leaving the on-scene experts to finish the job.

20.6.1 Integration within the United States (domestic operations)

Emergency response in the United States is managed at the lowest possible jurisdictional level—typically the local government—with the state government becoming involved when local resources (police, fire, public health and medical, emergency management, and other personnel) have been, or are expected to be, exhausted. A key partner for state and local governments is the American Red Cross, which responds to thousands of disasters, large and small, every year.

There are four categories of operations inside the United States and its territories:

1. *Homeland Defense:* For the protection of U.S. sovereignty, territory, domestic population, and critical infrastructure against external threats and aggression or other threats as directed by the president. The homeland is the physical region that includes the continental United States, Alaska, Hawaii, U.S. territories and possessions, and the surrounding territorial waters and airspace.
2. *Emergency Preparedness:* The measures taken in advance of an emergency to reduce the loss of life and property and to protect a nation's institutions from all types of hazards through a comprehensive emergency management program of preparedness, mitigation, response, and recovery.
3. *Civil Support:* DOD support to U.S. civil authorities for domestic emergencies and for designated law enforcement and other activities.
4. *Defense Support to Civil Authorities:* Support provided by U.S. military forces (Regular, Reserve, and National Guard), DOD civilians, DOD contract personnel, and DOD agency and component assets in response to requests for assistance from civilian federal, state, and local authorities for domestic emergencies, designated law enforcement support, and other domestic activities.

20.6.2 Integration outside the United States (international/foreign operations)

2011 was the costliest year in terms of property damage due to natural disasters. International disasters in 2011 included floods and landslides in Thailand, Guatemala, El Salvador, Pakistan, and Brazil; earthquakes in New Zealand and Turkey; Cyclone Yasi and flooding in Australia; drought in Somalia; wildfires in Canada; and the severe winter storm Joachim that swept across Europe. The worst and most costly single natural disaster of the year was the earthquake in Japan, the ensuing tsunami, and the resulting nuclear disaster.

The DOD normally has a supporting role in humanitarian assistance and disaster relief operations outside the United States and its territories. The U.S. ambassadors or chiefs of mission, posted to the affected host nations, and the U.S. Department of State usually serve as the U.S. government lead agency. There are two exceptions: (1) when the U.S. mission is not functional because of damage, or (2) the host nation government collapses and the country descends into anarchy. In these instances, the operation often changes from humanitarian assistance and disaster relief to other forms of international intervention.

The DOD's initial objective is to achieve stabilization and provide security first for its own forces and then for others; its end-state includes an exit strategy and a defined—and limited—mission. The objective of many NGOs is to address the humanitarian needs brought on by the disaster or conflict, and their end-state is open. Most NGOs will seek to continue their activities until the humanitarian crisis is contained and longer-term development work can resume.

On the other hand, the DOD needs to understand that a linear mind set is not sufficient to account for the many possible scenarios that develop in complex humanitarian operations. However, DOD and NGO relief workers do not always enjoy an easy working relationship. The CMOC can function smoothly as an operational interface between the DOD and NGOs. At the start of the operation, the military needs to obtain a long-term understanding of the humanitarian problem and plan a military-to-civilian transition. In order to achieve this particular goal, the military and the NGOs must endeavor to develop a shared understanding of the operation and a shared strategy for its completion. Ultimately, the fundamental for success in any peacekeeping or disaster relief operations is the efficient coordination of all involved, both DOD and NGOs.

20.7 Tools for integration

20.7.1 Mutual aid agreements and memorandums of understanding

Disasters and emergencies can happen at any given moment and are capable of ranging in magnitude and severity, from a house call about carbon monoxide to unpredicted events. For the most part, the majority of emergencies and disasters are handled in a timely manner by the local community and government. However, there are events that are simply too large for the local government or community to handle. They request help through mutual aid agreements (MAAs) and memorandums of understanding (MOUs).

An MAA is an agreement among emergency services, emergency responders, governments, and, in some cases, countries to lend assistance across jurisdictional boundaries. These agreements are developed and signed among neighboring communities, bordering states, and, in the case of the United States and Canada, between countries. The MAAs are activated when the emergency response to an event exceeds its local resources. Subsequently, a call for aid is sent out from the community in distresses to the co-signers of the MAA. By the terms of the MAA, the co-signers send resources for aid. At the local level, examples of incidents that might be causes for MAA activation include riots, multiple alarm fires, and major accidents. At the state level, an example of the use of MAAs is 9/11. The response to the World Trade Center was extremely overwhelming for New York police and fire alone. Therefore, police, fire, and emergency medical services were dispatched for aid from Massachusetts, Connecticut, Rhode Island, Vermont, New Jersey, and Pennsylvania. The American Red Cross has been active in training with the military for years, building relationships it maintains with the U.S. Armed Forces. The American Red Cross signed MOUs with both the DOD and the FEMA. These agreements delineate each organization's responsibilities in disaster preparedness planning and in operations in the event of a national emergency or disaster. They also outline areas of mutual support and cooperation, and pave the way for similar cooperative agreements in the future.

MOUs are more or less the same as MAAs. They are documents that describe an agreement between parties that may be multilateral or bilateral. The main differences between the two are that MOUs are not considered legally binding and are more suited for agreements within organizations, agencies, or communities. Various templates are available online for free download.

Within MAAs and MOUs, funding issues must be outlined in detail to acknowledge who pays for the support being provided. The entity that is requesting the aid from its mutual aid co-signers is usually responsible for providing and monitoring the funds used during the emergency. Funding must be monitored closely to compensate the responding resources appropriately. With that in mind, it is the responsibility of the responding resources to keep a log of personnel, hours, injuries, equipment, equipment utilization costs, power utilization, and other factors that are agreed upon in the MAA.

20.7.2   Request for federal government aid

20.7.2.1   State of emergency declarations

Throughout the duration of a disaster or emergency, there exists a wide range of unfortunate and unforeseen events that may occur. When they do, the local responding resources can be quickly overwhelmed. The need for more personnel, vehicles (cars, trucks, boats, helicopters, etc.), food, and water becomes desperate and the largest suppliers for those needs are the DOD and the DHS. If resources become overrun, then the governor of the state involved can issue a Declaration for a State of Emergency. Through this declaration, the governor may then call upon the state's National Guard units (all branches of the military) to aid in the response to the disaster or emergency. The Army National Guard and the Air National Guard then fall under the command of the governor. On the state level, the federal government is allowed to provide for aid only for a Declared State of Emergency. To receive aid from active duty military and the other federal agencies and organizations, there is one more step to take; request for a Presidential Disaster Declaration (PDD).

20.7.2.2 Presidential Disaster Declarations

In the event that a disaster or emergency should reach a level of magnitude where the local and state response become overwhelmed and the federal government is needed, a special request must be submitted by the governor of the state, or states, in need. The federal government cannot participate in the response or recovery of an event until a few steps have been taken:

1. State and local resources become overwhelmed.
2. Governor declares the region is in a state of emergency.
3. Governor forwards the Declaration for a State of Emergency to its FEMA region headquarters.
4. The Declaration for a State of Emergency is reviewed and forwarded to the FEMA director who then brings it before the president and his advisors.
5. On the basis of factual information and reports provided to the president, the president may then issue a PDD.
6. A PDD allows for the federal government to step in and deploy its resources.

Very few disasters receive PDDs. It is the hope that through MAAs and MOUs local communities and entities can resolve the disaster or emergency before it may escalate to the point where the state or federal government or both get involved. However, with a PDD, federal agencies and organizations such as FEMA, active duty military, the Environmental Protection Agency, DHS, Department of Public Health, Department of Health and Human Services, and all others may then begin to supply the disaster-ridden region with the resources needed to protect and sustain the lives of the victims and those threatened. Although we cannot prevent another hurricane as destructive as Katrina, with continued active collaboration at all levels of response and across all shared missions, we can improve disaster response using a "whole-of-government" approach.

20.8   Conclusion

To implement mission requirements, directives state the DOD should be prepared to engage in planning and operations with a range of civilian agencies, international organizations, and NGOs. The directives noted in this handbook make clear that coordination must begin before NGOs and the DOD first interact on the ground. In light of the directives, both groups are gaining a greater understanding of their respective roles, motivations, and responsibilities. NGOs are now attending DOD briefings, have an active role in training exercises, and are involved in the distribution of information such as this handbook.

References

American National Red Cross, The 2012. *The federal charter of the American Red Cross.* http://www.redcross.org/portal/site/en/menuitem.d229a5f06620c6052b1e cfbf43181aa0/?vgnextoid = 39c2a8f21931f110VgnVCM10000089f0870a RCRD.

Byman, D.L. 2001. Uncertain Partners: NGOs and the Military. *Parameters* 43(2): 97–114.

DHS. 2008. *National response framework.* http://www.fema.gov/emergency/nrf/ aboutNRF.htm.

DHS. 2004. *Homeland Security Presidential Directive (HSPD)-5: Management of domestic incidents.* Washington, DC: U.S. Government Printing Office.

DOD. 2010. *Quadrennial defense review report.* Washington, DC: The Pentagon. February 2010.

FEMA. 2006. *Fact sheet: NIMS implementation for nongovernmental organizations.* http://www.fema.gov/pdf/emergency/nims/ngo_fs.pdf.

Joint Staff. 2006. *Interagency, intergovernmental organization, and nongovernmental organization coordination during joint operations.* vol. I–II. JP 3-08. http://www .dtic.mil/doctrine/new_pubs/jp3_08v1.pdf. http://www.bits .de/NRANEU/ others/jp-doctrine/jp3_08v2(96).pdf.

Joint Staff. 2008. *Civil–military operations.* JP 3-57. http://www.dtic.mil/doctrine/ new_pubs/jp3_57.pdf.

Mitchell, D.G. 2006. *Policy implications for NGOs and contractors in permissive and non-permissive environments.* Carlisle, PA: U.S. Army War College.

OCHA. 2001. *Use of Military or Armed Escorts for Humanitarian Convoys.* www.who .int/hac/network/interagency/GuidelinesonArmedEscorts_Sept2001.pdf.

OCHA. 2003. *General Guidance for Interaction between United Nations Personnel and Military and Civilian Representatives of the Occupying Power in Iraq.* www.coe-dmha.org/Media/Guidance/6GeneralGuidance.pdf.

Senate Bipartisan Committee. 2006. *The final report of the select bipartisan committee to investigate the preparation for and response to Hurricane Katrina.* Washington, DC: U.S. Government Printing Office.

Willitts, P. 2002. What Is a Non-Governmental Organization? In: *UNESCO encyclopedia of life support systems.* Oxford, UK: EOLSS Publishers.

chapter twenty-one

Time is a murderer

The cost of critical path drag in emergency response

Stephen A. Devaux

Contents

21.1 Introduction

In the 1953 John Huston movie *Beat the Devil*, Peter Lorre says to Humphrey Bogart: "Time is a crook!" But in many projects, Time is much more sociopathic than any mere crook. After a disaster, as victims wait for help to arrive, Time wanders through the chaos with Death at its side. With every passing hour, mortality is likely to rise at an increasing rate, as the trapped, the injured, and the infected slowly succumb.

As Dr. Adedeji B. Badiru wrote in Chapter 17 on the Coordinated Systems Approach, emergency responses are programs consisting of both projects and operations and "require conventional planning, organization, scheduling, control, and phaseout." Benjamin Franklin's truism that "Time is money" has long been given lip service in project work, even if the specific quantification of time in monetary units has been much more noticeable by its absence. But new techniques and metrics in critical path scheduling have allowed planners to identify the time/money impact of both specific work items and specific resource scarcities on project and program value. These allow for optimizing the work process on the basis of reducing time and thus deriving maximum value from the project. These same techniques need to be applied in planning emergency response, where the cost of time is measured in human suffering.

In this chapter, we shall see the following:

- How using standard business project management techniques such as the creation of work breakdown structure (WBS) templates that can be rapidly modified to specific situations can increase agility and reduce response time
- How using program management techniques to identify and sequence the specific value generating projects can provide the response team with a clearer route to reduced mortality and damage
- How new techniques in critical path scheduling, such as clear quantification of the value/cost of time and new metrics such as critical path drag and drag cost, can be used to optimize response time in such planning templates and greatly reduce response time, mortality, and damage

21.2   Of programs and projects

The Guide to the Project Management Body of Knowledge (PMBOK™ Guide)[1] defines a project as "a temporary endeavor undertaken to create a unique product, service, or result," and a program as "a group of related projects managed in a coordinated way to obtain benefits and control not available from managing them individually" (p. 442). These definitions apply to the many specific outcomes and services that must be generated in an emergency response effort. Each project, such as securing an area, searching for survivors, providing medical care and evacuation, and establishing mechanisms for delivery of food, water, and shelter, carries with it its own parcel of benefits. In Dr. Badiru's Triple C model, coordination of these projects and the operations they create is the final product of the preceding communication and cooperation, during both the planning phase and the initial actions of the implementation phase.

In managing time and its harmful side effects, for more than 50 years, projects have used a technique called critical path analysis. A critical path is the longest sequence of activities, delays, and constraints through a project schedule. The following two vital data inputs to this analysis are the constraints due to activity duration and precedence.

- Duration is an estimate of how long each planned task will take to complete.
- Precedence, or dependency information, is the logical order in which the work must be performed.

As much as possible, the sole delaying criterion in defining precedence constraints should be the nature of the work. We cannot paint the wall until we have plastered it, nor debug the software code until we have written it, nor search through the rubble until the building has collapsed.

Of course, resource constraints must eventually be taken into consideration. For example, use of amphibious vehicles for supply and evacuation is only possible to the extent that they are available. However, best practice suggests postponing consideration of resource unavailability until the impact of each specific lack can be isolated and measured. As we shall see, critical path analysis will allow us to compress our schedule by altering durations ("crashing the critical path" by increasing resource usage) and precedence ("fast tracking" by doing more work in parallel) to optimize a schedule that will give us the greatest value for our investment of time and money. The resources can then be fitted onto that schedule and the impact of shortages measured and perhaps ameliorated.

Although critical path analysis is designed as a project-level tool, a modified version can also play a crucial role in planning the program. At the program level, the individual projects have dependencies among other projects. Here the precedences are based not so much on the logical nature of the work (as they are at the project and work activity level), but on the basis of the benefit that the inclusion of each project at a specific point in the program is designed to generate.

An early parachute was designed by Leonardo da Vinci, and the first successful jump was made in France in 1785. But it generated very little value prior to the success of the Wright brothers' project. The airplane was an "enabler" for the parachute (not to mention for the largest branch of the travel industry). The internal combustion engine enhanced the value of oil drilling and refining, as well as the aviation industry. If mankind had deliberately planned its aviation history in a program, the sequence would have been as shown in Figure 21.1.

Notice in Figure 21.1 that

- Parachute invention did not *require* practical airplanes as in a critical path method (CPM) dependency, but ...
- The benefits generated by parachutes, as well as the expanded aviation and travel industries required airplanes as a preceding enabler.
- All the benefits generated by practical airplanes required the internal combustion engine as a preceding enabler. Therefore from a program point of view, the internal combustion engine was the most

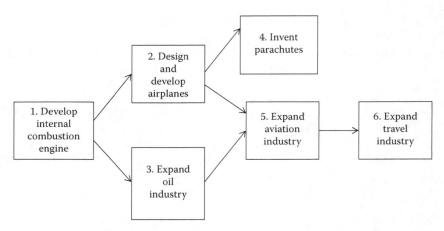

Figure 21.1 A flow chart for aviation history if it had been "properly planned."

valuable of the developments described earlier, with a value-added approximately equal to the value of the entire "program."

Figure 21.1 displays the logical order in which the projects would have been implemented as a program to generate the greatest value, if only human history were "properly planned." Although all the items described earlier can be seen as benefit generators in and of themselves, 1, 2, 3, and 5 are also enablers of other benefit generators, so that the cost of delay in those enablers is also magnified. In planning such a program, it would be crucial to understand that until we have a good internal combustion engine, the value of the rest of the program is close to nil. So the value-added of item 1 (develop internal combustion engine) is approximately equal to the value of the entire program.

Any delays in item 1 will also delay all of the other benefit generators. Delays in the early projects will delay (and thus almost always decrease) the generation of value in all the succeeding projects, just as a late project activity will delay successor activities and all of their successors (i.e., "descendants") in turn. The cost of time on enabler project delays will be magnified by the cost of time on succeeding delayed project value generation.

A program manager should not only take the time to plan the program in the logical order that will most enhance the program's benefit and value, but also recognize and monetize the huge costs of any delay on the early enabler projects.

21.3   The projects in emergency response programs

Emergency response planning suffers from an unfortunate dilemma. The life-and-death nature of an emergency, coupled with the enormous stress that it places on all decision makers, makes advance preparation, including planning, optimization of processes, and rehearsing to the point of second nature, absolutely indispensable.

However, the vast variation in emergency types makes it crucial to tailor the advance preparation to a wide variety of emergencies. For just a small sample, consider the variations in the following:

- A severe hurricane in Florida
- A severe hurricane and flooding in New Orleans
- An EF5 tornado hitting a major city
- An earthquake on the San Andreas Fault
- An earthquake and tsunami in the Puget Sound area
- Eruption of Mt. Rainier
- Mississippi River flooding of the 1993 type
- An Oklahoma City type of terrorist attack
- A 9/11 type of terrorist attack
- An emergency landing of an airliner on the Hudson River
- A city subway chemical, biological, or radiological attack
- A disease epidemic
- A crippling Northeast blizzard
- An electrical power blackout
- Meltdown of a nuclear power plant

These are only a tiny fraction of the potential emergencies, yet they cover a wide range in terms of amount of advanced warning, initial lethality, potential for extended lethality, geographical isolation or expansion, accessibility to aid, hazard to emergency responders, damage to property, and many other variables. Even within a specific emergency, specific circumstances can vary tremendously. An event such as that which struck Eastern Japan in March 2011, combining as it did earthquake, tsunami, and nuclear disaster, poses more magnified issues for emergency response than would any of those events individually. If project management methods are to be used to plan responses, how do we plan for such wide ranging catastrophes?

21.4   Project templates

In project-driven corporations that have mature processes in place, potential projects are often grouped by similar products and similar work. Template project plans are then developed for each group of projects. The template starts in the form of a WBS, with the final product decomposed into components, and the components decomposed into all the work activities necessary to create those components.

Delium Training, Inc. is a company that contracts to perform the development of employee training for large corporations. An example of a template WBS for one of its training development projects is shown in Figure 21.2. Needed resources and durations for all the work activities are

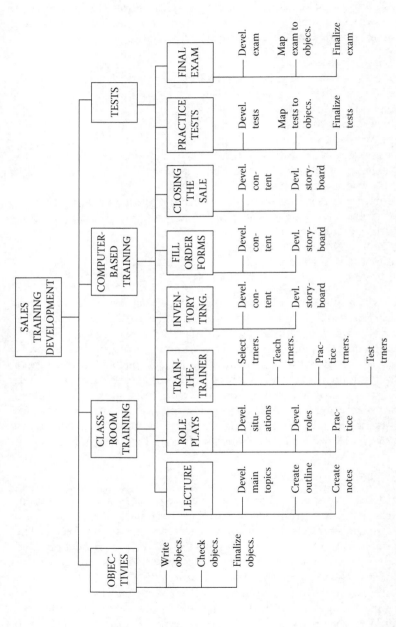

Figure 21.2 A work breakdown structure template for sales training on a contract basis.

then plugged into the WBS, and a schedule and budget created. And whenever a project of this type is performed, actual durations and costs are reported and captured in each element so as to further refine the template. Both schedule and budget can be constantly honed (e.g., through critical path analysis) to shorten duration and increase efficiency.

One of the greatest values of a WBS is the ability to manage scope changes by adding, removing, or modifying a branch, a twig, or a leaf of the structure. When the structure changes, so does whatever schedule or budgetary data that had been plugged into that portion of the WBS. For example, if a client decides that it does not want role plays and/or order form training, those two branches are lopped away from the template WBS, and the project plan is reduced by the amount of time, resources, and cost that would have been required for those work packages. And if another customer wants to expand the project to include training that is not part of the standard package (e.g., telemarketing), that branch can easily be added, either planned afresh or "borrowed" from a different template that includes telemarketing training.

21.5 Applying templates to emergency response plans

Just as the training development template is organized around the specific training products to be developed, emergency response templates should be organized around the products, services, and results (i.e., the outputs of each project) that are included in the response program. These outcomes should be based on the probable needs of the victim community: What things are likely, as much as possible, to decrease deaths, injuries, and property damage, and establish a basis for the earliest possible recovery of the community?

As mentioned earlier, the specifics of the emergency often evolve as the situation develops. For example, if a major earthquake strikes near a coastline, it may be an hour or more before it becomes clear whether it will be followed by a tsunami. Response planning and rehearsals must include dealing with such a devastating byproduct. When the earthquake occurs, response teams should prepare for the aftermath of both catastrophes until the tsunami risk has been eliminated. When that happens, the tsunami-specific portions of the response template can be discarded.

Perhaps the most significant difference between the training development project (or any commercial project) and an emergency response program is the vital importance of time. There are at least six ways in which time can take on huge importance:

1. The injured begin to die.
2. The trapped are without water, food, and shelter.

3. The ill or infirm lose access to medical aid.
4. One disaster spawns another (e.g., the 1906 San Francisco earthquake and fire).
5. An area of contamination spreads.
6. Collapse of governmental functions (e.g., post-Katrina New Orleans) begets criminality.

However, rapid response also can lead to high mortality among the first responders. The tremendous loss of life on 9/11 among New York City's police and firefighters, as well as the practice of many terrorists to follow up one explosion moments later with an even more deadly one, reminds us that sometimes first responders may need to take preliminary steps if they are to perform their functions safely. For example, securing the area must be a prerequisite for deploying many of the lifesaving measures that will follow.

Let us sequence the individual efforts within a possible emergency response plan into a flow chart (Figure 21.3). For illustration purposes, we will assume a response to a natural disaster where there has been

1. Probable injuries and loss of life.
2. Damage to property, including probable building collapses.

Key aspects of this flow chart are

1. Assembly of first responders occurs even before the full scope of the disaster is known. Thus, assembling first responders should be prepared with all necessary gear (hazardous material suits, automatic weapons, etc.) unless and until the details of the emergency are known.
2. Item 2 (DETERMINE SCOPE OF EMERGENCY) should include, in quantified terms that all leaders of the operation will understand, the extent of the damage, any continued or expanding threat to life and property, and the urgency of time in establishing three particular goals, as shown in Figure 21.4.

Items 10, 13, and 18 are all benefit generators, in that they provide benefits within themselves. For example, Item 3 (TRANSPORT IN FIRST RESPONDERS) and the setting up of the specific emergency operations (Items 15, 16, and 17) are crucial to much of the rest of the emergency response, but in and of themselves, they neither save lives nor protect property. Conversely, all of the highlighted activities are designed to save lives, treat casualties, and protect property. Reaching and accomplishing these activities as quickly as possible must be the clearly understood goal of the whole team. Also, the quantified benefits of saving time in

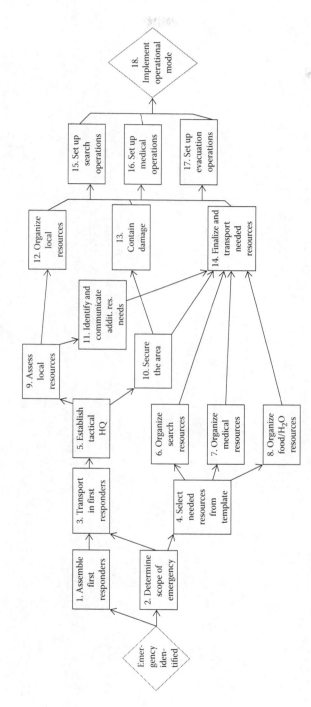

Figure 21.3 A flow chart for planned emergency response.

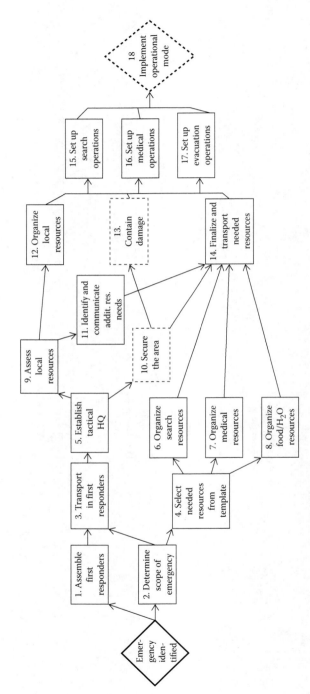

Figure 21.4 A flow chart highlighting the three benefit generators.

this process should have been communicated to all leaders back in Item 2 (DETERMINE SCOPE OF EMERGENCY): an estimate, for this specific emergency, of how many lives are likely to be lost, or how much property damage is likely to occur, for every hour that passes after the emergency has arisen.

However, it is crucial to recognize that, although the three items described earlier are the benefit-generator activities, all of their "ancestors" in the flow chart are enablers that must be performed in order even to reach the benefit generators. The value of the enabler is no less than that of the benefit generator it enables, any more than the goal scored in the first minute of a 2-1 soccer victory is less valuable than the goal scored in the final minute! Indeed, as we saw in the example of development of the internal combustion engine, the value of an enabler can often be greater than that of any single benefit generator if it is enabling more than one benefit generator. And, of course, Items 10 (SECURE THE AREA) and 13 (CONTAIN DAMAGE) are both benefit generators, as well as enablers of Item 18 (IMPLEMENT OPERATIONAL MODE).

21.6 Quantifying benefits in emergency response plans

Planning should tie to each benefit generator the expected quantified benefits (e.g., the expected monetary value of the project investment in business project management). In the case of almost every project, time is a factor that can greatly enhance or decrease value. In the case of emergency response, the benefit of faster projects is more lives saved and more injuries and damage prevented. The benefit of completing each of the generators should therefore be stated in terms of the incremental increase in damage to people or property for every additional hour. A Damage Control Time Chart such as the one shown in Figure 21.5 can be used with each emergency response template, with the estimated levels filled in as information about the precise nature of a specific disaster emerges. Notice that the specifics of such a chart might be completely different across a range of emergencies: an epidemic, a fire in a skyscraper, a car bomb in Manhattan, a sarin attack in a subway, a Mississippi flood, or a New Orleans hurricane. For our purposes, we are presuming an earthquake in a city where

- Many buildings have collapsed and people are trapped in the rubble.
- There is no danger of tsunami, but there is some danger of fires.
- Municipal communications and authority have largely disintegrated.

Hours	Secure the area			Contain the damage			Implement operational mode		
	Lives/h	Injuries/h	Damage/h	Lives/h	Injuries/h	Damage/h	Lives/h	Injuries/h	Damage/h
0–2	B	B	A	C	B	C	E	B	B
3–4	B	B	A	B	B	D	F	B	B
5–6	B	B	A	B	A	D	F	B	B
7–8	C	B	A	B	A	D	F	B	A
9–12	C	B	B	C	A	D	G	A	A
13–24	D	B	B	C	A	B	F	A	A

A = 0 lives/h., 0 Injuries/h., $0 damage/h.
B = 2 lives/h., 5 Injuries/h., $500K damage/h.
C = 5 lives/h., 15 Injuries/h., $1M damage/h.
D = 10 lives/h., 50 Injuries/h., $5M damage/h/
E = 20 lives/h., 100 Injuries/h., $20M damage/h.
F = 50 lives/h., 250 Injuries/h., $100M damage/h.
G = 100 lives/h., 500 Injuries/h., $500M damage/h.

Figure 21.5 A Damage Control Time Chart for a specific emergency.

The chart suggests that delays in securing the area and containing the damage will result in a steadily increasing mortality rate. Delays in implementing the operational mode for searching through the rubble, setting up triage and other medical processes, and evacuating acute cases will result in the greatest number of postevent casualties, which will peak at about 9 to 12 hours after the event and then decline slowly as the number of possible rescues that are still alive declines.

21.7 Using critical path method to plan a schedule for emergency response

We shall now plan and optimize our emergency response schedule. We will start by taking our flow chart and turning it into a network logic diagram (sometimes called a PERT Chart) by estimating the expected durations, in hours, of each of the efforts (Figure 21.6).

In a network logic diagram, an activity may not start until *all* of its immediate predecessors are finished. (There are such things as complex precedence dependencies, called start-to-start, finish-to-finish, and start-to-finish relationships, or SS, FF, and SF relationships for short. These can be important tools for fast tracking a schedule, and every schedule planner should be familiar with their use. However, for purposes of this chapter, we will deal only with finish-to-start (FS) dependencies, where a predecessor must finish before a successor can start.) For example, Item 14 (FINALIZE AND TRANSPORT NEEDED RESOURCES) cannot start until all five of its predecessors (6, 7, 8, 10, and 11) have finished. This means that whichever predecessor finishes last determines the earliest possible start of the successor.

The CPM of scheduling consists of two algorithms as follows:

- The forward pass algorithm traces network logic and durations from the first activity to the last and determines the earliest that each activity can start (ES) and finish (EF).
- The backward pass algorithm starts at the end, with the early dates or times determined for the last activity by the forward pass, and it traces network logic and durations backwards, from last activity to first, and determines the latest that each activity can finish (LF) and start (LS) *without delaying the last activity beyond its early dates*. These dates or times are then printed in a box for each activity using the convention shown in Figure 21.7.

Based on the durations and the dependencies, we can now calculate the CPM schedule as shown in Figure 21.8.

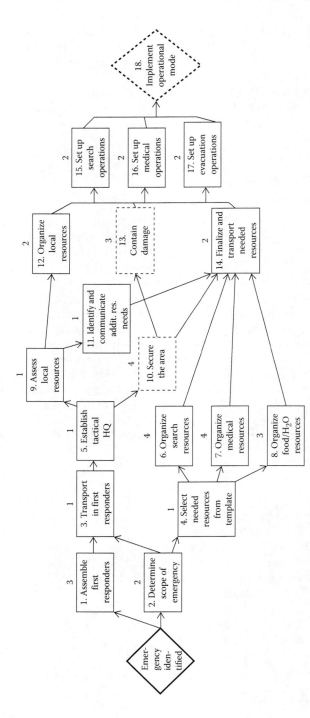

Figure 21.6 A network logic diagram showing durations of the efforts for this specific emergency.

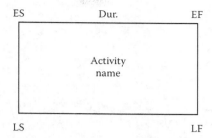

Figure 21.7 A network logic diagram activity box showing planned early and late starts and finishes.

Our project begins as soon as the emergency is identified. Certain items have exactly the same early numbers (above the box) as late numbers (below the box). These items are on the longest path through the project, which is called the *critical path* because

1. The duration of the longest path determines the duration of the project—every project is as long as its longest path.
2. Any delays on the longest path will make the project longer, whereas delays off the longest path will *only* make the project longer if the delays are so great as to make a new longest (and critical) path.
3. If we wish to shorten the total project duration, we have to shorten the longest path. This point is often overlooked, but is perhaps most important, especially on a project where hundreds of lives hang in the balance.

In Figure 21.8, we have highlighted the critical path with thick borders: Items 1, 3, 5, 10, 13, and 15 through 18. The fact that all three of our lifesaving benefit generators (Items 10, 13, and 18) are on the critical path will make the upcoming optimization process easier. If one were off the critical path, we would need to optimize the path to that item separately in order to pull it in earlier and thus save more lives.

CPM, which has been around for more than half a century, also quantifies the amount of time by which an activity is removed from the critical path into a metric called *total float* (TF) (or, in Microsoft Project software, *total slack*). TF is the amount of time an activity can slip without delaying the end of the project, that is, without becoming critical. It is calculated using the formula TF = LF–EF. In Figure 21.9, the TF of each noncritical item is printed directly under the middle of the activity box.

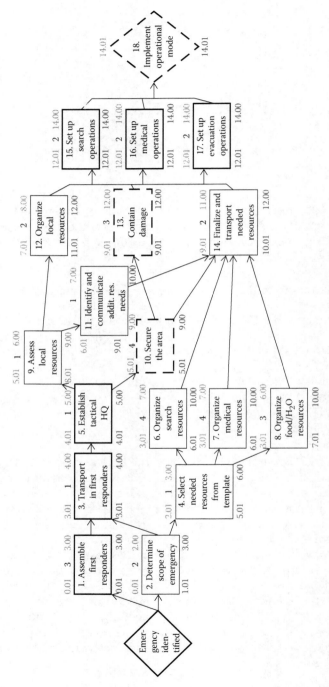

Figure 21.8 A critical path method network logic diagram with the critical path highlighted with thick borders.

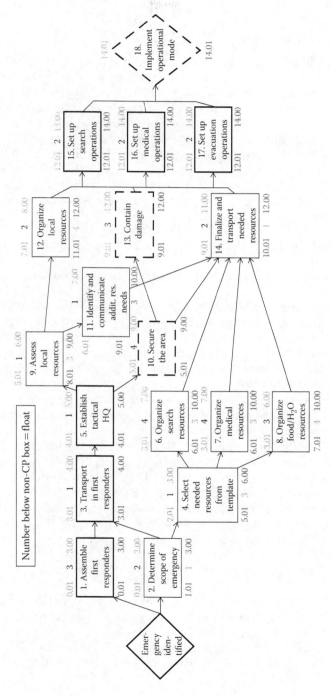

Figure 21.9 Critical path method network logic diagram with total float calculations.

21.8   Critical path drag: The missing quantification on the critical path

Although TF is an important metric, by itself it does nothing to help us shorten the project, because as we plan the project, float is always *off* the critical path, on activities that are not delaying the project at all and thus costing us neither time nor lives. Critical path theory, and all CPM software algorithms, computes TF. But what do they say about activities that are *on* the critical path? Answer: *zero!* They tell us they have zero TF, which is approximately the same as repeating to us that the activity is critical. When activities *are not* critical, CPM theory and the software quantify things nicely for us. But CPM tells us zero for the critical activities.

What *should* CPM theory and the software tell us about critical activities? We need to know

1. How much time each critical path activity is *adding* to the project's duration.
2. How much money, or pain, or human life that added time is costing.

In 1999, the concept of critical path drag was introduced by Devaux.[2] Drag is the amount of time an activity is adding to the project duration or, alternatively, the amount of time by which the project duration would be reduced if an activity were either removed from the project or its duration reduced to zero. Whereas float is always off the critical path, in order to have drag an activity must be critical.

In large schedules, and even in small schedules with complex dependencies (the previously mentioned start-to-start, finish-to-finish, and start-to-finish relationships, as well as delay or acceleration factors called lags and leads), computing drag can be very difficult. To date, only two software packages compute drag: Spider Project and Sumatra .com's Project Optimizer add-on to MS Project. But computing drag is important—human lives may depend on it! So we should learn to compute it in small networks with exclusively FS relationships (i.e., those where the successor cannot start until the predecessor is finished).

There are just two rules to follow in the computation:

1. If a CP activity has nothing else in parallel, its drag is equal to its duration.
2. If a CP activity has other activities in parallel, its drag is whichever is LESS: its duration or the TF of the parallel activity that has the LEAST TF (Figure 21.10).
   a. A and F have nothing in parallel (everything is either an ancestor or descendant), so their drags are each equal to their durations, 10 and 8 respectively.

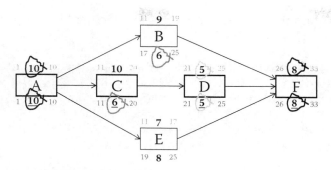

Figure 21.10 Simple critical path method network logic diagram with drag calculations underlined.

    b. C has drag constrained by parallel activity B's TF of 6.

    c. D is also parallel with B, but its duration is only 5 so its drag is 5.

If time is important, we can actually shorten the project more on the 8 days of Activity F than on the 10 days of Activity C.

A network logic diagram would be much larger and more complex for an emergency response than our diagrams in Figures 21.8 and 21.9. However, let us now compute critical path drag for the simple diagram we are using.

- Item 1 (ASSEMBLE FIRST RESPONDERS) has drag of 1 hour as it is parallel with Item 2, which has TF of 1.
- Item 3 (TRANSPORT IN FIRST RESPONDERS) has drag of 1 hour. Although it is parallel with many items (4, 6, 7, and 8) that have float of 3 and 4 hours, Item 3's drag is limited by its duration of 1 hour.
- Item 5 (ESTABLISH TACTICAL HQ) also has drag of 1 hour for the same reasons.
- Item 10 (SECURE THE AREA) has drag of 3 hours. It is parallel with many items (4, 6, 7, 8, 9, 11, and 12) that have float of either 3 or 4 hours, and it has a duration of 4 hours. (Note that Item 10 is NOT parallel with Item 2 and Item 14—it is a descendant of 2 and an ancestor of 14.)
- Item 13 (CONTAIN DAMAGE) has drag of 1 hour as it has a duration of 3 hours and is parallel with Item 14's float of 1 hour.
- Items 16, 17, and 18 are parallel critical path activities, each with zero float so that each has drag of zero. To pull in the end of the project on any one of them, we have to shorten all three. But note that the three of them together would have drag of 2 hours.

What does this all mean to the disaster victims? It means that, with our current schedule

- Those lives that are dependent on the completion of Item 10 (SECURE THE AREA) will be saved after Hour 9.
- Those lives that are dependent on the completion of Item 13 (CONTAIN DAMAGE) will be saved after Hour 12.
- Those lives that are dependent on Item 18 (IMPLEMENT OPERATIONAL MODE) will be saved after Hour 14.

Perhaps most important, it means we now have the needed information to optimize our plan by changing the schedule in ways that will allow faster response and thus save more lives.

21.9   Drag cost: The negative consequences of critical path drag

In business projects, time is money, often huge amounts of it. If we are building a luxury hotel on a tropical island for the tourist season and the hotel is not ready for opening until the fourth week of the season, the extra time on our schedule has cost 3 weeks of revenue, worth perhaps $1 million each. It is not just the last activity or two (landscaping, or electrical wiring, perhaps) that has caused the delay: *it is every activity that is on the final (or as-built) critical path*. And the *drag cost* of every critical path activity is equal to $1 million per week for up to 3 weeks of drag.

Similarly, it is not just the 2 hours required for each of Activities 15, 16, and 17, setting up the search, medical, and evacuation processes, that delay the saving of many lives until after Hour 14—it is every item on our critical path—and the drag cost of each such item is the lives, injuries, and damage that occur even as we rush to implement our lifesaving processes. It is not the individuals, who are doing their best to get the work done as swiftly as possible, who cause the deaths—it is Time itself that swings the scythe.

Let us look at two charts that show the loss of life in the first 16 hours after the catastrophe. Figure 21.12 shows the loss of life if there was no response, and Figure 21.13 corresponds to the response as scheduled in Figure 21.11. In Figure 21.11, the area is secured after Hour 9, the damage is contained after Hour 12, and search and rescue is fully operational after Hour 14. The information in these charts comes directly from the data in the Damage Control Time Chart in Figure 21.5.

The table in Figure 21.12 tells us that the total loss of life in 16 hours if there were no response would be 1084. However, if we follow the CPM schedule as shown in Figure 21.11, lives that otherwise would have been lost now start to be saved as early as Hour 10 by securing the area. The

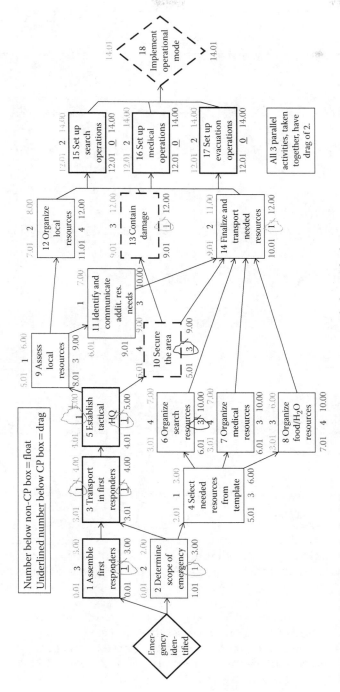

Figure 21.11 Critical path method diagram for emergency response with drag calculations.

	ITEM 10 (SECURE)	ITEM 13 (CONTAIN)	ITEM 18 (IMPLMNT)	HOURLY TOTAL
HOUR 1	B = 2	C = 5	E = 20	27
HOUR 2	B = 2	C = 5	E = 20	27
HOUR 3	B = 2	B = 2	F = 50	54
HOUR 4	B = 2	B = 2	F = 50	54
HOUR 5	B = 2	B = 2	F = 50	54
HOUR 6	B = 2	B = 2	F = 50	54
HOUR 7	C = 5	B = 2	F = 50	57
HOUR 8	C = 5	B = 2	F = 50	57
HOUR 9	C = 5	C = 5	G = 100	110
HOUR 10	C = 5	C = 5	G = 100	110
HOUR 11	C = 5	C = 5	G = 100	110
HOUR 12	C = 5	C = 5	G = 100	110
HOUR 13	D = 10	C = 5	F = 50	65
HOUR 14	D = 10	C = 5	F = 50	65
HOUR 15	D = 10	C = 5	F = 50	65
HOUR 16	D = 10	C = 5	F = 50	65
TOTAL DEATHS	82	62	940	1084

Figure 21.12 Loss of life in first 16 hours if there is no response.

result is that total expected deaths shrink by 175, from 1084 to 909, as shown in the table in Figure 21.13.

The highlighted cells in Figure 21.13 show how the planned response in Figure 21.11 reduces the mortality rate. Now we can look at our critical path from Figure 21.11, see what the drag cost is of each component, and examine whether there is a way to reduce the drag and, with it, the loss of life.

- The final trio of items (15, 16, and 17), setting up the search and rescue, medical, and evacuation processes, has, as a group, drag equal to the elapsed time of 2 hours for all three of these parallel activities. One hundred lives are estimated to expire in Hours 13 and 14. However, these setup processes are necessary to the future lifesaving, so they

	ITEM 10 (SECURE)	ITEM 13 (CONTAIN)	ITEM 18 (IMPLMNT)	HOURLY TOTAL
HOUR 1	B = 2	C = 5	E = 20	27
HOUR 2	B = 2	C = 5	E = 20	27
HOUR 3	B = 2	B = 2	F = 50	54
HOUR 4	B = 2	B = 2	F = 50	54
HOUR 5	B = 2	B = 2	F = 50	54
HOUR 6	B = 2	B = 2	F = 50	54
HOUR 7	C = 5	B = 2	F = 50	57
HOUR 8	C = 5	B = 2	F = 50	57
HOUR 9	C = 5	C = 5	G = 100	110
HOUR 10	0	C = 5	G = 100	105
HOUR 11	0	C = 5	G = 100	105
HOUR 12	0	C = 5	G = 100	105
HOUR 13	0	0	F = 50	50
HOUR 14	0	0	F = 50	50
HOUR 15	0	0	0	0
HOUR 16	0	0	0	0
TOTAL DEATHS	27	42	840	909

Figure 21.13 Loss of life in first 16 hours with planned response as in Figure 21.11.

cannot be eliminated. It might be possible to increase their resources and thus reduce their combined duration to 1 hour and their drag cost from 100 lives to 50 lives. But with three interrelated and parallel activities such as these, all efforts to shorten them will be in vain if any one of them slips back to 2 hours. The risk suggests that applying compression efforts elsewhere may be more beneficial. So, at least for the moment, let us accept that this trio has a drag cost of 100 lives.

- Item 13 (CONTAIN DAMAGE) is both a benefit generator in its own right and an enabler/predecessor of setting up the future lifesaving operations. It has drag of 1 hour, being parallel with Item 14 that has 1 hour of float. Being both a benefit generator and an enabler throws a wrinkle into the drag cost calculation for Item 13. In its enabler capacity, it is delaying the start of the main lifesaving operations by only 1 hour (i.e., 1 hour of drag). However, as a benefit generator, if its duration could be reduced from 3 hours to 1 hour, its completion at the end of Hour 10 would save 10 lives in and of itself as well as saving 50 lives by allowing Item 18 to start 1 hour earlier. Thus its drag cost as a benefit generator is 15 lives for its 3 days of duration and as an enabler its drag cost is 50 lives, for a total drag cost of 65 lives.

- Item 10 (SECURE THE AREA), like Item 13, is both a benefit generator in its own right and an enabler/predecessor of both containing damage and of setting up the future lifesaving operations. Its duration of 4 hours has a drag of 3 hours, being parallel with several items that have 3 hours of float. If the area is secure in 3 less hours, then
  - An estimated 15 lives would be saved just by completing Item 10 three hours earlier.
  - An additional 15 lives would be saved by enabling Item 13 to be completed three hours earlier.
  - A further 200 lives would be saved by launching the lifesaving processes at the beginning of Hour 12 rather than the beginning of Hour 15.

Thus the drag cost of Item 10 is 230 lives. (Notice that Item 10, in and of itself, is a benefit generator that saves lives. But the time it takes also costs lives.)

- Item 3 (TRANSPORT IN FIRST RESPONDERS) and Item 5 (ESTABLISH TACTICAL HQ) each has just 1 hour of drag due to their duration. But the entire path of Items 3, 5, and 10 (through SECURE THE AREA) also shares the drag-limiting parallel item float of 3 hours on Items 1, 6, and 7. What this means is that the combined drag on Items 3, 5, and 10 is not $1 + 1 + 3 = 5$ hours, since neither float nor drag is additive along a given path between merge points in the schedule. Any time gained by shortening any one of the three will reduce the drag on the other two items, so that the total combined drag of Items 3, 5, and 10 is only 3 hours, which may be gained by compressing any of these three items to a total of 3 hours.

The drag and drag cost of each of the critical path items are shown in Figure 21.14.

21.10   Using drag and drag cost to optimize the plan

As we get ready to show how to use critical path drag and drag cost to optimize the emergency response plan, we must caution the reader once

ITEM	DRAG (IN HOURS)	DRAG COST IN LIVES
1	1	60
3	1	60
5	1	60
10	3	230
13	1	65
15	0	0
16	0	0
17	0	0
15, 16, and 17	2	100

Figure 21.14 Drag and drag cost (in loss of life) for each critical path item.

more that we are showing this in a greatly simplified example. Any real project would need to be doing its computation and optimization on a much more granular and complex plan, involving perhaps thousands of activities and including the complex dependencies we mentioned previously. However, for showing how the concepts are designed to work, our simple network serves the purpose nicely without introducing confusing complexity.

After carefully analyzing the network in Figure 21.11 and the table in Figure 21.14, we work with the leaders of the individual efforts and those with access to additional resources to obtain a better assembly area for the first responders, both larger and closer to the likely disaster site. The benefits are

- The larger site allows both for assembly of a larger number of first responders and vehicles, as well as easy-to-load storage of all needed equipment and materials, so that responders just need to report to the site and retrieve them there. This shortens the needed time for assembly from 3 hours to 2 hours. The one hour of drag on Item 1 disappears, and the project duration becomes 13 hours instead of 14 hours.
- The greater proximity of the new site to the disaster area and the easy-loading design of the equipment and materials containers reduce the time necessary for Item 3 (TRANSPORT IN FIRST RESPONDERS) from 1 hour to 0.5 hours. This reduces the drag on Item 3 from 1 hour to 0.5 hours and the project duration becomes 12.5 hours instead of 13 hours.
- The expanded number of first responders that the new assembly site permits also reduces by half the amount of expected time necessary for Item 5 (ESTABLISH TACTICAL HQ) from 1 hour to 0.5 hours.

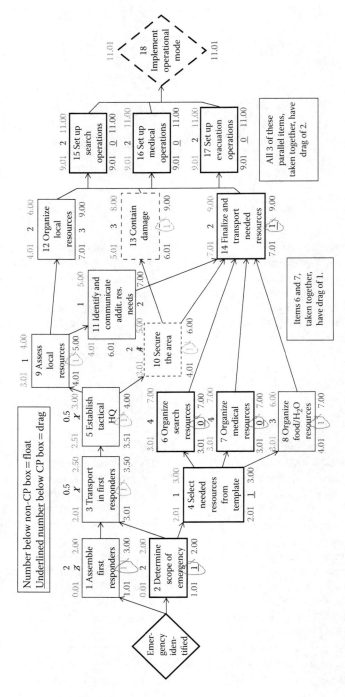

Figure 21.15 New durations, critical path, and drags for emergency response.

This reduces the drag on Item 5 from 1 hour to 0.5 hours and the project duration becomes 12 hours instead of 12.5 hours.
- Finally, the expanded number of first responders also reduces by half the amount of expected time necessary for Item 10 (SECURE THE AREA) from 4 hours to 2 hours. By this point, the earlier changes on Item 10's ancestors (1, 3, and 5) had reduced the drag on Item 10 from 3 hours to 1 hour, so this new two-hour compression changes Item 10's one remaining hour of drag to 1 hour of float as the critical path changes. Being a benefit generator, Item 10's earlier finish by 2 hours means that securing the area will start to save life 2 hours earlier. In addition, as Item 10 is the sole predecessor of benefit generator Item 13 (CONTAIN DAMAGE), it too will begin delivering its benefits 2 hours earlier. However, because the drag on Item 10 had previously been reduced to 1 hour, the end of the project and the implementation of the operational mode are only pulled in by additional 1 hour.

The CPM diagram for the modified schedule is shown in Figure 21.15. Notice that the critical path to the ultimate benefit generator has now changed to go through the items that involve determining, acquiring, and transporting in the additional resources (Items 2, 4, 6, 7, and 14).

What has our schedule compression done for the lethality of the disaster?

- If the new schedule is achievable, the deaths owing directly to a lack of security in the area should cease after the end of Hour 5. That is one huge modification. Establishing security 4 hours earlier reduces the mortality due to lack of security from 27 to 10.
- It also enables Item 13 (CONTAIN DAMAGE) to be completed by the end of Hour 8. This reduces the mortality from spreading damage (gas fires from ruptured pipelines, for instance) by containing it four hours earlier. The result is an estimated loss of just 22 lives from spreading damage, down from 42 when Item 13 was not scheduled for completion until Hour 12.
- The big payoff comes from being able to implement operational mode for search, medical, and evacuation operations 3 hours earlier. That enables this benefit generator to reduce expected mortality by 200 lives, from 840 to 640.

The chart (Figure 21.16) shows a reduction in the victims of Time by a total of 237, from 909 to 672. Similar analysis could be done for injuries and property damage.

The process of schedule optimization for this planned response does not have to be complete at this point. Again, planning at a more detailed level often shows additional opportunities for compression through fast tracking (i.e., doing more things in parallel) or crashing the critical path

	ITEM 10 (SECURE)	ITEM 13 (CONTAIN)	ITEM 18 (IMPLMNT)	HOURLY TOTAL
HOUR 1	B = 2	C = 5	E = 20	27
HOUR 2	B = 2	C = 5	E = 20	27
HOUR 3	B = 2	B = 2	F = 50	54
HOUR 4	B = 2	B = 2	F = 50	54
HOUR 5	B = 2	B = 2	F = 50	54
HOUR 6	0 (Was B = 2)	B = 2	F = 50	52 (Was 54)
HOUR 7	0 (Was C = 5)	B = 2	F = 50	52 (Was 57)
HOUR 8	0 (Was C = 5)	B = 2	F = 50	52 (Was 57)
HOUR 9	0 (Was C = 5)	0 (Was C = 5)	G = 100	100 (Was 110)
HOUR 10	0	0 (Was C = 5)	G = 100	100 (Was 105)
HOUR 11	0	0 (Was C = 5)	G = 100	100 (Was 105)
HOUR 12	0	0 (Was C = 5)	0 (Was G = 100)	0 (Was 105)
HOUR 13	0	0	0 (Was F = 50)	0 (Was 50)
HOUR 14	0	0	0 (Was F = 50)	0 (Was 50)
HOUR 15	0	0	0	0
HOUR 16	0	0	0	0
TOTAL DEATHS	10 (Was 27)	22 (Was 42)	640 (Was 840)	672 (Was 909)

Figure 21.16 Loss of life in first 16 hours with optimized response schedule as in Figure 21.15.

(adding resources). With the lives of hundreds of human beings on the line, we should look once more at the drag and drag cost totals on our new schedule in a chart (Figure 21.17) similar to the one in Figure 21.14.

The items that stand out are the two that have a duration of 2 hours *and* drag of 1 hour: Item 2 (DETERMINE SCOPE OF EMERGENCY) and Item 14 (FINALIZE AND TRANSPORT NEEDED RESOURCES). In practice, activities that have both longer durations and drag often provide opportunities for compression. Would it be possible to reduce the 2-hour durations of either or both of these items? If yes, then the total project duration could be reduced by up to 1 hour, with the operational mode starting after 10 hours. Of course, keep in mind that if we reduced Item 2 and Item 14 by 1 hour each, although that particular path would be reduced by 2 hours, the end of the project would only be pulled in by 1 hour and the critical path and drag would revert to the original critical path items.

Finally, the example we have used in this chapter is based on the model of a disaster that comes without warning. However, preventable loss of life and public frustration at what seems an inadequate level of

ITEM	DRAG (IN HOURS)	DRAG COST IN LIVES
2	1	100
4	1	100
6	0	0
7	0	0
6 and 7	1	100
14	1	100
15	0	0
16	0	0
17	0	0
15, 16, and 17	2	200

Figure 21.17 Drag and drag cost (in loss of life) of each item on the new critical path.

preparation often occurs following predicted disasters where there were 48 hours or more of clear warning that *something* bad was likely to happen. In situations such as these, the template model that is chosen can often be tweaked for the known specifics of the upcoming event. For example, in the CPM schedule we are using, not only could specifics of personnel and equipment be adjusted (e.g., amphibious vehicles and scuba gear brought into the assembly area if flooding is a possible hazard), but the preparatory time could also be used to provide a "head start" that would reduce the drag and drag cost of the early efforts. By performing Item 1 (ASSEMBLE FIRST RESPONDERS) even before the event has occurred, and simultaneously performing adequately detailed preparation for Item 2 (DETERMINE SCOPE OF EMERGENCY), we could turn the latter 2-hour effort into a half hour, thus starting Item 3 (TRANSPORT IN FIRST RESPONDERS) within 0.5 hours or less after the event has occurred, and thereby pulling in all the benefit generators by up to an additional 1.5 hours.

21.11 Conclusion

- Employment of modern project management techniques is essential in the planning of specific genres of emergency that seem likely to occur at some point and for which a response plan would have an ameliorating impact.
- Each of these plans should then be analyzed specifically with reference to the deleterious effects of the passage of time after the expected disaster. The impacts of time, in terms of mortality, injuries, and property damage should be clearly quantified, hour by hour if not in even smaller intervals, in a Damage Control Time Chart.

- Critical path analysis should then be used to assemble an initial schedule of response efforts. It should identify and focus on the specific actions/items/projects that will be the benefit generators, that is, those that will culminate in amelioration of some quantifiable amount of the negative impacts.
- Critical path drag and drag cost, with the latter measured in mortality, injuries, and property damage as quantified earlier in the Damage Control Time Chart, should be computed for the paths leading to each benefit generator. This information should then be used iteratively to make scheduling and resourcing decisions that compress the critical path and then, when it changes, focus on the new critical path. Careful track should be kept of the amount of benefit gained with each change until the plan seems to offer little or no further opportunity for improvement.
- In emergencies where the event is forecast in advance, all possible effort should be made to provide the template schedule with a head start by beginning the earliest items in advance.

References

1. The Project Management Institute, *A Guide to the Project Management Body of Knowledge, (PMBOK™ Guide)*, Fourth Edition, The Project Management Institute, Newtown Square, PA, 2008.

Bibliography

Devaux, Stephen A., "DRAG Racing on the Critical Path," *ProjectsatWork.com* (online magazine), Dec. 7, 2006.

Devaux, Stephen A., "DRAG Racing on the Critical Path," *ProjectsatWork.com* (online magazine), Dec. 14, 2006.

Devaux, Stephen A., "Paving the Critical Path," *ProjectsatWork.com* (online magazine), Dec. 4, 2006.

Devaux, Stephen A., "The Drag Efficient: The Missing Quantification of Time on the Critical Path." *Defense AT&L Magazine of Defense Acquisition University*, Vol. XLI, No. 1, DAU 224, Jan.–Feb. 2012, pp. 18–24.

Devaux, Stephen A., *Total Project Control: A Manager's Guide to Integrated Project Planning, Measuring and Tracking Systems*, John Wiley & Sons, New York, NY, 1999.

Duncan, William and Devaux S., "Scheduling Is a Drag," *ProjectsatWork.com* (online magazine), Jan. 15, 2009.

Sopko, J., and Strausser, G. *The Value of Organizational Project Management (OPM) Maturity Improvement – Understanding, Measuring, and Delivering the Benefits*. PMI Global Congress, Washington, DC, 2010.

Sopko, J., Yellayi, S., and Clark, S. *An Organization's Journey to Achieve Business Excellence through OPM Maturity*. PMI Global Congress, Marseille, France, 2012.

chapter twenty-two

Coordination and control in emergency response

John M. Flach, Debra Steele-Johnson, Valerie
L. Shalin, and Glenn C. Hamilton

Contents

22.1   Introduction

On the afternoon of Sunday, September 14, 2008, the remnants from the Gulf Coast's Hurricane Ike moved through Ohio. Although not as devastating as a tornado, hurricane force winds caused substantial storm damage and extensive loss of power. The high winds cut electrical service to more than half of the local power company's one-half million regional customers. The company's website initially indicated that service restoration would be a "multiday effort." However, full restoration was not reported until 2 weeks after the initial storm, on September 29, 2008. The result was that significant stress was placed on the regional medical emergency response resources. For example, 1100 new patients reported to local emergency departments in the first week of the power outage, and some ambulances were diverted from the nearest emergency rooms due to overflowing capacity.

In this chapter, we share lessons learned from extensive interviews with a wide range of participants in this emergency event. We summarize our observations in the context of theoretical work on the dynamics of "complex adaptive organizations." Our goals are to explicate theories of complex adaptive organizations by grounding them in the concrete events associated with this particular emergency situation. We also provide some practical suggestions that might be useful in preparing for future emergency situations. Section 22.2 will introduce the construct of a complex adaptive organization and will illustrate how this construct applies to the regional emergency response organization. This will be followed in Section 22.3 that considers the implications for coordination and control using Handy's (1992) recommendations for "federalism" as a general approach for managing complex adaptive organizations.

22.2   Complex adaptive organizations

22.2.1   Adaptive organizations

What is a complex adaptive organization? Let us begin by considering the term adaptive, which reflects a specific kind of control system. A simple control system (e.g., a servomechanism) adjusts its behavior based on error feedback as a function of fixed parameters (e.g., gains) that are optimized for a particular situation. An adaptive control system, however, is a system that is capable of adjusting its parameters to meet differential demands of changing situations. For example, a nonadaptive autopilot will have a fixed set of control parameters. These parameters will typically be optimized for some situations (e.g., altitudes), but these parameters will result in suboptimal performance or even instability in other situations (e.g., at other altitudes). In contrast, an adaptive autopilot is designed to adjust its parameters when situations change to achieve more satisfactory performance in a wider range of circumstances. Adaptive control will be necessary for systems that operate over a range of situations where no single control solution will yield stable performance. Although an adaptive control system may not achieve optimal performance in any particular situation, these systems will typically be more robust or resilient in their ability to achieve satisfactory performance over a wide range of situations.

Thus, an adaptive organization is an organization that is able to adapt its behaviors (tactics or processes) to yield satisfactory performance over a wide range of changing situations. For these organizations, rigid adherence to fixed procedures will not yield consistently satisfying outcomes due to the changing demands of situations. To achieve consistently satisfying outcomes in changing situations, it is necessary that these organizations adapt their processes to cope with the changing demands. For military organizations, the need for adaptation is reflected in observations such as "no plan survives contact with the enemy" and warnings against "fighting the last war." In the management literature, a term closely associated with the construct of adaptive organization is the construct "learning organization" (e.g., Senge, 1990). Adaptation and learning are essentially synonymous, reflecting the need for change to better fit the demands of changing circumstances.

Like military organizations, emergency organizations must be prepared to adapt to novel situations since no two emergency events will be exactly the same. The 2008 windstorm was a unique situation for the Ohio region. Tornados, which are more common to this region, are typically localized in both location and time. In contrast, this windstorm event impacted a large region, and the significant stresses on the emergency response system emerged gradually over time as a result of the cumulative effects of extensive power outages. Because of its uniqueness, it took significant time for many responders to fully appreciate the full scope of the emergency. Shortly after the precipitating event, interviewees varied widely in terms of threat recognition and in perceptions of both short- and long-term threats resulting from the windstorm.

Virtually all interviewees acknowledged the wind devastation on Sunday. Standby medical personnel attending an outdoor community event claimed early initial concern. One interviewee was off duty on Sunday but came into the office because of the storm. However, few others reported a proactive response. For example, most did not contact their emergency command centers or their employers. Although one interviewee acknowledged returning home from a trip to a "war zone," he did not contact his superiors. Another hospital interviewee reported that he noted the debris and wind but was not concerned and was not on call. A fire department interviewee also noted the wind and damage as he drove back into town from a weekend trip, but it was late and he went home. He initiated emergency procedures on Monday morning and began to assess the extent of the damage and loss of power. By Monday morning, some hospital administrators were concerned and went in early and/or contacted other hospitals to make plans for a surge in demand.

The insidious aspect of the windstorm, however, was not the initial physical damage of the wind but the cumulative effects of the resulting extended power outages. Few interviewees from the urban areas of the county claimed immediate recognition of the potential long-term threat to their communities. In contrast, one interviewee from a rural area did claim to recognize that the damage was likely to have long-term impacts due to prior experience with other weather events. This interviewee appreciated that for a low-priority rural community, the lack of power (electricity) would likely last weeks rather than days. However, the same interviewee reported that it was difficult to convince younger, less-experienced colleagues of the potential long-term consequences.

Since every emergency situation is unique in some way, there will be demands that cannot be anticipated in terms of planning or training exercises. Thus, it will be necessary for the organization to make adjustments in real time (i.e., to adapt) to achieve satisfactory performance.

22.2.2   Complex situations

In considering complex adaptive organizations, the term complex can reflect either properties of situation dynamics (e.g., degrees of uncertainty) or properties of the organization (e.g., the number of components or degrees of freedom) or both. The complexities of the situation dynamics associated with regional emergencies are easy to imagine. The county primarily impacted by the windstorm covered approximately 462 square miles and had a population over 500,000 in more than 200,000 households. Forty-three percent of the land was urban, supporting residential, commercial, and industrial usage. The county was home to eight hospitals, providing approximately 2500 beds.

The following list of initial consequences of the storm gives a sense of the scope of the problem:

Debris: The windstorm knocked down trees, obstructing public streets and private driveways. Unsafe debris removal practices sometimes resulted in accidents and injuries.

Downed power lines: The windstorm knocked down power lines, most of which were initially dead.

Refrigeration: The absence of electricity disabled home refrigerators, in some cases impacting the storage of temperature-sensitive medications and, more broadly, food storage. This, in turn, increased the risk of food-borne illnesses and caused economic hardship for low socioeconomic status citizens who lost food supplies. More than 19,000 people applied for replacement food stamps.

Gasoline: The absence of electricity prohibited the operation of gasoline pumps, resulting in gasoline shortages for first responders as well as the general public.

Traffic lights: The absence of electricity disabled traffic lights, resulting in both an increase in traffic accidents and the need for portable stop signs.

Water supply: Both of the city water supplies initially relied on emergency generators as a source of power.

Communications: The absence of electricity eliminated digital/wireless home telephones and traditional/cable television. Further, wind damage to cell towers, high demand for signals for calls, and reliance on car batteries for recharging phone batteries limited cell phone usage.

Electricity-dependent home and outpatient care: The absence of electricity affected many people who depended on electricity to live independently of the hospital healthcare facilities. This included people dependent on oxygen, dialysis patients, bariatric patients, and those dependent on electric wheelchairs for mobility. For example, the absence of electricity prohibited the use of in-home oxygen concentrators and delivery systems along with the electricity-based breathing treatments that some residents require. Otherwise stable and independent, oxygen-dependent residents had to seek alternative sources for oxygen and breathing treatments. Those unable to find alternative sources faced the potential of deteriorating health that in some cases required emergency medical treatment.

In the case of the windstorm, the complexity was not simply a function of the many dimensions, but it was magnified by the cumulative interactions within and across these factors over time. As noted in the Section 22.2.1, this complexity created a degree of ambiguity that made it difficult for many of the people involved to recognize the full extent of the emergency situation as it developed.

22.2.3   Complex organizations

The other facet of complexity has to do with the structure of the organization itself. Emergency response organizations are not monolithic organizations with clearly defined roles and hierarchical command structures. Rather, emergency response organizations are an example of what systems theorists call a federation of systems. Sage and Cuppan (2001) identify a federation of systems as a special case of a complex adaptive system within the more general class of "system of systems." A system of systems is a collection of component systems, each with independent organizational structure and function that collaborate to achieve some common function. The component systems are often geographically distributed, and typically the collaborations evolve over time. The goals and collaborative behaviors are emergent properties of this evolution. In other terms, a system of systems is a multiteam system (MTS). Mathieu et al. (2001) define an MTS as follows: "Two or more teams that interface directly and interdependently in response to environmental contingencies toward the accomplishment of collective goals. MTS boundaries are defined by virtue of the fact that all teams within the system, while pursuing different proximal goals, share at least one common distal goal, and in doing so exhibit input, process and outcome interdependence with at least one other team in the system" (p. 290).

Sage and Cuppan (2001) define the particular case of a federation of systems as a system of systems with "little central power or authority for 'command and control'" (p. 327), that is, a federation of systems is based on a rather loose coalition that emerges due to a common interest in achieving functions that require cooperation. Although the combination of teams within the federation of systems creates increased possibilities for meeting complex situation demands, it also adds to the complexity of the system control problem. Although each component system is a resource for meeting the variety of needs of a situation, each is also a source of uncertainty. In other words, for stable solutions to emerge the individual components must adapt to not only their local situation dynamics but also the changing behaviors of other components (that are also simultaneously adapting to their own local situation dynamics).

The federation of systems responding to the 2008 windstorm included multiple political and public service organizations including local (cities and towns and their associated police and fire units) and more global units (countywide organizations), multiple medical organizations (state and regional organizations, eight hospitals, an emergency operations center (EOC), an acute care center (ACC), and numerous distributed clinics and medical offices), the regional utility company, and civilian organizations (churches, neighborhood associations, and citizen band radio operators). Note that due to nesting and overlapping across these various teams, some individuals were members of multiple teams.

Consistent with the construct of a federation of systems, most of the component teams had primary functions other than regional emergency response, that is, they were not specifically organized for responding to regional emergencies. An exception to this was the EOC. The EOC was physically located in the major population center near the east side of the county. It had the dual mission of performing as both a planning (including training) agency and a response organization. During emergencies, the EOC is intended to perform a logistics function, coordinating resources in response to local requests and coordinating communications and resource requests with the state.

22.3   Coordination and control

Sage and Cuppan (2001) suggested five dimensions of federalism (attributed to Handy [1992]) that we will consider in relation to the challenge of coordination and control in federations of systems in general and for emergency operations in particular. These five dimensions are subsidiarity, interdependence, standardization, dual citizenship, and separation of powers. We will consider each in turn.

22.3.1   Subsidiarity

Under subsidiarity, power or authority resides with the lowest possible point within an organization. In other words, the component system elements have maximal flexibility and authority to respond to local contingencies. The flexibility for the components to adaptively respond to local situational demands without waiting for instructions from a higher authority is an attribute that allows federations of systems to adapt to complex dynamic situations more effectively than hierarchical, centrally controlled organizations.

In military systems, the need for subsidiarity is reflected in the construct of "command intent." This suggests that commanders should leave sufficient discretion to junior-level officers so that the system can adapt to changing contingencies that could not be anticipated in a fixed formal plan. Commanders should specify a general intent, but they should trust local subordinates to work out the details of implementing that intent based on local contingencies. Shattuck (2000) describes the tension that military commanders experience between centralization and flexibility: "The senior commander must make an inherent tradeoff which impacts the subordinate commander's ability to adapt to battlefield conditions. The battlefield is a highly complex, uncertain environment where a commander matches wits with his opponent while coping with such variables as terrain, weather, morale, fatigue and equipment. Providing subordinate commanders a large degree of flexibility is critical to success" (p. 68).

Thus, a value of federalism is that the component teams can function somewhat autonomously in dealing with contingencies that could not be anticipated in any a priori plan and where there is insufficient time to wait for directions from a centralized authority due to unavoidable communication delays. The subsidiarity property has two complementary implications. Centralized authority must avoid overconstraining their subordinates who are dealing with the chaotic demands at the front lines, as suggested by the construct of command intent. Additionally, subordinates must be willing to take the initiative to act in the absence of specific directives from above.

In emergency operations, the distributed and evolutionary nature of the organization that is composed of subteams, typically with a great deal of autonomy, helps to ensure that these teams are not hampered by a strict centralized authority. For example, the EOC typically functions as a logistics center, distributing resources where they are needed, rather than as a command center, that is, it is designed to be a liaison between local and state agencies. The EOC's role is to keep everyone on the same page, so to speak, and to direct resources to where they are needed most. This is more of a support and logistics function than a command function.

In our interviews after the windstorm, we observed large variabilities among the component units about whether they took local initiative. For example, some organizations (e.g., fire departments) took the initiative to canvas their local communities to identify where there were power outages, whereas other communities waited for the central utility company to tell them where the power outages were. Note that the communities that took the local initiative generally felt that they fared better than those that waited for direction from higher authorities. Similarly, as discussed earlier, some emergency personnel took initiative to report to their units, whereas others waited to be called.

22.3.2   Interdependence

Interdependence is a property of any organization. Thompson (1967) described three types of interdependence within organizations: "pooled," "sequential," and "reciprocal."

Pooled interdependence refers to situations in which each of the components within an organization operates with a high degree of autonomy, but the products must be combined to satisfy the shared goals. Sequential interdependence refers to situations in which there are precedence relations among the components in the organization such that processes of one component depend on the satisfactory completion of processes by another component. In this case, not only do shared functional goals require cooperation but also cooperation is required to achieve local subgoals. Finally, reciprocal interactions refer to situations in which the processes and functions of particular components must adapt based on how other components behave. In this case, there are not only functional dependencies but also process dependencies such that the processes within a component may need to change depending on the actions of other components.

Perhaps the distinct aspect of federations of systems in general and emergency operations in particular is the balance across these three types of dependencies. All three types of interdependence were evident in the windstorm event. Most of the component systems had specific independent functions (i.e., search and rescue, civil order, transport, medical treatment, logistics, etc.) that were necessary for satisfaction of the overall goals, but often these could be accomplished with a relatively high degree of autonomy. On the other hand, there were also obvious precedence constraints. For example, emergency departments depended on local resources to transport individuals and the EOC depended on information from hospitals to assess the regional needs. There were also obvious reciprocal relations, such as the utility company being able to work more efficiently in those communities that took the initiative to determine where power was out.

Ambulances having to reroute patients when their normal emergency departments became overpopulated is another example of reciprocal relations.

22.3.3   Uniform and standardized way of doing business

Sage and Cuppan (2001) suggest that due to the interdependencies of federations of systems, coordinated control is not possible without "agreement on basic rules of conduct, common traditions of communicating, and common units of measurement of progress and quality" (p. 330). Thompson (1967) differentiates three types of coordination (i.e., ways of doing business) that tend to be associated with the different types of interdependence: "standardization" (pooled), "planning" (sequential), and "mutual adjustment" (reciprocal).

Standardization reflects a minimal requirement for coordination. Namely, there must at least be some common language for communication so that the outputs from one component can be interpreted and combined with other components. In the windstorm event, the need for standardization was highlighted in association with the needs for oxygen. An important factor driving the emergency situation was the surprisingly large number of people who relied on in-home oxygen systems requiring electricity. Thus, significant demands were placed on emergency departments at local hospitals, as they became "oxygen bars."

As oxygen supplies became depleted, confusion arose because of ambiguities about terminology. For example, when a request for oxygen came in, did the requestor want large or small canisters with regulators? Did they want concentrators for personal or multipatient use? Perhaps the request referred to tanks of compressed gas or cryogenic liquid gas, which itself required special vaporizers to support inhalation. In some cases, those placing the request for oxygen did not understand the various types of oxygen delivery equipment and the specific resources this equipment required. Some emergency personnel needing oxygen suggested contacting welders, but this source of oxygen does not meet Food and Drug Administration standards. More typically, numerous oxygen-related confusions resulted from the fact that several key people who normally facilitated such requests were out of town. Interviewees reported confusion regarding the location of oxygen stores. In one case, both state resources and hospital association resources were stored in the same warehouses, but inventory-labeling problems and ambiguous requests made it unclear to managers whose oxygen resources were being requested, where those resources were located, and whether the requestor had the owner's authority to access these supplies. Persistent, though somewhat extended, communication eventually resolved this problem. However, more clearly established standards for labeling the oxygen supplies could have led to more efficient coordination.

Planning is typically needed when there are precedence dependencies among components. In this context, we are using the term plans to refer to a priori programs established prior to an emergency event to identify processes for coordinating activities. These typically provide a scheme or schedule for the appropriate sequence and timing of activities. The importance of a priori planning was illustrated by uncertainties regarding the EOC. The EOC was open and operating by 4:30 PM. on Sunday, the initial day of the windstorm event, as it should have been. However, there was widespread confusion about whether it was open and what its role was. For example, some saw it as a central node in the communication network for tracking resources. Thus, all requests and other information should have been routed through the EOC. Others saw the EOC as a resource of last resort. So, they did their best to access and monitor resources locally and only involved the EOC when all other sources failed.

Similar confusions were found with respect to the ACC, a state-level asset managed by the department of emergency medicine at a local university. The ACC was a lower level medical care facility designed by the state to be set up in a "building of opportunity." It required a physical building since it did not have an independent housing arrangement. The ACC was designed to provide four clinical care functions: intravenous hydration, antibiotics, pain management, and respiratory therapy. The ACC would not be used for trauma surgery but would be set up to handle less acute hospital care such as caring for oxygen-dependent patients. Several issues were highlighted in relation to the ACC, including understanding the asset, activation, and conflicts of interest.

Some interviewees, for example, hospital personnel and county EOC personnel, had a limited understanding of the capabilities of the ACC. Some did not realize that the ACC would have to be staffed by local medical personnel, that it required a physical location (i.e., a building in which the ACC could be set up), and that daylight and lead time would be required for deployment. Further, given the expense of deploying the ACC, some interviewees raised the possibility of the partial deployment of the ACC, namely, deploying only those components of the ACC needed to address the current event.

A specific protocol was required to activate the ACC, including a formal request to the state Department of Health. However, affected personnel, such as hospital administrators, were unclear on the protocol for activating the ACC. Further, there was uncertainty regarding who in the county had the authority to make the decision. Finally, interviewees reported that there were mixed and contradictory reports regarding whether the ACC had been activated and whether the ACC was needed.

Whereas at least one hospital advocated activation, another hospital-based interviewee commented that "we weren't out of supplies ... didn't need any of the stuff that they had." The association of area hospitals recommended against deployment. The EOC combined this recommendation with input from one of the local fire departments to decline pursuing activation. A question arose regarding whether the hospital association was the appropriate authority for making this decision or whether another group, such as the public health department or the Red Cross, would provide a more objective hand in deciding whether to activate the ACC.

These issues associated with both the EOC and the ACC reflect the kind of issues that should be addressed through planning and training prior to an emergency event. The lack of understanding of the appropriate protocols or standards added unnecessarily to the confusion and undermined effective coordination during the windstorm event.

Whereas planning involves a priori programs prior to an emergency, complex work typically also involves ad hoc planning or replanning. This leads to the third type of coordination in Thompson's model: mutual adjustment refers to the ability of components within an organization to adapt their processes and functions based on the activities of other components. This requires extensive communications for each component to know what the other components are doing. Thus, effective communications have been hypothesized as a key element that differentiates effective team performance (Entin and Serfaty, 1999; Serfaty et al., 1994). Entin and Entin (2001) have operationalized this in terms of the ratio of "push" to "pull" information. Pull refers to information that is elicited through requests. Push refers to unsolicited information that is made available or "pushed" to other components in the organization prior to any requests. Entin and Entin (2001) hypothesized that teams with high push-to-pull ratios anticipate the information needs of other components and push information to these components, making them more able to manage complexity.

SurgeNet was a critical communication resource among hospitals during the windstorm. It was an online database that could be used and accessed by the local hospitals. It was available and was regularly used by hospital and emergency personnel to manage patient distribution across hospitals. Emergency personnel used information from SurgeNet to determine which emergency departments were full and which could still accommodate patients. When an emergency department was full, it posted that information on SurgeNet so that emergency personnel could "reroute" patients to an emergency department in a different hospital.

According to news reports, area emergency departments treated record numbers of patients, and many were operating beyond their capacities at some points during the emergency period. The emergency departments were all on "reroutes" to other hospitals, and one interviewee reported that his hospital implemented its normal diversion plan due to the crowded emergency department.

SurgeNet was effective in communicating this information to emergency personnel. However, SurgeNet might have been expanded to push additional information about hospital resources other than beds in emergency departments. More process-oriented measures, including the availability of diagnostic and treatment devices (e.g., computer axial tomography scans and ventilators) as well as projected demand, may have been useful for allowing more effective mutual adjustments. For example, some hospitals were unable to provide cooling for diagnostic equipment due to the electrical outage, so this resource was not temporarily available.

Pushing information about resources should help the systems to anticipate future conditions. Knowledge of a future surge in resource requirements provides a valuable time cushion when the necessary mutual adjustments require time. Such an anticipation requires a richer model of the influences on resource demand, and technology can be an effective integrating mechanism. For example, if the surge in emergency department demands due to oxygen boarders, dialysis patients, and others in the community whose health depended on power could have been anticipated and documented on a tool such as SurgeNet, time would have been gained to coordinate a more efficient response to the surge.

An important implication of the advances in information technologies (e.g., SurgeNet or the Internet) is that they make it much easier to push information to others in the network. However, one factor that made adapting to the windstorm more difficult was the breakdown in the normal communication technologies resulting from the power outages. Several interviewees discovered belatedly that their home phones or cell phones were not usable due to lack of electricity or dead batteries. Assuming that supervisors would call with requests, emergency personnel at home may not have realized that their digital phones were not working.

Reliance on phones proved challenging for other reasons. Phone contact lists often failed to include cell phone numbers that were critical for administrators not at their desks. Indeed, some interviewees reported attempting to contact someone to ask whether the EOC had been activated, but the phone contact lists included only office telephone numbers and not cell phone numbers. Cell phone charging was limited to automobiles, requiring appropriate technology and sometimes separating the users, including emergency response personnel, from incoming calls during the charging process. In addition to breakdowns in phone communication systems, access to computers that also depended on electricity was compromised. One of the factors contributing to the uncertainty about the status of the EOC was the fact that e-mail was used to announce their opening.

For many of the teams identified as high-reliability systems (e.g., aircraft flight deck crews) there is an intact communication network (e.g., a voice loop). In these systems, the initiative to proactively push information may be important for team effectiveness. However, in the windstorm event, where there were significant compromises to the communication network, the key word seemed to be initiative rather than push. In other words, given the compromises to the communication infrastructure it was necessary that all the components take initiative to both push and pull the information that they needed. For example, it was clear that system response was hampered as some people passively waited for information to be pushed to them. However, failed equipment and lack of correct phone numbers prevented the dissemination of this information.

22.3.4   Dual citizenship

In federations of systems, individuals have at least dual citizenship, that is, they are both members of a component organization and members of the federation. In fact, in the windstorm event many people were affiliated with multiple component systems. For example, an individual representing a hospital may also be a member of the regional association of hospitals and may also participate on state-level bodies. Sage and Cuppan (2001) suggest that for federations of systems to work successfully, citizenship in the federations needs to be emphasized. When goal conflicts inevitably arise, the common goals of a federation need to take priority over the goals of individual component organizations.

A particular source of conflict between component and federation goals involved the activation of the ACC. Our interviews highlighted very real conflicts of interest for area hospitals in relation to the ACC. For example, if the ACC was activated, area hospitals might have lost revenue. On the other hand, area hospitals incurred increased costs, for example, from oxygen boarders. In addition, the cost of activating the ACC, as well as the cost burden related to activating the ACC, was unclear to participants, as was a clear means of reimbursement and restocking. Finally, several other issues arose that were related to relative costs, including how much the state would be charged for oxygen boarders and whether the ACC could accept health insurance.

Further, some interviewees substantiated hospital concerns regarding a potentially lower standard of care provided by the ACC, that is, the ACC is designed to perform only four functions rather than all of the functions of a hospital. This raised concerns in area hospitals regarding possible claims of physician or hospital malpractice resulting from referrals of patients to the ACC and questions on how and under what conditions such referrals would be made. Finally, some of the concerns regarding activating the ACC were related to the "can do" culture of hospital and medical personnel and long-range competition among hospitals. Healthcare providers felt that a request for help with their own mission might be interpreted as a weakness that they were reluctant to admit.

22.3.5   Separation of powers

Sage and Cuppan (2001) suggest that the functions of management, monitoring, and governance of federations of systems be "viewed as separate functions to be accomplished by separate bodies" (p. 330), as various functions have different time and value constraints. For example, issues associated with the standards for implementing the EOC and the ACC were governance issues that should have been addressed through long-term planning and training. However, the bodies that coordinate this long-range coordination are typically not the best vehicle for the process of real-time command and control during an emergency event. The fact that the EOC served a leadership role for planning and training may have been one source of confusion about its function during the actual windstorm event.

There was also separation of power issues associated with the ACC. The ACC generally was a state resource that was monitored by the local hospital association through a local medical school. However, in the context of the emergency situation, there has been a suggestion that the real-time management decisions about initiation and operation of this resource should be decoupled from the hospitals and should be made by a public health agency that has no revenue-based conflicts of interest and that can consider the larger emergency system, including the Red Cross operations.

With respect to the larger emergency context, many of the problems during the windstorm seemed to emerge as a function of a gap between the mission of the Red Cross to feed and clothe the healthy and the mission of hospitals to treat the sick and injured. The definition of good health is context dependent, and disasters change the context. Individuals who seem healthy with the assistance of technology, in this case electricity-based oxygen concentrators, are unhealthy in its absence. Weather or problems with the water supply could also differentially affect the same populations.

Issues arose when one of the Red Cross sites made it difficult to accommodate patients who were dependent on oxygen or other services that required access to electrical outlets. Thus, there was a need for a resource to meet the middle ground of people who were at the edges of health and required assistance for health support, including those with physical disabilities. The ACC, if it had been activated, may have been a resource for meeting this need. The disconnect between the hospital mission and the Red Cross mission suggests an opportunity to learn and adapt from the windstorm. This should be considered at the governance level in terms of potential implications for considering alternative management schemes and implementing these schemes in terms of planning and training. This, in turn, has obvious implications for changing the way in which the ACC resource might be managed in the future. It also has implications for the choice of sites for Red Cross–managed treatment centers (i.e., access to electricity and accessibility need to be considered).

22.4   Summary and conclusion

In the case of the 2008 windstorm, there is no evidence that any lives were lost due to failures of the emergency response systems. However, there is evidence to suggest that the system came dangerously close to its capacity limits. Some have wondered whether the region was saved from more catastrophic consequences because the windstorm was followed by unusually mild weather conditions. Further, there are lessons to be learned that might improve coordination in future emergencies. For example, the oxygen problem suggests the need to establish a clear taxonomy for identifying the different types of oxygen needs. It was also clear that there is a need for better standardization (through planning and training) with respect to the implementation and operation of the EOC and the ACC. Another important lesson is the need for people to better compensate for breakdowns in the normal communication infrastructures. Thus, they need to be more proactive in both pushing information from other components and pulling information from other components. Finally, there seems to be a clear gap between the mission of hospital emergency departments and the mission of the Red Cross treatment centers with respect to those at the edge of health who are dependent on electricity-based technology. This gap was an important factor in the excessive demand that was placed on emergency departments at local hospitals.

The construct of a federation of systems or a team of teams is relatively new in the organization literature. We doubt that the emergence of this construct reflects the emergence of a novel form of organization. In fact, we suspect that such an organization has always been true for regional emergencies (and probably also for military adventures). However, the increased attention on this type of organization may reflect the gradual extinction of more monolithic and hierarchical organizations that find it increasingly difficult to survive in a world where the pace of change is high and where there are increasing opportunities and demands for collaborations among globally distributed component teams. Thus, rather than being an anomaly, the federation of systems that has probably always been typical for emergency operations is increasingly becoming the standard for successful organizations (Galbraith, 1995).

A benefit of this evolution for emergency systems is that increasing attention is being paid to the problem of controlling these teams of teams. Therefore, we have used Sage and Cuppan's (2001) prescriptions for federalism as a framework for evaluating performance during the 2008 windstorm with the hope that this will provide insights toward potential improvements. Other literature that provides potential prescriptions for evaluating the effectiveness of emergency systems include research on "high-reliability organizations" (Weick et al., 1999) and the management literature on "learning organizations" (Senge, 1990). In all of this literature, the focus is on the ability of organizations to adapt to meet the demands of complex, dynamically changing situations.

We see larger overlaps in the prescriptions from the literature. Thus, these are not competing or conflicting views, but rather they seem to be converging insights about fundamental issues related to the stability of complex organizations operating in complex environments. To a large extent, this insight might be summarized in terms of Ashby's (1956) "law of requisite variety" (Flach, 2012). In essence, this law states that a controller must be at least as complex as the problem that is being controlled. In other words, it takes complexity to destroy complexity. The federalism described by Sage and Cuppan (2001) seems to be a particularly effective way for an organization to increase its complexity (i.e., degrees of freedom or flexibility) so that it can adaptively respond to solve complex problems like regional disasters. Thus, we have used that construct to organize some of our observations from the 2008 windstorm.

References

Ashby, R. (1956). *An introduction to cybernetics*. London: Chapman & Hall.

Entin, E. E., and Entin, E. B. (2001). Measures for evaluation of team processes and performance in experiments and exercises. Paper presented at the Proceedings of the 6th International Command and Control Research and Technology Symposium, Annapolis, MD.

Entin, E. E., and Serfaty, D. (1999). Adaptive team coordination. *Human Factors*, 41(2), 312–325.

Flach, J. M. (2012). Complexity: Learning to muddle through. *Cognition, Technology & Work*, 14, 187–197. DOI 10.007/s10111-011-0201-8.

Galbraith, J. R. (1995). *Designing organizations: An executive briefing on strategy, structure, and process*. San Francisco, CA: Jossey-Bass.

Handy, C. (1992). Balancing corporate power: A new federalist paper. *Harvard Business Review*, 70(6), 55–79.

Mathieu, J. E., Marks, M. A., and Zaccaro, S. J. (2001). Multi-team systems. In N. Anderson, D. Ones, H. K. Sinangil and C. Viswesvaran (Eds.), *International handbook of work and organizational psychology* (pp. 289–313). London: Sage.

Sage, A. P., and Cuppan, C. D. (2001). On the systems engineering and management of systems of systems and federations of systems. *Information, Knowledge Systems Management*, 2(4), 325–345.

Senge, P. M. (1990). *The fifth discipline: The art and practice of the learning organization*. London: Random House.

Serfaty, D. S., Entin, E. E., and Deckert, J. C. (1994). Implicit coordination in command teams. In A. H. Levis and I. S. Levis (Eds.), *Science of command and control: Part III coping with change* (AIP Information Systems) (pp. 87–94). Fairfax, VA: AFCEA International Press.

Shattuck, L. (2000). Communicating intent and imparting presence. *Military Review*, 66–72.

Thompson, J. (1967). *Organizations in action*. New York: McGraw-Hill.

Weick, K.E., Sutclife, K.M., and Obstfeld, D. (1999). Organizing for high reliability: Processes of collective mindfulness. In R.S. Sutton and B.M. Shaw (Eds.), *Research in organizational behavior*, Volume 1, (pp. 81–123). Stanford, CA: Jai Press.

chapter twenty-three

Managing the complexities of incident command

James R. Gruenberg

Contents

23.1 Introduction

Following the promulgation of Homeland Security Presidential Directive (HSPD)-5 on February 28, 2003, the U.S. government embarked on a mission to "ensure that all levels of government across the Nation have the capability to work efficiently and effectively together, using a national approach to domestic incident management" (White House 2003). The result was the arduous development and implementation of the National Incident Management System (NIMS). This system has emerged as the dominant construct by which civilian and military responders alike organize and execute responses to civil emergencies big and small. A significant component of NIMS is the Incident Command System (ICS). This system provides the nuts and bolts of response in the field and guides responding agencies on how they should structure response organizations under a defined system of terms and methodologies. Although ICS provides excellent structure for incidents up to and including large incident responses (Type I and II as defined in NIMS), it is argued that it does not provide adequate structure for the complexities of major, large-scale events (Type III and IV as defined in NIMS). Despite the complexities inherent within these disaster scenarios, strategies can be employed to address ICS challenges. This chapter will describe ICS, its history, and doctrine; identify the complexities that make the implantation of ICS diffi-cult in significant events; and suggest strategies by which these complexi-ties can be managed. Ultimately, this chapter should provide the reader new skills to address the challenges of a major response.

It would be tempting under the topic of this chapter to engage in a full dialogue about the effectiveness of the NIMS incident command methodology and its efficacy in a larger incident. Likewise, it would be interesting to parse the differences between incident management, as an organizational construct, versus incident leadership, as a human con-struct. Both these topics have been somewhat addressed in the literature, but need additional work.

British authors Katherine R. Devitt and Edward P. Borodzicz (2008) argued that incident complexity "is inexorably linked to the personal values, experience, drivers and expertise of the people responding to the incident." Although an in-depth examination of these human factors will not be conducted exhaustively, the strategies suggested in this chapter will provide the reader opportunities to understand and utilize practical methods by which a human can make a system work.

The value of ICS has been proven under the stressors of countless significant incidents; however, ICS also has weaknesses (Arbuthnot et al. 2002). This statement is meant to neither condemn nor criticize. In the con-text within which ICS was designed, it certainly has provided the nation's responders with a firm foundation by which incident objectives are orga-nized and executed. However, direct observation and experience indicate that ICS on paper varies greatly from that deployed in the field. This phe-nomenon exists because local social, political, organizational, and even human personality nuances vary so wildly that no single system, certainly not ICS, can address the infinite number of contingencies extant in grand-scale emergencies. The success or failure of incident command in the field is far more a function of the human rather than that of the system itself. More often than not, success and failure of incident command is a factor of the people who engage within it and make it work. No structure of man will be able to completely address the fury of nature or mankind, yet a sound examination of how responders can deploy strategies to better manage incident components is worthwhile and is the topic of this chapter.

23.2   The history of ICS

ICS was born out of necessity. As the American West's urban areas con-tinued to expand into otherwise open lands, the likelihood of wildfires impacting homes and businesses became an increasing problem to fire and other land management agencies. The fire season of 1970 high-lighted the wildland–urban interface problem and its subsequent risk to human life and property. Responders, specifically Southern California fire departments and state fire officials, decided to investigate the methods that could facilitate a better interaction between agencies as they responded to these large fire events. Specifically, a system needed to be developed that encumbered not only large numbers of responders to a widely distributed fire, but that also accommodated multiple layers of government: local, state, and federal.

In 1971, a system called FIRESCOPE (Firefighting Resources of California Organized for Potential Emergencies) was developed, a consortium authorized and funded through the U.S. Forest Service that spearheaded the development of a common incident command structure for wildland firefighting (FIRESCOPE n.d.). It is worth mentioning because the entire ICS evolution has had FIRESCOPE fingerprints on it and thus reflects a certain cultural framework: that of West Coast practices and experience in wildland fires.

For the purpose of this chapter, it is also important to understand that the ICS, throughout its development, has been a bureaucratic construct, both from the system's management and its field components. Its organization chart is linear and top down. Its process, the incident action planning process, is characterized by timelines and project management characteristics and results in the publication of a document: the incident action plan. This characterization may seem to have a negative connotation; however, these rigid structures do tend to bring order out of chaos. Unfortunately, its strength is also its weakness. Its rigidity also makes dynamic collaboration difficult, and it struggles to easily adjust to the multitude of complexities that exists in a large-scale incident environment.

Essentially, four iterations of ICS have emerged since the original founding meetings in the 1970s. For the most part, all four share a similar look and rationalizing doctrines: scalability, managed span of control, and a common language and organizational structure. The evolution process essentially provided some adaptation of the original system to accommodate a broader application. This adaptation process created some challenges to ICS's ultimate goal of a national framework.

The first version of ICS was organic to many Western fire agencies that had already managed incidents. Known as a Large Fire Organization (LFO), it provided the foundation for early ICS development. Terminology within LFO reflected the forest fire genesis of the system. Terms such as "crew boss," "sector boss," and others permeated the organization chart; however, "plans chief" and "operations chief," terms now employed within NIMS ICS, were also introduced (Society of American Foresters 2008). Perhaps the most significant aspect of the earliest ICS developments was the parallel development of the administrative machinery that would manage ICS doctrine and control fire policy. In 1976, the National Wildland Coordinating Group (NWCG) was instantiated to perform these very important roles. Even today, the NWCG name and logo appear on Federal

Emergency Management Agency (FEMA) ICS training materials. This group continues to promulgate policy as it relates to wildland fire management and greatly influences continued ICS development nationwide.

The early 1980s saw a further refinement of ICS terminology and the movement away from LFO and into what is known as ICS today. As fire departments in California began to voluntarily utilize this early ICS, naturally, the next step was to formalize it as a recognized entity. Both FIRESCOPE and the NWCG supported the "national adoption of a uniform emergency management organization" (FIRESCOPE n.d.). The new system, entitled National Interagency Incident Management System (NIIMS), was intended to meet the needs of fire service response, even beyond the confines of traditional firefighting. This concept was to be used in all hazards and was intended to apply ICS approaches to all types of emergency response including hazardous materials response, natural disasters, plane crashes, and so on.

Concurrently, the Phoenix, Arizona Fire Department, under the leadership of Fire Chief Alan Brunicini, developed a version of ICS called Phoenix Fireground Command System. Brunacini's (n.d.) book, *Fire Command*, was immensely popular within the municipal fire services and did much to improve this more streamlined approach. This command framework mirrored much of the ICS concepts that were emerging out of the West Coast. The primary difference was that *Fire Command* focused on a single-agency incident and thereby jettisoned some of the terminology and ICS organizational structure that reflected the campaign-style wildland incident management. For example, the NIIMS ICS organizational components of "branch," "division," and "group" were replaced by the single term "sector." Brunacini's work was polarizing. A municipal fire officer, who most often was fighting fires in buildings, could scarce understand the scale of NIIMS, was indifferent to fires in forests, and was thus endeared to the Phoenix model's simplicity and flexibility. Therefore, many municipal fire departments did not acknowledge NIIMS as their standard.

The final push to achieve a national incident management standard came in the aftermath of the events of September 11, 2001. Communications and organizational challenges plagued the response to New York City and led to significant urging from the federal government to adopt a common system. HSPD-5 was the impetus to take a national standard more seriously. The 2002 report by McKinsey and Company, requested by the New York City Fire Department (FDNY), reflects the reluctance of municipal fire departments (especially those larger, East Coast cities) to adopt ICS as part of their day-to-day routine. It is important to point out, however, that this does not mean that large municipalities, FDNY included, did not abide by an internal incident command procedure. The resistance was to adopt a system that they thought was too complicated when campaign fires were rare in the urbanized environment. The McKinsey Report recommended that the FDNY "review all its procedures to ensure consistency with ICS principles" (McKinsey and Company 2002).

The outcome of the September 11, 2001, events was a mandate for a common ICS system. Under this mandate, NIMS and its associated ICS became the regulated national incident management standard. This final form of ICS is essentially the NIIMS model with some updates. The challenge before the Department of Homeland Security (DHS), the agency sponsor of this system, was to determine how to enforce the standard across all levels of government and across the breadth of agencies involved in response. The standards debate raged in the immediate aftermath of the September 11th tragedy. One question asked among the emergency management and public safety communities revolved around whether federal funding would be attached to standardization compliance (Canada 2002).

Ostensibly, it was decided that the enforcement of NIMS would be through funding regulation. If a community did not comply with the NIMS rollout, then federal funding would not be provided to those communities (U.S. Department of Homeland Security 2004). Compliance was, and is, being accomplished through a Web-based tool called the National Incident Management System Compliance Assurance Tool. Each jurisdiction, whether state, county, or municipal, must annually meet certain adoption benchmarks and report their progress online. Thus, NIMS ICS has emerged as the national standard, but what are its basic tenets, and how do these tenets affect the complexities faced by responders in the field?

23.3   ICS doctrine and process

NIMS ICS has many nuances. Certain nuances are important to understand the inherent complexity of ICS as well as how to overcome them, namely, common terminology, unity of command, and unified command (U.S. Department of Homeland Security 2004). An overview will be given for each in order to offer background to the ICS complexities that will be discussed.

The need for common terminology arose from both the wildland and the municipal fire domains. Forest firefighters were plagued with asking for a resource and getting something unintended. For example, when needing a helicopter for a water drop on a remote portion of the fire, the request for a "helicopter" may yield a small helicopter for an aerial recon and therefore not equipped for water drops; thus, a helicopter was not a helicopter. This problem caused the early ICS developers to design a system of "type and kind." In its simplest form, the "helicopter" would be the "kind" of resource and that helicopter's capability would be its "type" (on a scale from Type I, the highest capacity helicopter, to a Type IV, the lowest capacity helicopter). This provided the field commanders the ability to easily and accurately request resources and ensured that they were requesting what they actually needed.

Another common terminology issue was brought to light in the immediate aftermath of the World Trade Center collapse in September 2001. This terminology problem, which was endemic throughout the United States, was that of the esoteric radio codes used by first responder agencies. Most commonly referred to as "10" codes, the purpose was to shorten radio chatter by associating a common radio message with a numerical code. For instance, a report of an auto accident may be "code 4." In the FDNY, it is "10–75" that communicates the initial conditions upon arrival to indicate the structure of fire. These unique identifiers became problematic in larger events where multiple jurisdictions would respond together to the same incident and not have a common code. The ensuing confusion is self-apparent.

Two initiatives have emerged within NIMS to overcome these terminology problems. First, a system of classifying types and kinds was developed. Essentially, the NWCG resource-typing document was adopted by NIMS. The challenge has been to expand this resource naming taxonomy beyond the scope of the original wildland context. Regional nuance interferes with a national standard. For example, in a single American market region, the vehicle that is staffed by emergency medical technicians to transport the sick and injured to the hospital may be called "ambulance," "medic," or "rescue." In other regions, an ambulance may be referred as a "bus."*

The second initiative to overcome common terminology problems was the demand for "plain English" communications. This mandate had better success and many smaller agencies welcomed the opportunity to end archaic radio codes. Large municipal agencies, the FDNY for example, whose radio codes were not only culturally significant but also had serious operational implications, refused to abide by this mandate. The city felt that the disruption of services that could be caused by a sizable paradigm shift presented more of a threat than the very rare need to communicate with outside resources. Although the FDNY has agreed to use plain English when communicating with outside departments, they continue to use the "10 codes" that were developed in the nineteenth century.

Another prominent ICS doctrine is the chain of command (Department of Homeland Security 2005). This doctrine states that there is a clear line of authority between the incident commander (IC) and the lowest organizational unit on the scene. Again, within the confines of a single jurisdiction, this doctrine makes sense and can be implemented relatively easily. For an incident that occurs in a single municipality, for instance, the IC is delegated

---

* The Dayton, Ohio region, still struggles with the resource-naming problem. In the given example, the immediate metroplex of Dayton has agreed to a certain naming lexicon but the surrounding rural communities continue to use traditional terms. The New York City area refers to ambulances as "buses." The problem is endemic across the county.

that authority by statute, and therefore by policy acts as the agent of the municipality during the incident. In a single municipality, for instance, the IC is likely to be deemed the IC by delegated authority of the chief elected official by statute. This statutory delegation gives him/her the "overall responsibility and complete authority for managing an incident" (National Fire Service Incident Management System 1998). Since the city may employ the IC, in this case a fire chief, and since the city owns the capital resources being utilized, in this case the fire trucks and personnel, the line of authority clearly exists and is quite clean. As will be shown, this clear line of authority ceases to exist once the confines of the incident exceed local boundaries.

The wildland firefighting community employs a unique concept to overcome authority challenges in large wildland fires. This concept, called delegation of authority, has been successfully used to allow outside commanders to manage a local incident. In the case of a rapidly expanding wildland fire, a local IC can request assistance of the National Forestry Service and the assistance of an Incident Management Team (IMT). Such a team would be deployed by the forestry service to the locale at which time the local IC would sign a document that would delegate his authority to the IC of the IMT. The signature authority granted to the incoming IC places all assigned resources under the control of the new IC, thus preserving the unity of command doctrine. The delegation process works cleanly as many of the firefighting resources called in on large-scale fires are federal resources, thus reinforcing the authority capacity of the federalized IC. This practice has been fully institutionalized in the American West and has been used with great success.

East of the Mississippi, however, such a practice is foreign and unlikely. Without the continual threat of large-scale incidents, such as forest fires, communities resist, both legally and culturally, the intrusion of outside authority. The use of local, regional, or state-level IMTs to assist in large-scale incidents is beginning to be more commonplace, but only under the proviso that authority is never actually transferred. In these cases, the IMT acts in an advisory role and fills in ICS positions where the local agencies have insufficient staff or expertise.

Unified command is another ICS doctrine worth mentioning. The idea is that the command can exist in the form of more than one person. In large incidents where many agencies or perhaps multiple jurisdictions are impacted, sufficient authority does not exist in one person. In such a case, select stakeholders may combine into a single command element that would together establish common incident objectives.

23.4   Large-scale complexities of ICS

In a large-scale incident, ICS reaches a certain threshold where its intuitive value is confused and its ability to sustain its doctrine is difficult if not impossible. This is where the operator is faced with numerous complexities and where incident success can be compromised. These complexities include complexity of authority and ownership, complexity of culture, and the complexity of the incident itself.

An old Scottish adage argues that possession is nine-tenths of the law. This is certainly the case in disaster response. True control of a resource lies with the owner of the resource. ICS provides a standardized means whereby one jurisdiction can borrow the resources of another under a commonly understood framework, but, at the end of the day, an external agency, the owner of a resource on loan to the incident, can decide to remove its resources from the scene without any recourse of the IC. This response challenge means that complex incident response requires careful and diplomatic relationship building. Authority of the IC and the ICS command and general staffs are thus a continuing negotiation with the agencies providing resources to the incident.

As an incident reaches such a scale that numerous state and federal resources are now deployed to the incident theatre, it becomes simply impossible for any single IC to have sufficient authority to direct all resources. Indeed, something as simple as internal rules and regulations dictates how certain resources can be used. For example, an IC cannot "order" an active duty military unit to conduct law enforcement activities due to federal regulation limitations of the Posse Comitatus Act. Similarly, private companies and some public agencies may have trade union limitations that dictate certain work conditions or even time limitations of their workers.

Another complexity of ICS in large-scale incidents is that of organizational culture. Ingrained cultural nuances of various organizations can make the command challenging. Nuances of practices common to specific organizations and agencies make it difficult for ICs to be able to navigate objectives. This challenge often creeps into the incident command post itself. One classic cultural difference lies between fire agencies and police agencies. Due to the quasi-classified nature of their normal work tasks, police agencies can sometimes rightfully be reluctant to merge their command activities with fire officials. This cultural clash was found to have existed in response to the World Trade Center collapse in the aftermath of September 11, 2001 (McKinsey and Company 2002). This is not to imply hostility in all cases, but simply a cultural difference that can impact large-scale incident cooperation and seamlessness.

Finally, and perhaps most obviously, the incident itself can create complexities that complicate the use of NIMS ICS in large-scale incidents. Incident characteristics such as poor communications, geographical space, and a never-ending shift in incident dynamics may result in the degradation of ICS.

Poor communications have plagued disasters since the dawn of time and, despite the revolution in information and communication technology, the problem persists. Communications devices, such as handheld public safety radios designed for close-quarters tactical communications, are inadequate to handle the volume and type of communications that are demanded in a post-disaster environment. Likewise, all electronic systems require a functioning infrastructure in order to operate. Service interruptions caused by direct damage or system overload make most communications links of no value.

Communications challenges also come in the form of both technical and cultural inoperability. Despite a national movement to overcome radio interoperability barriers, communications systems remain proprietary and cloaked in agency-specific nuance. Cell-based devices that are much more adept to disaster communications and information transfer are subject to the fragility of commercial systems and are often out-of-service in the immediate aftermath of an event. Interoperability bridges that have been formed create communications pinch points whereby the growing volumes of communications are forced to implode into a reduced bandwidth.*

Cultural inoperability occurs when organizations refuse to communicate, despite technology that ensures that the capability exists. There is no technology that can overcome the hesitancy to collaborate, rivalries, or agency chauvinism. The disaster theatre is riddled with agencies that, accustomed to working within their own paradigm, are not apt to reach out. This problem greatly confounds incident command entities that desperately need information but can be left out of the information loop by agencies who tend not to share vital data that they may possess. This behavior is exacerbated by agencies whose information is proprietary or is deemed sensitive and not for dissemination. The author directly observed this behavior during two nationally significant events where vital data with responder health and operational implications was withheld from responders. Although some level of cooperation is considered the norm in these events, intentional communications gaps can still be expected.

A commonly know axiom in emergency management and response is that "every incident is a local incident." This fact is certainly true. Another complexity brought about by the incident itself is simple geography. Whereby most emergency responses occur in a defined geographic boundary, many large-scale incidents cover a wide area and defy radio, foot, or even vehicle traffic. Often these incidents cross state lines, let alone county or municipal boundaries. It is virtually impossible for any IC to

---

* This is observed in the Dayton, Ohio region. The metropolitan area falls within two different counties each of which has its own proprietary radio system. To overcome the problem of cross-county mutual aid, a one-channel mutual aid radio bridge was created. The problem, of course, is that in a cross-county emergency (and this is tornado country), all communications between units from the two counties must collapse onto one channel.

have authority across great distances and thus across political boundaries. Brian J. Gallant (2008) admits that "the benefit of NIMS is most apparent at the local level, when a community as a whole prepares for and provides an integrated response to an incident." In a natural disaster, even when the scale of the incident requires a significant federal resource presence, there is no federal official, up to and including the President of the United States, that can control the lawful activities of a duly elected mayor. A construct exists in NIMS, called Area Command, that describes a geographically distributed incident command structure, but such an entity cannot possess adequate authority to execute its prescribed responsibilities. The Area Command established during Hurricane Katrina was used only for the direction of federal resources.

Perhaps the most profound challenge of the incident itself is its dynamic nature. Events change at an almost constant pace. As the incident unfolds, new information emerges, secondary emergencies occur, political priorities change, and many other variables can impact the operations flow. This ebb and flow defies ICS to try to predict objectives for the next operational period and beyond.

The Incident Action Plan (IAP), a document developed in the incident planning process, is the focal point of assigning incident objectives to time. Again, as a byproduct of the wildland firefighting domain, the IAP was designed to provide a common battle rhythm by which all agencies and resources were assigned to the incident abide. The process by which the IAP is developed is cyclical and tries to keep one operational period ahead. Incident dynamics, however, do not necessarily follow IAP cycles, and thus it is difficult for the IAP to keep up with the constant change. For example, a local elected official may change his priorities without regard to the IAP, thus confounding the planning process and frustrating ICS loyalists.

Given the background of NIMS and ICS, the core doctrines of ICS, and the complexities that ICS causes while addressing large-scale incidents, it is now time to turn attention to the strategies of overcoming some of these complexities.

23.5   Strategies for overcoming ICS complexities

It is important to point out here that although the below-suggested strategies offer a workaround to ICS challenges, they do not infer that the baby be thrown out with the bathwater. ICS has a clear role in managing incidents of appropriate scale and has been socialized enough that suggesting its demise, or even significant alteration, would be impractical. The suggested strategies are meant to address the shortfalls that have been already addressed and to assist responders in being effective despite them.

Three strategies are suggested. These three strategies include utilizing ICS for planned public events, horizontal communications, and the formation of *ad hoc* teams. Each uniquely recognizes that although challenges may face ICS in large-scale incidents, there are practical ways to overcome these challenges and succeed in incident management.

The use of ICS as a tool for large event organization and planning has been proven beneficial and is an excellent strategy for overcoming ICS challenges. At the heart of this strategy is the simple principle that responders work more collaboratively with people they know, and, perhaps more importantly, they will better understand ICS at a large scale when they experience its benefits and limitations in a controlled environment. Another wise saying that has been circulated around the emergency management profession states that "an emergency should not be where business cards are passed out." Essentially, this nugget of truth explains that the heat of chaos is not a good place to build relationships. By having community leaders, response agencies and community organizations adopt the ICS framework for event planning, they offer themselves the opportunity to get to know each other and their capabilities, along with how agency interactions affect the outcome of an event and how ICS is actually deployed community-wide. Major sporting events, community festivals, VIP visits, and other high-profile events become a unique training ground for collaboration and ICS implementation. There are two models by which this can be accomplished.

The first, and perhaps most aggressive model (but with the most reward), would be to use ICS and the incident planning process as the planning methodology for the event itself. The challenge to this model is to achieve buy in from organizations that are not accustomed to working under this construct and could possibly obstruct its use. Traditionally, special events are organized in a committee fashion with public safety agencies brought in as an appendage of the central planning committee. Although there is nothing wrong with this approach, it still creates a divide between community leaders and the response agencies that may need to intervene if an emergency were to occur. Also, the response agencies tend to not know all of the intricacies of the event and thus enter into an emergency situation with less than ideal situational awareness. This occurs especially when a third-party event management company is involved.

The second model for utilizing planned events as a framework to overcome ICS complexities is to simply broaden the scope of public safety planning, expand the planning group beyond the typical emergency response agencies, and incorporate other community organizations into the emergency contingency planning process. By utilizing the incident action planning process, more community members become familiar with how ICS is structured and how to effectively participate in the process.

Kettering, Ohio's annual "Holiday at Home" festival is a perfect example. In 2006, the city agreed to a more inclusive emergency planning team. An IC was named several months in advance of the festival and the planning cycle was initiated. The IAP became the central focus of planning and prompted the development of several plans that better refined the final product. A media plan, air operations plan, and a patient surge plan provided frameworks for even the most mundane activities of the event. Operational periods were divided into weeks at the beginning of the process, but as the event neared, the periods were narrowed to days. This model worked to the extent that a supervisor in the public works department emerged as the perfect Logistics Section Chief and his enthusiasm over the IAP process was infectious.

The value of using planned events as a strategy to overcome ICS challenges is that the problems of authority can be overcome by familiarity. The willingness to share resources can be improved by mere relationship building in a non-stressful atmosphere. The associated trust building, familiarity with ICS and the incident planning process, and enhanced communications will aid response agencies and many other community organizations to better collaborate in a real incident.

Another strategy to overcome ICS complexities is to embrace horizontal communications. This strategy is more difficult, especially as this will occur in a chaotic environment where communications are difficult to begin with. Essentially, this strategy suggests that responders should seek out communications pathways with other agencies. Often, ICS can encourage "stovepiped' communications, especially when the ICS is structured functionally. For instance, if the operations section is divided into the fire, emergency medical services (EMSs) and law enforcement branch paradigm, it is easier for these agencies to maintain their own communications pathways and stifle collaboration. This path of least resistance can lead to a sense of separation and reduction of important cross-communications. The preferred ICS structure in the large-scale incident is to establish a geographic breakdown, whereby agencies and functions are forced to cross-communicate within the geographic division or branch, and thus increase collaborative behaviors.

Failure to communicate horizontally between agencies can also degrade unified command. If various agencies continue to maintain a unique identity within a large-scale event, the ability for unified commanders to work together and identify common incident objectives is threatened. Horizontal communications also include communication between layers of government. Reaching out to various state and federal agencies, although difficult to establish initially, will reap significant benefits in situational awareness and creative problem solving. This level of communication will also help to shrink long distances as various agencies and organizations may have better communications infrastructure and thus are able to transcend greater distances. Tapping into these stronger systems has obvious benefits for everybody in ICS.

If anyone within ICS finds themselves talking only within their own agency, they need to reach out. Stovepiped communications can only lead to dangerous gaps in information and cross-communication. Taking time to figure out what agencies are in the incident and how to communicate with them will significantly overcome challenges of ICS in larger incidents.

The final strategy suggested is probably the most beneficial and most able to overcome ICS challenges. Building *ad hoc* teams is an excellent way to work around the complexities of ownership, authority, culture, and the incident itself. By joining with six to eight members of various agencies and levels of government, a tremendous synergy can occur and a large amount of important work can be accomplished. Furthermore, cultural complexities are very effectively dealt with through personal relationships rather than systematic constructs.

Industry has recognized the value of *ad hoc* teaming, thus its use in the theatre of a major emergency makes perfect sense. Author Joyce Osland draws parallelism between business and emergency response. Speaking of the challenge for global leaders to be able to effect significant change she relates, "global leaders deal with employees and stakeholders from a range of countries with distinct business and cultural practices," and, "the complexity of the global context in which they work forces them to develop what can be called metalevel leadership and cultural competencies" (Osland 2011). The parallelism is self-evident when you consider the complexities of large-scale incident environments and the challenge to responders to work around these complexities for the good of the impacted community. Kafner and Kerry (2012) explained the value of applying business practices to complex environments by asserting that "multiteam systems per se represent a potentially powerful ambient feature of the work environment."

A good example of this strategy was employed by the author in the response to Hurricane Katrina at the storm's epicenter: Hancock, Harrison, and Jackson Counties, Mississippi (Richardson et al. 2008). The Harrison County Coroner's Office was assisting its surrounding counties by coordinating the handling of the growing number of fatalities on top of coordinating missing person reports. As a Deputy Operations Section Chief for the FEMA's Urban Search and Rescue Response System, the author observed the volume of work trying to be accomplished by the coroner's office and their rapidly approaching point of saturation. A team was established quickly. The author, the Harrison County deputy coroner, the State of Mississippi search and rescue coordinator, officials of the Georgia Bureau of Investigation, and a lieutenant in the U.S. Coast Guard formed together to support the coroner and to coordinate search and rescue missions for the missing personnel between agencies. Although each member of this *ad hoc* team reported within their own ICS chain of command (remember that in large-scale incidents no single ICS will have sufficient authority to manage all resources), the team operated independently in order to streamline work and coordination. This was a notable success. The rapidity by which search priorities could be identified and search teams assigned allowed county officials to more quickly report to concerned family members, and thus assure the community that missing loved ones were being addressed. Each member of this team had each other's cell phone number on speed dial, and frequent meetings were arranged to address specific problems. This method was so successful that much of the *ad hoc* team stayed intact even as the incident transitioned from rescue to recovery. Together they worked on training programs for workers who were being contracted to remove debris while trying to still identify human remains during the removal process.

23.6   Conclusions

Although ICS has a rich history of development and implementation, and provides an excellent framework for the management of emergency response, there are complexities that make it difficult to navigate in large-scale, multi-jurisdictional incidents. Complexities such as authority and ownership, culture, and the incident itself can make ICS too rigid in these dynamic and politically diverse environments. The strategies suggested in this chapter allow the responder, and in some respects the planner, to be more effective in the large-scale event despite ICS challenges.

Using ICS as the framework for planning large-scale special events allows community agencies and organizations to work through ICS complexities without the stressor of an emergency. Here, leaders and managers can get to know one another and build trust that will go a long way to break down cultural and geo-political boundaries.

During the emergency, it is suggested that reaching out horizontally will help overcome gaps in information and coordination. The tendency for ICS to structure itself functionally (police, fire, EMS, etc.), rather than geographically, lends itself to stovepiped communications and information hoarding. Responders at all levels should recognize the need to look around them, identify support agencies and organizations, get their contact information, and start talking.

The strategy of *ad hoc* teaming has a lot of potential. A little more formalized than the above-suggested horizontal communications, this strategy has the ability to significantly overcome coordination gaps and get quantities of important work done for the betterment of the incident and its victims. Although this strategy may seem like an ICS workaround, it should be viewed as a force multiplier rather than an act of defiance. All players in ICS at an incident should still follow their chain of command and report accordingly. An *ad hoc* team simply allows a small group of people to more efficiently work through ICS complexities.

The study of ICS and its abilities to encompass large-scale events needs to be continued. Although such a study falls outside the content of this chapter, it is imperative that objective and scholarly attention be drawn to the ICS construct to adjust ICS as necessary to minimize these complexities and challenges. By employing the above strategies, the response community can enhance their effectiveness in large-scale events and build synergies that can improve the outcomes of devastating events. At the end of the day, it is the people who make this system work, and the people who recognize the complexities and deploy these and other creative strategies to plan for and respond to significant emergencies.

References

Arbuthnot, Kevin, and Rhona Flin. "Introduction." In *Incident Command: Tales from the Hot Seat*, by Kevin Arbuthnot and Rhona Flin, 3–9. Burlington, VT: Ashgate Publishing Company, 2002.

Brunacini, Alan. *Fire Command*, 1st Edition. Quincy, MA: National Fire Protection Association, n.d.

Canada, Ben. *First Responder Initiative: Policy Issues and Options.* Report for Congress, Congressional Research Service, The Library of Congress, 2002.

Department of Homeland Security. *FEMA ICS-100/200: Instructor Manual.* Washington, DC: Center for Domestic Preparedness, 2005.

Devitt, Katherine R., and Edward P. Borodzicz. "Interwoven Leadership: The Missing Link in Multi-Agency Major Incident Response." *Journal of Contingencies and Crisis Management* 16, no. 4: 208–216 (December 2008).

FIRESCOPE. "FIRESCOPE: Our Beginning." *FIRESCOPE.org.* n.d. http://firescope .org/firescope-history/Some%Highlights%20of%Evolution%20of%20the% 20ICS.pdf (accessed December 15, 2011).

Gallant, Brian J. *Essentials in Emergency Management: Including the All-Hazards Approach.* Lanham, MD: The Scarecrow Press, 2008.

Kafner, Ruth, and Matthew Kerry. "Motivation in Miltiteam Systems." In *Multiteam Systems: An Orgaizational Form for Dynamic and Complex Environments*, by Stephen J. Zaccaro, Michelle A. Marks and Leslie A. DeChurch. New York: Routledge, 2012.

McKinsey and Company. *Report on the New York City Fire Department Response to the World Trade Center.* n.p., 2002.

National Fire Service Incident Management System. *Structural Collapse and US&R Operations.* Stillwater, OK: Fire Protection Publications, 1998.

Osland, Joyce. "Expert Cognition and Sensemaking in the Global Organization Leadesrhip Context." In *Informed by Knowledge: Expert Performance in Complex Situations*, by Kathleen L. Mosier and Ute M. Fischer. New York: Psychology Press, 2011.

Richardson, Harry W., Peter Gordon, and James E. II Moore. "Introduction." In *Natural Disaster Analysis after Hurricane Katrina*, by Harry W. Richardson, Peter Gordon and James E. II Moore, 1–7. Northampton, MA: Edward Elgar, 2008.

Society of American Foresters. "The Dictionary of Forestry." *Society of National Foresters.* 2008. http://dictionaryofforestry.org/dict/term/large_fire_organization (accessed January 10, 2012).

U.S Department of Homeland Security. *National Incident Management System Complaince Assurance Support Tool.* Washington, DC: U.S. Department of Homeland Security, 2004.

White House, The. "About DHS: Laws and Regulations." *Department of Homeland Security.* February 28, 2003. www.fas.org/irp/offdocs/nspd/hspd-5.html (accessed January 9, 2012).

chapter twenty-four

Begin with the end in mind

An all-hazards systems approach to waste management planning for homeland security incidents

Mario E. Ierardi

Contents

24.1   Introduction

Waste management during homeland security (HS) incidents is a complex field that involves many organizations at the international, federal, state, local, territorial, and tribal areas of government, in addition to elected officials, the private sector, nongovernmental organizations, associations, academia, the media, and the public. It also involves specialists in many different fields of expertise across those organizations that may not have worked with each other before. In addition, there has been and continues to be an evolution in waste management during a HS incident since the terror attacks of 2001 and the subsequent formation of the Department of Homeland Security (DHS) in 2003. The DHS was formed to develop and coordinate a national response framework (NRF) across the country in response to HS incidents including large-scale natural (e.g., Hurricane Katrina) and man-made (e.g., Deepwater Horizon oil spill) incidents. The DHS has developed frameworks for National Response, Recovery and Mitigation, as well as National Planning Scenarios (NPSs) to assist emergency planners and managers in developing capabilities needed in response to HS incidents. Experience from real incidents (e.g., Hurricane Katrina, Fukushima Nuclear Power Plant, Deepwater Horizon oil spill) and national level HS exercises have shown the importance of waste management as a capability necessary for an effective and timely response and recovery to HS incidents. This paper explores applying an "all-hazards systems approach" to waste management planning related to large-scale HS incidents and Spills of National Significance (SONS) to assist emergency managers and planners at all levels of the public and the private sector in developing and/or enhancing this capability.

24.2   Why is an "all-hazards systems approach" to waste management planning needed?

The simple answer is because waste management during HS incidents involve a great deal of complexity and integration of people, processes, and systems that exhibit many commonalities that are independent of the type of incident and the nature of the contaminants involved. Waste management planning involves a multihazard, multistakeholder, multidisciplinary, multimedia, multiscale, and multiobjective decision-making process. Waste management frequently involves social, economic, environmental, political, regulatory/statutory, policy, and technical considerations in making sound, safe, cost-effective, risk-informed, and consensus-orientated decisions (Chang, 2003). Experience during several HS incidents and national level HS exercises has shown that waste management planning is not something that should be left to do during the incident itself. A systems approach is needed in waste management planning so that there is not a fragmented, piecemeal approach to waste management during HS incidents.

In a systems approach, the concentration is on the analysis and design of the whole, as distinct from the total focus on the parts (Figure 24.1). This is where the concept of *beginning with the end in mind* (Covey, 1989) and understanding the interdependencies of people, processes, and systems becomes important to effective planning for HS incidents. This same concept is why waste management cannot be separated from the response or recovery, just as it cannot be separated from cleanup, decontamination, sampling/analysis, or even evidence gathering. This is because waste is generated throughout the entire response system and therefore must have a "cradle to grave" plan ahead of time because *waste management starts on day 1!*

It should be noted here that we are not referring to the many responses that occur on a daily basis across the country. The Environmental Protection Agency (EPA), US Coast Guard (USCG), the Federal Emergency Management Agency (FEMA), states, and local emergency response personnel have a very good handle on these types of responses. What distinguishes HS incidents from a typical response from a waste management perspective is the magnitude, scope, unique threat agents, economic impacts, resource requirements, and logistics required for this type of response. In addition, many of the NPSs identified within the NRF (Department of Homeland Security, 2008) identify wide area releases of threat agents that are not encountered frequently by emergency response personnel or waste management officials. There are 15 all-hazards NPSs that are reflective of the all-hazards approach. These 15 NPSs are intended to be utilized by emergency planners and managers across the country in development of their emergency response plans. These planning scenarios

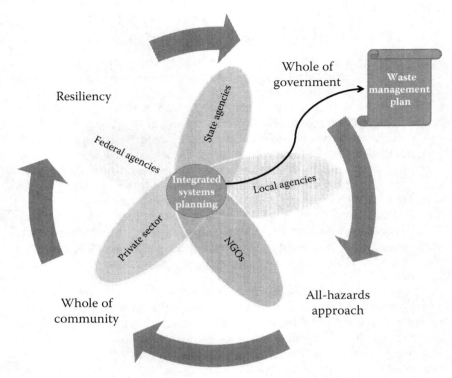

Figure 24.1 Integrated systems planning. (From Ierardi, EPA Office of Resource Conservation and Recovery, Homeland Security Team Leader, 2012.)

provide valuable assumptions on what to expect and, therefore, what to plan for during HS incidents. They identify unique chemical, biological, and radiological (CBR) threat agents and wide area release scenarios that are not encountered during a typical spill or emergency response. From a waste management perspective, these NPSs identify some important planning considerations that are not part of a typical emergency response, including

1. *Larger quantity of waste*
   The amount of waste generated may be greater than the amount of waste many communities typically handle in a year, overwhelming state, local, and territorial resources.
2. *Wider variety and uniqueness of wastes encountered*
   The incident may generate a broader range of waste streams, including CBR-contaminated wastes not typically handled by communities or waste management facilities.

3. *Wider area of impact*

In a HS incident, the area of impact may be extremely large (e.g., several states or regions). Multiple regulatory jurisdictions may be involved with varying requirements and approaches to waste management. Several of the NPSs identify wide area release scenarios.

4. *The need for an "all-hazards systems planning" waste management approach*

The NPSs present a wide spectrum of scenarios, threat agents, release scenarios, impacts, and complexity, with many commonalities that span across multiple scenarios and drive the need for an "all-hazards systems planning" waste management approach.

5. *The need to engage the public and manage public perception*

The high visibility of a HS incident may result in communities resisting the treatment or disposal of generated wastes into their local facilities (e.g., landfills), including wastes that would otherwise be managed at those facilities under normal conditions.

6. *Large-scale incidents quickly become politicized*

Information gathered/developed in advance assists decision makers in weighing political/social issues during an incident. Planning in advance and incorporating local plans into state, regional, and federal plans ensure that political decisions take local considerations into account.

24.3   What is waste management and where does it fit into a homeland security incident?

Waste management is usually undefined in most national HS guidance and planning documents. Instead, the term waste disposal is typically used. Although waste disposal is a commonly used catch-all term that encompasses many aspects of waste management, waste disposal only refers to the ultimate disposition of waste in a landfill. Waste disposal is an event, and a very narrow one, whereas waste management is a process. From a systems approach, it is quite a different concept to plan for an event, especially one as narrow as waste disposal (usually seen as the last thing on the waste management spectrum), versus waste management that covers the entire spectrum of a response. Terminology and definitions are quite important when there are so many different stakeholders involved in the process. Currently, there are a host of terms utilized in describing waste, such as debris, waste materials, solid waste, hazardous waste, household hazardous waste, mixed waste, low-level radiological waste (LLRW), biohazardous waste, infectious waste, special waste, and so on. It is important that each of the stakeholders in the planning process understand the terms and what they mean from a statutory and regulatory standpoint. In addition to the many terms utilized in describing waste, there are also many terms involved in describing waste management activities. Disposal is typically the endpoint of the waste management process, a process that also includes collection, handling, packaging, labeling, sampling/analysis, characterization, interim and final storage, various treatment methods, various source-reduction methods, recycling, waste minimization, and as mentioned, disposal. Waste management can involve some or all of these activities in any given HS incident. To communicate consistently and improve the meaning of this discussion, we are using an operational waste management definition as follows:

> **Waste Management** is the collection, transport, processing or disposal, managing and monitoring of waste materials. The term usually relates to materials produced by accidental or intentional human activity or natural disasters, and the process is generally undertaken to reduce their effect on public health and the environment. The management of wastes treats all materials as a single class, whether solid, liquid, or gaseous, chemical, biological or

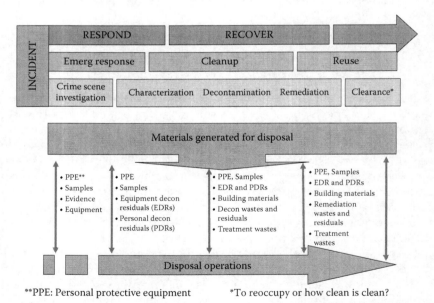

Figure 24.2 Role of waste management in a homeland security incident response. (From Ierardi, EPA Office of Resource Conservation and Recovery, Homeland Security Team Leader, 2010.)

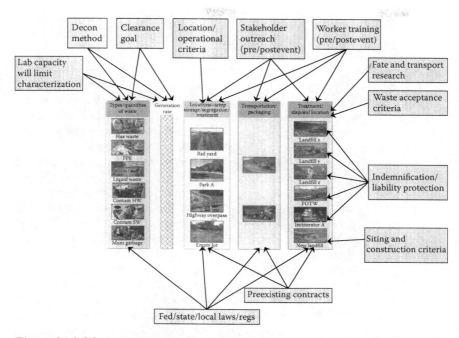

Figure 24.3 Waste management considerations involved in homeland security incidents. (From Parrish, *Waste Management Workshop*. EPA Office of Homeland Security, 2012.)

radioactive substances, and tries to reduce the harmful environmental impacts of each through different management methods.

(Wikipedia, 2012)

Waste management fits in almost every NPS because waste is generated from all of them. In addition, waste can be generated immediately, even during the evidence collection process, if the incident is intentional. As a result, a waste management plan (WMP) should be in place prior to the event and subsequent response and recovery. Figure 24.2 illustrates how waste management and operations fit into the overall response and recovery to a HS incident.

As previously mentioned, the waste management decision-making process is complex due to the many different considerations that are present during HS incidents. This is captured in Figure 24.3, which was created during a waste management workshop conducted by the DHS and EPA as part of an effort to improve WMP for wide area releases and HS incidents such as those described within the DHS NPSs (EPA Office of Homeland Security, 2012).

24.4 What is the framework for waste management decision making?

Decision making within complex systems involving so many factors is a good candidate for applying a systems approach to assist in the facilitation of these decisions (Ramo and St. Clair, 1998). Delaying so many of these decisions until the incident may result in a long, drawn-out and costly response and recovery. As an alternative, a preevent systems approach provides for a description (what we are referring to as a WMP) of the waste management process in its entirety, taking into account the planning assumptions outlined within the NPSs as well as the considerations shown in Figure 24.3. Conducting this process prior to an incident provides the opportunity to understand the needs of an incident, of the people involved or impacted by the incident, and to understand the interrelationship of people who would be involved at the different levels of government (vertical alignment) with those involved across functions and expertise within the same level of government (horizontal alignment). It provides the opportunity to optimize the utilization of resources by coordinating and aligning emergency response plans across different agencies and entities involved. It also provides opportunities to identify gaps in resources ahead of time and develop mutual aid agreements to fill those gaps instead of scrambling for them in the middle of an incident. Where additional needs exist, it provides an opportunity for development of preexisting/prenegotiated contracts for services in support of the waste management activities. By taking a systems approach, one is able to begin to see a common framework of decision making regardless of the incident, threat agent, or location. Identifying these similarities ahead of time reduces the number of decisions that need to be made during the incident. In fact, it allows those involved in the response to focus on tailoring these existing frameworks to site or incident-specific conditions versus starting from the beginning of the planning process. Figure 24.4 is an example of a waste management decision framework that can help visualize the overall decision process made from information as it is obtained during an incident.

24.5 A systems approach to develop a waste management plan within the national planning process

Why do we need to put a WMP together when we have responded to incidents before without one? The answer is because this is not a typical incident! According to the DHS, the death, damage, and therefore the

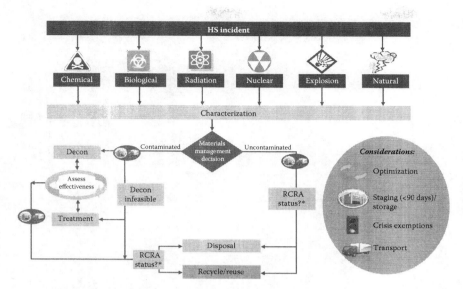

Figure 24.4 Homeland security incident waste management decision framework. (From Ierardi, EPA Office of Resource Conservation and Recovery, Homeland Security Team Leader, 2010.)

waste associated with the NPSs may be greater than anything we have ever seen from a natural or man-made disaster (Figure 24.5).

In addition, HS incidents may require a large number of waste management facilities across several states, regions, and perhaps the country. Fortunately, there is a national planning process through which planning for HS incidents should take place. The National Contingency Plan (NCP) (EPA, 1980) is the federal government's emergency response blueprint for responding to oil spills and hazardous substance releases and to promote coordination among emergency responders and contingency plans. The NCP describes the National Response System as an organized network of agencies, programs, and resources with authorities and responsibilities in oil and hazardous materials response at the federal, state, territorial, local, and tribal levels of government as well as key private sector stakeholders (e.g., waste management facility owner/operators). These representatives are involved in the development of Regional Contingency Plans and Area Contingency Plans (ACPs) in accordance with the NCP. The ACPs are the focal point of response planning, providing detailed information on response procedures, priorities, and appropriate countermeasures. WMPs should be developed for HS incidents and oil SONS, and thereby become part of the ACP.

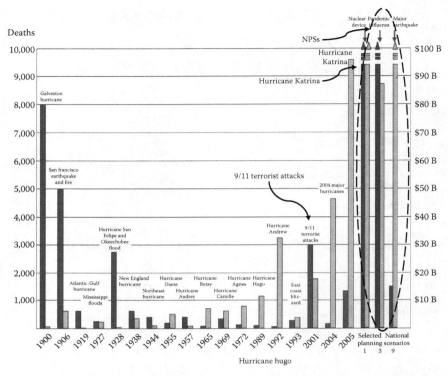

Figure 24.5 Death/damage estimates of NPSs compared with other disasters. (From National Commission, *Deep Water, The Gulf Oil Disaster and the Future of Offshore Drilling*, National Commission on the BP Deepwater Horizon oil spill and Offshore Drilling, 2011.)

24.6   Improving the waste management planning process

Lessons learned from The Report to the President on The Gulf Oil Disaster and the Future of Offshore Drilling identified needed improvements in the area of waste management planning within the ACP development process. These recommendations included a need to establish distinct plans and procedures for responding to a "Spill of National Significance" as well as to strengthen state and local involvement in the oil contingency planning process (National Commission 2011). Additional findings from the report indicated that the ACPs and WMPs in existence before the Deepwater Horizon oil spill were inadequate for responding to the magnitude of the spill. As a result, WMPs were still being prepared by BP with federal and state input several months into the oil spill response. This resulted in delays in waste management decision making, adding costs and time to the overall response, and attracting public concern. Listed below are the overarching waste management lessons learned from recent domestic (e.g., Deepwater Horizon Oil SONS) and international (e.g., Fukushima Nuclear Power Plant) incidents.

1. They involve large quantities of waste and extensive waste management activities that exceed state and local capabilities.
2. Having no or inadequate WMPs prior to HS incidents or SONS can negatively impact the cost, as well as the response and recovery timeframe.
3. The NCP identifies waste management elements as part of the ACP planning process.
4. These responses are complex and often require a regional or national response involving resources and coordination at the national and regional level.

24.7   Key decisions before developing a waste management plan

Before starting the process of developing a WMP, it is useful to identify what key decisions need to be made.

- Who needs to be part of the waste management planning team?
- How much and what types of waste will be generated from the NPSs?
- On the basis of past experience (process knowledge) and available information, what will the waste be characterized or designated as? Are there different waste management requirements in the states, territories, or tribes impacted by the response?
- How will samples be analyzed, collected, labeled, segregated, and waste minimized?
- How many waste samples are anticipated? Is sufficient laboratory capacity available to handle these samples along with those samples that will be needed for contaminant characterization, cleanup, and clearance? What arrangements need to be made to ensure laboratory capacity can be provided?
- How and where will the waste be managed? What will the process be for assessing what facilities are appropriate for the waste to be managed at? With whom will this be coordinated?
- How will the waste get to the waste management facilities? With whom will this be coordinated?

- How will the waste be tracked to ensure it is managed appropriately? How will the result of the waste management process be reported to keep the public and other stakeholders informed?
- What safety measures will need to be taken for those handling and managing the waste?
- What resources (people, equipment, contracts, funding, etc.) are needed to execute the WMP per the NPSs? Do any gaps in resources exist? If so, what mutual aid agreements or other backup agreements are in place to support the implementation of the WMP?
- What will need to be communicated to the public and other stakeholders about this WMP and how will that be performed?
- How will the WMP be exercised and updated? Who will review and approve of the document and when will it need to be reviewed and updated if necessary?
- Who will be trained on the WMP and how will that be documented?
- How will lessons learned from exercises and real events be captured and incorporated into subsequent versions of the WMP?

24.8 Framework of an "all-hazards" waste management plan for homeland security incidents

Experience from previous HS incidents as well as national level HS exercises have enabled the formation of a common framework for what should be included in a WMP that has begun to emerge. The elements of this framework, listed and discussed here, are consistent across many types of incidents that are reflective of both a systems and "all-hazards" approach. The sections of the plan should include

1. Purpose, scope, and key contacts
2. Waste management requirements, waste types, and quantities
3. Strategies, facilities, and assets
4. Waste/material tracking and reporting
5. Oversight and exit strategy
6. Community outreach and communications
7. Health and safety considerations

Attachments to the WMP are as follows:

- Maps and lists of approved waste management contractors, facilities, and assets
- Waste staging/storage/decontamination area location maps
- Waste management documentation/tracking forms

- Waste management oversight reporting and cost reimbursement forms
- State requirements, emergency declaration/orders, mutual assistance agreements

24.8.1   Section A: Purpose, scope, and key contacts

A WMP for HS incidents should be a dynamic document that is able to expand with the incident as it unfolds and robust enough to address worst case scenarios. It should incorporate appendices to address individual state-specific requirements. Planning for this ahead of time and working out the logistics and procedures to address this within the incident command (IC) structure can save a lot of time during an actual event. The purpose and scope of a WMP for HS incidents is to

- Reduce the overall cost of waste management activities through effective waste management planning across the impacted area
- Identify waste management regulatory, staffing, and resource requirements across all states/regions impacted
- Identify the types, quantities, and designation of wastes that are likely to be generated for each of the states impacted by the HS incident (this can be estimated from past spill experiences and by utilizing waste estimating tools)
- Develop a waste sampling/analysis process for characterizing/ designating the wastes generated
- Establish waste management strategies for each waste type, including collection, segregation, staging, temporary storage, minimization, treatment, and disposal
- Identify and assess waste treatment, storage, and disposal facilities that will manage the waste
- Identify waste acceptance criteria, sampling, labeling, and tracking requirements for each waste management facility
- Develop transportation plans for the waste across all regions that will handle the waste generated
- Establish waste tracking/reporting processes to keep the public and IC informed of waste management operations
- Establish waste management oversight and exit strategy for waste management activities associated with the HS incident
- Establish a community outreach/communications process in support of waste management operations
- Establish health and safety requirements in support of waste management activities
- Identify the key points of contact, subject matter experts, agency representatives, and owner/operators of public and private sector waste management facilities, including phone numbers and e-mails of the appropriate federal, state, local, and tribal waste management officials

24.8.2 Section B: Waste management requirements, waste types, and quantities

24.8.2.1 Identify waste management requirements of all stakeholders

This section of the WMP involves the preidentification of regulatory requirements at the federal, state, local, tribal, and territorial levels of government as well as private sector requirements for the waste types anticipated with a HS incident. Although there are federal standards for management of solid and hazardous wastes, state programs have generally been authorized to operate in lieu of the federal program. States may have regulations or requirements that are broader in scope or more stringent than federal requirements. As a result, it is important that these differences in waste management are understood, reflected, and integrated into the WMP and become part of emergency operation plans for HS incidents. The WMP should identify where federal and/or state requirements and/or exemptions from hazardous waste requirements for CBR agents apply and where they do not to ensure that waste generated from the response is handled properly and in accordance with federal, state, local, and tribal requirements. Some CBR wastes may need to be disposed of at facilities that meet certain criteria (e.g., liners, leachate collection systems, groundwater or air monitoring, financial assurance, closure, and postclosure care). Other CBR wastes generated during cleanup efforts may be hazardous wastes under the Resource Conservation and Recovery Act (RCRA), which must be disposed of in a RCRA Subtitle C facility. The WMP should anticipate both situations and plan accordingly.

State, local, tribal, and territorial waste management resources may be overwhelmed in a HS incident due to the magnitude of wastes generated, the wide area impacted, and the uniqueness of the type of wastes generated (i.e., difficulty in finding waste management facilities [both public and private sector facilities] that can or will accept the waste). Therefore, it is important that WMPs are robust enough to address the magnitude of response needed for these incidents. Because state solid and hazardous waste programs have generally been approved or authorized to operate in lieu of the federal program, states may have regulations or requirements that are broader in scope or more stringent than federal requirements. For example, some states designate anthrax wastes as hazardous waste, whereas others may designate it as medical or special wastes, or existing designations may not address this waste type at all. This may result in very few public or private sector facilities that are capable or permitted to handle this type of waste. As a result, it is important that state, local, tribal, and territorial waste management officials have their specific waste management requirements and any private sector requirements preestablished within a WMP as part of the ACP for that state and region.

24.8.2.2   Identify waste types and quantities anticipated

Using process knowledge from previous incidents, we can anticipate the type of materials and waste streams that will be generated from a HS incident response. The waste types from Hurricane Katrina (Table 24.1) provide a useful example of waste types and quantities associated with this incident.

The NPSs developed by the DHS also provide planning assumptions that can be utilized in determining the types and quantities of waste anticipated. These scenarios provide planning information about the type of release, the agent involved, the area over which the release occurred, and the amount of structural damage. In addition, there have been some tools developed to assist in estimating waste streams and quantities. The EPA's National Homeland Security Research Center (NHSRC) has developed an online "Incident Waste Assessment System and Tonnage Estimator (I-Waste)" (EPA National Homeland Security Research Center, 2012) tool to assist emergency and waste management planners involved in developing the WMP for handling, transport, treatment, and disposal of waste from different HS incidents. The I-Waste tool can be tailored to different CBR scenarios as well as to incident-specific information. This allows the user to identify specific waste management facilities and contacts for facilitating waste management decisions.

24.8.2.3   Identify waste sampling and analysis requirements

Waste streams will need to be sampled to determine their characteristics and designations for use by receiving waste management facilities.

Table 24.1 Waste Streams and Quantities from Hurricane Katrina

Type of waste/debris	Amount
Curbside debris (construction, demolition, and vegetative/wood debris)	53 million cubic yards
White goods (refrigerators, ranges, water heaters, freezers, air conditioning units, washer/dryers)	~892,000 units
Freon removal	~325,000 units[a]
Electronic goods	~603,000 units
Waste containers (drums, propane tanks, fuel tanks, etc.)	~3,740,000 containers
Household hazardous waste (batteries, oil, automotive products, paint, cleaners, pool chemicals, pesticides, etc.)	~16,114,495 lbs
Nonhazardous household waste (furniture, mattresses, carpets, textiles)	~3,645,025 lbs
Putrescible waste (meats, fruits, vegetables from grocery stores and residents)	~36 million lbs
Vehicles and vessels (cars, boats, etc.)	~410,000 units

Source: From, CRS Report for Congress, 2008.

[a] Units are the number of refrigerators that had freon removed.

This is important because the owners/operators of these waste management facilities are permitted by state environmental agencies with facility-specific waste acceptance criteria. Waste generators must complete facility-specific waste profiles prior to the facility's acceptance of the waste. Sampling and analysis of waste streams also provide additional information to guide health and safety plans (HASP) for response workers, as well as waste management facility operators, waste haulers, and the general public. Details related to the standard sampling and analysis methodologies, frequencies of sampling, as well as data reporting of each of the anticipated waste streams are captured in a waste sampling and analysis plan.

A waste management quality assurance project plan (QAPP) should also be prepared due to the amount of waste management data to be generated from sampling and analysis efforts for the various waste streams, as well as for the long period of time over which these data will be collected. A QAPP lays out a description of waste management sampling and analysis objectives, the organization and responsibility of waste management sampling and analysis efforts, data quality objectives, quality control measures, sampling and analysis procedures, calibration procedures, preventive maintenance, data reduction, validation and reporting, and performance and system audits.

24.8.3   Section C: Strategies, facilities, and assets

This section of the WMP builds upon the results of the waste sampling and analysis performed and the waste designations made in Section B of the WMP. From those results, potential waste management facilities and assets need to be identified and assessed to ensure that specific waste management facilities can meet the anticipated waste types and quantities to be generated from cleanup operations as well as meet all federal, state, local, and tribal waste management requirements.

24.8.3.1   The waste management hierarchy serves as a waste management strategy

The waste management hierarchy (Figure 24.6) is well known as a waste management framework for generators that manage wastes. It is also a hierarchy that is helpful in planning for HS incidents. The waste management hierarchy uses principles of waste reduction as well as reuse and recycling to minimize the amount of waste produced during cleanup operations, thus reducing the environmental and economic costs, and expediting recovery. It provides a useful tool for structuring a waste management strategy and can be utilized as a model for waste management operations.

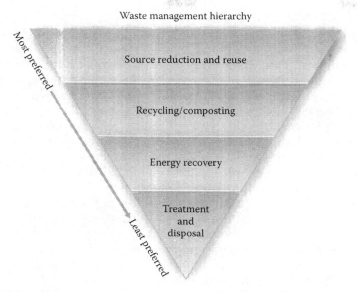

Waste management hierarchy

Figure 24.6 Waste management hierarchy. (From EPA, *Waste Management Hierarchy*, Retrieved February 2012, from Waste Management for Homeland Security Incidents, http://www.epa.gov/osw/homeland/options.htm, 2012.)

On the basis of the waste management hierarchy discussed earlier, the following general waste management guidelines can be adopted for waste generated during HS incidents (Note: Adjustments to these general guidelines may be necessary to tailor these recommendations to a specific CBR agent.):

1. Segregate the different types of wastes (liquid, solid, debris, personal protective equipment, hazardous, nonhazardous, etc.) at the point of generation wherever possible.
2. Waste management sites (temporary storage, staging, decontamination sites, etc.) should have procedures to prevent spills, rainwater infiltration, and runoff.
3. Label all waste containers and identify the source and date of generation.
4. Equipment should be cleaned and reused rather than discarded when possible.
5. Use on-site management to reduce the waste amount requiring further transport and/or treatment.
6. Use reusable personal protective equipment whenever possible.
7. Prevent the mixing of hazardous and nonhazardous wastes.
8. Prevent wastes from contaminating soil, surface waters, or underground aquifers by using liners underneath drums, tanks, and within decontamination/cleaning sites.

24.8.3.2   Waste management options/facilities and assets

Many of the wastes associated with a CBR agent release may not be a waste stream for which current waste management facilities were built and/or designed or for which they are permitted to be used. States are authorized to operate the management of hazardous and solid waste management programs in lieu of federal programs because they may have requirements that are broader in scope or more stringent than federal requirements for some threat agents (e.g., anthrax). Therefore, there can be differences between how the waste is regulated and designated between different states. For some wastes (e.g., LLRW), there are only a few commercially available waste management sites in the entire country. In addition, waste management facility owners and operators may be reluctant to accept waste from HS incidents, even if the waste management facility is considered an effective method for the waste to be managed. Therefore, making arrangements with public and private sector waste management facilities prior to a HS incident will help eliminate uncertainty and potential delays to the cleanup and recovery. The WMP should consider the full capability and capacity of waste management facilities for temporary storage, mobile and fixed treatment systems, as well as final disposal facilities.

It is likely that a variety of options will be necessary to include in a WMP to be able to address the magnitude and extent of waste generated.

24.8.3.3   Develop a strategic waste/materials management process flow diagram (Know the flow!)

This section of the WMP is considered the cornerstone of the overarching waste management response strategy. It lays out a visual display of the waste and material flow through the waste management strategy by identifying what, where, and when the proper waste management actions are to be taken. The process flow of the waste management actions reveals where the waste will be generated, what waste will be generated, how much will be generated, the rate of waste generation, and the physical state (solid, liquid, and gas) of the waste. This information can be used to determine the waste designation (e.g., RCRA status), and from that determine appropriate management approaches that meet federal, state, local, tribal, and territorial waste management requirements.

From previous HS incidents, we have some process knowledge as to what type of wastes to expect as well as the typical approaches utilized in response to a HS incident. With that in mind, it is important to understand that these incidents have unique elements that need to be factored into waste management decisions. These elements include differences in the physical and chemical properties of the CBR agent involved, the characteristics of how and where (i.e., urban versus rural area) the CBR agent was released or dispersed, weather conditions, and the environmental conditions. These elements also support waste sampling and characterization efforts. We also know from experience that many of the waste management planning elements are the same and be planned ahead of time, and then tailored to the site-specific conditions at the time of the HS incident. Figure 24.7 is an example of waste and material handling flow diagram utilized during Hurricane Katrina. A flow diagram like this is important for any HS incident to ensure waste management decisions are based on an understanding of the incident, the release, types of wastes generated, and the overall response approach (i.e., knowing the flow of your waste and material streams and matching them up with the best management practice or practices). It also gives a sense of the number of different waste streams you are likely to encounter during a HS incident. For HS incidents involving a CBR agent, one can assume that there would be similar materials that are contaminated with the agent involved and would need to be cleaned up, decontaminated, or designated as waste without any further treatment. In addition, the cleanup and decontamination processes utilized will also produce waste streams that would need to be managed as well.

In addition to the type of waste management facility needed, there is a need to determine where the facility exists and make appropriate plans

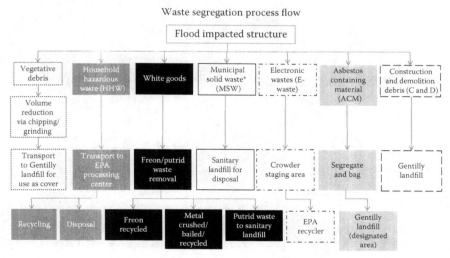

*Non-recoverable municipal solid waste includes tires and motorized devices. These items are segregated, but not currently picked up.

Figure 24.7 Waste flow diagram utilized after Hurricane Katrina. (CRS Report for Congress, 2008.)

with the facility to handle the waste streams expected from the incident. As can be seen from Figure 24.8, there are not many waste management facilities that can handle hazardous waste across the country. So, depending upon where the incident occurs, transportation may be a significant issue. In addition, each waste management facility has operating permits issued by the appropriate state waste management authority in which that facility is located. Each facility has certain waste acceptance criteria and profiling requirements that must be met before the waste may be shipped to it.

In addition to the hazardous waste management facilities in Figure 24.9, there are many nonhazardous waste management facilities across the country, which are also permitted and overseen by state and local waste management officials. The utilization of a nonhazardous waste management facility for a HS incident involving a CBR agent would require even more coordination, approval, and perhaps emergency authorizations from these state and local waste management officials.

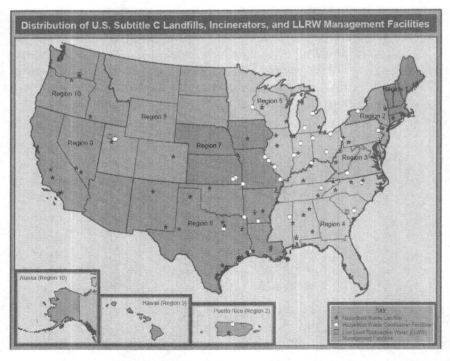

Figure 24.8 Hazardous waste landfills, incinerators, and low-level radioactive waste (LLRW) facilities in the United States. (From U.S. Environmental Protection Agency, Office of Homeland Security, Wide Area Recovery and Resiliency Program Waste Management Workshop Report, Denver, CO, March 15–16, 2012.)

24.8.3.4   *Assessment of waste management facilities for use during a homeland security incident*

A HS incident is a highly visible event that will draw intense attention from the public, media, and federal, state, local, and tribal waste management officials. As a result, it is important that a thorough vetting of potential waste management facilities be performed before recommended approval to the IC for use during a HS incident. This involves assessing potential waste management facilities against a number of different potential considerations, some of which are listed in Table 24.2.

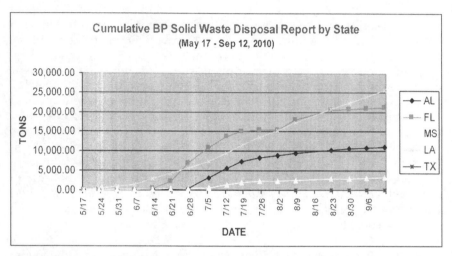

Figure 24.9 Solid waste disposal by state during the Deepwater Horizon oil spill. (From Michael, *Case Studies BP Oil Spill, WARRP - Waste Management Workshop,* 2012.)

Table 24.2 Considerations for Evaluation of Potential Waste Management Facilities

• Waste compatibility	• Insurance	• Site closure planning
• Facility capacity	• Safety record	• Adjacent properties
• Facility acceptance	• Community relations	• Storm-water controls
• Financial status	• Public acceptance	• Spill control plans
• Distance	• State concerns	• Drum/container storage
• Costs	• Environmental monitoring	• On-site laboratory
• Compliance history	• Environmental justice	• Record keeping

During the Deepwater Horizon oil spill of 2010, many waste management treatment, storage, or disposal facilities were evaluated in support of the amount and different types of wastes generated during that spill response. These waste management facilities were located across five states, which included facilities outside the impacted area because of the need for specialized facilities (e.g., deep well injection) to handle some of the waste streams.

24.8.3.5 *Transportation considerations*

The magnitude and scope of a HS incident will involve a large amount of transportation of different types of wastes in many different kinds of vehicles, vessels, and containers. The transportation of waste has its own considerations and requirements (e.g., Department of Transportation, State Department of Transportation) that require preplanning, logistics, and oversight. Transportation plans to support the WMP need to be robust enough to identify the appropriate types and amount of vehicles, vessels, or other transportation methods to handle the different types of waste based upon expected generation rates, points of generation, and transportation routes to waste management facilities that will accept the waste. A HS incident may involve the use of dump trucks, front-end loaders, pump trucks, all-terrain vehicles, aircraft, boats, skimmers, helicopters, oil tankers, private boats, railroads, and other vehicles. Sources of these vehicles should be identified in a WMP, and agreements for their use should be made in advance. The transportation plan may include the following considerations:

1. Type of vehicles to be utilized categorized by waste type, the media (i.e., land, water, and air), and areas (e.g., states, tribal, and ACP region) in which they will be operating
2. Waste hauler permit, placard, and documentation (e.g., EPA or state ID numbers, waste manifests, and waste profiles) by vehicle, waste type, and area of operation
3. Insurance requirements for vehicles; spill response plan for vehicles
4. Inter- and intrastate transportation routes
5. Identification of inspection requirements
6. Decontamination of vehicles during the response and at the end of the response

24.8.4 *Section D: Waste/material tracking and reporting*

The national waste management framework is based upon a "cradle to grave" system designed to ensure the proper management of wastes and the protection of public health and the environment. In addition, the IC set up in response to a HS incident also requires the tracking and documentation of waste management activities to keep the IC informed of the progress of cleanup and waste management operations. As a result, a waste/material tracking and reporting system is an important element of a WMP for HS incidents.

In addition, this waste is being managed over a very large geographic area, involving many state and local governments, as well as a number of waste management facilities. In order to assure the public, the media, and federal, state, local, and tribal officials that the waste from a HS incident is being managed appropriately, it is important to be able to demonstrate that a cradle to grave waste tracking and reporting system is in place. The importance of this capability was highlighted during the Deepwater Horizon oil spill of 2010, where BP was required to track and report waste and materials to inform the public, regulators, the media, and others about the status of waste management activities. This requires careful planning to ensure that there is someone identified with keeping the tracking system updated with accurate information, that there are data management and data quality control methods in place to ensure the accuracy of a reporting system, and that this system will provide decision makers at the federal, state, local, and tribal levels the information necessary to effectively manage and oversee a HS incident response. The considerations of a waste management tracking and reporting system should include where feasible

1. Reporting of waste sampling and analysis results by location
2. Uniform tracking of waste/materials of recovered product as well as liquid and solid wastes
3. The quantity and volume handled at each waste management location, including where the waste is temporarily (e.g., staging areas) or permanently (e.g., disposal site and recovery operation) located
4. Reporting on the status of waste management activities on a daily and cumulative basis for each type of waste/material
5. Online Web posting of flow charts showing the quantity and volume of how and where each category of waste type is being managed
6. Archived posting of the IC system reports (e.g., ICS 209, "Oil Spill Incident Status Summary")

There has also been a greater push for transparency across the government, including the emergency management arena. As a result, there has been a push to provide information out to the public as it becomes available. Doing this in the middle of an emergency takes quite a bit of up-front planning. During the Deepwater Horizon oil spill, there were hundreds of thousands of waste management data points that were developed and reported as a result of sampling, performance of site visits, and waste staged and stored, treated, and disposed of across five states and involving over 100 waste management facilities. Having a system in place to track and report that much information is complex. A tracking and reporting system should allow for "cradle to grave" management of waste from points of generation to staging/storage, treatment, recycling, and disposal. It also should allow for a visualization of the waste management process for those involved in the response as well as federal, state, local, and tribal waste management officials, the media, and the public. A waste management tracking and reporting system should provide the opportunity to track various aspects of the waste management process in as close to real time as possible (Figures 24.9 and 24.10).

24.8.4.1   *Data management associated with waste management activities*

Data management is an important element of the waste management planning process. In addition, because waste management requires a cradle to grave system of management, it generates large quantities of data by its very nature. This includes sampling and analysis data; staging and storage data; transportation tracking data; waste manifest data; waste acceptance data; chain of custody data; treatment and disposal data; health and safety data; site visit data; waste type; designation, and categorization data; waste shipment log data; waste weight ticket data; and so on. This amount of data can easily overwhelm a system that has not been thought out and sized appropriately ahead of an event. A particular concern for waste management data is to define common units of measurement (tons, cy^3, etc.) in which data will be reported to facilitate timely posting of data from different sources. The suggested content for a data management plan includes the following:

1. Description of the data generation, chain of custody, and management process
2. Record-keeping procedures/document control, data storage, and retrieval/security systems

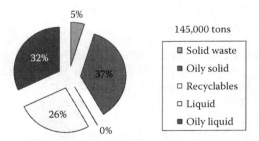

Figure 24.10 Waste generated by waste type during the Deepwater Horizon oil spill. (From Michael, *Case Studies BP Oil Spill, WARRP - Waste Management Workshop*, 2012.)

3. Identification of data handling equipment and procedures to process, compile, and analyze data
4. Discussion of data handling procedures to detect/correct errors and loss during data processing
5. Examples of any forms or checklists to be used
6. Identification of any specific computer hardware/software performance requirements and how configuration acceptance will be determined
7. Description of how resource management requirements will be satisfied
8. Identification of the individuals responsible for data validation, verification, and data posting
9. Identification of the time-frame requirement from data generation to data posting
10. Identification of website formats, reporting formats, frequency of reports, and archiving of data

24.8.5   Section E: Oversight and exit strategy

A HS incident may or may not involve a responsible party who would be designated as the waste generator and, therefore, be responsible for waste management activities. Where no responsible party exists, federal and/or state on-scene coordinators may actually be performing the waste management functions as part of the response. In either case, such large-scale waste management functions will likely involve many contracted waste management services, involving contracts and financial resources. Federal, state, local, tribal, and territorial waste management officials may find part of their response role being the oversight of waste management activities, contractors' performance, as well as cost accounting associated with these efforts. During the Deepwater Horizon Oil SONS of 2010, federal and state regulatory agencies became involved in many different types of waste management oversight activities. These activities included

1. Review and approval of the responsible party's WMPs
2. Review and approval of the proposed waste management facilities
3. Sampling and analysis of waste streams
4. Site visits/inspection of facilities/sites (staging/storage, decontamination, landfills, etc.)
5. Reporting and posting of waste sampling results on an oil spill website

These waste management oversight activities required the development of plans in support of these activities, taking into account quality assurance and data management considerations, development of forms for the information gathered, and reporting formats to make the information collected publicaly available. States have authorized waste management programs and already have some waste management oversight activities in place that occur on a routine basis. Although the waste management activities described earlier serve as an example of waste management oversight functions, any waste management operation may have over-sight needs during SONS. These plans should be developed prior to a HS incident or SONS and made part of the WMP. The reporting of results was often linked to each of the impacted states' spill response websites so that there was consistency in reporting these activities. The EPA websites also allowed for users to download any specific waste management sampling results conducted by EPA.

Because waste management activities associated with a HS incident or SONS can go on for an extended period of time (e.g., >18 months for the Deepwater Horizon Oil SONS), it is also important to plan an exit strategy ahead of time for the waste management oversight activities. Basically, this is a transition plan from the increased oversight activities associated with the spill back to the routine oversight activities conducted by state agencies. The waste management oversight exit strategy should address the current oversight activities performed, who performs them and the frequency, and then the proposed transition of these activities back to their normal functions prior to the SONS. Typically, this involves a transi-tion of all activities back to the state in which these facilities are located and overseen on a normal basis.

24.8.6   Section F: Community outreach and communications

Community outreach and communications are a key part of a HS incident response for effective waste management operations during a HS incident. There are cases where effective response actions have been delayed due to poor community outreach and engagement. This is magnified when there is a HS incident that draws national attention from federal, state, local and tribal officials, the media, and the public. Waste management–related community outreach issues may include environmental justice, inform-ing the public about waste management operations, having processes in place allowing the public to communicate concerns about waste manage-ment operations, and addressing those concerns. Although a responsible party can be expected to conduct some community outreach and commu-nications, all parties involved can expect to be involved in these activities and should have plans in place ahead of time that address these needs.

The EPA has decades of experience in community relations associ-ated with hazardous waste cleanups around the county. This experience has shown the importance of this function to the overall success of a cleanup. The same is true for a HS incident with the only difference being that the time frame is greatly condensed, the number of people involved is much greater, and the need to be able to respond to requests for information from officials, the media, and public is tremendous. Having processes and plans in place to handle this magnitude of effort is essential to the success of a response. This waste management community outreach plan should apply to all impacted states and include the following objectives:

1. Identify and assess individual community information needs related to waste management operations and respond appropriately with relevant information and actions
2. Work with local elected/community leaders to seek their support and participation in community outreach efforts related to waste management operations
3. Identify stakeholders who need to be kept informed and engaged on an ongoing basis during waste management operations
4. Develop accurate and timely waste management information to be disseminated to the community in a variety of methods and languages, including print, website, and public venues
5. Identify the right tools and/or personnel needed to reach out to key stakeholders
6. Implement appropriate processes to respond in a timely manner to any potential concerns or complaints from the communities affected in each state about waste management operations

During its waste management operations during the Deepwater Horizon SONS of 2010, BP reached out to more than 20,000 people using town hall meetings, beach information booths, briefings to emergency operations centers, as well as briefings to governors.

It is recommended that a waste management communications plan be developed jointly with the public information officer and be documented within the WMP. Past HS incidents have shown that federal, state, local, and tribal waste management officials spend thousands of hours in emergency operations centers, IC posts, joint field offices, public meetings, and other venues responding to numerous requests for information regarding waste management issues. The magnitude of this effort requires a plan for communications ahead of time to ensure proper coordination and flow of timely and accurate information. Communication procedures and processes should be developed, documented, and demonstrated to work during training and exercises to ensure that they will work in a real incident. In addition, we would recommend developing frequently asked questions and answers for your specific WMP. Fact sheets that address specific waste streams, waste facilities, or waste management operations are another effective communications tool. These would be useful to have prescripted and coordinated ahead of time and attached to the waste management communications plan.

24.8.7   Section G: Health and safety considerations

Health and safety take on an increased importance during waste management operations involving a HS incident due to the amount of waste generated, the potential hazards from unique CBR agents, as well as dispersants, detergents, disinfectants, fumigants, and degreasers, the length of time of the response, the large geographic area, and the large number or facilities, contractors, and people involved. In addition, response actions conducted under the NCP must comply with the provisions of the Occupational Health and Safety Administration's (OSHA) Hazardous Waste Operations and Emergency Response (HAZWOPER) Standard, under 29 CFR 1910.120 and 1926.65. These health and safety provisions apply even though some of the oil waste may not be considered federal hazardous waste under the RCRA.

OSHA's HAZWOPER regulations require employers to have a detailed HASP to protect workers involved in cleanup operations in accordance with the NCP. As a result, a waste management HASP needs to be prepared that addresses the potential hazards from handling waste that is contaminated with CBR threat agents or byproducts of them. The OSHA goals include ensuring that these workers receive appropriate training and protective equipment for the potential hazards and exposures that exist during waste management operations. A HASP for waste management operations should consider the following items:

1. Situational overview of the HS incident
2. Key personnel involved and contact information
3. Hazard assessment of all waste management operations and potential exposures
4. Training requirements associated with waste management operations
5. Personal protective equipment requirements associated with waste management operations
6. Monitoring and management of temperature extremes, medical surveillance requirements
7. Exposure monitoring and air sampling
8. Site safety control measures, confined space entry
9. Decontamination and waste management operations, spill containment requirements
10. Emergency response/contingency plan, documentation requirements
11. Material safety data sheets for chemicals utilized during waste management operations
12. HASP coordination and approval

24.9   Putting it all together: "A system of systems" all-hazards waste management planning approach

In the end, it takes a "system of systems" all-hazards waste management planning approach (Figure 24.11) to effectively plan for the complexity and integration of people, processes, and systems involved with a HS incident. The NPSs developed by DHS identify scenarios involving CBR threat agents under various release scenarios, including wide area releases. Waste management planning involves a multihazard, multistakeholder, multidisciplinary, multimedia, multiscale, and multiobjective decision-making process. Waste management frequently involves social, economic, environmental, political, regulatory/statutory, policy, and technical considerations in making sound, safe, cost-effective, risk-informed, and consensus-orientated decisions. The need for this planning to be done prior to an event is critical if there is to be a timely and cost-effective response to a HS incident. This document presents a systems approach to do so and to document this within a WMP so that when the incident occurs we are *truly ready to respond.*

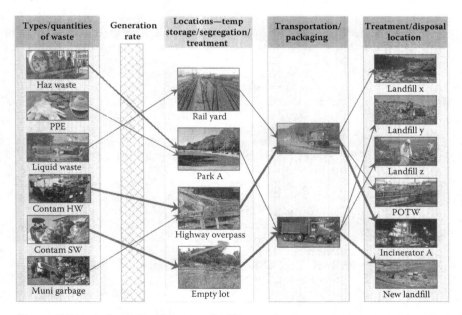

Figure 24.11 A "system of systems" all-hazards waste management planning approach. (From Parrish, *Waste Management Workshop.* EPA Office of Homeland Security, 2012.)

Acknowledgments and disclaimer

The following individuals made valuable contributions to this chapter for which the author is grateful: *James Michael*, Chief, Waste Characterization Branch, Materials Recovery and Waste Management Division, Office of Resource Conservation and Recovery, US EPA; *Melissa Kaps*, Office of Resource Conservation and Recovery, US EPA; *Anna Tschursin*, Office of Resource Conservation and Recovery, US EPA; *Paul M. Lemieux*, Associate Division Director, Decontamination and Consequence Management Division, National HS Research Center, Office of Research and Development, US EPA; *Cayce Parrish*, Office of HS, US EPA.

Reference herein to any specific commercial products, process, or service by trade name, trademark, manufacturer, or otherwise, does not necessarily constitute or imply its endorsement, recommendation, or favoring by the United States Government. The views and opinions of authors expressed herein do not necessarily state or reflect those of the United States Government and shall not be used for advertising or product endorsement purposes.

References

Chang, N. (2003). Planning, Design, and Management of Hazardous Waste Disposal by Environmental Systems Engineering Approach. *ASCE Pactical Periodical of Hazardous, Toxic, and Radioactive Waste Management.* 7(2), 67–67.

CRS Report for Congress. (2008). Disaster Debris Removal After Hurricane Katrina: Status and Associated Issues. Order Code RL33477.

Covey, S. R. (1989). *Seven Habits of Highly Effective People.* New York: Free Press.

Department of Homeland Security. (2008, January). *National Response Framework.* Retrieved May 23, 2012, from National Response Framework Resource Center: http://www.fema.gov/NRF.

EPA National Homeland Security Research Center. (2012). *Incident Waste Assessment System and Tonnage Estimator.* Retrieved February 2012, from I-Waste: http://www2.ergweb.com/bdrtool/login.asp.

EPA. (1980). *National Oil & Hazadous Substances Pollution Contingency Plan (NCP).* EPA.

EPA Office of Homeland Security. (2012). WARRP Waste Management Workshop, March 15–16, 2012. Denver, Colorado: EPA Office of Homeland Security.

EPA. (2012). *Waste Management Hierarchy.* Retrieved February 2012, from Waste Management for Homeland Security Incidents: http://www.epa.gov/osw/homeland/options.htm.

Ierardi, M. E. (2010). EPA Office of Resource Conservation and Recovery, Homeland Security Team Leader.

Ierardi, M. E. (2012). EPA Office of Resource Conservation and Recovery, Homeland Security Team Leader.

Michael, J. (2012, March). *Case Studies BP Oil Spill, WARRP—Waste Management Workshop.* Retrieved from http://www.warrp.org/events/event_info/2012_wgms/waste_management/docs/march_15/11%20michael_WARRP_Case%20Studies_JM_BP%20Oil_v5.pdf.

National Commission. (2011). *Deep Water, The Gulf Oil Disaster and the Future of Offshore Drilling*. National Commission on the BP Deepwater Horizon Oil Spill and Offshore Drilling. Retrieved from http://www.oilspillcommission .gov/final-report.

Parrish, C. (2012, March). *Waste Management Workshop*. EPA Office of Homeland Security. Retrieved from http://www.epa.gov/wastes/homeland/docs/ warrp_report.pdf.

Ramo S. and Robin K. St.Clair. (1998). *The Systems Approach: Fresh Solutions to Complex Problems Through Combining Science and Practical Common Sense*. Anaheim, CA: KNI, Incorporated.

Wikipedia. (2012). *Definition of Waste Management*. Retrived From http://en .wikipedia.org/wiki/Waste_management.

chapter twenty-five

Creating effective response communications*

Tay W. Johannes

Contents

This chapter describes crisis and emergency response communication considerations and offers guidelines for strategy, details elements of response communications systems, and discusses the effects of social networks and new media. Its details and explanations are not meant to be prescriptive, because various situations and organizational factors introduce too much variability. The intent here is to establish a baseline of ideas and concepts necessary to tailor an organization's plan for communicating in a response.

An emergency or a crisis requires timely and accurate communication with first responders, government officials, the public, and the media. Organizational emergencies or crises range from fire alarms to earthquakes or from armed robbery to terrorism. Management and control of these incidents require a sound communications approach to ensure the best possible outcome.

* The material contained in this chapter was prepared as an account of work sponsored by an agency of the United States Government. Neither the United States Government nor any agent thereof, nor any of their employees, makes any warranty, express or implied, or assumes any legal liability or responsibility for the accuracy, completeness, or usefulness of any information, apparatus, product, or process disclosed, or represents that its use would not infringe upon privately owned rights. Reference herein to any specific commercial product, process, or service by trade name, trademark, manufacturer, or otherwise does not necessarily constitute or imply its endorsement, recommendation, or favoring by the United States Government or any agency thereof. The views and opinions of the author expressed herein do not necessarily state or reflect those of the United States Government or any agency thereof.

25.1   Strategic elements of response communication

The basis for an effective response communications strategy depends on the following five factors:

1. Stakeholders
2. Organizational leadership
3. Planning
4. Situational awareness
5. Common operating picture

Response communications process incoming and outgoing information to stakeholders who are either information producers or consumers or both. In an emergency response situation, the emergency manager must consider the context in terms of implications, which may be subject to diverse and possibly competing needs. Basic stakeholders in emergency response include the following:

- First responders—fire, police, and emergency medical
- Survivors—those directly affected or impacted by specific event
- Public officials—department heads, mayors, councils, and other elected officials
- Volunteer groups—American Red Cross, Salvation Army, and so on
- Business community—typically service providers (food or lodging) and public utilities
- Media—news services (strongest link to general public)
- General public—all others (especially vulnerable population groups such as elderly and non-English speaking) (Haddow et al. 2011, p. 139)

Committed organizational leadership is the key to successful response communications both internally and externally. The organizational leader provides vision, focus, and endorsement for establishing, expanding, and adjusting lines of communication with organization staff, partners, and the public. A successful response requires that the leader is effective and decisive not only in the response but also in the planning and preparedness leading to a response. In fact, communication planning may be the most important part of the leadership commitment to the organization.

The leadership must inculcate several critical elements in emergency planning and response for successful communications operations. Factors to be considered are integration, training, a crisis information center, and hardware and infrastructure systems. These factors span the communications spectrum for all phases of emergency management and require thorough communications planning to be effective.

Beginning with integration, having comprehensive, written operating procedures and policies to capture organizational goals and objectives is an imperative. This information will assist the organization with timely and accurate interfaces with internal and external entities before, during, and after a crisis. Risks of confusion and miscommunication will be much higher when procedures and policies are not standardized. Training also becomes more difficult without standardized training objectives and guidance.

Consideration for training in emergency response communication would include exercises, drills, and tabletop discussions. Because experience is fundamental to building capability for an organization and its responders, the organization needs to cultivate opportunities for experiential growth in emergency response. Training and exercises fulfill the need for experience in a controlled setting, but only if it is valid and realistic. Additionally, training should focus on realistic communication issues that would drive the need for redundant methods of delivery.

The organization should designate an emergency information center; this may be integral to the emergency operations center. A distinction to be made is that an emergency information center makes information available to all other entities that are not directly involved in response operations. Fundamentally, the information center's primary mission would be to provide information support focused on response operations. Redundant forms of voice, data, and audio and video information systems to the extent that the organization is able to support would allow for unimpeded incoming and outgoing information. Telephones, computers, teleconferencing, whiteboards (or smart boards), and television with public broadcast reception are essential information tools and systems.

The fundamental reason for the existence of communications hardware and infrastructure is to provide flexible, dispersed access to critical response data and information. The organization should evaluate the mechanisms for creating redundant linkages to information that can accommodate multiple users and responders with access to the most accurate and useful data as fast as possible. Additionally, the organization's evaluation should consider current inventory lists of available information systems and resources. Finally, information itself (especially regarding the response) should be backed up in remote locations in the event that the crisis or emergency creates a disruption of information and communications systems.

Advances in information technology create a dynamic situation in the hardware and infrastructure capabilities that challenge organizational leadership with investment and resource allocation decisions. The key to keeping abreast of advancing technology is incorporating interoperability into all planning considerations. Homeland Security's program, called SAFECOM, describes the elements of interoperability for public safety service and support providers as the ability to communicate via voice and data on demand, in real time, when needed, and when authorized (Department of Homeland Security, n.d.). Advances in cell phone technology and new fourth-generation (4G) networks offer flexibility and interoperability for organizational response efforts. Cell phone technology may potentially transform cell phone users into remote sensors capable of transmitting informative situational information to central locations. Unfortunately, many types of hazards affect cell phone operations and drive the need for backup communications.

Situational awareness has been defined in many different ways. First identified as a U.S. Air Force concept for fighter pilots, courtesy of Colonel John Boyd's "observe-orient-decide-act loop" or OODA loop, situational awareness is defined as making sense of what's happening and, based on that analysis, making predictions about what will happen (as referenced in von Lubitz 2008, p. 562). In a more general context, situational awareness by Human Factors practitioners and researchers in the Enhanced Safety through Situational Awareness Integration in training (ESSAI) consortium is

> the perception of the elements in the environment
> within a volume of time and space, the comprehen-
> sion of their meaning, and the projection of their
> status in the near future. (Endsley 1988, p. 98)

The National Response Framework (NRF) describes the response process for gaining and maintaining situational awareness. This action enables follow-on actions of activating and deploying key resources and capabilities, coordinating response actions, and then demobilizing. One of the baseline priorities for achieving situational awareness hinges on the public and private sectors sharing information to develop a common operating picture. Standardized reports and documents offer a consistency in shared, critical information—a necessary element to enhance situational awareness.

A lessons-learned report by the White House on the national response to Hurricane Katrina made special mention of situational awareness in the findings. The White House committee found that inadequate situational awareness during the response resulted in decision makers using incorrect and incomplete information (Townsend 2006, pp. 97–98). The major contributing factor of the inadequate situational awareness was the impacted communications infrastructure. Communications failures resulting from Hurricane Katrina included

- Communication facilities and backup power supply system's destruction
- Over 3 million phone lines down in Louisiana, Mississippi, and Alabama
- Thirty-eight 911 call centers down
- Over 2000 wireless network cell sites offline
- 37 of 41 radio stations off-the-air in New Orleans (Townsend 2006, pp. 5–10)

Impaired response, command, control, and situational awareness were all cited as the results of the massive communication damage and the failure to adequately plan for alternatives (Townsend 2006, p. 3). This indicates a strong need for alternative methods of creating situational awareness and drives an assumption of major losses in communication infrastructure.

The common operating picture is a key concept and the principle is described in the National Incident Management System (NIMS), which calls it an overview of an incident made through collecting and organizing various types of information from various responsible sources to support decision making. This facilitates consistent information regarding the incident for on-scene responders and off-scene command and support personnel including asset availability, location, and status of requests. The development of a common operating picture and common situational awareness for first responders is considered a key issue by the National Academy of Science in addressing the issue of providing the most effective assistance to those affected by disasters (National Academy of Science 2007, pp. 49, 50). The common operating picture is listed first under NIMS concepts and principles. Intentionally or not, this apparent preeminence of common operating picture is corroborated in the NRF:

> The continued development and rapid integration at the Federal, State, tribal, and local levels of electronic reporting and information-sharing tools supporting the [National Operations Center] common operating picture is a very high priority of the [NRF]. (NRF 2008, pp. 55, 56)

A common operating picture will be shared across many different functional perspectives. The NRF identifies, defines, and elaborates on 15 emergency support functions, each with a separate annex that includes purpose, capability, members, and operational concepts. The emergency support functions are identified in Table 25.1.

The common operating picture must also be able to address information needs that directly result from the disaster response. Incident management under NIMS offers some general response-based questions such as what resource and how much is needed; where and when is this resource needed; and who gets it. At closer inspection of the common operating picture, many different facets begin to appear. The level of detail needed varies due to specific mission tasks, type of event, location, phase of response, demographics, time period, and so forth. Although the common operating picture serves the response system's overall unity of effort, variation in detail focus is apparent across various information users.

Table 25.1 Emergency Support Functions (ESFs) of the National Response
Framework

ESF#1 Transportation	ESF#2 Communications	ESF#3 Public Works and Engineering	ESF#4 Firefighting
ESF#5 Emergency Management	ESF#6 Mass Care, Emergency Assist., Housing and Human Services	ESF#7 Logistics Management and Resource Support	ESF#8 Public Health and Medical Services
ESF#9 Search and Rescue	ESF#10 Oil and HazMat Response	ESF#11 Agriculture and Natural Resources	ESF#12 Energy
ESF#13 Public Safety and Security	ESF#14[a] Long-Term Community Recovery	ESF#15 External Affairs	

[a] Superseded by National Disaster Recovery Framework (FEMA 2011).

25.2   Elements of response communication systems

Successful emergency responses require communication systems that are adaptable to a wide variety of situations. These communication systems must convey messages that are timely, accurate, routed properly, and in a format that ensures clarity and efficient understanding. The most effective communications systems in response are characterized by five basic elements:

1. Communications team
2. Communications plan
3. Information exchange
4. Training and exercise
5. Communications center (Haddow et al. 2011, pp. 151–152)

Orchestrating a balance of these elements will produce a capable communications system.

The communications team is made up of select individuals oriented to higher organizational perspectives relative to the organization's vision and strategy. Ideally, the team members would include senior representatives vested with some decision authorities. Where possible, the most senior personnel (along with their enabling authorities) would come from the following functions:

- Corporate communications (such as a chief information officer [CIO])
- Public affairs
- Operations
- Legal
- Security and/or safety
- Human resources
- IT

The team leader (typically the CIO) would assemble the team for meetings and act as the conduit to the senior organizational leader. During a response, the team forms the core of the strategic response capability. One of the most important functions is developing the emergency communications plan.

When considering the various hazards and threats to an organization, it is almost impossible to establish the range of events and predict how and when they will occur. Regardless, an emergency communications plan must focus on delivering information to stakeholders and identifying the mechanisms of transfer. As the team builds this plan, the organization's strategic objectives and vision play a leading role in establishing policy, identifying stakeholders, and characterizing desired outcomes. The communications planning process tracks the flow of information from sources to end-users and how the information evolves, changes, and is transmitted. The team must maintain a current list of contacts that are sources and consumers of information including the media and key personnel within the organization. The emergency communications team should keep this list readily available for emergency responses.

For effective planning, the planners should address common communications issues in emergency response. Five typical issues to address are as follows:

1. Loss of primary communications networks: direct and indirect (i.e., sometimes the network operation is knocked out due to power, physical damage, etc.). Direct loss of communications is often due to overload. Responses improve with more realistic exercises and training that account for overload.
2. Failure to plan and operate alternatives for such outages. In a scenario where a disaster destroys a communication tower, it impacts not just a single system, but rather multiple systems such as landline phones (microwave links), cell phones, and radios all at once. Various communication system components tend to cluster (communication towers,

duct banks, and other network nodes) as a method for reducing operating and maintenance costs.

3. In organizations that use handheld communication devices (i.e., radios) in emergency response, loss of communications discipline—open microphones, radio discipline, "walking" over other transmissions, proximity of transmitters causing interference, microphone too close, yelling causing overdrive conditions, slurring, and so on.

4. Ignorance of radio signal propagation (low-frequency signals, which propagate through a ground wave, will not travel very far), ignorance of distance/interference issues, and so on.

5. Interoperability issues with external equipment resources (including things such as different encryption and different frequencies) (United States Council of Mayors 2004).

Sources of information are critically important for the emergency communications team and must be credible and resilient. Sources will not necessarily present consistent information, and the communications team will need to establish working relationships with each source before an emergency occurs. Familiarity with the sources not only improves awareness of their point of view but also enables the team to anticipate potential organizational biases. The emergency communications team must also be prepared to identify and work with new sources of information that emerge after an emergency occurs.

With several sources of information, the team must establish information procedures and formats that will effectively contribute to the organization's response efforts and decision making. The team must standardize the procedures to ensure a proper understanding of the type, content, and timing of information needed. In terms of media where interaction might be limited or not possible, the team should prepare in advance a process that screens or selects relevant sources depending on the situation.

In addition to working with sources, the emergency management team should work with information customers. Information leaving the organization should be reviewed by the emergency communications team for consistency with organizational goals and objectives with respect to the recipient of the information. This is best worked during planning and not during an emergency. Again format, content, and timing are key considerations. Various recipients would include the general public, organization personnel, emergency first responders, government officials, and so on.

Effective response communications require well-trained personnel throughout the system in the processes that generate, transmit, and receive information. Training exercises, using various emergency-related events, are practical and effective in testing emergency communication plans and ensuring people know and understand the information flow process. The emergency communications team should review the following major developments in a crisis situation:

- *Assessment:* A quick initial assessment is critical in knowing the what, when, and where of an event. Collect basic facts such as whether the event is over, developing, or ongoing and its severity to initiate emergency response actions.
- *Identify key groups:* In a given situation, various groups will emerge in a hierarchy of response involvement. Distinguish information priorities and potentially conflicting needs.
- *Course of action development:* Determine warning and/or emergency notification content with respect to mission and anticipated response. In addition to content, method of delivery must also consider the specifics of the event and the imposed constraints.
- *Evaluation:* Determine response success by recording how well the communication system met short-term and long-term needs of key groups. Immediate feedback allows the system to adapt to dynamic conditions. Overall feedback factors into systemic improvements in planning and training.

Ideally, an emergency communications center should be close to the organization's emergency operations center. The center should be equipped to support internal and external information flow including considerations for redundancy based on system priorities and the nature of the emergency event. As determined by the various threat scenarios, the center should have backups both physical and informational backups.

25.3   Social networks and new media

Growth and capability in new communication methods such as social media and social networks have dramatically impacted public access to information. In crisis and emergency management communities, past and often ardent resistance is giving way to mainstream incorporation into communication plans and operations. Connectivity and versatility continue to increase the widespread adoption along with the revolutionary aspects of personal/private interaction brought on by social media. The latest estimates for large applications such as Facebook and Twitter are at 750 million and 100 million users, respectively; 230 million tweets are sent per day (Serrano 2011). The public's reflexive response in emergencies now is to grab their smart phone and hit video record.

Gerald Baron, a commentator on the website of *Emergency Management* magazine, wrote a piece entitled "Five Ways Social Media is Changing

Emergency Management" that summarizes several of the main impacts this media has on the emergency management profession:

1. How we get our news
2. How we start and do business
3. How we meet and stay in touch with people
4. What we reveal
5. What we can influence (Baron 2009)

Transportation and movement are problematic during disaster response, but technology available today on smart phones allows for mobile and dispersed operations. Owing to today's higher expectations for nationwide connectivity, communication delays with stakeholders and those directly controlling or influencing the response operations are not tolerated for very long. Transparency and the official (or unofficial) release of information, coupled with national-level commitment for open government, have led to an expectation of immediate and almost instantaneous release of good or bad information before being preempted by other external sources. Social media have influenced emergency response by truncating the span of control for communication. This is an age of unprecedented access and ability to inform in real time through the manipulation of new media capabilities.

Across the Arab world in 2011 (the "Arab Spring") and continuing into 2012, even strict government sanctions could not control social media content or span of influence. "Facebook has served as a medium to begin civil war, Twitter has been a means for people lost in a massive disaster to let their family know they survived, and YouTube has been a venue to show the entire world scenes that you could have only witnessed individually before" (Becker 2011). New technology has also allowed millions of people to be receptors and feed current information into information centers. These "first informers" armed with photo- and video-enabled cell phones capture and transfer events as they occur. An example of a poorly controlled scene in Ohio in 2011 was the killing of the exotic animals by the local authorities in Zanesville. The incident led to major public outcry that distracted responders in an already stressed situation with limited resources (Jarman et al. 2011).

The U.S. government is increasingly becoming more engaged with social media by integrating various social media tools into website and communications strategies. Through the Federal Emergency Management Agency's (FEMA) Office of External Affairs, the government is establishing direct communication links with stakeholders, the public, and news centers through online interfaces to enhance outreach objectives. Since June 2008, FEMA has been incorporating social media into its communications plans to improve transparency in response and recovery operations along with opening channels for receiving unofficial data and information inputs.

The news media have established an avenue to receive and manage incoming knowledge from first informers. The next step is for emergency managers to follow suit. Currently, 911 call centers are not equipped to be able to receive short message service (SMS) capabilities despite the fact that often during an emergency the data-intensive phone network is overloaded and an SMS-format text message may be the only viable option.

Some researchers are looking to capture more relevant information and piece it together to maximize situational awareness. In a project called "Tweak the Tweet," researchers are attempting to explore historical records of tweets and text messages to look back at past events such as the Virginia Tech shooting, the tsunami in Japan, and the earthquake in Haiti (University of Colorado-Boulder 2009b). Hidden behind the tweets is metadata that identify a person, the time, and their location (Pisano-Pedigo 2011). If this information were to be captured, it could form a powerful collaborative tool for situational awareness. Tweak the Tweet is a project within a larger project known as Project EPIC (Empowering the Public with Information in Crisis). Project EPIC, started in September 2009, was funded by the National Science Foundation. It is a "multi-disciplinary, multi-university, multi-lingual research effort to support the information needs by members of the public during times of mass emergency" (University of Colorado-Boulder 2009a). It is a group of students and faculty from the University of Colorado, Boulder and University of California, Irvine. In the age of social media, the project brings "behavioral and technical knowledge of computer mediated communication to the world of crisis studies and emergency response ... studying massive wide scale coordination across the internet, and conducting action research and employing participatory design oriented approaches" (Project EPIC 2009). Looking past current technology to anticipate future socio-technical changes with broader and expanded sources of information is one of the main objectives.

IBM (2011) has developed an artificial intelligence machine, Watson, which has been suggested to help the medical field assess patients' symptoms and search all available medical information to conclude the most likely cause. In a similar way, the machine might be able to process available communication patterns and threats and aid an emergency manager, or even publish public service warnings. As another example, Facebook's relevancy engines recognize common events that people post such as a person's birthday, Christmas, news events, and then group them by topic for expedient review. This technology could be harnessed to focus on events such as disasters and feed information into an emergency response center.

The YouTube platform offers shared access to host video materials to support response objectives. During the response, this platform could offer directions for the public in explaining missions and locations of services and resources.

As social media grows in popularity and demand for resources, this new communication channel presents an additional risk for responders—a dependency on information networks that may not reach all sections of society. Of particular concern are vulnerable and disadvantaged segments of the population who are uninformed or unable to access social media. In the same way that past considerations have been made for the deaf and blind, new considerations arise with each new method or capability. As an example, the homeless are not likely to be most effectively informed of critical emergency messages via social media interaction if they do not have regular access to a computer or portable data device such as a smart phone.

25.4   Conclusion

Creating effective communication in emergency response takes concerted effort before, during, and after a crisis. Organizations will need to commit substantial resources (primarily intellectual) to planning, assessing, and testing their response system. The external interfaces necessary to quickly progress into the recovery phase demand flexibility offered through parallel lines of communication. The resulting synergies of robust response communications offer great potential for organizations to sustain operations and increase resiliency.

Technology advances continue to change and shape emergency response. The demand placed on communications systems drives a need for agile and innovative planning within the organization. The interface with multiple, external parties with competing priorities and conflicting areas of responsibility challenges an organization's ability to pierce the fog and friction of an emergency or a crisis. Integrating new media forms and incorporating new technology to leverage evolving networks helps organizations meet response objectives. Finally, response success requires skill and acumen to overcome inherent difficulties.

References

Baron, G. 2009. Five Ways Social Media is Changing Emergency Management. http://www.emergencymgmt.com/emergency-blogs/crisis-comm/Five-Ways-Social-Media.html (accessed on November 8, 2011).

Becker, D. 2011. Social Media: A Responsibility and Habit in Emergency Management Communications. *The CUSEC Journal*, 15(3), 1–3. http://www.cusec.org/publications/cusec-newsletter.html (accessed on March 28, 2012).

Department of Homeland Security. n.d. SAFECOM Interoperability. http://www.safecomprogram.gov/interoperability/Default.aspx (accessed on April 3, 2012).

Endsley, M.R. 1988. Design and Evaluations for Situational Awareness Enhancement. *Proceedings of the Human Factors Society 32nd Annual Meeting*, 1, 97–101. Santa Monica, CA: Human Factors Society.

Haddow, G.D., J.A. Bullock, and D.P. Coppola. 2011. *Introduction to Emergency Management, 4th Ed.*, Burlington, MA: Elsevier Butterworth-Heinemann, 132–152.

IBM. 2011. *IBM Watson.* http://www-03.ibm.com/innovation/us/watson/? cn=agus_watson-20100712&cm=k&csr=google&cr=jeopardy_ watson&ct=USJWK002&S_TACT=USJWK002&ck=jeopardy_watson&cmp= 00000&mkwid=sRONsUpeR_15714878373_432n0d3749 (accessed on December 12, 2011).

Jarman, J., Q. Truong, J. Woods, and B. Jackson. 2011. Sheriff: 56 Exotic Animals Escaped from Farm Near Zanesville; 49 Killed by Authorities. http://www .dispatch.com/content/stories/local/2011/10/18/Wild-animals-loose-in-Muskingum-County.html (accessed on March 28, 2012).

National Academy of Sciences. 2007. *Successful Response Starts with a Map: Improving Geospatial Support for Disaster Management.* Washington, DC: National Academies Press. http://www.nap.edu/catalog/11793.html (accessed on September 29, 2011).

National Disaster Recovery Framework. 2011. http://www.fema.gov/pdf/ recoveryframework/ndrf.pdf (accessed on March 28, 2012).

National Response Framework. 2008. http://www.fema.gov/pdf/emergency/ nrf/nrf-core.pdf (accessed on March 24, 2012).

Pisano-Pedigo, L. 2011. Social Media and Disaster Communication. *Regional Interagency Steering Committee.* http://www.fema.gov/pdf/about/regions/ regionviii/risc_0711.pdf (accessed on March 28, 2012).

Serrano, A.F. 2011. The Social Media Explosion: By the Numbers. http://www .thefiscaltimes.com/Articles/2011/09/12/The-Social-Media-Explosion-By-the-Numbers.aspx#page1 (accessed on March 19, 2012).

Townsend, F.F. 2006. *The Federal Response to Hurricane Katrina: Lessons Learned.* Washington, DC: Office of the Assistant to the President for Homeland Security and Counterterrorism. http://georgewbush-whitehouse.archives. gov/reports/katrina-lessons-learned/ (accessed on March 27, 2012).

United States Council of Mayors. 2004. *Interoperability Survey: A 192-City Survey.* http://www.usmayors.org/72ndannualmeeting/interoperability-report_062804.pdf (accessed on April 3, 2012).

University of Colorado-Boulder. 2009a. Project EPIC. http://epic.cs.colorado. edu/ (accessed on December 12, 2011).

University of Colorado-Boulder. 2009b. Tweak the Tweet Earns CU Graduate Student Second Place in National Technology Competition. http://www .colorado.edu/news/releases/2009/12/03/tweak-tweet-earns-cu-graduate-student-second-place-national-technology (accessed on March 20, 2012).

von Lubitz, Dag K.J.E. 2008. Medical Readiness for Operations other than War: Boyd's OODA Loop and Training Using Advanced Distributed Simulation Technology. *International Journal of Risk Assessment and Management*, 9(4), 409–432.

chapter twenty-six

Assessing the state of knowledge about emergency management in other countries

David A. McEntire

Contents

26.1 Introduction

Disaster scholars and practitioners have long recognized the lack of knowledge about disasters at the international level and the need for additional studies about emergency management in other countries. Two well-known disaster researchers, Thomas Drabek (1986) and Russell Dynes (1988), pointed these weaknesses out over 25 years ago. Despite increased attention during the United Nations International Decade for Natural Disaster Reduction, the call for such studies was repeated in the late 1990s (Peacock, 1997; McEntire, 1997). Even in recent years, the request for additional studies from a global and/or comparative perspective has not diminished (Britton, 2006; National Academy of Sciences, 2006; McEntire, 2007).

While progress has undoubtedly been made (see, e.g., Alexander, 2006; Coppola, 2007), there is still ample room for improvement in our understanding of disasters at the global level. Fortunately, academia and practicing emergency planners are responding to the exigency and are making important contributions in this regard. For this reason, it is essential to continually develop new knowledge, assess what we have learned, and take stock of future needs in disaster management.

This chapter provides background about the Comparative Emergency Management Project initiated by the author for the Federal Emergency Management Agency (FEMA) Higher Education Program. It identifies key propositions emanating from this work and discusses the implications of such findings for the future. The goal of this chapter, therefore, is to better understand the challenges nations face and how they might be best overcome as we move forward in our efforts to successfully manage disasters.

26.2   Propositions from the Comparative Emergency Management Project

In order to support professors who teach emergency management, the FEMA Higher Education Program has been actively involved in the development of educational materials. Many of these resources are instructor guides and electronic books that cover various course subjects that are of interest to faculty and students (FEMA, 2011). One of these texts is the Comparative Emergency Management Project, and the findings are summarized here.

At least 10 propositions about global and comparative disasters may be gleaned from the FEMA Higher Education Comparative Emergency Management Project. Some of these relate to hazards, vulnerability, and disasters while others deal with laws, organization, and activity in emergency management. All of the propositions may help advance comprehension of disasters around the world and facilitate in some small way the improvement of emergency management domestically or abroad.

26.2.1   Nations face severe risks

Research from the Comparative Emergency Management Project confirms that the world is indeed a hazardous place. For instance:

- In 2005, the World Bank report entitled *Natural Disaster Hot Spots—A Global Risk Analysis* indicated that "Taiwan might be the most vulnerable [country] to natural hazards on Earth, with 73% of land and population exposed to three or more hazards" (McEntire and Tso, 2011).
- "Costa Rica ranks second in the world among countries most vulnerable to hazards based on land area, with 36.8 percent of the total area exposed to three or more adverse natural events" (Afedzie et al., 2011).
- In Turkey, "92% of the country's 780,580 km² is prone to earthquakes" due to the Anatolian faults (Ural, 2011).
- Sir Geoffrey Palmer, the Former Prime Minister of New Zealand commented, "We live on two volcanic rocks where two tectonic plates meet, in a somewhat lonely stretch of windswept ocean just above the Roaring Forties. If you want drama—you've come to the right place" (Webb and McEntire, 2011).
- Peru has "always been described as a very hazard-prone region of the world" (McEntire et al., 2011, citing Oliver-Smith, 1985).
- "The Virgin Islands are among the most vulnerable societies in the world" (McEntire and Samuel, 2011).

Taking these facts into consideration, it appears that countries almost seem to be vying for the title of who is most at risk. While disagreement about individual status is almost certain to abound, one may conclude with confidence that the world is undoubtedly full of significant disaster risks.

26.2.2 The hazard profile is both similar and different in each nation

Although all nations face at least some degree of risk, the hazards they are confronted with are obviously not uniform around the world. Some, like the United States, face a broad spectrum of natural, industrial, and anthropogenic hazards. Several others also fall into this category. As an example, "India is one of the most disaster prone countries of the world. It has had some of the world's most severe droughts, famines, cyclones, earthquakes, chemical disasters, mid-air head-on air collisions, rail accidents, and road accidents. India is also one of the most terrorist prone countries" (Gupta, 2011). "China is [likewise] one of the countries that is most affected by natural disasters.... This includes floods, droughts, meteorological, seismic, geological, maritime and ecological disasters as well as forestry and grassland fires" (Bai, 2011). Finally, "one thing is certain about Mexico: hazards of virtually every kind abound!" (McEntire and Urby, 2011a).

Even though nations are affected by diverse hazards, they must also deal with commonly recurring events like flooding. Excessive precipitation and runoff (associated with any number of meteorological and even other hazards) creates major problems for most regions of the world. France is among the countless nations struggling to deal with this hazard. "The second half of the 20th century has seen an elevated pattern of water-related disasters in France. The nature of these events changed from slow water risings of the rivers Louire and Seine, to highly destructive flash floods in recently urbanized mountainous areas south of France" (Mancebo and Renda-Tanali, 2011). The Netherlands is also well-known for its flooding risk and received its name because 60% of the population lives below sea level (Engel et al., 2011).

This brings up another important point about nations, that is, they may also be extremely prone to a specific hazard in particular. As a case in point, Denmark suffers from landslides due to its topography (Wyman, 2011). Australia is prone to drought and brushfires (McEntire and Peters, 2011). "Israel has faced more than its share of emergency management challenges, primarily originating from warfare and terrorist attacks" (Rozdilsky, 2011). Malawi must cope with a major HIV AIDS epidemic (Misomali, 2011). South Korea has had some major structural fires throughout its history (Ha, 2011). New Zealand is threatened by geological hazards such as volcanic and seismic activity (Webb and McEntire, 2011). Thus, no two nations are fully alike in their hazard profile, although parallels do exist around the world.

26.2.3   The social construction of disasters is painfully evident

Disasters do not occur as a result of natural processes alone. Vulnerability to hazards is produced through social factors such as poverty. For instance, Malawi ranks 162 out of 177 countries in the Human Development Index, and over 65% of the citizens live on less than $1.00 per day. Under these dire conditions, it is extremely difficult to mitigate disasters and react effectively to disasters. Although such circumstances are recognized most often in the developing world, wealthy nations are not immune from this social vulnerability. Hurricane Katrina is a visible reminder that even countries like the United States face serious disaster challenges when hazards and poverty collide. Many, and perhaps even most, of the victims were from lower economic classes.

The social construction of vulnerability is not only related to income disparities, however. There are many other factors that also lead to disaster. Overcrowding in urban areas (as much as 800 people per hectare) is a major problem in Syria (Alqusairi, 2011), while a sparse population poses other emergency management challenges in Canada perhaps due to scarce resources (Lindsay, 2011). Improper land use in Thailand (e.g., people living along the Chao Phraya River in Bangkok) has resulted in major flood impacts (Khunwishit and McEntire, 2011). Construction materials and practices augment damages in severe storms in Oman and earthquakes in Chile (Al-Shaqsi, 2011; Aguirre, 2011). Insufficient building codes or a lack of enforcement for fire hydrants and sprinkler systems has contributed to major fires in Paraguay (McEntire and Urby, 2011b). The lack of seismic monitoring in Chile limits awareness of earthquake risks (Aguirre, 2011). Social, cultural, religious, and political conflicts have made Ireland prone to terrorist attacks (McMullan, 2011). Gupta (2011) has illustrated that factors such as apathy, corruption, illiteracy, and technological misuse also have an impact on vulnerability. Even the pilgrimage to Mecca is associated with heat exhaustion, possible trampling, public health concerns, and fires (Alamri, 2011). Human attitudes and behaviors indisputably play a role in the occurrence of disasters.

26.2.4 Past and current disasters have significant impacts

Disasters around the world have serious consequences for the countries affected by them. The devastating effects in recent years are striking and troublesome.

- Canada's ice storm in 1998 was the most expensive disaster in the nation's history and necessitated that 250 communities declare a state of emergency (Lindsay, 2011).
- In Turkey, the Kocaeli earthquake in 1999 resulted in major damage to 73,342 structures (Ural, 2011).
- The Gujarat earthquake in 2001 killed about 25,000 people (Gupta, 2011).
- Terrorist attacks on 9/11 killed nearly 3000 people and produced billions of dollars in economic losses (McEntire, 2011).
- Drought and food shortage in 2001 adversely impacted approximately 3 million people in Malawi (Misomali, 2011).
- In 2001, Typhoon Nari resulted in the loss of power to 650,000 people and the loss of water to another 350,000 (McEntire, and Tso, 2011).
- The 2004 shopping center fire in Paraguay claimed the life of over 2000 people (McEntire and Urby, 2011b).
- Also, in 2004, the 9.3 magnitude earthquake in the Indian Ocean produced a tsunami that resulted in 14 billion Baht in damages and the loss of 30 billion Baht (Khunwishit and McEntire, 2011).
- The limited provision of health care during a 2005 holiday season in France resulted in the death of 14,000 people (predominantly the elderly) (Mancebo and Renda-Tanali, 2011).
- Cyclone Gonu affected 90% of all roads in Oman in 2007 (Al-Shaqsi, 2011).
- The fires that ravished San Diego in 2007 forced the evacuation of 500,000 people (McEntire, 2011).

- The collision of an oil tanker with another ship in 2007 resulted in a spill of 12,547 kL of oil and tainted 70 km of shoreline in Chungnam province in South Korea (Ha, 2011).
- The 2008 Sichuan earthquake killed nearly 70,000 people and injured almost 400,000 more (Bai, 2011).
- In Saudi Arabia, there were almost 500,000 motor vehicle crashes in 2008 alone, resulting in over 6000 deaths (Alamri, 2011).
- Over 2000 homes were lost in the 2009 South Eastern Australia Heat Wave (McEntire and Peters, 2011).

It is clear that disasters around the world produce substantial physical damage and are accompanied by economic losses and social disruption. While some events are concentrated in cities and provinces, others have national and international dimensions (e.g., the Indian Ocean tsunami affected multiple nations directly and countless others indirectly).

26.2.5   The most common time to promote change is immediately after a disaster

While it is always best to implement emergency management policies before disasters occur, this is oftentimes difficult due to political, budgetary, and cultural constraints. The most common time to create or alter laws is directly after disasters occur. For instance, in 2003, the government in China passed new regulations on the Handling of Public Health in Emergencies owing to the Severe Acute Respiratory Syndrome (SARS) epidemic (Bai, 2011). The 2003 subway fire in Daegu, South Korea, resulted in the passing of the Emergency and Safety Management Basic Act to better coordinate civil defense and natural disaster activities (Ha, 2011). The Gujarat Disaster Management Act of 2003, India's first comprehensive law regarding emergency management, was passed subsequent to the devastating earthquake in 2001 (Gupta, 2011). After the 2004 shopping center fire in Paraguay, the Secretaria de Emergencia Nacional created new laws to promote a culture of prevention (McEntire and Urby, 2011b). The United Kingdom introduced the Civil Contingencies Act after the 2004 bombings in London in order to replace and update prior civil defense powers (Kapucu, 2011). The first legislation relating to warnings in Thailand was created in 2005 after the tsunami ravaged much of the country (Khuwinshit and McEntire, 2011). When Typhoon Morakot dissipated in 2010, the public demanded and succeeded in getting new laws to professionalize emergency management in Taiwan (McEntire and Tso, 2011).

Incidentally, these types of policies and regulations sometimes have significant impact upon the nation. The event on 9/11 is a perfect example. Virtually all areas of life in the United States were altered due to the new laws relating to the Department of Homeland Security. Immigration and customs, banking and trade, and travel and transportation all were altered dramatically by the new regulations. Many of these homeland security policies have shaped activities abroad as well.

26.2.6 *Nations have common and distinct forms of organization*

There are analogous organizations for emergency management with major variance in systems as well. For instance, emergency management was originally seen by many nations as a military responsibility due to the threat of air raids during World War II or nuclear conflict during the Cold War. Countries such as Australia, Canada, Thailand, the United Kingdom, the United States, and many others established civil defense organizations with direct or strong ties to the military. In many cases, this tradition remains strong and is still evident in places like Cuba (Aguirre and Trainor, 2011) and Saudi Arabia (Alamri, 2011). Perhaps because of the threat of war and terrorism, there is also a very close connection today between the military and emergency management in Israel (Rozdilsky, 2011) and Peru (McEntire et al., 2011). Many countries, therefore, have based (and may currently base) their emergency management programs on or around the military or departments of defense.

The Incident Command System (ICS) is another common organizing principle for governments around the world. This structure was initially developed for firefighting in the 1970s in an attempt to improve leadership, decision making, and resource management in multi-organizational response operations. The system aims to facilitate coordination of first responders with unified leadership and organization based on the functions of planning, operations, logistics, and finance/administration. In the United States, ICS has morphed into the National Incident Management System with an additional emphasis on preparedness, training, exercises, and evaluation. Similar systems have been developed in Canada (McEntire and Lindsay, 2012), Australia (McEntire and Peters, 2011), and elsewhere.

Despite these mutually shared elements, there are dramatic distinctions among national emergency management organizations. Although many countries rely heavily on the military, this is not the case in Costa Rica. In fact, this Central American nation has no armed forces whatsoever. Such a situation at times has limited response operations and requires reliance on foreign military services (Afedzie et al., 2011), but it has permitted additional investment in other government programs such as health, education, and even emergency management. Also, even though many nations around the world are adopting ICSs, this practice is not universal and is less common in developing nations.

There are a number of other organizational differences in national emergency management systems. Emergency management in large countries like Australia and Canada is decentralized, giving power to the states and provinces instead of the central government (McEntire and Peters, 2011; Lindsay, 2011). In contrast, others, like Paraguay, have a more top-down approach to disaster management (McEntire and Urby, 2011b). Some nations, including Taiwan and South Korea (McEntire and Tso, 2011; Ha, 2011), have dedicated emergency management organizations like FEMA in the United States. Others, including China (Bai, 2011), have no such organization and instead spread responsibility among all government organizations.

Finally, emergency management may be located in a variety of government departments and agencies. As an example, the National Committee for Civil Defense is largely attached to the police in Oman (Al-Shaqsi, 2011). Emergency management is placed under the Department of Poverty and Disaster Management Affairs (Misomali, 2011). Turkey's emergency management arrangement has a close relation with the Prime Minister and the Ministry of Public Works and Settlements (Ural, 2011). The Department of Disaster Prevention and Mitigation is in the Ministry of Interior in Thailand (Khuwinshit and McEntire, 2011). And, in New Zealand, the Ministry of Civil Defense and Emergency Management is closely tied to the Prime Minister (Webb and McEntire, 2011). Thus, while many nations favor organization based on the military and ICS, there are no universally agreed standards for governmental structure.

26.2.7 The participants in emergency management are expanding

Emergency management in most countries was initiated by the government at the central or federal level. Nevertheless, this important function is growing horizontally across departments as well as vertically among lower levels of government. In addition, emergency management is being addressed by others in the private and nonprofit sectors.

In Syria, for example, emergency management activities are performed by the Department of Seismology, the National Committee on Hazardous Waste Management, the Chemical Safety Committee, the Ministry of Irrigation, and the Ministry of Agriculture and Agrarian Reform (Alqusairi, 2011). Similarly, the United States has a great deal of redundancy owing to numerous federal partners, FEMA regional offices, the Interstate Emergency Management Assistance Compact, state emergency management organizations, regional council of governments, urban mutual aid agreements, and local emergency management offices (McEntire, 2011). Furthermore, in the Netherlands, responsibilities are delegated to fire, emergency medical services, and police (Engel et al., 2011), which is typical of most nations.

These and other nations are likewise doing more to engage the private and nonprofit sectors. "In Taiwan, the role of public–private cooperation in responding to disasters has become more important in the emergency management framework" (McEntire and Tso, 2011). "There are many other public organizations, non-profit organizations, businesses and civil networks that also actively participate in disasters in Thailand" (Khunwishit and McEntire, 2011). Oman, too, is pushing for further private sector involvement (Al-Shaqsi, 2011) while more volunteers are participating in emergency management in South Korea (Ha, 2011). Thus, it seems as if emergency management principles and practices are infiltrating all levels and departments of government, and other sectors in society. In fact, the concept "Total Defense" in Denmark (Wyman, 2011) and "Whole Community" in the United States (McEntire, 2011) are manifestations of this trend in emergency management in many nations.

26.2.8 Countries have been reactive, but are becoming more proactive

Historically, nations have focused emergency management on response operations and, to a lesser extent, preparedness activities. Such a reactive approach is now being questioned in policy, if not entirely in practice. Although some nations like the Netherlands are rethinking their primary focus on mitigation, given the unique nature of the threats they face, many countries are also attempting to embrace prevention and capacity-building activities. A number of examples can be given.

- The United States implemented new legislation under the Disaster Mitigation Act to improve risk assessments and post-disaster mitigation (McEntire, 2011).
- Much more government oversight of the construction process is expected in the U.S. Virgin Islands (McEntire and Samuel, 2011).
- The 2007 Emergency Management Act in Canada made planning a responsibility of every government department (Lindsay, 2011).
- The Communal Safeguard Plan in France authorizes the mayor to take needed measures regarding risk reduction and mitigation (Mancebo and Renda-Tanali, 2011).
- Saudi Arabia is placing increased attention on planning for mass religious gatherings (Alamri, 2011).
- Turkey is giving more attention to building codes to prevent mass casualties in major earthquakes (Ural, 2011).
- India is developing five-year national plans to guide emergency management in a strategic manner (Gupta, 2011).

These and other proactive approaches indicate that a broader and more anticipatory form of emergency management is required to meet future disasters. For this reason, New Zealand and many others regard resilience as a guiding concept for their nation's emergency management system (Webb and McEntire, 2011).

26.2.9   There is a global focus on terrorism

There is an international trend illustrating that most nations are now increasingly concerned about deadly, destructive, and disruptive terrorist attacks. Countries like Ireland and Israel have long had experience in dealing with terrorism throughout their histories (McMullen, 2011; Rozdilsky, 2011) while others like Saudi Arabia have seen increased terrorist activity over time (Alamri, 2011). Although more recent attacks have often been the work of Islamic radicals, this is not always the case. France has faced Islamic terrorism as well as attacks from internal separatists (Mancebo and Renda-Tanali, 2011). There are an estimated 174 insurgent groups in India and these represent diverse interests and political ambitions (Gupta, 2011). Peru has faced terrorism from drug-related organizations including the Sendero Luminso (McEntire and Urby, 2011a).

However, the Al Qaeda terrorist attacks in the United States on 9/11 seem to have had the most influence on emergency management in this country and elsewhere. The reasoning is that if terrorists could strike with such a fury on American soil, they could inflict similar damages and casualties elsewhere. Unfortunately, the potential for terrorism has become a frightening reality. The 2002 bombings in Bali got the attention of the government in Australia as citizens frequent these tourist locations for vacations (McEntire and Peters, 2011). The United Kingdom experienced simultaneous attacks on its transportation system in London in 2005 (Kapucu, 2011). In response to some satirical cartoons that appeared in 2005 in a Danish newspaper, Denmark has received countless threats and has been on heightened alert ever since (Wyman, 2011).

In light of this resultant heightened state of security, countries have undertaken major governmental reforms. The United States created the Department of Homeland Security, a massive conglomerate of 22 federal agencies, and over 122,000 employees. Other countries have also experienced major reorganizations to better deal with terrorism, although some have retreated like Canada since its Office of Critical Infrastructure Protection and Emergency Preparedness has been changed in a more recent reform (Lindsay, 2011). Nevertheless, the overall trend is to consider terrorism to be a major threat and react accordingly.

26.2.10 Countries must learn from one another

Because countries must gain knowledge from the experience of others, these findings are important for practitioners around the world. There are many negative lessons from abroad that emergency managers will not wish to repeat in their own nations. However, success in emergency management is most likely to be achieved when governments can emulate the achievements of their counterparts wherever appropriate. There are many positive examples that can be gleaned from this study:

- The United States has developed standards (e.g., NFPA 1600) that may help guide or shape emergency management in other countries (McEntire, 2011).
- The Virgin Islands have elevated emergency management to a cabinet-level position (McEntire and Samuel, 2011).
- The United Kingdom has identified different levels of responses based on the nature of the disaster or catastrophe (Kapucu, 2011).
- Ireland has established its very first university program in emergency management to increase professionalism in the field (McMullen, 2011).
- Canada has passed policies to clarify the distribution of federal funds to facilitate recovery (Lindsay, 2011).
- France has recognized the vital importance of risk assessment as a foundation of emergency management programs (Mancebo and Renda-Tanali, 2011).
- The Dutch can provide a great deal of advice and expertise on managing the flood hazard (Engel et al., 2011).
- Denmark reminds public officials to plan for disasters that affect their own citizens in other countries (e.g., tourists impacted by the 2004 tsunami) (Wyman, 2011).
- New Zealand has a clear plan to pursue the goal of disaster resilience (Webb and McEntire, 2011).
- Australia provides a great model of engaging volunteers in all types of emergency management activities (McEntire and Peters, 2011).
- Israel is well-known for its intelligence operations and the "Defense of the Rear" concept (Rozdilsky, 2011).
- Syria helps emergency managers to consider sometimes unanticipated emergency management functions including sheltering of refugees (Alqusairi, 2011).
- Saudi Arabia offers a good model for dealing with mass gatherings (Alamri, 2011).
- Oman encourages emergency managers to determine how development practices may impact future disasters (Al-Sahqsi, 2011).

- Malawi has planned on how best to integrate international donors into their emergency management system (Misomali, 2011).
- Turkey is aggressively implementing more stringent construction practices to minimize vulnerability to future earthquakes (Ural, 2011).
- India is giving a great deal of attention to disaster education for the rising generation (Gupta, 2011).
- Thailand has one of the most active nonprofits, The Fajaprajanugroh Foundation, involved in emergency management (Khuwishit and McEntire, 2011).
- China is employing critical technology to improve its capabilities in emergency operations centers around the nation (Bai, 2011).
- Taiwan has an active faith-based community, The Buddhist Compassion Relief Tzu Chi Foundation, which does much to respond to disasters of all types (McEntire and Tso, 2011).
- The government in South Korea has done much to improve business continuity planning in the private sector (Ha, 2011).
- Mexico has one of the first earthquake warning systems in the world (McEntire and Urby, 2011b).
- The Cuban experience with disasters provides several recommendations for successful dealing with international aid (Aguirre, 2011).
- Costa Rica requires that all government departments and agencies allocate money for emergency management purposes (Afedzie et al., 2011).
- Paraguay is attempting to train their emergency management professionals to better deal with the impacts of disasters (McEntire and Urby, 2011b).
- The Peruvian government is working closely with the nonprofit sector to link relief activities with development and environmental conservation (McEntire et al., 2011).
- Chile is beginning to recognize the need to alter culture to improve how it deals with vulnerability and disasters (Aguirre, 2011).

26.3   *Discussion and conclusions*

Because there is a dearth of information about emergency management in other countries, such studies are desperately needed. The Comparative Emergency Management Project has attempted to add to our knowledge base by examining hazards, vulnerability, and disasters around the world as well as national emergency management laws and practices. While it is unreasonable to assume that the propositions discussed in this chapter will hold true in every particular national context, there are several recommendations that may be drawn from the findings of this project. Ten of these recommendations are listed as follows.

1. Emergency managers must recognize all types of hazards and take them seriously. It is a grave mistake to downplay potential risks.
2. More expertise on hazards and vulnerability assessments are needed. While emergency managers are knowledgeable in many areas, it is likely that risk assessment deserves additional improvement.
3. Practitioners should give additional attention to public disaster education. Ways must be found to reverse the social construction of disasters.
4. Professionals in emergency management must anticipate broad consequences that result from disasters. Besides physical damage, deaths, and injuries, there are many other far-reaching effects that have to be taken into account (e.g., economic loss, psychological impacts).
5. Emergency managers must be ready to take advantage of windows of opportunity when disasters strike. Failing to have clearly defined mitigation goals after disaster strikes will limit the amount of change that can take place when disasters occur.
6. Experts involved in emergency management should strive to understand the implications of various organizational arrangements. Currently, there is a lack of information on this subject.
7. Those involved in disaster planning should identify the best way to engage others in emergency management. If the number and scope of participants are expanded, it will be important to clearly define roles and responsibilities.
8. The individuals working in emergency management should consider which methods are most likely to advance our goals in the future. Determining the best methods to achieve proactive approaches is required in this crucial profession.
9. Practitioners must not only acknowledge the threat of terrorism but also consider how homeland security activities might impact emergency management. Countries like the United States have had painful experiences when terrorism overshadows other types of hazards (e.g., Hurricane Katrina).
10. Emergency managers should become followers of the emergency management issues in other countries. Learning from others is a great way to avoid mistakes and capitalize on successful practices.

In conclusion, it should be recognized that emergency managers around the world have difficult jobs that require a great deal of knowledge and expertise. Learning more about the disaster systems in other nations only adds to the significant responsibilities that already exist. However, incorporating the lessons drawn from other countries may improve emergency management internationally and ultimately make the careers of such professionals easier. It is hoped that this chapter may help practitioners reach these important goals now or in the future.

References

Afedzie, R., McEntire, D.A., and Urby, H. (2011), "Emergency Management in Costa Rica: A Unique Model for Developing and Developed Nations," *Comparison Emergency Management Book Project*, http://training.fema.gov/EMIWeb/edu/CompEmMgmtBookProject.asp (accessed September 1, 2011).

Aguirre, B.E. (2011), "Chile's Civil Defense and Lack of Mitigation," *Comparison Emergency Management Book Project*, http://training.fema.gov/EMIWeb/edu/CompEmMgmtBookProject.asp (accessed September 1, 2011).

Aguirre, B.E., and Trainor, J.E. (2011), "Emergency Management in Cuba: Disasters Experienced, Lessons Learned, and Recommendations for the Future," *Comparison Emergency Management Book Project*, http://training.fema.gov/EMIWeb/edu/CompEmMgmtBookProject.asp (accessed September 1, 2011).

Alamri, Y.A. (2011), "Emergency and Crisis Management in Saudi Arabia: Past, Present and Future," *Comparison Emergency Management Book Project*, http://training.fema.gov/EMIWeb/edu/CompEmMgmtBookProject.asp (accessed September 1, 2011).

Alexander, D. (2006), "Globalization of Disaster: Trends, Problems and Dilemmas," *Journal of International Affairs*, Volume 59, Number 2, pp. 1–24.

Alqusairi, D. (2011), "Risk Profile and Emergency Management of Syria," *Comparison Emergency Management Book Project*, http://training.fema.gov/EMIWeb/edu/CompEmMgmtBookProject.asp (accessed September 1, 2011).

Al-Shaqsi, S.Z. (2011), "Emergency Management in the Arabian Peninsula: A Case Study from the Sultanate of Oman," *Comparison Emergency Management Book Project*,http://training.fema.gov/EMIWeb/edu/CompEmMgmtBookProject.asp (accessed September 1, 2011).

Bai, V. (2011), "Emergency Management in China," *Comparison Emergency Management Book Project*,http://training.fema.gov/EMIWeb/edu/CompEmMgmtBookProject.asp (accessed September 1, 2011).

Britton, N.R. (2006), "National Planning and Response," in Rodriguez, H., E.L. Quarantelli and R.R. Dynes (eds.), *Handbook of Disaster Research*, Springer, New York, pp. 347–367.

Coppola, D.P. (2007), *Introduction to International Disaster Management*, Butterworth-Heinemann, Boston.

Drabek, T.E. (1986), *Human System Responses to Disaster*, Springer, New York.

Dynes, R.R. (1988), "Cross-Cultural International Research: Sociology and Disaster," *International Journal of Mass Emergencies and Disasters*, Volume 6, Number 2, pp. 101–129.

Engel, K., Harrald, J.R., McNeil, S., Shaw, G., Trainor, J.E., and Zannoni, M. (2011), "Flood and Disaster Management in the NL: God Created the World, but the Dutch Created the NL," *Comparison Emergency Management Book Project*, http://training.fema.gov/EMIWeb/edu/CompEmMgmtBookProject.asp (accessed September 1, 2011).

FEMA. (2011), *Comparative Emergency Management Book Project*, http://training.fema.gov/EMIWeb/edu/CompEmMgmtBookProject.asp (accessed September 1, 2011).

Gupta, K. (2011), "Disaster Management and India: Responding Internally and Simultaneously in Neighboring Countries," *Comparison Emergency Management Book Project*, http://training.fema.gov/EMIWeb/edu/CompEmMgmtBookProject.asp (accessed September 1, 2011).

Ha, K. (2011), "Emergency Management in Korea: Just Started, but Rapidly Evolving," *Comparison Emergency Management Book Project*, http://training.fema.gov/EMIWeb/edu/CompEmMgmtBookProject.asp (accessed September 1, 2011).

Kapucu, N. (2011), "Emergency and Crisis Management in the United Kingdom: Disasters Experienced, Lesson Learned, and Recommendations for the Future," *Comparison Emergency Management Book Project*, http://training.fema.gov/EMIWeb/edu/CompEmMgmtBookProject.asp (accessed September 1, 2011).

Khunwishit, S., and McEntire, D.A. (2011), "Emergency Management in Thailand: On the Way to Creating a More Systematic Approach to Disasters," *Comparison Emergency Management Book Project*, http://training.fema.gov/EMIWeb/edu/CompEmMgmtBookProject.asp (accessed September 1, 2011).

Lindsay, J. (2011), "Emergency Management in Canada: Near Misses and Moving Target," *Comparison Emergency Management Book Project*, http://training.fema.gov/EMIWeb/edu/CompEmMgmtBookProject.asp (accessed September 1, 2011).

Lindsay, J., and McEntire D.A. (2012), "One Neighborhood, Two Families: A Comparison of Intergovernmental Emergency Management Relationships." *Journal of Emergency Management*. Volume 10, Number 2, pp. 93–107.

Mancebo, F., and Renda-Tanali, I. (2011), "French Emergency Management System: Moving Toward an Integrated Risk Management Policy," *Comparison Emergency Management Book Project*, http://training.fema.gov/EMIWeb/edu/CompEmMgmtBookProject.asp (accessed September 1, 2011).

McEntire, D.A. (1997), "Reflecting on the Weaknesses of the International Community during the IDNDR: Some Implications for Research and Application," *Disaster Prevention and Management*, Volume 6, Number 4, pp. 221–233.

McEntire, D.A. (2007), *Disciplines, Disasters and Emergency Management: The Convergence and Divergence of Concepts, Issues and Trends from the Research Literature*, C.C. Thomas, Springfield, IL.

McEntire, D.A. (2011), "Emergency Management in the United States: Disasters Experienced, Lessons Learned, and Recommendations for the Future," *Comparison Emergency Management Book Project*, http://training.fema.gov/EMIWeb/edu/CompEmMgmtBookProject.asp (accessed September 1, 2011).

McEntire, D.A., and Peters, E.J. (2011), "Emergency Management in Australia: An Innovative Progressive and Committed Sector," *Comparison Emergency Management Book Project*, http://training.fema.gov/EMIWeb/edu/CompEmMgmtBookProject.asp (accessed September 1, 2011).

McEntire, D.A., and Samuel, C. (2011), "Emergency Management in the U.S. Virgin Islands: A Small Island Territory with a Developing Program," *Comparison Emergency Management Book Project*, http://training.fema.gov/EMIWeb/edu/CompEmMgmtBookProject.asp (accessed September 1, 2011).

McEntire, D.A., and Tso, Y. (2011), "Emergency Management in Taiwan: Learning from Past and Current Experiences," *Comparison Emergency Management Book Project*, http://training.fema.gov/EMIWeb/edu/CompEmMgmtBookProject.asp (accessed September 1, 2011).

McEntire, D.A., Peters, E.J., and Urby, H. (2011), "Peru: An Andean Country with Significant Disaster and Emergency Management Challenges," *Comparison Emergency Management Book Project*, http://training.fema.gov/EMIWeb/edu/CompEmMgmtBookProject.asp (accessed September 1, 2011).

McEntire, D.A., and Urby, H. (2011a), "Emergency Management in Mexico: A Good Beginning but Additional Progress Needed," *Comparison Emergency Management Book Project*, http://training.fema.gov/EMIWeb/edu/CompEmMgmtBookProject.asp (accessed September 1, 2011).

McEntire, D.A., and Urby, H. (2011b), "Emergency Management in Paraguay: A Landlocked Country Not Without Disasters and Accidents," *Comparison Emergency Management Book Project*, http://training.fema.gov/EMIWeb/edu/CompEmMgmtBookProject.asp (accessed September 1, 2011).

McMullan, C. (2011), "Emergency Management in the Republic of Ireland: A Rising Tide Has Lifted All Boats," *Comparison Emergency Management Book Project*, http://training.fema.gov/EMIWeb/edu/CompEmMgmtBookProject.asp (accessed September 1, 2011).

Misomali, R. (2011), "Emergency Management in Malawi: A Work in Progress with a Strong Foundation," *Comparison Emergency Management Book Project*, http://training.fema.gov/EMIWeb/edu/CompEmMgmtBookProject.asp (accessed September 1, 2011).

National Academy of Sciences. (2006), *Facing Hazards and Disasters: Understanding Human Dimensions*, The National Academies Press, Washington, D.C.

Peacock, W.G. (1997), "Cross-national and Comparative Disaster Research," *International Journal of Mass Emergencies and Disasters*, Volume 15, Number 1, pp. 117–133.

Rozdilsky, J.L. (2011), "Emergency Management in Israel: Context and Characteristics," *Comparison Emergency Management Book Project*, http://training.fema.gov/EMIWeb/edu/CompEmMgmtBookProject.asp (accessed September 1, 2011).

Ural, D.N. (2011), "Emergency Management in Turkey: Disasters Experienced, Lessons Learned, and Recommendations for the Future," *Comparison Emergency Management Book Project*, http://training.fema.gov/EMIWeb/edu/CompEmMgmtBookProject.asp (accessed September 1, 2011).

Webb, C., and McEntire, D.A. (2011), "Emergency Management in New Zealand: Potential Disasters and Opportunities for Resilience," *Comparison Emergency Management Book Project*, http://training.fema.gov/EMIWeb/edu/CompEmMgmtBookProject.asp (accessed September 1, 2011).

Wyman, J.S. (2011), "Emergency Management in Scandinavia: Lessons Learned at Home and Abroad," *Comparison Emergency Management Book Project*, http://training.fema.gov/EMIWeb/edu/CompEmMgmtBookProject.asp (accessed September 1, 2011).

chapter twenty-seven

Framework for real-time, all-hazards global situational awareness*

Olufemi A. Omitaomu, Steven J. Fernandez, and Budhendra L. Bhaduri

Contents

* The material contained in this chapter was prepared as an account of work sponsored by an agency of the U.S. government. Neither the U.S. government nor any agent thereof, nor any of their employees, makes any warranty, express or implied, or assumes any legal liability or responsibility for the accuracy, completeness, or usefulness of any information, apparatus, product, or process disclosed, or represents that its use would not infringe upon privately owned rights. Reference herein to any specific commercial product, process, or service by trade name, trademark, manufacturer, or otherwise does not necessarily constitute or imply its endorsement, recommendation, or favoring by the U.S. government or any agency thereof. The views and opinions of authors expressed herein do not necessarily state or reflect those of the U.S. government or any agency thereof.

27.1   Introduction

When disaster strikes, effective incident management and response coordination is essential to ensuring the resilience of critical infrastructure. This, in turn, depends on the availability of critical infrastructure data, as well as geospatial modeling and simulation capabilities, that can complement the decision-making process at various stages of the disaster. Hence, disaster consequence management organizations should have access to the best available geospatial technical expertise, global and regional datasets, and modeling and analytical tools. However, an optimal combination of data assets and modeling expertise is often beyond the resources available internally within a single organization but can be accessed through collaboration with other organizations. This provides an opportunity to develop a network of solutions for emergency response. Such a network becomes a platform for sharing data and model outputs. For such a network to be successful, though, there are at least three necessary conditions that must be fulfilled.

The first condition is that members of this network should be able to share data and information with one another without any proprietary restrictions. The reality, however, is that the sharing of information is usually limited by license agreements. As a result, some members may only have access to hard copies of results and not the actual data and/or models. This restriction hinders the true benefits of the network. Emergency responders can base their decisions only on credible information to minimize false alarms. If they cannot obtain such information from credible sources, they are limited in what they can do and/or provide during an emergency. This tends to unnecessarily prolong disaster response time and recovery. To remove this restriction, the question then is, "Can infrastructure data be built from open content sources so that it is amenable to sharing among credible organizations or members of the network?"

Another necessary condition is the need for real-time updates of data and models as the disaster evolves. This condition depends solely on the first condition—unrestricted access to data and model outputs. If the first condition is not met, there is little chance that this second condition can be of value. Furthermore, there should be a mechanism for real-time updates of data and models and for subsequent sharing of those updates with all members of the network. The lack of a true information-sharing mechanism has limited the benefits of real-time updates of disaster data and models. In case of hazards such as hurricanes, consequence analyses are usually completed hours before a landfall. When disaster strikes, emergency response and management is based on information previously generated with little or no modification using real-time field data. However, the emergence of social networking sites and advances in consumer electronics have changed the way we report and receive news; therefore, disaster models should take advantage of an abundance of information that is shared every second through social networking sites. Again, this is possible only if there is a true real-time situational awareness system that uses no proprietary data. Based on this description, the next question then is, "Can we build upon the first necessary condition by updating and sharing new data and models in real time during a disaster?"

The third necessary condition for a successful network is the development of an all-hazards framework for disaster impact analysis. Such a framework should cover all countries in the world as well as all known hazards. Several emergency management communities have developed internal systems for some hazards, but not all hazards are covered. Therefore, communities may rely on their own network of colleagues for information not available at their disposal. With the establishment of a network of solutions, members of the network can bring their expertise to bear in developing an all-hazards tool. However, if there is no mechanism for data and information sharing among members, some of the members may be reluctant to contribute their expertise in their strong areas because they cannot get similar expertise in their deficient areas. With the first and second necessary conditions satisfied, this condition should be easier to formalize. This brings us to our third question, "Can we build an all-hazards platform for situational awareness?"

We present in this chapter a framework developed to answer these three questions and to satisfy the three necessary conditions. The Energy Awareness and Resiliency Standardized Services (EARSS) framework was developed as a system of systems for real-time, all-hazards global situational awareness without any restrictions in information sharing among members of the network. In this chapter, we will describe the EARSS framework, discuss some of the methodologies behind the system, and highlight the EARSS application to emergency management.

27.2   EARSS framework

The EARSS framework is a parallel framework in which each hazard is a separate, functional, and complete system. The overall system is developed as a framework so that algorithms and models can easily be integrated into the existing systems. Each hazard is treated as a system; therefore, there are as many systems as the number of known hazards. However, four major hazards—tropical cyclone, flood, wildfire, and earthquake—are usually used for illustration purposes. For each system, there are three specific subsystems—analytic models (automatic event detection and experts' evaluation of scale of disaster), real-time notification, and real-time model updates—as depicted in Figure 27.1. There is also a common subsystem, called the geodatabase of global energy data.

The specific subsystems use data within this common subsystem to identify assets and infrastructures affected by a particular disaster. It should be noted that the EARSS outputs are provided as services, and members can subscribe to the outputs of interests to their organization.

27.2.1 Geodatabase of global energy data

The geodatabase of global energy data contains critical infrastructure data. To develop this system, datasets from open content sources are used. The data sources include images, pictures, and other online resources. The static data are mapped (or digitized) into editable data layers using a geographic information system (GIS) as the human–machine interface. The output of this system informs the analytic models. The data layers provide the geographic location, size, capacity, and other pertinent information about critical infrastructures such as electric lines, electrical substations, power-generating stations, and gas pipelines. The quality control of the mapped data is achieved using high-resolution imagery. The data layers are developed for all countries. Since the data layers are created from open

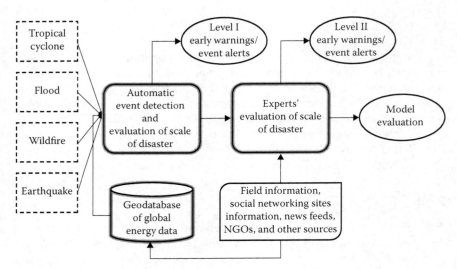

Figure 27.1 Depiction of the Energy Awareness and Resiliency Standardized Services (EARSS) framework.

sources, the output can be easily shared with members of the EARSS network without any restrictions. In addition to the mapping of the data, the layers are digitally connected for real-time disruption analysis using the analytic models. Furthermore, the status of the layers is monitored in real time to determine which of the infrastructures have been impacted. Once an impacted infrastructure is identified, the downstream implications of such an impact, in terms of the number of people without power and potential cascading effects of the failure, are analyzed and made available to the EARSS members. It should be noted that this geodatabase is much more than a database. It also helps overcome some of the compatibility issues with data that use mixed coordinate systems. Since maps cannot display data with mixed coordinates, users tend to be restricted in their analyses of data with different coordinates. The EARSS geodatabase helps overcome such incompatibilities by using the same coordinate system for all the data layers. The EARSS geodatabase can change the paradigm of how information is treated and shared by making data available for all to use.

27.2.2   Analytic models

The analytic models consist of several submodels, some of which are offline models (models used for analysis before a disaster) and real-time models (models for automatic tuning of the predictions as more information about the disaster becomes available). This is the engine for the EARSS system. Some of the methodologies within the EARSS analytic models are described in Section 27.3.

27.2.3   Real-time detection and notification system

As part of the EARSS system, we monitor in near, real-time natural disasters all over the world. However, some of the hazards including earthquakes and wildfires are monitored in real time because they provide no sufficient time for alerts and advisories. When any of these events are detected, detection information is automatically disseminated to all members via a short message service (text messaging). In addition to the detection notification, some automated initial analyses are also performed, and the results are emailed to all members in real time. These services are often disseminated before news about the events even gets to the media. A typical text message reads "Earthquake: M6.1, Alaska Peninsula – 74 people potentially affected (see email for report)." This message means that an earthquake of magnitude 6.1 has been detected in the Alaska Peninsula, and about 74 people are potentially within a 20 mi radius of the earthquake epicenter. Text messages like this guarantee that emergency management personnel are informed of events as soon as they occur. Furthermore, the email report gives the recipients some results of the initial analysis. The results can be used for some initial planning before detailed results become available. Analyses on hazards with sufficient time for advisories, such as hurricanes, are updated every day for about 3–5 days before a landfall and every hour for about 6 hours before a landfall.

27.2.4   Real-time model updates

The real-time model updates use data from news feeds, blogs, social networking sites, and Twitter to characterize conditions in the affected areas. The data are then used with the analytic models to update the predicted impact areas and information about recovery.

27.3   EARSS methodologies

In this section, we describe the approaches that drive some of the subsystems presented in Section 27.2.

27.3.1   Developing a geodatabase of global critical infrastructure

As stated previously, one of the challenges with the present system is the use of proprietary data, which has limited the sharing of information among users. Therefore, the first step in the EARSS framework is the development of global critical infrastructure data layers using nonproprietary sources. The steps involved include identifying the infrastructure of interest and searching for data sources, which may include static images, reports, spreadsheet-based datasets, and satellite imagery. Once the data sources are identified and verified for completion, the data are converted into a geodatabase layer using the georeferencing technique. A data source is georeferenced by defining its existence in physical space, that is, reproducing the data by establishing its location on a map. This process basically determines the spatial location of all the geographical features in these sources. In addition to establishing the spatial location of features, this technique also helps to develop a common projection or coordinate systems for all the data sources, that is, mapping all the data sources into one common projection for easy analyses. Georeferencing can be used with a set of points, lines, polygons, images, or three-dimensional structures. The georeferencing technique will automatically record the latitude and longitude of the features. Other information can be manually documented in an attribute table. A unique identifier can be created for each feature for use within a data analysis framework. There are various GIS tools available that can transform data sources to a common geographic mapping framework. The tool used in this case is ArcMap—a product by the Environmental Systems Research Institute. Once the georeferencing process is completed, the static features can be overlaid on a virtual globe, such as Google Earth, for verification (using high-resolution imagery data) and visualization purposes.

27.3.2 Modeling electrical substation service areas

The problem of estimating electrical substation service areas (SAs) can be considered as a variant of a transportation model. The problem seeks the determination of a transportation (or electrical distribution) plan for a single commodity (in this case, electric power) from several sources (or electrical substations) to several destinations (or utility customers). One difference between this model and the traditional transportation model is that we assume that each destination can only receive its demand from the closest source (a single substation). In optimization, we can describe this model as an integer optimization model. However, our problem cannot be solved as an optimization model because we are not trying to build new networks of SAs for the existing electrical substations, but to make inference about the existing distribution network. Therefore, we are interested in heuristic solutions (not optimum solutions) that make sound engineering judgment. Nevertheless, the data for the model must include the level of supply (power capacity) at each substation and the demand (electricity required) at each destination. Since the substations are already in existence, we also assume a balanced model—the total power capacity equals the total power demand.

27.3.2.1 Approach for estimating electrical substation service area

The driver of all activities during an emergency is the availability of electric power. The loss of power for a prolonged time will bring life as we know it to a halt. Therefore, the door to ensuring the resilience of other critical infrastructure is the availability of electric power. To estimate the impact of a disaster when there is no power, there is a need for understanding the SAs for electrical substations. When an electrical substation is taken down by a storm, for example, we then have an idea of where people may be without power, what other critical infrastructures may be impacted, and what downstream impacts could be expected. However, data about substation SAs are proprietary data that cannot be shared. Therefore, our approach is to develop a methodology that accounts for principled engineering methods for estimating substation SAs all over the world. In this section, our methodology for estimating substation SA is presented.

The methodology is a directional nearest-neighbor approach for acquiring cells based on the availability of power at a substation and the closeness of this substation to where power is actually needed. The approach creates SAs by assigning demand cells to each supply cell. The first step is to create supply and demand matrices for the input datasets. For the supply matrix, the spatial locations and capacities of existing electric substations are obtained from the geodatabase subsystem described in Section 27.3.1. To obtain the number of electrical customers, we used the LandScan Global population data as described in Section 27.3.2.2. The cells in the demand and supply matrices measured approximately 1 km as depicted in Figure 27.2.

The demand in each square cell $\left(D_{ijk}\right)$ is defined as follows:

$$D_{ijk} = \frac{A_{ijk}}{\sum\limits_{i,j} P_k}\left(\sum\limits_{i,j} S_k\right)$$

where A_{ijk} is the total population in a cell with position i, j and state k, P_k is the total population in state k, and S_k is the total available substation capacity in state k. The supply in each cell $\left(C_{ijk}^T\right)$ is defined as follows:

$$C_{ijk}^T = \sum C_{ijk}$$

where C_{ijk} is the available capacity of each substation in position i, j and state k and T is the step of iteration. If there is more than one substation in a cell, the capacity of the substations is summed together. This introduces some uncertainties into the methodology, but since only a few cases

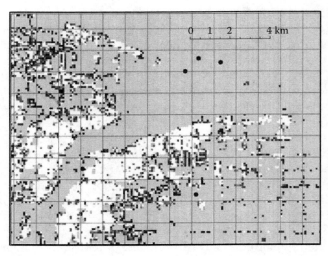

Figure 27.2 Matrix with population and substation data.

are encountered in each country, this does not affect the accuracy of the models. In addition to the demand and supply matrices, we also create taboo regions. The taboo regions are areas that violate common engineering principles for the power distribution system. As an illustration, power cannot be supplied from one side of a mountain to the other side to avoid excessive power loss. The same constraint holds for customers on different sides of rivers and other reserved areas. Furthermore, substations cannot supply power to customers that are located at a distance beyond a set distance threshold to minimize power loss because of long distance of power transportation. With the input data estimated and the threshold and taboo regions set, our methodology assigns demand cells to each supply cell until there is no available capability to assign or one of the other constraints is met.

The neighborhood of range r for implementing the algorithm is defined as follows:

$$N^M_{(i_0, j_0)} = \left\{(i, j): |i - i_0| = r, |j - j_0| = r\right\}$$

where $r = 0, 1, 2, \ldots, M$ is the stage of iteration, i_0, j_0 is the location of the supply source of interest, and N is the number of acquired cells in a specific range. The substation cells acquire demand cells based on their ranked proximity and the available capacity. Based on prior knowledge of the guiding engineering principles, the acquisition starts west of each substation and proceeds in the clockwise direction. For each run of acquisition, the proximity of the available demand cells are ranked and assigned to the substation that is closest. This trend continues until all demand cells have been assigned or all supply capacity has been used up. In cases where there is more than one feasible substation for a demand cell, the demand cell is randomly assigned to one of the substations. The steps for implementing the algorithm are presented in pictorial form in Figure 27.3. This algorithm has been implemented for several countries, and the outputs have been validated using recent disasters. The results of some of the validation exercises are presented in Section 27.4.

27.3.2.2   The state of the art

The problem of estimating electrical substation SAs has received prior attention in the research community. One approach to estimate substation SAs uses the Voronoi method of weighted distances (Newton and Schirmer, 1997). This approach is based on substation load and distance from the substations. It produces a geometric partitioning of the area around substations. This approach does not include crucial inputs such as land use, population, and terrain features; hence, there is much uncertainty in the results. As an improvement to the Voronoi-based

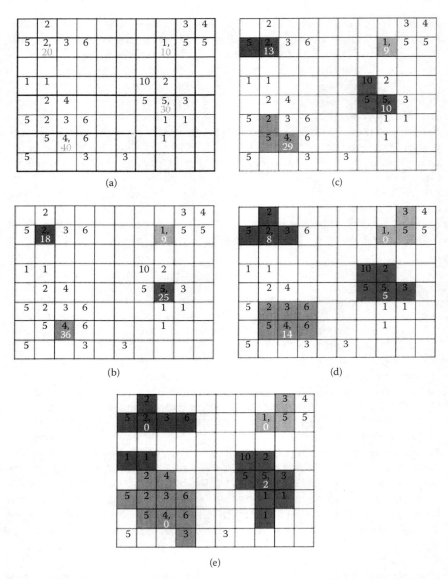

Figure 27.3 A step-by-step implementation of the SA algorithm. (a) Combined demand and supply matrix. Demand is shown in dark gray color and supply points are the four other colors. (b) At the first iteration, demand cells are assigned to supply cells and available capacity for supply reduced. The same color indicates cells assigned to the same supply source. (c) More cells assigned as iteration progresses. (d) At this time, the available capacity of one supply source becomes zero. (e) At the end of the iteration, the capacity of three of the four sources is zero. One source has a capacity of 2, but no demand point that meets all the constraints. So, the algorithm ends.

approach, Fenwick and Dowell (1998) developed a Cellular automata–based approach for estimating substation SA. Their approach is similar to the concept of Conway's Game of Life (Gilbert and Troitzsch, 1999). The approach divides an area into a discrete number of cells, each of which represents a SA portion. The cellular automata (CA) algorithm evolves by acquiring the cells that surround each substation much like a growing organism in the Game of Life. The cells that each substation acquires constitute its SA. The idea of using CA for constructing SA polygons is intuitive and has potential for accounting for the engineering assumptions of assigning utility customers to electrical substations. As a result, our approach is an extension of the CA approach by Fenwick and Dowell (1998).

The differences between the state of the art and our approach are in the sources and types of input data and the method of cell acquisition. Our methodology uses the LandScan Global Population Distribution Database, a high-resolution population dataset that the majority of key U.S. and foreign nongovernmental organizations working in disaster response use as their database system for worldwide geospatial distribution of populations. Furthermore, it is based on the nearest-neighbor method (Webb, 2002) in which the demand cells are acquired based on their proximity to each substation within a fixed spatial radius and the available capability of the respective substation. In some cases, a cell may be assigned to more than one substation to account for incremental development to the distribution system, which is a more realistic assumption. In the following subsection, we have described the input data for our approach.

27.3.2.3   LandScan population model

The LandScan Global population distribution model (Bhaduri et al., 2002; Dobson et al., 2000) involves a collection of best available census counts for each country and other primary geospatial input datasets such as land cover, roads, and slope. LandScan is at a higher resolution of approximately 1 km. There are many applications of the LandScan Global population data. The data have significantly enhanced the utility and impact of various applications such as counterterrorism, homeland security, emergency planning and management, consequence analysis, epidemiology, exposure analysis, and urban sprawl detection (Bhaduri et al., 2002). LandScan is extensively used by national and international organizations including the United Nations, the World Health Organization, the Food and Agriculture Organization, and several federal agencies in the United States and other countries. Oak Ridge National Laboratory has distributed LandScan Global population data to over 200 different organizations across the world. To inquire about how to obtain a copy of the latest LandScan Global population data, please visit http://www.ornl.gov/sci/landscan/ for additional information. Another variant of the LandScan

Global model is LandScan USA. The LandScan USA model uses the U.S. Census population counts at the block level to produce an output at a more detailed spatial resolution (~90 m resolution) than LandScan Global. LandScan USA also includes a temporal component in which the database contains both a nighttime residential as well as a baseline daytime population distribution.

27.3.2.4   Estimating electrical outage restoration time

In addition to knowing how many people may be without power after a disaster and where those people are located, we also need to know when power may be restored to those people. Therefore, to estimate restoration time, we rely on historical data to predict the number of repair crews that may be needed based on the assessment of the event (Lubkeman and Julian, 2004). Therefore, the restoration estimator is event dependent. The prediction changes as more information about the storm becomes available. However, our methodology assumes the following:

1. Restoration starts from areas of least damage and then moves to areas of heaviest damage. This assumption allows the cleaning crew to clear areas of heaviest damage before the repair crew movies in.
2. Restoration starts from densely populated areas and then moves to summer homes. Substations in the cities are restored before substations in the rural areas.
3. Restoration proceeds nonlinearly, and it is a function of the severity of the storm and the available resources.
4. It takes a crew of 8 technicians and 8 hours to restore 1 substation, which is called the 8-8-1 rule.
5. A crew restores only one substation in a day. This is a conservative assumption but is a reflection of historical data on restoration.
6. Restoration starts several hours after an event (usually 24 hours) depending on how large and strong the storm is. The 24-hour time lag allows the crew to clear any debris on the road and to get a better sense for the direction of the storm.

The methodology for the agent-based restoration estimator is summarized in Table 27.1.

27.3.3   Visualizing national electricity outage data

In addition to estimating the number of customers potentially without power and estimating when power may be restored to those customers, we also need a system for monitoring in real time the electricity outage situation. As a result, we also develop a visual analytics system for reporting

Table 27.1 Algorithm for Agent-Based Simulation of Power Outage Restoration Time

---

GIVEN *the number of affected substations and where they are located*
GIVEN *the number of electrical customers without power and where they are located*
GIVEN *the number of utility companies that serve these customers*
GIVEN *the number of available repair crew from each utility*
GIVEN *how many hours each repair crew member works*
 for each utility company $i = 1, ..., p$ do

- Develop a ramp and travel time table for the expected repair crews (ramp time is the time it will take each crew to travel to the damage area, and travel time is the time taken to move between repairs)
- Determine substations that serve critical assets such as police and fire stations, hospitals, and government facilities; these are critical substations
- Assign the first available repair crew to repair critical substations
- Develop a substation priority list by ranking the remaining substations with respect to the number of customers they serve
- Start the restoration process by assigning crews to substations at the beginning of each day based on the substation priority list
- Repeat restoration process until all substations are restored

end for

---

real-time electricity outages. Some of the U.S. utility companies maintain a version of electricity outage reporting system on their website. Visiting utility websites sequentially for situational awareness about power outage situation is not a solution. Our solution is to develop a process of extracting outage information from utility websites. This is a Web data extraction process. It is an implementation of low-level Hypertext Transfer Protocol. It basically transforms unstructured data on utility websites into structured data that can be stored, analyzed, and visualized in a central local database. This is an automated process that runs every 5 minutes. Figure 27.4a shows an example of the visualization system for the national outage map using Google Earth as a visualization platform. The outage data are aggregated at the county level.

The lighter shaded counties in Figure 27.4a are counties with no outage as of when this image was captured. A higher resolution image and other information about one area, Hamilton, are shown in Figure 27.4b. It should be noted that not all utility companies maintain a website of their power outage information. As a result, we are developing a mechanism for extracting outage information from social networking sites and Twitter feeds. The stages involved in this data-filtering process are much more intense. This is designed to be intense to avoid double counting and to guarantee the credibility of the extracted information.

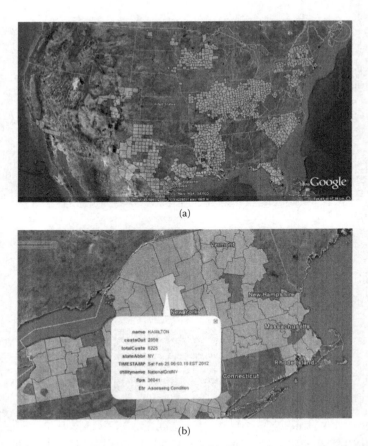

Figure 27.4 (a) The national outage map and (b) the number of customers without power in Hamilton.

27.4   Application of the EARSS system

In this section, we present two applications of the EARSS system. The first application was during the 2008 hurricane season in the United States, whereas the second application was during the 2011 tsunami disaster in Japan.

27.4.1   Application of EARSS during the hurricane season

Four of the tropical storms that made landfall in the United States in 2008—Dolly, Gustav, Hannah, and Ike—are used as examples in this section. In case of Hurricane Dolly, some 96 hours before landfall, we estimated that about 93 substations could be impacted by the hurricane wind using the projected damage area shown in Figure 27.5a. The spatial locations of the identified substations are also shown in Figure 27.5a. Using the identified substations, we also estimated that about 200,000 electrical customers could be without power for some hours or days. The spatial distribution of people that may lose power with respect to the locations of the substations is shown in Figure 27.5b. Furthermore, a projected restoration time is also estimated as shown in Figure 27.5c. After the storm, the observed restoration time is compared with the projected restoration time as shown in Figure 27.5c.

Similar analyses were also generated for the other three storms. For example, an overlay of Hurricane Gustav and the predicted damage contour is shown in Figure 27.6. An overlay of observed outages and projected damage contour for Hurricane Ike are shown in Figure 27.7.

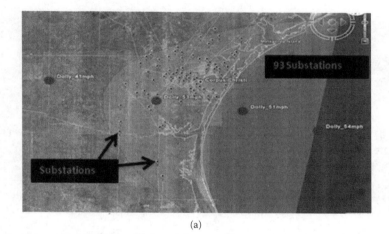

(a)

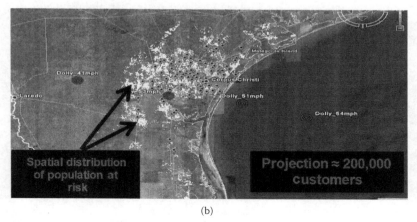

(b)

Figure 27.5 (a) The projected damage contour for Hurricane Dolly. (b) The projected spatial distribution of population at risk because of Hurricane Dolly. (c) The projected and actual restoration time for Hurricane Dolly.

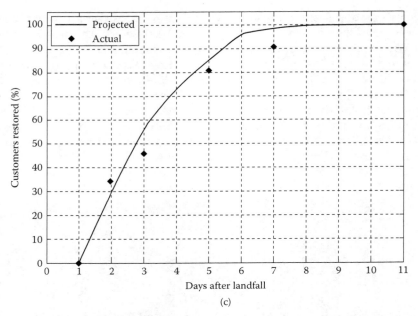

(c)

Figure 27.5 (*Continued*) (a) The projected damage contour for Hurricane Dolly. (b) The projected spatial distribution of population at risk because of Hurricane Dolly. (c) The projected and actual restoration time for Hurricane Dolly.

The comparison of the projected and observed number of customers without power for these hurricanes is shown in Table 27.2. One insight from the results is that it is usually difficult to predict accurately around the edge of a storm because storm direction changes as it makes landfall (see Figure 27.6 for an example). Even so, the projected number is a useful resource for the utility companies and emergency responders to plan for response and recovery.

27.4.2   Application of EARSS during an earthquake/tsunami event

As part of the real-time monitoring capability with EARSS, earthquake activities are monitored continuously using data from the U.S. Geological Survey. Once an earthquake that could cause damage to people and infrastructure (an earthquake with magnitude above a set threshold) is detected, real-time alerts are sent out as described in Section 27.2.

Once the 9.0 magnitude earthquake was detected about 43 mi east of the Oshika Peninsula of Tohoku on March 11, 2011, the EARSS data products were made available for different types of analyses. Some of the captured images of the data products that were available to the members of the EARSS network are shown in Figure 27.8. In addition to the data

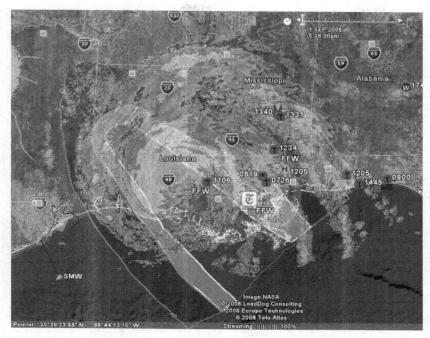

Figure 27.6 Projected damage contour for Hurricane Gustav before landfall (shaded area) and during the storm.

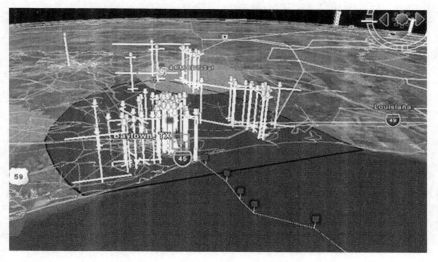

Figure 27.7 Projected damage contour for Hurricane Ike (dark shaded area) and observed outages (white poles).

Table 27.2 Comparison of the Projected and Observed Number of Customers
without Power

Hurricane	Projected	Observed	Absolute difference	Percent difference (%)
Dolly	200,000	236,000	36,000	−18.0
Gustav	1,830,000	1,820,000	10,000	0.5
Hannah (North Carolina)	210,000	187,000	23,000	11.0
Ike	3,820,000	3,770,000	50,000	1.3

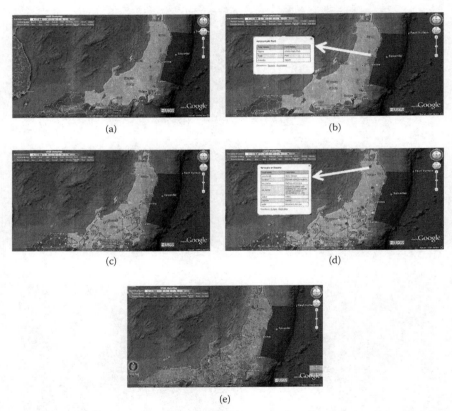

(a)　　　　　　　　　　　　　　　　(b)

(c)　　　　　　　　　　　　　　　　(d)

(e)

Figure 27.8 Application of Energy Awareness and Resiliency Standardized
Services (EARSS) during an earthquake event in Japan. (a) The shakemap for
the earthquake. (b) An overlay of ports on the shakemap for impact assessment.
(c) An overlay of the power transmission lines for downstream analysis. (d) An
overlay of refineries for further analysis. (e) An overlay of population distribution
for outrage assessment.

products, some of the model outputs were also shared. The detailed analyses performed during this disaster are beyond the scope of this chapter.

27.5 Conclusions

A real-time, all-hazards situational awareness framework cannot be anything other than a system of systems. The EARSS system is an example of such a system of systems. The EARSS system uses many systems, including the models for mapping and georeferencing assets and infrastructures, the models for assessing the impact of hazards on things and people within their affected areas, the computing resources for implementing complex algorithms, the manpower to keep the operational system functioning and alive, and the sensor infrastructures to be deployed and maintained. All these functions cannot exist in a single organization. The development of a system such as EARSS allows organizations to collaborate and share data, tools, models, and expertise. Furthermore, the use of data from open content sources guarantees the sharing and real-time updating of all data and model outputs.

Acknowledgments

This chapter has been authored by employees of UT-Battelle, LLC, under contract DE-AC05-00OR22725 with the U.S. Department of Energy. Accordingly, the U.S. government retains and the publisher, by accepting the article for publication, acknowledges that the U.S. government retains a nonexclusive, paid-up, irrevocable, worldwide license to publish or reproduce the published form of this manuscript, or allow others to do so, for U.S. government purposes.

References

Bhaduri, B., Bright, E., Coleman, P., and Dobson, J. (2002). LandScan: locating people is what matters. *Geoinformatics*, 5(2):34–37.

Dobson, J.E., Bright, E.A., Coleman, P.R., Durfee, R.C., and Worley, B.A. (2000). LandScan: a global population database for estimating populations at risk. *Photogrammetric Engineering and Remote Sensing*, 66(7):849–857.

Fenwick, J.W. and Dowell, L.J. (1998). *Electrical Substation Service-Area Estimation Using Cellular Automata: An Initial Report*. Available online http://www.osti.gov/energycitations/servlets/purl/319895-EnJlo0/webviewable/319895.pdf. (accessed July 24, 2010).

Gilbert, N. and Troitzsch, K.G. (1999). *Simulation for the Social Scientist*. Open University Press, Buckingham, Philadelphia, PA.

Lubkeman, D. and Julian, D.E. (2004). Large scale storm outage management. Power Engineering Society General Meeting, 6–10 June, Denver, Colorado, USA, *IEEE*, 1:16–22.

Newton, K.Q. and Schirmer, D.E. (1997). On the methodology of defining substation spheres of influence within an electric vehicle project framework. Environmental Systems Research Institute (ESRI) User Conference , July 8–11, 1997, San Diego, California, USA. Available online at http://proceedings.esri .com/library/userconf/proc97/proc97/to450/pap445/p445.htm (accessed January 5, 2007).

Webb, A. (2002). *Statistical Pattern Recognition*, 2nd Edition. Wiley, England, UK.

chapter twenty-eight

All-hazards response team preparation
Planning and training concepts*

Paul S. Pirkle III

Contents

---

* The material contained in this chapter was prepared as an account of work sponsored by an agency of the U.S. Government. Neither the U.S. Government nor any agent thereof, or any of their employees, makes any warranty, express or implied, or assumes any legal liability or responsibility for the accuracy, completeness, or usefulness of any information, apparatus, product, or process disclosed, or represents that its use would not infringe on privately owned rights. Reference herein to any specific commercial product, process, or service by trade name, trademark, manufacturer, or otherwise does not necessarily constitute or imply its endorsement, recommendation, or favoring by the U.S. Government or any agency thereof. The views and opinions of the author expressed herein do not necessarily state or reflect those of the U.S. Government or any agency thereof.

28.1   Introduction

A response team leader's primary incident response goal should be to prepare the team to perform its assigned duties competently and confidently in a complex, physically demanding, and time-limited environment. This is achieved by safely employing response team capabilities and coordinating efforts with multiple agencies. The response team leader must ensure that preparations are made in accordance with governing policies and guidance. Since the term "all-hazards" is prevalent in the incident response community, let us consider its origins and applications.

28.2   Origins and applications of the all-hazards approach

The term "all-hazards" is rooted in Federal Emergency Management Agency (FEMA) lore. FEMA can trace its beginnings to the Congressional Act of 1803. This act, generally considered the first piece of disaster legislation, provided assistance to a New Hampshire town following an extensive fire. More than 100 pieces of ad hoc legislation following natural disasters were passed over the next 100 years.

Emergency response activities were still disjointed into the 1970s. Over 100 federal agencies were involved in some aspect of emergency response. The proliferation of parallel state and local programs only exacerbated the convolution.

In response to a request from the National Governor's Association to clarify emergency response roles at all levels of government, President Carter issued Executive Order 12127 in 1979 to consolidate the fractured disaster-related responsibilities into FEMA.

During the tenure of FEMA's first director, John Macy, FEMA instigated the Integrated Emergency Management System with an all-hazards approach that included, "direction, control and warning systems which are common to the full range of emergencies from small isolated events to the ultimate emergency—war."

Following 9-11, the agency coordinated its activities with the newly formed Office of Homeland Security, and FEMA's Office of National Preparedness was given responsibility for helping to ensure that the nation's first responders were trained and equipped to deal with weapons of mass destruction.

Just a few years past its 20th anniversary, FEMA was actively directing its "all-hazards" approach to disasters toward homeland security issues. In March 2003, FEMA joined 22 other federal agencies, programs, and offices in becoming the Department of Homeland Security. The new department brought a coordinated approach to national security from emergencies and disasters—both natural and man-made (FEMA 2012).

There are many plans, programs, systems, and reports that have "all-hazards" inculcated in them. The National Response Framework and Homeland Security Presidential Directive 5 use this term. But what does "all-hazards" mean to an emergency responder? What does it mean to the various emergency responder communities? What does it mean to specific response teams? Let us bring "all-hazards" from the strategic to the tactical level and look at emerging changes in the U.S. Air Force.

The concept of all-hazards incident response nests well with the Air Force's capabilities-based planning, programming, and budgeting system. A threat-based viewpoint, which served its purpose for a long time, was predicated on a conventional nation-state enemy. The Cold War allowed the United States and the Soviet Union to produce rigid doctrines, plans, programs, and weapon systems over decades to deter and combat a relatively well-known adversary. Because of the nature of today's warfare, a shift from threat-based to capabilities-based thinking has occurred at the Department of Defense level and has subsequently permeated down through the services. Cold War relics such as "mutually-assured destruction" mean nothing to fanatical, nonstate terrorist tribes.

This approach describes how the Air Force operates on a macro level—to deliver air power anywhere in the world. But the same logic applies to incident response. The mentality for decades was to build response programs for specific events, such as wildfires, hurricanes, and acid spills.

The days of incident-specific checklists such as the "chemical incident checklist" are gone. Prepackaging response actions based on a presumed cause are relatively inefficient, ineffective, and inadequately capture reality. Being able to pinpoint the exact type or nature of the hazards associated with a response is frequently not immediately possible. Often, the incident unfolds like peeling layers of an onion, so the planned response actions should be organized accordingly. The prospect of perpetrated disasters (e.g., terrorist attack) fundamentally shifted incident preparedness, response, and recovery schemes.

The all-hazards planning approach is a more efficient and effective way to prepare for emergencies. Rather than managing planning initiatives for a multitude of threat scenarios, the all-hazards planning approach focuses on developing capacities and capabilities that are critical to preparedness for a full spectrum of emergencies or disasters (Ishikawa et al. 2007).

Penn State University follows an "all-hazards" approach to emergency preparedness, an idea that acknowledges the many similarities in responding to any threatened or actual emergency, regardless of the cause. Although there is a wide range of potential emergencies that could befall a campus—from fires, pandemics, and floods to power outages, cyber disruptions, and beyond—procedures can be put in place now that will help a campus community respond quickly and recover from any potential threat (Powers 2010).

Experience over the past few decades has taught us that attempting to identify the cause of a mass casualty incident (MCI) cause, especially in the early stages of an incident, is an exercise in futility. The Oklahoma City Fire Department thought it was responding to a gas explosion on April 19, 1994 when heading to the bombing site at the Alfred P. Murrow Federal Building. Firefighters responding to the World Trade Center bombing in 1993 believed the incident was a transformer fire. In both cases, the initial assumptions of cause led to decisions and actions that incurred more risk to responders and the public. An all-hazards approach would have mitigated that risk. Incident commanders who direct actions based on observed conditions and effects without attribution of cause are exhibiting best practice (Kanarian 2010).

The bottom line is that effective incident response capabilities are designed, built, and exercised so that responders can apply them during a situation, independent of immediate knowledge of the incident's cause.

28.3 Basic incident response preparedness and execution concepts

There are many ways to successfully plan, train, and exercise, but best practices have been observed during numerous incident response exercises and competitions. The following is a generalized approach based on those best practices that have proven effective for a hazardous materials response team. It is not all-inclusive or detailed enough to build a complete program but provides a sound framework to do so. Depending on your response team's mission, the content described below may be very different. The main point is to plan to a sufficient level of detail to positively influence the training and exercise program.

- Initial information gathering
  - Take the initial call
  - Employ a checklist to capture as much relevant information as possible
- Organizing the team
  - Gather all personnel present in the office
  - Start assigning team roles
  - Consider recalling other personnel
- Initial response team activation
  - Take initial response team kit

- Know location of/directions to Incident Command Post
- Use safe route
- Equipment preparation
  - Use checklist to pull equipment off the shelf and get ready for deployment
  - Perform calibrations/operations checks
  - Conduct background measurements
  - Prepare response vehicle and trailer
  - Ensure auxiliary equipment (e.g., generators) is ready
- Initial response team arrival
  - Report to incident commander
  - Obtain critical incident details using a checklist
  - Begin shaping remainder of team's response
  - Initial response team chief assumes role of response team chief
- Follow-on team arrival/entry
  - Report to response team chief
  - Stage equipment/supplies
  - Receive situation briefing from response team chief
  - Establish entry objectives and specify associated tasks
  - Don personal protective equipment
  - Receive safety briefing
  - Enter the hot zone
  - Perform tasks to fulfill entry objectives
- Follow-on team exit
  - Decontamination
  - Doff personal protective equipment
  - Brief response team chief
- Reconstitution
  - Repack materiel onto trailer/response vehicle
  - Return materiel to permanent storage location
  - Plug in batteries to recharge them
  - Refill self-contained breathing apparatus cylinders
  - Order replacement supplies
- Reporting
  - Complete postexercise/incident summary (PEIS) report within a specified time
  - Submit the PEIS report to medical readiness
  - Comply with other local reporting requirements

This approach begins to tackle the planning aspects of incident response. However, there are two major areas—equipment management and training/exercising—that must also be addressed to fully develop an incident response capability.

28.4   Optimum incident response equipment management and proficiency practices

Many incident response teams utilize equipment to accomplish assigned tasks. For illustrative purposes, the following discussion involves equipment needs for a notional hazardous materials response team. Equipment preparedness is fairly straightforward and breaks down to the following:

- Is the equipment present and ready to use?
- Are the operators proficient in using the equipment?

Regarding the first question, there are three main areas: (1) inventory, (2) calibration and maintenance, and (3) operations checks.

Although a centralized function within the organization may be responsible for the official equipment inventory, the response team must be actively engaged in the process. This is especially true for expendable supplies.

Not only do equipment items need to be present, they also need to be properly calibrated and maintained. One key feature of your calibration and maintenance program needs to be staggering like pieces of equipment and capabilities. For example, if there are four like-equipment items, schedule two for annual calibration and the other two with a 6-month offset from the original two. The same should be done for two equipment items having similar capabilities. This strategy will balance availability of capability with the needs of calibration and maintenance.

Operations checks need to be done on a routine schedule. The response team should choose a schedule that makes the most sense for the equipment items and the nature of the response team mission. This may sound relatively simple, but there are some learned lessons that will enhance this component of your equipment preparedness efforts. First and foremost, you need to define what "operationally ready" means for each piece of equipment and then build a step-by-step review process for a technician to accomplish the operations check. Dimensions of "operationally ready" include the following:

- Batteries are charged.
- All required items—calibration source, carrier gas, screwdriver, gasoline for generator, and so on—are present and ready to operate the equipment.
- Equipment appropriately responds to challenge (e.g., calibration, blank run).

Note that a sound operations check program cross-links with a strong training program. In other words, insight gained during a thorough hands-on training session can improve operations check procedures for that piece of equipment.

Like many aspects of managing and leading a response team, incident response equipment preparedness is primarily an issue of priority and diligence. It is a classical example of "what gets measured, gets done" in that you need to establish a system of expectations and accountability. The price for not doing so is too high, given that mission performance is critically tied to availability and operational readiness of response team tools and the proficiency of the operators who use them.

28.5   Novel incident response training and exercise techniques

28.5.1   Purpose

The purpose of a response team's training and exercise program is to prepare team members to competently and confidently execute the team's mission in a complex, physically demanding, and time-limited operational environment. This can best be fulfilled with a diligent, targeted, and integrated approach that incorporates maximum realism and skill demonstration (verification).

28.5.2   Background

Some response team training programs could be characterized as mid to junior team members taking turns presenting prepackaged, outdated slide shows. The result is often drudgery for the presenter because of last minute preparation and lack of oversight. There is a general sense of simply "filling a square." The training topic and delivery miss the mark, which means that team members are still not competent or confident in how to execute many facets of the team's mission. There can also be a negative impact on morale.

28.5.3   Overview

Break free from the old way of flipping PowerPoint slides! There are basic tenets of effective training and exercise programs. They include the following:

- Training objectives: Systematically identify what you need to know and be able to do.
- Block training structure: Different levels of learning for specific training objectives.

- Preparation/review cycle: Instructor expectations, presentation fine tuning, and hands-on logistics.
- The "take-away": What you want the student to take away from the training event. This should be closely aligned with training objectives.
- Make-up training: Since some training sessions are hands-on, you will need a novel approach.
- Quarterly schedule: Set training objectives and a mix of training/ exercise events on a quarterly schedule.
- Maximum reality exercises: Simulate as little as possible.

28.5.4 Training objectives

Establish annual training objectives based on overarching directives, policies, and guidance for the response team and the local assessment of operational needs. Identify topics, take-aways, training blocks, and frequency.

Training blocks are identified in Table 28.1. Training frequency is determined by assessing the complexity and significance of the skill. In other words, skills that are more difficult to master and contribute more significantly to the response team mission should be covered more frequently than those that are relatively simple to perform and have a secondary importance to mission accomplishment.

28.5.5 Quarterly schedule

Build a quarterly schedule using a checklist that reminds the response team's training manager which topics should be covered and how often. Assign instructors based on guidance in Table 28.1 and on instructor experience and skills. Obtain approval from the response team leader or designated representative approximately 3 weeks prior to the start of the next quarter.

28.5.6 Preparation

One of the primary deficiencies with the "old way" is a lack of communication on training expectations from the instructor and a subsequent lack of accountability. To counter that, a preparation and review scheme should be devised. As a reference point, training day is designated $T = 0$. In the example below, Friday was assigned as "training day."

At T–9 days, the Wednesday of the week prior to training, the instructor is required to brief the training manager on the training topic and proposed methodology for achieving the assigned training objectives.

Table 28.1 Training Blocks

Block no.	Training goals	Method	Plain language	Minimum level of instructor	Time allotment (estimate)
1	Gain knowledge: Students presented with knowledge through concepts, procedures, techniques, and so on.	Teaching	I tell you	Technician	30 min. to 1 h
2	Demonstrate knowledge: Students prove knowledge gained by demonstration.	Testing	You tell me	Senior technician	1 h
3	Show and tell: Students introduced to tools such as equipment and computer software	Presentation to group	I show you	Technician	30 min. to 1 h
4	Task/skill performance: Students taught how to operate tools and yield a result. Students competent and confident in utilizing tools via demonstration. Students interpret results and provide analysis.	Hands-on demo by individual	You show me	Senior technician	1 to 2 h
5	Capability performance: Students demonstrate Block 4 skills in a physical near real-world environment.	Scenario	Do in the field	Response team leader	3 to 4 h

At T–3 days, the Tuesday of the week of training, the instructor provides a "dry run" to a review panel. Instructors do not give a complete lesson. Instead, focus on the take-aways, primary teaching points, methods, logistics, adjunct instructors, and the like.

At T–1 day, the day prior to training, the instructor provides the final version of the presentation to the response team training manager. After final review with the training manager, the instructor provides the final outline for filing in the team members' training records.

28.5.7   Training day

Whether training occurs every day, week, month, or quarter, the team needs to have a clear expectation so that other duties can be scheduled. This frequency will be determined by many factors, including the team's mission, personnel, and culture.

One team declared every Friday as training day. This was more a mindset than an unbreakable rule. However, response team members were prohibited from setting any work or personal appointments on Fridays, except under rare, approved circumstances. Vacations and business travel were scheduled, to the maximum degree possible, on days other than Fridays because the training experience could often not be sufficiently reproduced.

Although every Friday was designated as a training day, there were cases when the scheduled training did not comprise the entire duty day. When that happened, supervisors ensured their subordinates returned to routine assigned work.

28.5.8   Make-up training

Make-up training is actually more challenging to manage because of the difficulty of re-creating the training experience (no longer flipping PowerPoint slides). Therefore, consider videorecording sessions for those making up training. Hands-on portions can be done one-on-one with the trainer during the make-up session.

28.5.9   Response team exercises

The cadre for a particular quarter also plans and executes the response team exercises. When the schedule allows and there is training value in doing so, it is best to conduct a tabletop exercise one week and a field exercise on the same topic the next week. Start with a training objective and build the scenario and evaluation criteria accordingly.

Logistics is a primary consideration for exercises. Work out in advance who (cadre and players) and what (equipment/supplies) are supposed to be where and when. Identify an adequate field site and then contact the owner to ensure proper access and authorization. Artificial inputs and conditions should be kept to an absolute minimum, and simulants should be used to register a response from equipment whenever feasible. Be sure to label and store simulants properly.

Invite other response team partners at least 2 weeks in advance as appropriate, depending on the objective/scenario. Submit an exercise summary report and send it to the appropriate oversight office to ensure you receive exercise credit.

28.5.10 Response team competition

Some response team communities might be advanced enough in their preparations that they are ready to install the next level of preparedness—competition. Examples include multiple national-level firefighter competitions and military events such as the Air Force's Global Strike Challenge for security forces and the Integrated CBRN Exercise & Training (ICE-T) event for bioenvironmental engineering and emergency management personnel.

A response team can set up its own quarterly competition based on the same principles used to design and execute large-scale Olympics-style competitions. One way to do that is to establish competitive sub-teams that will be shuffled every quarter to foster overall response team cohesion and avoid parochialism. The number of competition sub-teams will be dictated by the overall size of the response team. The response team leader should assign sub-teams and consider excused absences such as projected vacation and business trips when doing so. One successful approach is to hold the competition for six consecutive quarters, take a break one quarter to reconstitute, and then return to six consecutive quarters of competition.

The response team leader should serve as permanent cadre and select other response team members, based on past performance and their ability to carry out cadre tasks, to serve on his or her staff for that quarter. Cadre members should be expected to perform their duties in addition to their regularly assigned work from their supervisors.

Jeopardy games (Block 2 in Table 28.1) and relay races (Block 5 in Table 28.1) may comprise the competition events. Scoring the events is a combination of art and science. The response team leader must gauge each event's weighted contribution to the overall effort. In general, the scoring is stringent to provide distinction in performance and to give players a high goal to achieve. For example, a person who grades out with a 75% might actually be a star performer. This score would not necessarily be considered "C" average work.

Table 28.2 presents an example relay race scoring sheet. Each relay race event is assigned a station number, which is described below the table. Below each station number at the top of Table 28.2 is a station point value. This is the highest score that an individual can achieve for that station. The individuals' names and their respective scores for each station are recorded in the table. The station total for that team is annotated below Magellan's row of scores. Below that, the maximum station value (i.e., station max, calculated by multiplying the station point value by the number of team members) is posted. The station percentage below that is simply the sub-team's station total divided by its station max.

Scanning the right side of Table 28.2, the person total value is a summation of that person's scores for all of the stations. The person max number is a summation of the station point values and the person percentage is person total divided by person max.

Table 28.3 is an example consolidated scoring sheet to track points accumulated during a given quarter. The raw point value is multiplied by a weight to come up with a score for each event. Each jeopardy is typically assigned a base weight of 1. Most relay races are assigned a weight of 3, but not all relay races are equally challenging. So, there is some flexibility consistent with the art/science comment made earlier. Because of the arbitrary nature of assigning weights, the response team leader should always

Table 28.2 Example Relay Race Scoring Sheet

Station	1	2	3	4			
Team member	20	100	90	125	Person total	Person max	Person (%)
Jackson	20	55	80	40	195	335	58
Smith	20	75	75	90	260	335	78
Jones	5	35	70	30	140	335	42
Cortez	20	20	90	40	170	335	51
DeGama	15	45	80	20	160	335	48
Magellan	15	10	70	0	95	335	28
Station total	95	240	465	220	1020	2010	51
Station max	120	600	540	750	Team total	Team max	
Station (%)	79	40	86	29			

Station	Description
1	Gamma spectrometer—identify isotope
2	PID/FID instrument quiz
3	pH/M-8 (bleach)—30 each phys prop, ID (15 pH, 15 M-8), interpretation
4	Electronic personal dosimeter—recognize alarms

Table 28.3 Example Consolidated Scoring Sheet

Date	Event	Weight	Sub-team Alpha		Sub-team Bravo		Sub-team Charlie	
			Points	Score	Points	Score	Points	Score
April 25, 2011	Jeopardy	1	0	0	0	0	380	380
April 25, 2011	Relay race	3	480	1439	534	1603	482	1447
May 30, 2011	Jeopardy	1	540	540	240	240	0	0
May 30, 2011	Relay race	2.5	581	1453	495	1238	480	1200
June 27, 2011	Jeopardy	1	258	258	100	100	220	220
June 27, 2011	Relay race	2	0	0	100	200	200	400
			1859	3690	1469	3381	1762	3647

do a "sanity check" to make sure that the final results are consistent with the overall performance during the quarter. To help with that, the score for the winning team for each event is highlighted in gray.

This being a competition, the response team leader might consider awarding the winning sub-team members. The individual contributing the most points to his or her sub-team also could be recognized as the most valuable player.

At the end of a quarter, the response team leader then analyzes the results primarily in two ways. First, he or she identifies top performers over time in multiple situations using many tools. It becomes apparent who excels and can execute under some time and environmental stressors. This helps with selecting cadre members and recognizing who can be relied on for real-world events. The person percentage values on the right side of Table 28.2 help with this assessment. In that example, Smith might be a star performer. Track his or her performance in other quarters.

Second, the analysis affords the opportunity to determine what the response team–wide training improvement opportunities are. If a large percentage of personnel did not understand how to employ a particular tool, it indicates the need to readdress either the content of the original training or to increase the frequency of training on this topic. The station percentage values at the bottom of Table 28.2 help with this assessment. In that example, electronic personal dosimeter (EPD) alarm recognition and photoionization detector/flame ionization detector (PID/FID) knowledge probably need to be considered for retraining.

28.5.11   Response team competition logistics

Ensure all equipment and supplies needed for your response team competition are properly used before, during, and after employment. This means ensuring needed items are available, operationally ready, and calibrated prior to planning a competition event or exercise. Cadre members must be familiar enough with the tools used to perform field services to keep them running properly. After they are used, tools must be returned in an operationally ready state for future use. This may require ordering supplies to reconstitute your response team capability.

28.6   Conclusion

The odds of knowing when the bad day will come and what it will bring are very low. More likely, the bad day comes and your response team will be left with managing and mitigating its effects, not knowing right away what its cause is. Your response team will counter with its capabilities and desired effects using an all-hazards approach. The tools and their operators must be ready today.*

References

Federal Emergency Management Agency (FEMA) History, site last reviewed June 2012. Available at http://www.fema.gov/about.

Ishikawa, C.K.T., Simonsen G.W., Ceconi B., Kuss K. "Universal Design: The All-Hazards Approach to Vulnerable Populations Planning" (abstract), American Public Health Association Conference, November 5, 2007. Available at https://apha.confex.com/apha/135am/techprogram/paper_160527.htm.

Kanarian, S. "An All-Hazards Approach to Terrorism," Fire Engineering, March 1, 2010. Available at http://www.fireengineering.com/articles/print/volume-163/issue-3/departments/fire-service_ems/an-all-hazards-approach-to-terrorism.html.

Powers, L. "University Continues Emergency Response Work at Campuses," Gant Daily, August 17, 2010. Available at http://gantdaily.com/2010/08/17/university-continues-emergency-response-work-at-campuses/.

---

* Any opinions expressed in this draft document are solely those of the author and do not necessarily represent those of Battelle Memorial Institute or the U.S. Air Force. No endorsement of any product is offered or implied.

chapter twenty-nine

Medical supply chain resiliency in disasters

Jack E. Smith II

Contents

29.1 Introduction

The primary objective of emergency responders, public agencies, and nongovernmental organizations (NGOs) during the emergency response phase of a disaster is to reduce the mortality, morbidity, and suffering of the victims. In addition to search and rescue operations and initial medical care, the rapid delivery of vital goods such as food, water, medical supplies, and shelter has always been an essential component of an effective disaster response. While responders and humanitarians from around the world are well acquainted with this doctrine, and the devastating consequences of its limitation, the gravity and importance of this concept were demonstrated for the American people during the aftermath of Hurricane Katrina. As the surviving inhabitants of New Orleans sat for days sweltering in the Superdome, the Convention Center, and on the elevated highway overpasses, the world watched as they waited for the most basic supplies including food, water, and medicines. Owing to poor planning, weak leadership, and failed coordination of response efforts at all levels of government, the same post-disaster chaos that responders and humanitarians are so familiar with in developing countries unfolded in one of the richest countries in the world.

While many suffered in the heat, this failure of initiative kept convoys of Red Cross and commercial haulers loaded with food, water, hygiene kits, and ice sitting on the roadside just outside the city limits. Mismanagement and supply chain breakdowns led to 110 million pounds of ice (60% of the total ordered) and $900 million in manufactured homes being wasted (the infamous Federal Emergency Management Agency [FEMA] trailer debacle) (Dowty and Wallace, 2010). In two separate cases, truck drivers for Universe Truck Lines were dispatched to haul ice for FEMA to the Katrina victims. After they were redirected five and six times over a 7- to 10-day period, they logged 3,282 and 4,100 miles, respectively, and never delivered their ice to a single hurricane victim. The total bill to FEMA was $15,000 when the ice was worth only $5,000 (Dowty and Wallace, 2010).

29.2   Supply chain management

Supply chain management (SCM) involves more than delivering products to a destination. It requires close coordination for the transportation, distribution, storage, and timely delivery of inventory, while minimizing costs and serving the maximum number of consumers. While this is important in private enterprise, it is critical when human life is at stake. Some of the concepts and lessons from commercial SCM are applicable to the disaster relief supply chain. However, the dynamics of a disaster relief supply chain would confound most commercial supply chain managers. Emergency managers and logisticians, unlike their counterparts in the private sector, are constantly faced with ever-evolving and unanticipated situations. Information on when, where, what, how much, where from, and how often are the basic parameters of an efficient supply chain, yet they are completely uncertain in the post-disaster situation. On the other hand, military logisticians have long established principles and procedures for constructing and operating long supply chains in high risk, dynamic environments. Their experience bridges the need to support the warfighter, rapidly evacuate and treat the wounded, as well as support humanitarian missions in developing and failed states. Therefore, the military-oriented logistics and supply chain is likely to have significant commonality with the challenges and complications of the disaster supply chain.

One strategy for improving supply chain efficiency is forecasting demand of both the types of products and consumption rate. However, in the post-disaster environment, even historical data are not always reliable in forecasting demand for relief supplies and resources. While some data have been compiled from past events by various NGOs and government agencies such as the Emergency Events Database by the Center for Research on the Epidemiology of Disasters, they tend to be largely inaccurate due to inconsistent and/or insufficient data collection and reporting problems. Moreover, each disaster is unique even if it occurs in the exact same location. Although injury and illness profiles can be forecast for a given type of disaster, the number of victims can be affected by time of day, variations in evacuation compliance, seasonal transiency, virulence of a contagion, economic conditions, and other factors (Ergun et al., 2010).

While focusing on the effects of short notice natural disasters on hospitals and community health-care systems, this chapter will compare disaster supply chains with commercial supply chains, identify strategies for enhancing supply chain resilience, and apply those strategies to existing hospital supply chains.

29.3   Disaster versus commercial supply chains

The disaster relief supply chain operates under conditions that would overwhelm most commercial supply chain managers. Different objectives and mechanisms drive disaster relief supply chains. Their operating environment is extremely uncertain and dynamic, life safety and assistance are prioritized over cost, unique solutions are often used to overcome obstacles, and unusual management principles are often employed. Disaster relief supply chains are considered a subcategory of humanitarian supply chains. The primary difference between these categories is that disasters tend to be sudden onset events with zero or very little notice, whereas humanitarian assistance can be an ongoing operation as in the case of famine relief or refugee support. A disaster relief supply chain must be established with very little notice; requires good information from the end user as to the resources needed; depends on well-identified, improvised delivery locations, travel routes, and so on; and relies on a coordination of efforts between organizations that have disparate organizational cultures, utilize different jargon and communication methods, and probably have never previously worked together.

Within the United States, disaster response is organized under the FEMA in accordance with the National Response Framework (NRF) (FEMA, 2008). The NRF outlines the roles and responsibilities of federal, state, local, and tribal government agencies and how they should coordinate among themselves during the response phase of a disaster, as well as the mitigation, preparedness, and recovery phases. The NRF utilizes the National Incident Management System (NIMS) as the national standard based on "accepted best practices" for emergency management and response. During the response phase of a disaster, local emergency management and first response agencies should establish an Incident Command System (ICS), as outlined by NIMS, to provide the framework and structure necessary to effectively manage the resources required to respond to the incident. While this construct is the national standard within the United States and has been widely adopted by state and local governments across the country, proficiency levels in its application and utilization can vary widely from community to community. Both NIMS and any local incident-specific ICS provide officials and responders with a clear chain of command/unity of command, clear lines of communication, and efficient division of labor/span of control. Under the incident commander, the ICS is structured into five functional areas including command, operations, planning, logistics, and finance/administration (FEMA, 2008).

Disasters by definition are events that overwhelm the capability of local authorities (or a specific organization) to adequately respond to and recover from the situation. When outside resources such as manpower, equipment, and supplies are required to adequately manage the response and ultimately recover from the impact of the event, disaster relief supply chains are established under the logistics section of the ICS. Because life-saving resources are often scarce during these types of events, the ability to manage them effectively and efficiently is paramount. It has been said that the goal of disaster supply chains and logistics is to "get the right supplies to the right people, at the right time, in the right place". In order to do this successfully, emergency managers and responders must accurately size up the situation and develop an Incident Action Plan that includes identifying the critical factors of the incident (e.g., immediate rescues), and developing SMART objectives (objectives that are Specific, Measurable, Action oriented, Realistic, and Time sensitive) (FEMA, 2010). With regard to logistics and supply chain, this strategy helps incident commanders, planners, and logisticians to develop good situational awareness and assists them in anticipating what resources will be needed to mitigate the incident.

Commercial supply chains provide the end user (demand) with the goods and services they require from various vendors (suppliers). In a typical disaster supply chain, the logistics section of the ICS functions much like a third-party logistics provider by attempting to identify and forecast demand of the end customers (responders and victims) and then coordinating government resources (search and rescue teams, or National Guard units) and NGO resources (Red Cross, volunteers) to fulfill those needs (supply). However, as Table 29.1 illustrates, the disaster situation creates a number of dynamic situations that are very different from the well-established and relatively static environment in which a commercial supply chain operates.

Domestically, state, local, and tribal governments retain the ultimate authority and oversight of the emergency response and recovery activities. Ignorance of both the NRF and NIMS can lead to poor proficiency in implementation of a local ICS and the disaster supply chain as was demonstrated during the Hurricane Katrina response. Internationally, sovereign nations can complicate the response and recovery efforts due to their foreign relations and internal policies and the decisions they make directly affect the supply chain. In 2004, during the aftermath of the massive Indonesian tsunami, the Indian government prohibited international

Table 29.1 Disaster Domains and Corresponding Supply Chains

Domain	Disaster supply chain	Commercial supply chain
Operating environment	Dynamic	Static
System priority	Life	Cost
Demand	Difficult to forecast	Easier to forecast
Supply	Some unknowns	Standard order filling
Operating procedure	Impromptu	Standardized
Cost efficiency	Low	High
Parties	Disparate/unfamiliar	Consistent/well established
Goods	Inconsistent	Consistent
Delivery routes	Fluctuating	Well established
Delivery points	Fluctuating	Well established
Transportation modes	Various/improvised	Standardized

aid agencies from participating at all in the first 60 days of the relief effort, and relied solely on local resources (Ergun et al., 2010). After Cyclone Nargis hit Myanmar (Burma) in 2008, the ruling military government refused to allow a number of relief organizations to enter the country. After remaining anchored offshore for more than 2 weeks, several French and American ships finally had to return home (Whybark et al., 2010).

Another challenge unique to disaster and humanitarian supplies chains is the supplies that come from in-kind donations. In the regular supply chain, a standardized order fulfillment process is followed, whereas in the disaster supply chain, a portion of supplies comes from voluntary donations rather than being driven by the specifications of the end user. Because speed is of the essence immediately after a disaster and real-time information can be vague, incorrect, or nonexistent, both government agencies and NGOs often "push" anticipated resources to the impact area. This influx of supplies, equipment, and volunteers all requires manpower and effort to manage and coordinate. Furthermore, the dynamic environment and the uncertainty of demand can cause significant inconsistencies between supply and demand in terms of mix, quantity, timing, and location (Whybark et al., 2010). This creates a high level of uncertainty of what is going to be received and when. For example, consumables that arrive too early and cannot be stored for a long time and non-consumables that arrive too late are wasted. Nearly every disaster both domestically and internationally in recent history has accounts where donated items were not needed and were not distributed to the people affected by the disaster. After the 1988 Armenian earthquake, international relief operations provided over 5000 tons of drugs and consumable medical supplies. However, only 30% were immediately usable (sorted, relevant for the type of emergency, and easy to identify), and 20% of these supplies were unable to be used and ultimately destroyed. Supplies that are inappropriate and incompatible demand management effort, tax scarce storage and transportation resources, and cause bottlenecks in the supply chain (Ergun et al., 2010).

Closely related to the influx of unrequested donations is the equally frustrating and confounding issue of "self-dispatched" volunteers (individuals, faith-based and community service groups, or contractors) that surge on a community and begin operations, sometimes without obtaining proper credentials or permission from local emergency managers or incident commanders. Regardless of whether these volunteers are altruistically motivated or are opportunistic scammers, anyone operating outside the established ICS create additional workload for emergency managers and local officials who have to validate their credentials; provide food, water, and shelter for them; coordinate their work efforts; treat their injuries; and prevent fraud and price gouging.

Another obstacle not typically encountered in a well-established commercial supply chain is the utilization of ad hoc warehouses that are influenced by factors such as building availability based on the extent of damage sustained; condition and level of damage to the nearby transportation infrastructure; availability of transportation, fuel, and drivers; condition of the communication infrastructure; and other dynamic and cascading events. The nature of sudden onset disasters along with the inability to forecast the impact of most disasters creates a plethora of problems that are common in disaster relief supply chains. These problems include the challenge of developing rapid and accurate situational awareness to determine and anticipate necessary resources, the lack of predetermined resource lists and redundant suppliers, high expediting costs, choice of wrong transportation mode and/or provider, unidentified or poorly chosen staging areas and warehouses, inability to identify and clear bottlenecks throughout the supply chain, and incomplete execution. Therefore, the importance of adequate planning to address these issues during the preparedness phase of emergency management cannot be stressed enough (Ergun et al., 2010).

Perhaps the most significant challenge of the disaster supply chain is the management and coordination of the numerous agencies and organizations (government, military, NGOs, volunteers, and suppliers) along with all the resources that need to be delivered. Despite the different cultural, political, geographical, and historical differences among them, the specialized capability of each is becoming increasingly needed in the disaster supply chain. Further complicating this issue is the fact that people in charge of logistics or SCM in most ad hoc ICS structures and NGOs typically are not specialized in this field. Therefore, the level of proficiency in solving problems that occur during the operations is inconsistent. Additionally, domestic barriers such as required permits, credentialing, specific policies of the region, or excessive paperwork may cause additional delays, as well as external complications due to foreign relations.

29.4 Hospital supply chain

Not only did Hurricane Katrina expose the weaknesses of federal, state, and local government agencies to successfully establish and coordinate disaster relief supply chains, it also demonstrated that as institutions, hospitals must do their own preparedness and planning with regard to business and supply chain continuity. Because of the critical nature of their business, it is imperative that hospitals collaborate with their regional planning partners (public and private), group purchasing organizations (GPOs), and even competitors to develop their own SCM strategies and infrastructure to ensure that essential provisions are available during a disaster.

To reduce a hospital's vulnerability to supply chain disruption, supply chain managers and emergency managers often seek to balance commercial SCM efficiency strategies with those that develop robustness and resiliency. Challenges to efficient SCM include developing supply chain strategies that support other business systems and overall organizational objectives, minimizing costs while maximizing service levels, and managing uncertainty and risk. This section focuses on managing the uncertainties and risks that are inherent in every supply chain, including those that serve hospitals, and how to develop an efficient, yet resilient supply chain.

One foundational element of an efficient or a lean supply chain is the ability to forecast demand and, in turn, flex or surge the supply chain to respond to increases in demand. Small surges are typically handled through internal buffers of extra supplies or simply increasing the next order. Larger surges in demand require the supply chain to have a degree of resilience in order to accommodate those surges. A hospital's daily demand can be forecasted to a degree by considering patient throughput, elective surgeries, scheduled visits, and historical census data for emergency department visits. However, unforeseen surges in patient volumes resulting from communicable disease outbreaks, extreme weather, mass casualty events, and obviously disasters cannot be predicted. These uncertainties necessitate that medical supply chains must be designed to enhance both resiliency and reliability (Day et al., 2009).

A Government Accountability Office (2003) report on hospital bioterrorism preparedness found that of the 2,000 urban hospitals surveyed, half had fewer than six ventilators per 100 staffed beds, three or fewer personal protective equipment suits, and fewer than four isolation beds. Since it is cost prohibitive to stockpile and maintain enough supplies or equipment for every disaster or mass casualty event that a hospital could face, it is crucial that each participate in collaborative regional planning. For example, states can access resources such as the Centers for Disease Control and Prevention's Strategic National Stockpile (SNS) of antibiotics, antidotes, or the Federal Medical Station for patient surge. However, according to testimony in 2004 before the House of Representatives, most states had not at that time developed plans to access the SNS, and only about a third had outlined how they would distribute the drugs (Dimitruk, 2005).

Currently, the cost of health care in the United States is around 18% of gross domestic product (The White House, 2009). Because of increased competition, dwindling reimbursement, greater regulatory constraints, and higher customer expectations, hospitals, like any other business, must manage costs and improve efficiencies. As part of their overall cost containment strategies, hospitals have adopted the same "Just-in-Time" and lean supply chain strategies used by private industry. These cost reduction strategies cause most hospitals to typically stock only 24–72 hours worth of medical supplies, pharmaceuticals, food, and water (Day et al., 2009). However, this strategy of reduced overhead costs increases a hospital's vulnerability to supply chain disruption and demand spikes from patient surge levels secondary to a large-scale emergency or disaster.

The Joint Commission on Accreditation of Healthcare Organizations (JCAHO) requires hospitals and health-care organizations to be able to maintain operations for 96 hours without outside assistance from the community (Chisholm, 2010). This does not mean that health-care organizations are required to stockpile 96 hours worth of supplies. The 96-hour standard can be met through the utilization of several strategies including stockpiling of supplies, vendor-managed inventory (VMI) where increased consumption rates trigger increased ordering of supplies, prime vendors leveraging their own supply chain resiliency to pull supplies from unaffected regions, and memorandums of agreements/understanding with other organizations for the sharing of supplies and equipment during shortages.

In emergency management parlance, the terms "robust" and "resilient" are often used interchangeably when discussing community or organizational mitigation and preparedness initiatives. However, in the context of supply chains, the two terms have different implications for for-profit businesses. The term "robust" is defined as "strong, and sturdy." Thus, a robust supply chain would be expected to falter little when impacted by a disaster. The primary way to build a robust hospital supply chain is through layers of redundancy (e.g., stockpiling of supplies on site and in extra warehouses, or utilizing multiple vendors), none of which supports the lean supply chain theory nor is it typically an economically viable option for for-profit businesses (Christopher and Rutherford, 2004).

"Resilience" is the ability of a community or organization to return to its original state (or newly desired state that incorporates improvements to enhance future disaster resilience) in a timely manner following the impact of a disaster. A resilient supply chain must also be adaptable, especially if the desired state is different from the original. A robust supply chain may be strong but if it is not adaptable, it is not necessarily resilient. Conversely, a resilient supply chain can be robust, but it is also able to respond to sudden and unexpected spikes in demand or interruption from extreme weather, for example (Christopher and Rutherford, 2004). A hospital that has a robust (sturdy but inflexible) supply chain may find all of its off-site redundancy useless when surrounded by flood waters, blanketed by a severe winter storm, or cut off by blocked or damaged roadways. The hospital with a resilient supply chain that utilizes strategies such as VMI, helicopter air drops, for example, or multiple sources for scarce resources like ventilators has a more resilient and adaptable plan.

Health-care administrators seeking to utilize best practices for SCM to develop a resilient supply chain should consider the following four primary components before a crisis occurs:

1. Utilize the organization's hazard and vulnerability analysis (HVA) to identify the required supplies and equipment needed for general and specific "all-hazards" threats.
2. Determine whether stockpiling supplies should be done on-site or elsewhere. Again the HVA will be useful in determining vulnerability and accessibility of those supplies in either location.
3. Identify the role of the hospital's GPO, supply vendors, and public first response agencies including police, fire, emergency medical services, public health, emergency management, public utilities, and so on.
4. Recognize that most community health-care systems are designed to collapse back on the hospital and that the hospital must be prepared to receive and treat a surge of patients—many with special medical and functional needs, which will further impact the consumption rate of supplies and equipment.

During disaster planning and preparedness activities, a hospital's disaster or emergency management team must consider some of the cascading events that can impact the supply chain. First, it must be assumed that a disaster will interrupt a hospital's normal supply chain, which in turn will impact daily operations. Therefore, a medical facility must determine its minimum supply levels in order to maintain its core service and treatment capabilities, until the supply chain can be reestablished. Second, a disaster will result in a patient surge, many of whom will have special functional or medical needs. Using the all-hazards approach, hospitals must be able to anticipate and identify the special supplies and equipment necessary to treat the injuries or illnesses related to any possible disaster scenarios faced by the facility. Third, the hospital's disaster or emergency management team must determine the specific resources necessary to maintain the hospital's core capabilities including contingency plans for loss of utilities and the safety and health of hospital personnel. For instance, medical providers may need additional personal protective equipment or antidotes in case of a bioterrorism attack (Dimitruk, 2005). Some hospitals even plan for staff members' families to seek shelter at the hospital in order to allow the staff to better focus on their jobs.

Once the critical supplies and equipment have been determined, several factors must be considered to determine the best location to store them including the hospital's physical vulnerability to the impacts of various disasters. For example, a hospital located in a flood plain may not want to store all the disaster supplies in the basement, whereas one susceptible to tornados may find the basement to be the safest place. Physical vulnerability can also include impacts to the supply chain. A health-care facility can quickly become isolated when its supply chain is broken due to roads, bridges, and airports that are rendered impassible due to physical damage, flooding, debris, or snow and ice. Another consideration is the likely geographic perimeter of areas affected by possible disasters. Tornados are fairly localized incidents whereas hurricanes and earthquakes can be more regional in scope.

Sustainability costs are a significant consideration when considering the shelf life of perishable supplies, the conditions in which they can be safely stored, and the ongoing maintenance of biomedical equipment. Some antidotes and pharmaceuticals require refrigeration whereas equipment such ventilators and oxygen concentrators require regular inspection, testing, and maintenance. If the hospital decides to store items off-site or they are part of a regional preparedness stockpile, special transportation needs must be considered, especially in light of hospital locations that are vulnerable to damaged roadways and bridges. The stockpile of supplies must be prioritized in the event that transportation is limited and who will load and transport the supplies must also be determined prior to the event. Finally, the health and safety of the staff must be considered as well. Supplies such as flashlights, water, protective masks, and personal hygiene supplies for staff should be readily available on-site (Dimitruk, 2005).

Most community health-care systems are designed to collapse back onto the hospital in much the same way that the medieval castle or frontier fort provided a safe haven for the villagers. This is a key point to recognize for planning and preparedness initiatives. Consider how many nursing homes, dialysis centers, family practice physicians, pharmacies, and home health oxygen providers exist in a community to support the medical needs of the population, the majority of which do not have backup electrical generators. Even a simple electrical power outage lasting more than a day can easily result in a surge on a hospital with patients needing dialysis, oxygen, refrigeration for insulin, or a myriad of other medical support needs. Too often this scenario is not considered in emergency planning at either the organizational or community level. Proactive planners in several regions have collaborated with their stakeholders to outfit family practice clinics, dialysis centers, and pharmacies with backup generators. Keeping these facilities operational helps reduce the surge of patients that would otherwise be forced to rely on the hospitals for these services. This mitigation strategy enhances the community's resilience by creating a "concentric ring" of protection around the hospital similar to a moat around the castle.

It is essential that these and other considerations about the supply chain be integrated into all aspects of a hospital's Emergency Operations Plan. This means that these issues must be forefront in the minds of those who will fill leadership roles in the Hospital Incident Command System for operations, planning, logistics, and finance. Incident-specific assignments of responsibility and job action sheets (usually found in the appendices of the Emergency Operations Plan) should address these issues as well. Furthermore, since hospitals do not exist in a vacuum but are an integral part of the community's overall health-care system, emergency planning should not take place in a vacuum either. As with all emergency preparedness planning, hospitals should integrate their SCM plan with their regional stakeholders including response agencies, emergency management, other regional hospital association partners, and their own GPOs. These organizations all have a stake in issues such as regional SCM, stockpiling of disaster supplies and equipment, coordination of scarce resources, current bed status, and surge capacity. More specifically, the GPO and regional hospital association can help balance the costs associated with stockpiling critical supplies and equipment with mitigating the vulnerabilities and risks of supply chain interruption.

Hospital administrators must be committed to developing a resilient and robust supply chain that will be reliable under the duress and rapid tempo of a real crisis. Furthermore, developing a supply chain that is resilient enough to perform under any of the potential impacts identified by the HVA is a considerable challenge. Dimitruk (2005) suggested that "the complexity of disaster management, combined with the heightened demands for best practices and a robust response capability from JCAHO and our communities, supports an effort to bring a more systematic approach to SCM for disasters in advance. Technology, in the form of online disaster management tools, can provide that extra, essential measure of preparedness."

29.5 Conclusion

Strategies for developing supply chain resilience include the following:

- Stockpiling of critical supplies and equipment based on a thorough HVA
- Partnering with vendors and suppliers to understand their disaster and contingency plans
- Determining what mix of on-site stockpiling, off-site stockpiling, and VMI strategies provide the optimal balance of cost efficiency and preparedness
- Coordinating with regional partners like other hospitals and the county or regional Emergency Management Agency to identify gaps and redundancies
- Establishing alternate delivery methods with agencies such as the state's National Guard units to provide trucks with all-wheel drive or high water-fording capabilities, or helicopters for airlift of supplies

One cautionary note: health-care administrators and emergency planners should also carefully consider and assess the capabilities and resiliency of their vendors' supply chains. While it may be cost-effective to streamline purchasing by using a few dedicated vendors, sole reliance on prime vendors reduces supply chain resiliency and makes the hospital or health-care facility highly dependent on those vendors. This situation is magnified when multiple hospitals relying on the same suppliers are impacted by a large-scale disaster that can quickly overwhelm the ability of a prime vendor to meet a spike in demand from multiple customers within the impact zone. Prudent administrators and planners will not simply take a vendor's word that they are resilient enough to absorb any emergency impact. The same holds true for other finite resources such as ambulances or buses for evacuation. If all eight hospitals in the city and all 147 nursing homes are depending on the same 53 ambulances to be available in the event of a total evacuation, the scarcity of resources will quickly be realized, albeit too late. Coordinating this type of planning on a regional basis and testing the plans through exercises designed to stress the system to the point of failure is essential to identifying the weakness in disaster plans.

References

Chisholm, M. October 15, 2010. "The Joint Commission 2010 EM and EC Update." Joint Commission on Accreditation of Healthcare Organizations.

Christopher, M., and Rutherford, C. August 2004. "Creating Supply Chain Resilience through Agile Six Sigma." *CriticalEye.net*. http://martin-christopher.info/wp-content/uploads/2009/12/critical_eye.pdf.

Day, J., Junglas, I., and Silva, L. August 2009. "Information Flow Impediments in Disaster Relief Supply Chains." *Journal of the Association for Informational Systems*. Volume 10, Issue 8, pp. 637–660. http://student.bus.olemiss.edu/files/conlon/Others/SCM/supplyChain_categories_np/impact%20and%20mgmt%20of%20disruption/disaster%20relief%20supply%20chain.pdf.

Dowty, R., and Wallace, W. July 2010. "Implications of Organizational Culture for Supply Chain Disruption and Restoration." *International Journal of Production Economics*. Volume 126, Issue 1, pp. 57–65. http://www.sciencedirect.com/science/article/pii/S0925527309003934.

Dimitruk, P. November 2005. "Hospital Preparedness: Most Urban Hospitals Have Emergency Plans but Lack Certain Capacities for Bioterrorism Response." *Healthcare Purchasing News*. http://www.hpnonline.com/inside/December%2005/0512HavingMySay.html.

Ergun, O., Karakus, G., Keskinocak, P., Swann, J., and Villarreal, M. June 2010. *Operations Research to Improve Disaster Supply Chain Management*. H. Milton Stewart School of Industrial and Systems Engineering, Georgia Institute of Technology, Atlanta, Georgia. http://www.wiley.com/WileyCDA/Section/id-397133.html.

Federal Emergency Management Agency (FEMA). 2008. *National Response Framework*. http://www.fema.gov/pdf/emergency/nrf/nrf-core.pdf.

Federal Emergency Management Agency (FEMA). 2010. *ICS 200*. http://emilms.fema.gov/IS200b/index.htm.

Government Accountability Office. August 2003. *Hospital Preparedness*. http://www.gao.gov/new.items/d03924.pdf.

White House, The 2009. *The Economic Case for Health Care Reform*. http://www.whitehouse.gov/administration/eop/cea/TheEconomicCaseforHealthCareReform.

Whybark, D., Melnyk, S., Day, J., and Davis E. May 2010. "Disaster Relief Supply Chain Management: New Realities, Management Challenges, Emerging Opportunities." *Decision Line*. http://www.decisionsciences.org/decisionline/Vol41/41_3/dsi-dl41_3fea.pdf.

chapter thirty

Decision support for inland waterways emergency response

Heather Nachtmann, Edward A. Pohl, and
Leily Farrokhvar

Contents

30.1   Introduction

In emergency management, it is generally assumed that the standard means of transportation will be available and feasible when a disaster occurs. In severe cases, however, the incident may disable emergency vehicles or destroy the roads and bridges that are vital to providing emergency response. As emergency management teams prepare contingency plans for emergency responses, it is important to recognize the transportation resource offered by the inland waterways in conjunction with or as an alternative to the highway system. More than 12,000 miles of federally maintained, navigable inland waterways could serve as a contingency transportation mode in response to a variety of disasters across much of the United States. For many communities, including those in rural areas, inland waterways can provide access to equipment and services when other modes of transportation are unavailable due to capacity overload or destruction. Barges have great capacity to transport heavy loads of emergency equipment and supplies, although they are limited to certain types of emergencies because of their relatively slow response time. Communities may spend weeks or even months recovering from catastrophic incidents such as tornadoes or earthquakes. In such cases, the potential to benefit from an Emergency Response Barge System (ERBS) is clear.

This research contributes the first known ERBS decision support for using barges on inland waterways to provide emergency response. We provide insight into the inland waterway transportation system as a contingency mode of emergency response. A Waterway Emergency Services (WES) Index enables emergency management teams to evaluate the feasibility and potential benefit level of using inland waterways for local emergency response. After identifying the potential benefit from WES, a multi-objective ERBS approach also helps emergency management teams to determine the starting location of available barges to ensure that communities with the potential to benefit from emergency response via inland waterways have the maximum possible coverage. Our demonstration on the lower Mississippi River region provides insight into the potential of communities within this region to benefit from inland waterway emergency response and demonstrates the application of our research.

30.2   Background

Inland waterways are a tremendous asset to the United States, providing an economical and environmentally sound mode for moving cargo. The U.S. Army Corps of Engineers (USACE) is responsible for nearly 12,000 miles of commercial, navigable U.S. inland and intra-coastal waterways—the Mississippi/Ohio River System, the Gulf Intracoastal

Waterway, the Intracoastal Waterway along the Atlantic Coast, and the Columbia-Snake River System in the Pacific Northwest. Inland and intracoastal waterways serve 38 states with 192 commercially active lock sites (USACE, 2009). Domestic waterborne trade over inland waterways amounted to 522.5 million short tons in 2009 alone (USACE, 2010). The nation's waterways are used to transport approximately 20% of America's coal, 22% of its petroleum, and 60% of its farm exports (USACE, 2009).

History reveals that barges have been used to provide services relevant to emergency response. In New York City, a barge served as a floating hospital providing free medical and dental care to low-income families from 1866 until fairly recently (Anonymous, 1988). Barges have also been used to provide medical services to the military. During World War I, British troop casualties were evacuated via floating hospital barges. The slow speed of the vessel actually proved to be useful for the injured troops, allowing them to recover before arriving at their destination (Quaranc, 2009). The New Hampshire Public Service has used a floating power plant in one of the discarded hulls of World War I, the "Jacona," to supplement power output at various points of its system in the northeast United States (Wecksler, 1942). The floating power barge could move over the Great Lakes, Illinois River, Mississippi River, and along the intercoastal canal system of the Gulf states to firm up the power at regions along these waterways. Also, as part of a recovery efforts after the January 2010 Haiti earthquake, tugs and barges participated in the vast international relief operation, carrying large volume supplies of food and aid to help ease some of the shortage (Ewing, 2010).

The Federal Emergency Management Agency (FEMA, 2008) states that effective hazard preparedness requires strategies to obtain and deploy supplies in sufficient quantities to perform emergency response missions and tasks. Resilience is the "ability to resist, absorb, recover from, or adapt to an adverse occurrence" (FEMA, 2010). At the local and state level, community resiliency is increased by knowing the community and its demographics (FEMA, 2010). This motivates the need to consider community attributes while developing emergency operations plans.

30.3 WES Index

We developed a WES Index to assess the potential of a given community to benefit from emergency services via inland waterways. The WES Index is an additive value function consisting of seven factors: accessibility to navigable waterways, population demand, social vulnerability, risk of disaster, limited access to medical services, limited access to resources, and limited access to transportation modes (Farrokhvar et al., 2011) as described in Table 30.1. The selection of an additive model was made to allow for ease of interpretation and implementation by practitioners. The

Table 30.1 WES Index Factors

Factor	Description	Metric	Scale	Value
Accessibility to navigable inland waterway	Proximity of a community to a navigable inland waterway. Emergency response is not feasible for communities located too far from a navigable inland waterway.	Distance between county population centroid and closest inland port/terminal (U.S. Census Bureau, 2011)	Accessible (≤3-h drive at 35 mph) = 1 Inaccessible (>3-h drive at 35 mph) = 0	1 0
Population demands	Size of population and its proximity to metropolitan areas. Important for identifying the level of services that may be needed during an emergency.	Rural–Urban Continuum Code (USDA ERS, 2004)	Low (7–9) Medium (4–6) High (1–3)	1 2 3
Social vulnerability	Social, economic, demographic, and housing characteristics that influence a community's ability to respond to, cope with, recover from, and adapt to environmental hazards. Useful for identifying which counties may need the greatest assistance during an emergency.	National percentile ranking of the Social Vulnerability Index (SoVI) (Hazards and Vulnerability Research Institute, 2008)	Low (0.01–33.33) Medium (33.34–66.66) High (66.67–99.99)	1 2 3

			Total
Risk of disaster	The risk of tornado, earthquake, flood, or terrorist attack. Useful for identifying which counties are most likely to need inland waterway-based emergency assistance.	Combined risk level of tornado (The Tornado Project, 1999), earthquake (U.S. Geological Survey, 2009), flood (Federal Emergency Management Association, 2009), and terrorism	
		Tornado: Low (<2.5) = 1, Medium (2.5–4.99) = 2, High (≥5) = 3	Low (4–6) 1
		Earthquake: Low (<20) = 1, Medium (20–79.9) = 2, High (≥80) = 3	Medium (7–9) 2
		Flood: Low (<3) = 1, Medium (3–4) = 2, High (>4) = 3	High (10–12) 3
		Terrorism: Low = 1, Medium = 2, High = 3	
Limited access to medical services	Number of community hospital beds per 100,000 people available in the area. Important for identifying the necessity of medical services that may be brought to the area during an emergency.	Number of community hospital beds per 100,000 people (U.S. Census Bureau, 2007)	
		Low (>317)	1
		Medium (1–317)	2
		High (0)	3

(Continued)

Table 30.1 (Continued) WES Index Factors

Factor	Description	Metric	Scale	Value
Limited access to resources	Availability of resources including clean water supply, power supply, temporary housing, and fuel supplies. This factor is important in identifying the necessity of providing resources via barge.	Combined availability level of water supply and irrigation systems; electric power generation, transmission, and distribution; number of hotels, motels, B&B, other travel accommodations, RV parks and camps, and rooming and boarding houses; and number of gasoline station establishments. To be consistent, all the metrics are measured per 100,000 people (U.S. Census Bureau, 2008a and 2008b)	**Clean Water:** Low (>8) = 1, Medium (1–8) = 2, High (0) = 3 **Power:** Low (>7) = 1, Medium (1–7) = 2, High (0) = 3 **Temporary Housing:** Low (>23) = 1, Medium (1–23) = 2, High (0) = 3 **Fuel:** Low (>67) = 1, Medium (1–67) = 2, High (0) = 3	Total Low (4–6) 1 Medium (7–9) 2 High (10–12) 3
Limited access to transportation modes	Accessibility to railroad system or airports. If a county does not have easy access to other modes of transportation it has higher potential to benefit from waterway-based transportation.	Railroad passes through the county and/or at least one public airport is located in the county (U.S. Department of Transportation, 2010)	Both railroad and airport(s) are accessible Railroad or airport is accessible Neither railroad nor airport is accessible	1 2 3

seven factors were selected based on our discussions with emergency managers and the availability of public-level data at the county level that would be necessary for emergency management teams with limited resources to be able to access the necessary data to implement the system. In the basic model presented here, we assume that all factors are weighted equally; however, it is certainly conceivable that extensions of the model could allow for a weighting scheme to be applied to the WES Index function. Table 30.1 also contains a description of each factor's corresponding metric, scale, and data source used to compute a county's WES Index. The WES Index is computed using Equation 30.1.

$$\text{WES Index Value} = A(PD + V + R + M + LR + T) \tag{30.1}$$

where

A = Accessibility to navigable waterway score
PD = Population demands score
V = Social vulnerability score
R = Risk of disaster score
M = Limited access to medical services score
LR = Limited access to resources score
T = Limited access to transportation modes score

The possible values for the WES Index are 0, 6, 7, 8, 9, 10, 11, 12, 13, 14, 15, 16, 17, and 18. A WES Index value of 0 indicates that the county is not located within an assumed reasonable 3-hour drive of a public port on a navigable inland waterway, and, therefore, WES coverage is inaccessible. A WES Index value of 6, 7, 8, or 9 indicates that the county has low potential to benefit from inland WES. A WES Index value of 10, 11, 12, or 13 indicates that the county has medium potential to benefit from inland WES; and a WES Index value of 14, 15, 16, 17, or 18 indicates that the county has high potential to benefit from inland WES.

30.4   Demonstration of WES Index

We use the lower Mississippi River region to demonstrate the use of the WES Index to evaluate the extent to which a given county in this region can benefit from WES. An output of this demonstration is an assessment of the WES Index values for 316 counties and parishes within the four states of Arkansas, Louisiana, Mississippi, and Tennessee.

30.4.1   Data collection

In the first step of our demonstration, we collected the data necessary to compute the WES Index factor values for each county in our region.

30.4.2 *Accessibility to navigable inland waterway*

In order to calculate the accessibility to navigable inland waterway factor values, we estimated the drive times between the origin and destination points for residents of the counties within our region to get to the nearest public port located on the lower Mississippi River. We assumed that emergency response barges can only have access to the 16 public ports along the Mississippi River within our four-state region as shown in Table 30.2.

The origin point is defined as the county's population centroid and is retrieved for each county in the region (U.S. Census Bureau, 2011). After identifying the origin and destination points for each county, we used Google Maps (maps.google.com) to find the distances between these points. The drive time was computed by dividing the distance by an assumed average travel speed of 35 miles per hour. If one of the 16 public ports is located within a 3-hour drive of a given county, it is considered feasible for that county to benefit from inland WES, and the county is assigned an accessibility to navigable inland waterway factor value of 1. There are 145 counties within our four-state region that have access to one of the 16 public ports.

30.4.3 *Population demands*

The rural–urban continuum codes for each county in the region were collected from the Economic Research Service (USDA ERS, 2004) and are classified as high, medium, or low population demands according to their

Table 30.2 Public Ports on Lower Mississippi River

Port number	Port name	State	Port number	Port name	State
1	Plaquemine	LA	9	Madison Parish	LA
2	St. Bernard	LA	10	Lake Providence	LA
3	New Orleans	LA	11	Greenville	MS
4	South Louisiana	LA	12	Yellow Bend	AR
5	Greater Baton Rouge	LA	13	Rosedale	MS
6	Natchez	MS	14	Helena	AR
7	Claiborne County	MS	15	Memphis	TN
8	Vicksburg	MS	16	Osceola	AR

rural–urban continuum code as defined in Table 30.1. Counties with high, medium, and low population demand factor values received scores of 3, 2, and 1, respectively.

30.4.4   Social vulnerability

A county's social vulnerability (SoVI) represents its "ability to respond to, cope with, recover from, and adapt to environmental hazards" (Hazards and Vulnerabilities Research Institute, 2011). The SoVI value for each county can be obtained from the Hazards and Vulnerability Research Institute. In addition to the SoVI values, the database also provides the national percentile ranking for each county (Hazards and Vulnerabilities Research Institute, 2008). We categorized the counties based on their national percentile. For the purposes of calculating the social vulnerability factor value, the low, medium, and high vulnerabilities were based on national percentiles from 0.01 to 33.33, 33.34 to 66.66, and 66.67 to 99.99, respectively. Counties with low, medium, and high percentiles were given values of 1, 2, and 3, respectively.

30.4.5   Risk of disaster

To determine the risk of disaster factor value for each county in the region, data on risk of tornadoes, earthquakes, floods/hurricanes/tropical storms, and terrorist attacks are needed. We used historical tornado data to determine each county's risk level for violent tornadoes. A tornado's intensity is measured by its rating on the Fujita Scale (The Tornado Project, 1999). Using data from www.tornadoproject.com, we identified the total number of tornadoes and their Fujita Scale ratings for each county in the region from 1950 to 1995. This source indicates that 67% of tornado-related deaths are caused by F4 and F5 tornadoes, 29% are caused by F2 and F3 tornadoes, and only 4% are caused by F0 and F1 tornadoes (The Tornado Project, 1999). Using this information about tornado-related deaths, we weighted the total number of F0 and F1 tornadoes, F2 and F3 tornadoes, and F4 and F5 tornadoes by 4%, 29%, and 67%, respectively, and then summed to obtain a Tornado Risk Score for each county, as described in Equation 30.2. As an example, Table 30.3 gives the historical tornado data for Howard County, Arkansas.

$$\lceil 0.04(F0+F1) \rceil + \lceil 0.29(F2+F3) \rceil + \lceil 0.67(F4+F5) \rceil = \text{Tornado Risk Score}$$
(30.2)

Table 30.3 Historical Tornado Data for Howard County, Arkansas

County	Total	F0	F1	F2	F3	F4	F5
Howard	18	6	5	4	1	2	0

where F0, F1, F2, F3, F4, and F5 represent the county's total number of F0, F1, F2, F3, F4, and F5 tornadoes, respectively.

In order to calculate the Tornado Risk Score for Howard County, we used the historical tornado data from Table 30.3 and applied it to Equation 30.2 to obtain a Tornado Risk Score of 3.23.

$$[0.04(6+5)] + [0.29(4+1)] + [0.67(2+0)] = 3.23$$

The Tornado Risk Scores for each county were then categorized as low risk (0–2.49), medium risk (2.50–4.99), or high risk (≥5.00). Counties with low-, medium-, and high-risk levels were given subfactor values of 1, 2, and 3, respectively.

Earthquakes are capable of causing significant damage to ground structures and roads and are also known to initiate other natural disasters including landslides and tsunamis. A powerful earthquake can easily disrupt standard means of transportation, inhibiting emergency response teams from reaching victims of the disaster. In order to determine each county's risk of earthquake, we gathered information on the seismicity of the four states. The U.S. Geological Survey (2009) measures seismicity in terms of peak acceleration during an earthquake. By overlaying the seismicity map with a map of each state's counties, we estimated the seismicity level for each county. The seismicity was then categorized into three risk levels based on peak acceleration as expressed as a percentage of the acceleration due to gravity: low risk (0–19.9), medium risk (20–79.9), and high risk (≥80). Counties with low-, medium-, and high-risk levels were given subfactor values of 1, 2, and 3, respectively.

The next risk of disaster subfactor we considered was flooding. Floods are dangerous because they cause damage through inundation and soaking as well as the powerful force of moving water. FEMA (2011) provides a description of all major disasters that have occurred in each state. Since flooding often occurs during hurricanes and tropical storms, we studied the number of disaster declarations that included flooding, hurricane, or a tropical storm over the last 10 years. Based on the total number of flood-related disaster declarations made over the last 10 years, the counties were categorized as having a low (<3), medium (3–4), or high (>4) risk of flood. Counties with a low, medium, and high risk of flood were given subfactor values of 1, 2, and 3, respectively.

There is no sure way to predict future terrorism events. We naively assumed a set of locations in the four-state region that could be targets of a terrorist attack, including active military bases and nuclear power plants. An attack on any of these targets would pose a threat to the surrounding areas. Using this information, we assigned a low, medium, or high risk for terrorist attack to each county in the region based on its proximity to one or both these locations. The counties containing one of the targets and all of their adjacent counties were categorized as being at high risk for terrorist attack. Non–high-risk counties adjacent to a high-risk county were categorized as having a medium risk for terrorist attack. All other counties were categorized as the low risk for terrorist attack. A county categorized as low, medium, or high risk was given a subfactor value of 1, 2, or 3, respectively. These target areas are easily adjusted in practice using the emergency management team's assessment of locations at risk of a terrorist attack.

After the subfactor values for tornado, earthquake, flood, and terrorist attack were determined for each county, we calculated the overall risk of disaster factor value by summing the four subfactor values for each county. Counties with values in the ranges 4–6, 7–9, and 10–12 were classified as having low, medium, and high risk of disaster, respectively. Disaster risk levels of low, medium, and high were given risk of disaster factor values of 1, 2, and 3, respectively.

30.4.6 Limited access to medical services

To measure each county's limited access to medical services, we used the number of community hospital beds per 100,000 people (U.S. Census Bureau, 2007). In the year 2004, the average number of hospital beds per 100,000 people for the four-state region was 287 per county. The counties with zero hospital beds per 100,000 people are considered to have a high potential of benefiting from an emergency response barge, counties with 1–287 (the county-level average number of beds per 100,000 people in the case study region) are considered to have medium potential, and counties with more than 287 are considered to have low potential. Counties with a low, medium, and high potential were given values of 1, 2, or 3, respectively.

30.4.7 Limited access to resources

We identified four types of resources that are feasible for transport by a barge via inland waterways in case of disaster. These resources are clean water, power, temporary housing, and fuel. We defined a subfactor to measure the limited access of a county to each of these resources and then combined them to obtain a limited access to resources factor value for each county.

To measure the level of available clean water supplies in each county, we computed the number of water supply and irrigation systems establishments per 100,000 people (U.S. Census Bureau, 2008a,b). The counties with zero water supply and irrigation systems establishments per 100,000 people are considered to have a high potential of benefiting from clean water supplies and water treatment equipment provided by an emergency response barge, counties with one to eight (the average number of systems per county in the case study region) water supply and irrigation systems are considered to have medium potential, and counties with more than eight are considered to have low potential. Counties with a high, medium, and low potential to benefit were given subfactor values of 3, 2, and 1, respectively, for limited access to clean water.

When measuring each county's level of access to power sources, we used the number of available electric power generation, transmission, and distribution establishments (U.S. Census Bureau, 2008a) per 100,000 people in a county. Counties were then categorized as having high potential to benefit from power generating supplies via barge if there are zero establishments, medium potential if there are one to seven (the average numbers of establishments per county in the case study region) establishments, or low potential if it has more than seven establishments. Counties with high, medium, and low potential received a subfactor value of 3, 2, and 1, respectively, for limited access to power supplies.

The number of hotels, motels, bed and breakfast establishments, RV parks and camps, other travel accommodations, and rooming and boarding houses (U.S. Census Bureau, 2008a) per 100,000 people was used to measure a county's limited access to temporary housing. A county with zero establishments has high potential to require waterway-based assistance for temporary housing and receives a subfactor value of 3. If the county has between 1 and 23 (the average numbers of establishments per county in the case study region) establishments, the county has medium potential to benefit and receives a subfactor value of 2. Finally, if the county has more than 23 establishments, the county has low potential to benefit from temporary housing supplies via barge and receives a subfactor value of 1.

We considered the number of gas station establishments in each county (U.S. Census Bureau, 2008a) per 100,000 people as the measure for a county's access to fuel. If a county has zero establishments, it receives a subfactor value of 3 to represent its high potential to benefit from waterway-based assistance for fuel supplies. If the county has between 1 and 67 (the average numbers of establishments per county in the case study region) gas station establishments, it receives a subfactor value of 2, which represents its medium potential to benefit. Finally, if the county has more than 67 establishments, it receives a subfactor value of 1 for its low potential to benefit.

After the subfactor values for limited access to clean water, power, temporary housing, and fuel were determined for each county, we calculated the overall limited access to resources factor value for each county by summing its subfactor value on the four resource subfactors. A county with a value in the range 4–6, 7–9, or 10–12 was classified as having a low, medium, or high potential to require assistance via inland waterways for these resources. Low, medium, and high levels of access to these resources were given a limited access to resources value of 1, 2, and 3, respectively.

30.4.8 Limited access to transportation modes

There are four common modes of cargo transportation: highway, rail, air, and water. In addition to ground transportation via the highways, which is available in all counties within the region, we examined two alternative transportation modes—rail and air. We used the National Transportation Atlas Database (U.S. Department of Transportation, 2010) to find the number of public use airport facilities in each county and to identify the counties that have access to the rail system. The counties that do not have a rail system passing through them and do not have any public airports were considered to have high potential to benefit from waterway-based emergency response. If a county has access to either rail or air transportation, the county was categorized as having medium potential, and if a county has access to both modes, the county has low potential to benefit. Counties with high, medium, and low potential were assigned values of 3, 2, and 1, respectively.

30.5 WES Index computation

After the seven factor values are determined for each county in the region, the overall WES Index value for each county was calculated using Equation 30.1. Figure 30.1 graphically depicts the WES Index value of each county in the four-state region, which indicates its potential to benefit from inland waterway emergency response.

There are 171 counties in the four-state region that are more than a 3-hour drive from the public ports on the lower Mississippi River, making the use of inland WES infeasible for these counties. These counties have a WES Index of 0. Thirty-nine counties (12%) in the four-state region have a WES Index value of less than 10 and, therefore, have low potential to benefit from inland waterway-based emergency response. Ninety-seven counties (31%) in the region have medium potential, and nine counties (3%) have high potential. Overall, based on WES Index values, there are 106 counties (73%) among 145 that have access to the Mississippi River public ports with at least medium potential to benefit from inland waterway-based emergency response. If additional inland waterways are taken into consideration, the counties that currently have a WES Index value of 0 could potentially have access to a navigable inland waterway other than the Mississippi River and, therefore, benefit from WES. Once the WES Index was created, we were interested in providing decision support that would assist regional emergency management teams whose counties have potential to benefit from WES in determining the required number of emergency response barges along with their pre-positioning locations of the team would need to provide acceptable coverage to their region. The developed approaches are presented in the next section.

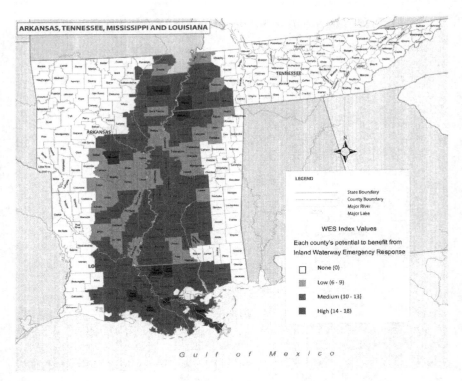

Figure 30.1 WES Index values for lower Mississippi River region.

30.6   ERBS approaches

The basic question of "Where are we going to put things?" is modeled as a facility location problem. The classic facility location problem was first described by Weber (1929) who studied the location of a single facility. Hotelling (1929) expanded this initial work by studying the location of two facilities. The problem has been expanded to cover many realistic location decisions such as emergency medical service bases, fire stations, and wireless communication towers. Daskin (2008) provided a thorough review of the facility location literature and a taxonomy for classifying location problems (see Drezner and Hamacher [2002], and Francis et al. [1992]). The proposed ERBS problem can be classified as a dynamic, multiple facility location problem based on actual transport distances. It can be solved using covering-based models, which allow for demand to be "covered" within a critical coverage distance or time. Specifically, we model the ERBS problem as both a set covering problem and a maximal covering problem.

To use our methodology, the emergency management team needs to set a coverage range for the emergency response barges. Each emergency response barge is assumed to provide emergency response service to the ports that are within the defined coverage range. In this research, we define the base coverage range for an emergency response barge to be 12 hours, which means that a county is considered to have WES coverage if a barge can get to an accessible port for that county in less than 12 hours, assuming a barge travel speed of 115 river miles per day. Here, we present three different approaches toward our goal of developing an optimal ERBS via inland waterways. First, we developed the Minimum Required Emergency Response Barges (MRERB) model, a set covering model to determine the number of required barges; second, we developed the maximum WES coverage model, a maximal covering model to locate the response barges in order to provide maximum WES coverage while considering resource limitations; and third, we combined these two single-objective approaches into a multi-objective optimization model using goal programming.

30.6.1   Model notation

Here, we define the notation that is used in formulation of the ERBS optimization models. There are two sets defined in this problem:

C Set of counties, indexed by i
P Set of ports, indexed by j and k

Parameters defined for the ERBS problem are

$a_{ij} = 1$ if county i has access to port j (less than 3 hours drive), 0 otherwise
$d_{jk} = 1$ if port j is within the coverage range of port k (12 hours), 0 otherwise
n is number of available emergency response barges
m is number of ports
w_i is WES Index value for county i

Variables defined in order to model the ERBS problem are

$x_i = 1$ if county i is covered, 0 otherwise
$y_j = 1$ if there is a barge at port j, 0 otherwise

30.6.2   MRERB model

A set covering model was formulated to determine the minimum number of emergency response barges required to provide a desired level of WES coverage to determine the starting ports to locate the barges. The goal is to use as few barges as possible in order to provide a desired level of WES coverage to counties that have access to at least one port on an inland waterway. The objective Function 30.3 minimizes the number of required barges. Constraint 30.4 relates the upper bound of variable x_i to variable y_j. It makes sure if a county is not covered under the defined rules, x_i must be less than or equal to zero. Constraint 30.5 relates the lower bound of variable x_i to variable y_j. It guarantees that variable x_i must be strictly greater than zero when a county is covered under the defined conditions. Since variable x_i is a binary variable, when it is strictly greater than zero, it is set to 1. In addition to Constraints 30.4 and 30.5, additional optional Constraints can be added to the model to address scenario-specific performance goals. As examples, we have provided optional Constraints 30.6 through 30.8. Optional Constraint 30.6 verifies that all the counties must be covered. This constraint is included in the model when the emergency management team requires all of the counties to be covered. Otherwise, coverage requirements can be specified by other factors of the WES Index. Optional Constraints 30.7 and 30.8 are presented as examples of additional factor-specific coverage requirement constraints. Optional Constraint 30.7 ensures that if a county has a value of 2 or 3 for the risk of disaster factor, then the county is covered by the inland waterway emergency response service. R_i is defined as the value of the risk of disaster factor for county i. Optional Constraint 30.8 ensures that if a county has a value of 3 for the social vulnerability factor, then the county is covered by the inland waterway emergency response service. S_i is defined as the value of the social vulnerability factor for county i. Constraint 30.9 ensures that the binary variables only obtain values of 0 or 1. Additional coverage constraints similar to Constraints 30.7 and 30.8 can be added to the model for any factor of the WES Index depending on the needs and priorities of the emergency management team.

$$Min \sum_{j \in P} y_j \tag{30.3}$$

$$x_i \le \sum_{j \in P}\sum_{k \in P} y_j d_{jk} a_{ik} \quad \forall i \in C \tag{30.4}$$

$$\sum_{j \in P}\sum_{k \in P} y_j d_{jk} a_{ik} \le m x_i \quad \forall i \in C \tag{30.5}$$

$$x_i = 1 \quad \forall i \in C \tag{30.6}$$

$$R_i \le 2\left(x_i + \frac{1}{2}\right) \quad \forall i \in C \tag{30.7}$$

$$S_i \le 2(x_i + 1) \quad \forall i \in C \tag{30.8}$$

$$x_i, y_j \in \{0, 1\} \quad \forall i \in C, \forall j \in P \tag{30.9}$$

30.6.3   Maximum WES coverage model

In a separate maximal covering location model, we take the resource limitation on the number of available barges into account while determining the optimal location for the barges to be pre-positioned in order to provide maximum WES coverage to the counties that have access to an inland waterway. The goal here is to provide WES coverage to as many counties as possible given that there are a limited number of emergency response barges available. Objective Function 30.10 maximizes the number of counties covered while giving priority to counties that have higher WES Index values. Constraints 30.11 and 30.12 are identical to Constraints 30.4 and 30.5. Constraint 30.11 relates the upper bound of variable x_i to variable y_j. It makes sure if a county is not covered under the defined rules, x_i must be less than or equal to zero. Constraint 30.12 relates the lower bound of variable x_i to variable y_j. It guarantees that variable x_i must be strictly greater than zero when a county is covered under the defined conditions. Since variable x_i is a binary variable, when it is strictly greater than zero, it is set to 1. Constraint 30.13 verifies that the number of the barges used is less than or equal to the number of the available barges. Constraint 30.14 ensures that the binary variables only obtain values of 0 or 1.

$$Max \sum_{i \in C} w_i x_i \tag{30.10}$$

$$x_i \leq \sum_{j \in P} \sum_{k \in P} y_j d_{jk} a_{ik} \quad \forall i \in C \tag{30.11}$$

$$\sum_{j \in P} \sum_{k \in P} y_j d_{jk} a_{ik} \leq m x_i \quad \forall i \in C \tag{30.12}$$

$$\sum_{j \in P} y_j \leq n \tag{30.13}$$

$$x_i, y_j \in \{0, 1\} \quad \forall i \in C, \forall j \in P \tag{30.14}$$

30.6.4   Multi-objective ERBS approach

In our third approach, we employed a goal programming approach to jointly achieve the maximum possible WES coverage with the minimum number of required emergency response barges, independent of how much we can achieve on these single objectives by defining target values for each objective function and adding soft constraints to enforce goal achievement. We defined deficiency variables that are non-negative variables that equal the difference between the target values and our objective function values. The objective of the goal programming model is to minimize the weighted deficiency. Assigning various weights to each criteria of the multi-objective function enables the decision makers to differentiate between the importance of each objective. Since our objective functions do not have the same scale, we used scaling factors to simultaneously study multiple objectives. In order to introduce target levels and deficiency variables, we define additional notations as follows:

t_1 = target value for MRERB model
t_2 = target value for maximum WES coverage model
v_1 = scaling factor for MRERB model
v_2 = scaling factor for maximum WES coverage model
α = weight assigned to MRERB objective function
d_1 = deficiency variable for MRERB model
d_2 = deficiency variable for maximum WES coverage model

Using the defined notation, we present the model as follows. Objective Function 30.15 minimizes the difference between the target values and the objective function values. Since the objective function values for two objective functions are not on the same scale, we multiply each of the objective functions by a scaling factor. By assigning weights to each objective function, emergency management teams are able to assign their important level to the two single objective functions by varying the value of α. As shown below in Equation 30.15, higher levels of α will place a greater weight on the MRERB objective function that represents the team placing more importance on this objective. Constraint 30.16 relates the upper bound of variable x_i to variable y_j. It makes sure that if a county is not covered under the defined rules, x_i must be less than or equal to zero. Constraint 30.17 relates the lower bound of variable x_i to variable y_j. It guarantees that variable x_i must be strictly greater than zero when a county is covered under the defined conditions. Since variable x_i is a binary variable, when it is strictly greater than zero, it is set to 1. Constraint 30.18 defines the deficiency variable for the MRERB objective function. The difference between the target value for the number of required barges and the total number of required barges that the model assigns is our first deficiency variable. Constraint 30.19 defines the deficiency variable for the Maximize WES Coverage objective function. The difference between the target value for the WES coverage and the value of the WES coverage achieved in the model is our second deficiency variable. Constraint 30.20 ensures that the binary variables only obtain values of 0 or 1. Constraint 30.21 ensures the non-negativity of the deficiency variables.

$$Min \ \alpha v_1 d_1 + (1 - \alpha) v_2 d_2 \tag{30.15}$$

$$x_i \leq \sum_{j \in P} \sum_{k \in P} y_j d_{jk} a_{ik} \quad \forall i \in C \tag{30.16}$$

$$\sum_{j \in P} \sum_{k \in P} y_j d_{jk} a_{ik} \leq m x_i \quad \forall i \in C \tag{30.17}$$

$$\sum_{j \in P} y_j - d_1 \leq t_1 \tag{30.18}$$

$$\sum_{i \in C} w_i x_i + d_2 \geq t_2 \tag{30.19}$$

$$x_i, y_j \in \{0, 1\} \quad \forall i \in C, \forall j \in P \tag{30.20}$$

$$d_1, d_2 \geq 0 \tag{30.21}$$

30.7   Demonstration of ERBS approaches

We used the WES Index values resulting from the lower Mississippi River region to implement our three ERBS approaches on the counties within this region that have the potential to benefit from inland WES. The ERBS approaches were implemented by considering the public ports located along the lower Mississippi River.

30.7.1   Designing ERBS

We implemented our ERBS approaches utilizing the WES Index results of the lower Mississippi River region. As previously mentioned, we considered 16 public ports listed in Table 30.2 as potential starting locations of emergency response barges. Based on the WES Index results, there are 145 counties in the region that have access to at least one of the 16 public ports. We implemented our decision support methodology on these 145 counties. The models were run in AMPL (A Mathematical Programming Language) on a notebook computer and analyzed using CPLEX (IBM ILOG CPLEX Optimizer) in a negligible amount of time.

30.7.1.1   MRERB results

The goal here was to determine the minimum number of emergency response barges required to satisfy the coverage criteria that is defined for the region. As discussed earlier, we can define optional performance constraints for the model to enforce the desired WES coverage criteria. Here, we require that all the 145 counties must be covered (Constraint 30.5).

In our base model, we assumed that a barge can provide emergency response coverage to a county if the barge can travel from its starting location to an accessible port for that county in less than 12 hours. Emergency management teams may assume a different coverage range that is reflected in the value of d_{jk}. The results for multiple barge coverage ranges for the emergency barges to completely cover the 145 counties are presented in Table 30.4.

30.7.1.2   Maximum WES coverage results

Here, we assumed that the number of barges available for emergency response is predetermined. We implemented the model considering multiple barge coverage ranges (d_{jk} defined for 3, 6, 12, 24, and 48 hours coverage ranges) for various numbers of available barges (1, 2, 3, and 4). Table 30.5 contains the results for all combinations of barge coverage ranges and numbers of available barges. As an example, for a barge coverage range of 12 hours with two barges available, Ports 7 and 15 are the selected ports and 110 counties (76%) with access to Mississippi River public ports are covered. As expected, when we have a limit on the number of available barges with shorter barge coverage ranges, fewer counties are covered by WES.

30.7.1.3   Multi-objective ERBS approach

The first step in using a goal programming approach is setting the target values for each decision criteria. Here, we used the results from the single objective models to help us define our target levels to be achieved. For the MRERB model, the ideal situation is to have only one barge and still be able to cover the entire region, so the target value for this objective function is set to 1 ($t_1 = 1$). In the maximize WES coverage model, the ideal situation is to locate the barges in a way that all the counties with access to the Lower Mississippi River are covered. For our region of interest, results from the previous model show that the maximum objective function value is 1553, which is set as our second target value ($t_2 = 1553$).

In order to scale the two objective functions appropriately, we need to find appropriate scaling factors for each objective function. The lower

Table 30.4 MRERB Model Results

Barge coverage range (hours)	Minimum number of required barges	Origin ports
3	8	2, 5, 6, 8, 9, 12, 15, 16
6	7	2, 5, 6, 8, 12, 15, 16
12	5	1, 5, 7, 12, 15
24	3	3, 9, 15
48	2	4, 13

Table 30.5 Maximum WES Coverage Model Results

Barge coverage range (hours)	Number of available barges	Origin ports	Number of covered counties (% covered)	Coverage scores (objective function values)
3	1	9	50 (34%)	501
	2	9, 15	89 (61%)	917
	3	5, 9, 15	122 (84%)	1313
	4	2, 5, 9, 15	129 (89%)	1393
6	1	8	58 (40%)	582
	2	8, 15	97 (67%)	998
	3	5, 8, 15	127 (88%)	1362
	4	2, 5, 8, 15	134 (92%)	1442
12	1	7	69 (48%)	712
	2	7, 15	110 (76%)	1148
	3	5, 7, 15	133 (92%)	1423
	4	1, 5, 7, 15	139 (96%)	1494
24	1	9	75 (52%)	772
	2	3, 12	115 (79%)	1248
	3	3, 9, 15	145 (100%)	1553
	4	3, 9, 15	145 (100%)	1553
48	1	13	116 (80%)	1207
	2	1, 13	145 (100%)	1553
	3	1, 13	145 (100%)	1553
	4	1, 13	145 (100%)	1553

bound for the Minimum Required Emergency Barges objective function is one and, for the upper bound, we consider the minimum number of barges required to cover the entire region with the shortest coverage range of 3 hours. Table 30.4 shows that when we assume d_{jk} was defined for 3-hour barge coverage range, we need eight barges to cover the entire region. Therefore, the upper bound on the number of available barges is set to eight. To set the lower bound on the Maximum WES Coverage objective, we looked at the coverage score for d_{jk} defined for 3-hour coverage range with only one barge available (Table 30.5), which is 501. The upper bound for the Maximum WES Coverage objective is the coverage score for the case when all the 145 counties are covered, which is 1553. We used these bounds to compute the normalized scaling factors for the goal programming objective function. By solving Equations 30.22 and 30.23, we obtained 0.9934 as the scaling factor value for the MRERB objective function (v_1) and 0.0066 as the scaling factor value for Maximum WES Coverage objective function (v_2).

$$(8-1)v_1 = (1553 - 501)v_2 \tag{30.22}$$

$$v_1 + v_2 = 1 \tag{30.23}$$

We solved the goal programming model given in Equations 30.15 through 30.21 by giving equal weight to each single objective function ($\alpha = 0.5$). This weight can be adjusted in practice to reflect the emergency management team's priorities. We tested the models using the following parameters and multiple barge coverage ranges:

$$t_1 = 1 \qquad t_2 = 1553 \qquad v_1 = 0.9934 \qquad v_2 = 0.0066 \qquad \alpha = 0.5$$

Based on the results of the goal programming analysis shown in Table 30.6 with d_{jk} defined based on 3, 6, 12, and 24 hours coverage range, there are three barges required in the optimal solution. Only in the case where d_{jk} is defined based on a 48-hour coverage range does the minimum number of required barges reduce to 2.

30.7.2   Sensitivity analysis

The goal programming model is initially solved assuming that both objective functions have the same weight ($\alpha = 0.5$). In some cases, the emergency management team may decide to give higher priority to one objective function by assigning a greater weight to α. To study the effects of various function weights, we solved the model with multiple values of α. Assigning greater values to α places greater importance on the Minimum Required Emergency Barges objective over the Maximum WES Coverage objective. Table 30.7 shows the results for various values of α for different barge coverage ranges.

Table 30.6 Multi-Objective ERBS Model Results

Barge coverage range (hours)	Number of available barges	Origin ports	Number of covered counties (% covered)	Coverage scores (objective function values)
3	3	5, 9, 15	122 (84%)	1312
6	3	5, 8, 15	127 (88%)	1362
12	3	5, 7, 15	133 (92%)	1423
24	3	3, 9, 15	145 (100%)	1553
48	2	1, 13	145 (100%)	1553

Table 30.7 Sensitivity Analysis of Goal Programming Model for Various α Levels and Various Barge Coverage Ranges

Results	Barge coverage range	Weights				
		(α = 0.9)	(α = 0.7)	(α = 0.5)	(α = 0.3)	(α = 0.1)
Number of required barges	3	1	3	3	4	8
Origin ports		9	5, 9, 15	5, 9, 15	2, 5, 9, 15	2, 5, 6, 8, 9, 12, 15, 16
Number (%) of covered counties		50 (34%)	122 (84%)	122 (84%)	129 (89%)	145 (100%)
Coverage score		501	1313	1313	1393	1553
Number of required barges	6	1	3	3	4	7
Origin ports		8	5, 8, 15	5, 8, 15	2, 5, 8, 15	2, 5, 6, 8, 12, 15, 16
Number (%) of covered counties		58 (40%)	127 (88%)	127 (88%)	134 (92%)	145 (100%)
Coverage score		582	1362	1362	1442	1553
Number of required barges	12	1	2	3	4	5
Origin ports		7	7, 15	5, 7, 15	1, 5, 7, 15	1, 5, 7, 12, 15
Number (%) of covered counties		69 (48%)	110 (76%)	133 (92%)	139 (96%)	145 (100%)
Coverage Score		712	1148	1423	1494	1553
Number of required barges	24	1	2	3	3	3
Origin ports		12	3, 12	3, 9, 15	3, 9, 15	3, 9, 15
Number (%) of covered counties		75 (52%)	115 (79%)	145 (100%)	145 (100%)	145 (100%)
Coverage score		772	1248	1553	1553	1553
Number of required barges	48	2	2	2	2	2
Origin ports		1, 13	1, 13	1, 13	1, 13	1, 13
Number (%) of covered counties		145 (100%)	145 (100%)	145 (100%)	145 (100%)	145 (100%)
Coverage score		1553	1553	1553	1553	1553

The results of the sensitivity analysis show the tradeoffs between desiring fewer number of barges and higher WES coverage. Sensitivity analysis results confirmed that the model behaves as expected. When assigning higher weight to the Minimize Required Emergency Barges objective, fewer barges are recommended; however, the WES coverage is then lower in the optimal solution. Placing a higher weight on the Maximize WES Coverage objective results in covering more counties while requiring an increase in the number of required barges. In our base model (d_{jk} defined for 12-hour coverage range), full coverage of all counties is achieved when the Maximize WES Coverage objective is weighted nine times greater than the weight of the Minimize Required Emergency Barges objective. In contrast, when the Minimize Required Emergency Barges objective is weighted nine times greater than the Maximize WES Coverage objective, only 48% (69) of the counties are covered.

30.8   Implications for engineering managers

Anderson et al. (2004) recognized the important role that engineering managers play in protecting our nation from natural and unnatural threats, and in providing decision support for emergency situations. These authors specifically mention the importance of systems engineering, project management, and organizational structure development in successful emergency management. Critical transportation infrastructure systems are complex, comprised of many components that act and interact to achieve the system's purpose. Historically, engineering managers strive to improve the technological capability and reliability of these components—and will rightly continue to do so; however, a growing critical need relates to defining, modeling, and optimizing interdependencies between the components within a system, and the interdependencies between systems. It is vital that the transportation infrastructure and associated systems of the future—and the workforce charged with maintaining, enhancing, and securing such—be developed and examined from a systems-level perspective to support and secure the nation's transportation needs.

30.9   Conclusions and future work

This research provides emergency management teams with insights into the benefits of inland waterway emergency response. First, a WES Index was developed to measure the potential of individual counties to benefit from inland waterway emergency response. The WES Index consists of seven factors: accessibility to navigable waterways, population demand, social vulnerability, risk of disaster, limited access to medical services, limited access to resources, and limited access to transportation modes. We obtained the WES Index values for four states along the lower Mississippi River. The results show that, among counties with public port access along Mississippi River, more than 73% have at least a medium level of potential to benefit from emergency response via this river.

We then present three ERBS optimization approaches to aid emergency management teams in designing the most efficient and effective inland waterway-based emergency response system. We developed an MRERB model to help emergency management teams determine the minimum number of barges required to provide a predefined level of emergency response coverage. Then, considering resource limitations, we also formulated a maximize WES coverage model to determine the optimal starting location for the available barges in order to provide maximum WES coverage. Finally, we developed a multi-objective ERBS optimization model that combines these two single objectives. Our demonstration of these three approaches shows that, with an assumed 12-hour barge coverage range, three barges are required to provide emergency response coverage to 92% of the accessible counties in our four-state region. While some general assumptions are made in our demonstration, emergency management teams are likely to be more knowledgeable about their available resources via inland waterways and are encouraged to adjust the WES Index and to adopt the methodology according to their specific needs. In future efforts, we plan to collect systematic feedback from the emergency management community in order to ensure the usability of the WES Index and ERBS approaches in practice.

This research investigates the feasibility of emergency response via inland waterways and provides a framework for finding the optimal starting locations of emergency response barges. Further research could include determining which resources should be supplied by an emergency response barge and how these resources should be allocated. Available funding and specifications of the barge may limit the number and type of emergency services that could be provided. It may be useful to explore the layout, capacity, and potential capabilities of various barge configurations in order to identify what level of service could be provided. The economic feasibility of emergency response via inland waterways is another area of future research. Because all emergency operations plans are limited by a budget, estimating the costs of equipment, personnel, supplies, maintenance, and daily operations of a response barge would prove useful to emergency management teams.

Acknowledgement and disclaimer

This material is based upon work supported by the U.S. Department of Homeland Security under Grant Award Number 2008-ST-061-TS003. The work was conducted through the Mack-Blackwell Rural Transportation

Center at the University of Arkansas. The views and conclusions contained in this document are those of the authors and should not be interpreted as necessarily representing the official policies, either expressed or implied, of the U.S. Department of Homeland Security.

References

Anderson, Anice I., Dennis Compton, and Tom Mason, "Managing in a Dangerous World – The National Incident Management System," *Engineering Management Journal* (December 2004), 16, pp. 3–9.

Anonymous, "Healthful Cruise on a Hospital Barge," *New York Times* (August 07, 1988), p. 44.

Farrokhvar, Leily, Heather Nachtmann, and Edward A. Pohl, "Measuring the Feasibility of Inland Waterway Emergency Response," *Proceedings of the 2011 Industrial Engineering Research Conference* (May 2011).

Daskin, Mark S. "What You Should Know about Location Modeling," *Naval Research Logistics* (2008), 55, pp. 283–294.

Drezner, Zvi, and Horst W. Hamacher, *Facility Location: Applications and Theory*, Berlin, Germany: Springer (2002).

Ewing, Philip, "U.S Sends 5 More Ships to Help Haiti," *Navy Times* (January 18, 2010).

Federal Emergency Management Agency (FEMA), U.S. Department of Homeland Security, *National Response Framework*, Washington, DC: FEMA Publication P-682 (Catalog Number 08011-1) (2008).

Federal Emergency Management Agency (FEMA), U.S. Department of Homeland Security, *Developing and Maintaining Emergency Operations Plans: Comprehensive Preparedness Guide*, Washington, DC: FEMA Publication CPG 101 V2 (November 2010).

Federal Emergency Management Agency (FEMA), U.S. Department of Homeland Security, *Declared Disasters by Year or State*, http://www.fema.gov/news/disaster_totals_annual.fema#markS, accessed April 9, 2011 (2011).

Francis, Richard L., Leon F. McGinnis, Jr., and John A. White, *Facility Layout and Location: An Analytical Approach*, Engelwood Cliffs, NJ: Prentice-Hall, Inc. (1992).

Hazards and Vulnerabilities Research Institute, *Social Vulnerability Index for the United States 2000*, http://webra.cas.sc.edu/hvriapps/SOVI_Access/SoVI_Access_Page.htm, accessed April 2, 2011 (2008).

Hazards and Vulnerabilities Research Institute, Department of Geology, University of South Carolina, *Social Vulnerability Index: Frequently Asked Questions*, http://webra.cas.sc.edu/hvri/products/sovifaq.aspx#score, accessed March 9, 2011 (2011).

Hotelling, Harold, "Stability in Competition," *The Economy Journal* (1929), 39, pp. 41–57.

Quaranc, *Hospital Barges*, http://www.qaranc.co.uk/hospitalbarges.php, accessed March 2, 2009 (2009).

Tornado Project, The, *Tornadoes 1950–1995*, http://www.tornadoproject.com/, accessed April 2, 2009 (1999).

U.S. Army Corps of Engineers (USACE), *Inland Waterway Navigation: Value to the Nation*, http://www.corpsresults.us/, accessed July 10, 2011 (2009).

U.S. Army Corps of Engineers (USACE), Navigation Data Center, *The U.S. Waterway System, Transportation Facts and Information*, http://www.ndc .iwr.usace.army.mil/factcard/temp/factcard10.pdf, accessed July 10, 2011 (November 2010).

U.S. Census Bureau, *Centers of Population Computation for 1950, 1960, 1970, 1980, 1990, 2000, and 2010*, http://www.census.gov/geo/www/2010census/ centerpop2010/county/CenPop2010_Mean_CO05.txt, accessed October 30, 2011 (2011).

U.S. Census Bureau, *County and City Data Book: 2007, Table B-6. Counties – Physicians, Community Hospitals, Medicare, Social Security, and Supplemental Security Income*, http://www.census.gov/statab/ccdb/cc07_tabB6.pdf., accessed October 1, 2009 (2007).

U.S. Census Bureau, *County Business Patterns (NAICS)*, http://censtats.census .gov/cbpnaic/cbpnaic.shtml, accessed October 10, 2011 (2008a).

U.S. Census Bureau, *Population Estimates*, http://www.census.gov/popest/ estimates.html, accessed October 21, 2010 (2008b).

U.S. Department of Agriculture Economic Research Service (USDA ERS), "Measuring Rurality: Rural-Urban Continuum Codes" http://www.ers .usda.gov/briefing/rurality/ruralurbcon/, accessed April 21, 2011 (2004).

U.S. Department of Transportation, Research and Innovative Technology Administration, *National Transportation Atlas Database*, http://www.bts .gov/publications/national_transportation_atlas_database/2010/, accessed October 12, 2011 (2010).

U.S. Geological Survey, *Seismic Hazard Maps*, http://earthquake.usgs.gov/ earthquakes/states/, accessed July 19, 2010 (2009).

Weber, Alfred, Uber den standort der industrien, tubingen, English translation, by C.J. Friedrich, *Theory of the Location of Industries*, Chicago, IL: University of Chicago Press (1929).

Wecksler, A, "Floating Power Barges Planned to Provide Emergency Power Source," in *A Handbook of Production & Maintenance Know-How. To Win the War, American Industry Must Product: More-Faster-With less*, New York: Mill & Factory (1942), pp. 102–103.

Index